Exclusive Papers of the Editorial Board Members and Topical Advisory Panel Members of *Catalysts* in Section "Catalytic Materials"

Exclusive Papers of the Editorial Board Members and Topical Advisory Panel Members of *Catalysts* in Section "Catalytic Materials"

Guest Editors

Leonarda Liotta

Narendra Kumar

Konstantin Ivanov Hadjiivanov

MDPI

Basel • Beijing • Wuhan • Barcelona • Belgrade • Novi Sad • Cluj • Manchester

Guest Editors
Leonarda Liotta
Institute of Nanostructured Materials
Palermo Research Division
Palermo
Italy

Narendra Kumar
Faculty of Science and Engineering
Åbo Akademi University
Turku
Finland

Konstantin Ivanov Hadjiivanov
Institute of General and Inorganic Chemistry
Bulgarian Academy of Sciences
Sofia
Bulgaria

Editorial Office
MDPI AG
Grosspeteranlage 5
4052 Basel, Switzerland

This is a reprint of the Special Issue, published open access by the journal *Catalysts* (ISSN 2073-4344), freely accessible at: www.mdpi.com/journal/catalysts/special_issues/0Y7NQC649X.

For citation purposes, cite each article independently as indicated on the article page online and using the guide below:

Lastname, A.A.; Lastname, B.B. Article Title. *Journal Name* **Year**, *Volume Number*, Page Range.

ISBN 978-3-7258-4014-4 (Hbk)
ISBN 978-3-7258-4013-7 (PDF)
https://doi.org/10.3390/books978-3-7258-4013-7

Contents

About the Editors

Leonarda Liotta

Dr. Leonarda Liotta works at the Institute of Nanostructured Materials in Palermo, Italy. Hew research interests are in the fields of supported noble metals; nanostructured and mesoporous materials; inorganic perovskites for application in NO SCR from exhaust gases (stationary and mobile sources); VOCs oxidation; dry/steam hydrocarbons reaction; CO_2 methanation.

Narendra Kumar

Dr. Narendra Kumar works at the Faculty of Science and Engineering, Laboratory of Industrial Chemistry and Reaction Engineering, Åbo Akademi University, Finland. His research interests are in the fields of heterogeneous catalysis; catalyst synthesis; nanoporous materials; catalyst characterization; zeolite catalysis; refinery processes; hydrocarbon conversion; reaction mechanisms; biomass transformations; environmental catalysis; exhaust emission control; water purification.

Konstantin Ivanov Hadjiivanov

Prof. Dr. Konstantin Ivanov Hadjiivanov works at the Institute of General and Inorganic Chemistry, Bulgarian Academy of Sciences. His research interests are in the fields of vibrational spectroscopy; surfaces; environmental catalysis; adsorption; porous materials.

Preface

This reprint is a compilation of contributions originally published in the *Catalysts* Special Issue entitled "Exclusive Papers of the Editorial Board Members and Topical Advisory Panel Members of *Catalysts*", within the Section "Catalytic Materials".

It features a collection of 15 papers—1 review and 14 original research articles—spanning a wide range of topics in both homogeneous and heterogeneous catalysis. These works reflect the diverse expertise of selected members of the journal's Editorial Board and Topical Advisory Panel. The volume opens with a brief Editorial from the Guest Editors, summarizing the key scientific achievements presented in the Special Issue.

The contributing authors represent a broad international community, with affiliations in Bulgaria, China, Italy, Iran, Poland, Portugal, the Republic of Korea, Russia, Slovakia, Sweden, and Switzerland.

We are confident that the research presented in this reprint will be a valuable resource for scientists, researchers, and academics working in both academic and industrial settings. The collection offers in-depth insights, innovative methodologies, and comprehensive data, all of which can inspire new ideas, support ongoing research, and contribute to practical advancements in the field of catalysis.

Leonarda Liotta, Narendra Kumar, and Konstantin Ivanov Hadjiivanov
Guest Editors

catalysts

Editorial

Exclusive Papers of the Editorial Board Members and Topical Advisory Panel Members of Catalysts in Section "Catalytic Materials"

Leonarda Francesca Liotta [1,*], Narendra Kumar [2,*] and Konstantin Ivanov Hadjiivanov [3,*]

1 Institute for the Study of Nanostructured Materials (ISMN)-CNR, Via Ugo La Malfa 153, 90146 Palermo, Italy
2 Faculty of Science and Engineering, Industrial Chemistry and Reaction Engineering, Åbo Akademi University, Henriksgatan 2, FI-20500 Turku, Finland
3 Institute of General and Inorganic Chemistry, Bulgarian Academy of Sciences, 1113 Sofia, Bulgaria
* Correspondence: leonardafrancesca.liotta@cnr.it (L.F.L.); narendra.kumar@abo.fi (N.K.); kih@svr.igic.bas.bg (K.I.H.)

Citation: Liotta, L.F.; Kumar, N.; Hadjiivanov, K.I. Exclusive Papers of the Editorial Board Members and Topical Advisory Panel Members of Catalysts in Section "Catalytic Materials". *Catalysts* **2024**, *14*, 564. https://doi.org/10.3390/catal14090564

Received: 12 August 2024
Accepted: 21 August 2024
Published: 26 August 2024

The Special Issue "Exclusive Papers of the Editorial Board Members and Topical Advisory Panel Members of Catalysts in Section "Catalytic Materials" contains 14 peer-reviewed research articles and 1 review paper (Contributions 1–15), which broadly focus on the field of homogeneous and heterogeneous catalysis, with an emphasis on synthesis, physico-chemical characterizations, and applications in several environmental protection reactions, such as CO_2 valorization, NOx SCR, removal of VOCs, photocatalysis. These fifteen studies offer in-depth explanations of reactions taking place in gas and liquid phases using different types of reactors—such as fixed bed, plug flow, continuous, and batch reactors (Contributions 1,2,4,6–9,14,15). Moreover, analyses of the reaction products were carried out using gas chromatography, mass spectrometer and high-power liquid chromatography.

The physico-chemical characterization of the catalytic materials is performed using scanning electron microscopy, transmission electron microscopy, nitrogen physisorption, high-resolution transmission electron microscopy, X-ray powder diffraction, X-ray photo electron spectroscopy, Fourier-transform infrared spectroscopy, atomic absorption and X-ray fluorescence (Contributions 1,3,8,14). Notably, most of the research papers offer detailed descriptions of the correlation between physicochemical characterization results and catalytic performance, thus contributing in-depth interdisciplinary knowledge to this field.

Furthermore, proposed reaction mechanisms, kinetic studies, and catalyst deactivation and regeneration are supported by reliable evidence of catalyst physico-chemical characterizations and the synthesis of catalytic materials. The catalytic materials studied in the published research papers are transition metal oxides, porous oxides, alumino-silicate zeolites, silica, alumina, mesoporous materials, structured micro- and mesoporous materials, and carbon-based materials (Contributions 5,6,8,9,13). Some of these catalytic materials have been synthesized in academic and industrial research laboratories, and their novel methods of synthesis and characterizations are described in these papers, thus enhancing the scientific quality of the published research papers.

In conclusion, the synthesis and characterization of catalytic materials play a pivotal role in advancing our understanding of reaction mechanisms, kinetic behaviors, and strategies for dealing with catalyst deactivation. The different types of catalysts studied in this Special Issue highlight ongoing efforts to develop more efficient, durable, and versatile catalysts for industrial and environmental applications.

We are confident that these research articles will be immensely beneficial to distinguished scientists, researchers, and academics engaged in both, academic and industrial laboratories, by providing in-depth insights, innovative methodologies, and comprehensive data that can advance their understanding, foster new research ideas, and enhance practical applications in their respective fields.

Author Contributions: Conceptualization, L.F.L., N.K. and K.I.H.; methodology, L.F.L., N.K. and K.I.H.; software, L.F.L., N.K. and K.I.H.; validation, L.F.L., N.K. and K.I.H.; formal analysis, L.F.L., N.K. and K.I.H.; investigation, L.F.L., N.K. and K.I.H.; resources, L.F.L., N.K. and K.I.H.; data curation, L.F.L., N.K. and K.I.H.; writing—original draft preparation, N.K.; writing—review and editing, L.F.L., N.K. and K.I.H.; visualization, L.F.L., N.K. and K.I.H.; supervision, L.F.L., N.K. and K.I.H.; project administration, L.F.L., N.K. and K.I.H. All authors have read and agreed to the published version of the manuscript.

Funding: This research received no external funding.

Acknowledgments: We would like to thank all of the authors who shared their research and the referees for their invaluable contributions.

Conflicts of Interest: The authors declare no conflicts of interest.

List of Contributions:

1. Liu, P.; Qiao, H.; Su, X.; Peirong; Yang, F. Visible Light-Photocatalyzed C5-H Nitration of 8 Aminoquinoline Amides. *Catalysts* **2024**, *14*, 263. https://doi.org/10.3390/Catal14040263.
2. Neves, P.; Simoes, G.; Naprus, B.D.; Pamin, K. Versatile Polyoxo Metalates of Different Structural Dimensionalities for Liquid Phase Catalytic Oxidations. *Catalysts* **2024**, *14*, 251. https://doi.org/10.3390/Catal4040251.
3. Sidorenko, N.D.; Topchiyan, P.A.; Saraev, A.A.; Yu, E.; Gerasimov; Zhurenok, A.V.; Vasilchenko, D.B.; Kozlova, E.A. Bimettalic Pt-IrOx/g-C_3N_4 Photocatalysts for the Highly Efficient Overall Water Splitting Under Visible Light. *Catalysts* **2024**, *14*, 225. https://doi.org/10.3390/Catal14040225.
4. Marks, K.; Erbing, A.; Hohmann, L.; Chien, T.-E.; Yazdi, M.G.; Muntwiler, M.; Hansson, T.; Engvall, K.; Harding, D.J.; Öström, H.; Odellus, M.; Göthelid, M. Naphthalene Dehydrogenation on Ni(111) in the Presence of Chemisorbed Oxygen and Nickel Oxide. *Catalysts* **2024**, *14*, 124. https://doi.org/10.3390/Catal14020124.
5. Cepollaro, E.M.; Cimino, S.; Marco D'Agostini; Gargiulo, N. Giorgia Franchin, and Luciana Lisi, 3D-Printed Monoliths Based on Cu-Exchanged SSZ-13 as Catalystsfor SCR of NOx. *Catalysts* **2024**, *14*, 85. https://doi.org/10.3390/Catal14010085.
6. Chakarova, K.; Drenchev, N.; Mihaylov, M.; Hadjiivanov, K. Interaction of O_2 with Reduced Ceria nanoparticles at 100–400K: Fast Oxidation of Ce^{3+} ions and dissolved H_2. *Catalysts* **2024**, *14*, 45. https://doi.org/10.3390/Catal14010045.
7. Mohammadpour, N.; Uznanski, H.K.-P.; Tyczkowski, J. Plasma Deposited CoO-(Carbon Matrix) Thin-Film Nanocatalysts: The Impact of Nanoscale p-n Heterogunctions on Activity in CO_2 Methanations. *Catalysts* **2024**, *14*, 38. https://doi.org/10.3390/Catal14010038.
8. Zhu, Q.; Luo, Y.; Yang, K.; Che, G.; Wang, H.; Qi, J. Construction of Spinel/Perovskite Heterojunction for Boosting Photocatalytic Performance for Polyacrylamide. *Catalysts* **2023**, *13*, 1424. https://10.33901/Catal13111424.
9. Chernov, A.N.; Cherepanova, S.V.; Gerasimov, E.Y.; Prosvitin, I.P.; Zenkovets, G.A.; Shutilov, A.A.; Gorbunova, A.S.; Koltunov, K.Y.; Sobolev, V.L. Propane Dehydrogenation over Cobalt Aluminates: Evaluation of Potential Catalytic Active Sites. *Catalysts* **2023**, *13*, 1419. https://doi.org/10.3390/Catal13111419.
10. Rizzato, L.; Cavazzani, J.; Osti, A.; Scavini, M.; Glsenti, A. Cu-Doped $SrTiO_3$ Nanostructured Catalysts for CO_2 Conversion into Solar Fuels Using Localiged Surface Plasmon Resonance. *Catalysts* **2023**, *13*, 1377. https://doi.org/10.3390/Catal13101377.
11. Panek, B.; Kierzkowska-Pawlak, H.; Uznanski, P.; Nagy, S.; Nagy-Trembosova, V.; Tyczkowski, J. The role of Carbon Nanotube Deposit in Catalytic Activity of FeOx-Based PECVD Thin Films Tested in RWGS Reaction. *Catalysts* **2023**, *13*, 1302. https://doi.org/10.3390/Catal13091302.
12. Osti, A.; Rizzato, L.; Canazzani, J.; Gilsenti, A. Optimizing Citrate Combustion Synthesi of A-Site Defficient La, Mn-Based Perovskite: Application of Catalytic CH_4 Combustion in Stoichiometric Conditions. *Catalysts* **2023**, *13*, 1177. https://doi.org/10.3390/Catal13081177.
13. Afshari, M.; Carabineiro, S.A.C.; Gorjizadeh, M. Sulfonated Silica Coated $CoFe_2O_4$ Magnetic Nanoparticles for the Synthesis of 3, 4-Dihydropyrimidin-2(1H)-one and octahydroquinazoline Derivatives. *Catalysts* **2023**, *13*, 989. https://doi.org/10.3390/Catal13060989.

14. Lee, J.W.; Jeong, R.H.; Shin, I.; Boo, J.-H. Fabrication of New TiO_2 Photo Catalyst for removing organic dyes and Harzadous VOCs in Air Purifier System. *Catalysts* **2023**, *13*, 935. https://doi.org/10.3390/Catal13060935.
15. Peng, A.; Huang, Z.; Li, G. Ethylene Oligomerization Catalyzed by Different Homogeneous or Heterogeneous Catalysts. *Catalysts* **2024**, *14*, 268. https://doi.org/10.3390/Catal14040268.

catalysts

Review

Ethylene Oligomerization Catalyzed by Different Homogeneous or Heterogeneous Catalysts

Anfeng Peng †, Zheng Huang † and Gang Li *

State Key Laboratory of Fine Chemicals, School of Chemical Engineering, Dalian University of Technology, Dalian 116024, China; pengaf@mail.dlut.edu.cn (A.P.); hz22245199@mail.dlut.edu.cn (Z.H.)
* Correspondence: liganghg@dlut.edu.cn
† These authors contributed equally to this work.

Abstract: Linear α-olefins (LAOs) are linear alkenes with double bonds at the ends of the molecular chains. LAOs with different chain lengths can be widely applied in various fields. Ethylene oligomerization has become the main process for producing LAOs. In this review, different homogeneous or heterogeneous catalysts recently reported in ethylene oligomerization with Ni, Fe, Co, Cr, etc., as active centers will be discussed. In the homogeneous catalytic system, we mainly discuss the effects of the molecular structure and the electronic and coordination states of complexes on their catalytic activity and selectivity. The Ni, Fe, and Co homogeneous catalysts are discussed separately based on different ligand types, while the Cr-based homogeneous catalysts are discussed separately for ethylene trimerization, tetramerization, and non-selective oligomerization. In heterogeneous catalytic systems, we mainly concentrate on the influence of various supports (metal–organic frameworks, covalent organic frameworks, molecular sieves, etc.) and different ways to introduce active centers to affect the performance in ethylene oligomerization. Finally, a summary and outlook on ethylene oligomerization catalysts are provided based on the current research. The development of highly selective α-olefin formation processes remains a major challenge for academia and industry.

Keywords: ethylene oligomerization; linear α-olefins; homogeneous catalysts; heterogeneous catalysts; Ni; Fe; Co; Cr

Citation: Peng, A.; Huang, Z.; Li, G. Ethylene Oligomerization Catalyzed by Different Homogeneous or Heterogeneous Catalysts. *Catalysts* **2024**, *14*, 268. https://doi.org/10.3390/catal14040268

Academic Editors: Narendra Kumar, Leonarda Liotta and Konstantin Ivanov Hadjiivanov

Received: 27 March 2024
Revised: 13 April 2024
Accepted: 13 April 2024
Published: 17 April 2024

1. Introduction

Linear α-olefins (LAOs) are linear alkenes with double bonds at the ends of the molecular chains. Due to their unique structure, they are applied widely in various fields depending on the chain length. For example, 1-butene and 1-octene are often used as copolymerization monomers for linear low-density polyethylene (LLDPE) and polyolefin elastomers (POE) [1]. LAOs with carbon numbers of 6–30 are used as important raw materials for the synthesis of plasticizers, detergents, surfactants, and other chemical products. LAOs have developed extremely rapidly in recent years, with the demand growing year after year around the world [2]. Most of the currently commercialized LAO units mainly produce full-fraction α-olefins, which can meet a variety of downstream markets, with product carbon number distributions ranging from C_4 to $C_{30}{}^+$. However, owing to the different demand in markets for various fractions of LAO products, the technology of producing linear α-olefins with specific carbon numbers has also received attention.

Compared with wax pyrolysis [3], Fischer–Tropsch synthesis [4–6], and alkane dehydrogenation [7–9], the ethylene oligomerization method has become the main process for producing LAOs due to its relatively mild operating conditions, flexible product distribution, and high linearity. According to literature reports, there are a wide variety of catalysts for ethylene oligomerization, mainly Ni, Fe, Co, and Cr as metal active centers [10–13], and some catalysts have been industrialized to produce linearity α-olefin products. For example, the Shell higher olefin process (SHOP) is an important catalytic system for ethylene oligomerization, with more than one million tons of α-olefins prepared industrially via this method [14]. Chevron Phillips realized the industrial production of 1-hexene using a Cr catalyst in 2003 [15].

The mechanism of ethylene oligomerization can be divided into the Cossee–Arlman mechanism and the metallacyclic mechanism. As early as 1966, Cosse [16] proposed the mechanism of cis-ligand insertion. After the enrichment by Arlman [17], the Cossee–Arlman mechanism (Figure 1) successfully demonstrated the process of catalyzing ethylene oligomerization to α-olefins. The mechanism is divided into four steps: (i) coordination between ethylene monomer and metal active centers; (ii) ethylene insertion into metal hydride or alkyl intermediate; (iii) repeated coordination and insertion to promote chain growth; and (iv) β-H elimination reaction to obtain α-olefins and regain active centers. The products obtained from the catalytic process following the Cossee–Arlman mechanism tend to follow a Schulz–Flory or Poisson distribution, with a wide distribution range and no specific carbon number.

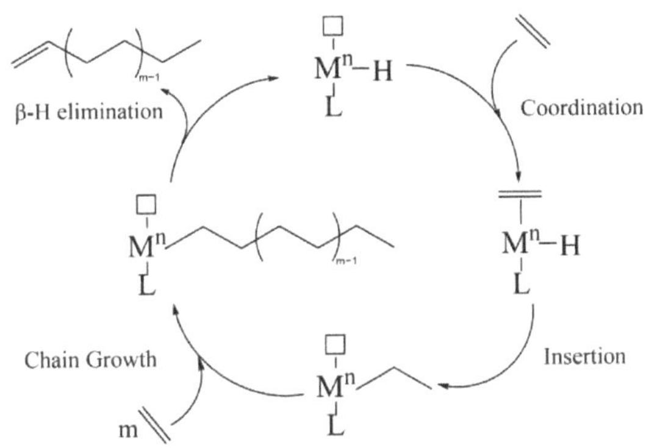

Figure 1. Cossee–Arlman mechanism.

However, Cr-based catalysts can produce 1-hexene and 1-octene through the catalytic tri/tetramerization of ethylene, which cannot be explained by the Cossee–Arlman mechanism. In 1989, Briggs [18] optimized the 2-ethylhexanoate chromium system and validated the metallacyclic mechanism (as shown in Figure 2) to explain the trimerization phenomenon. Chabbra et al. [19] used kinetics and deuterium isotope labeling techniques to identify intermediate species in the catalytic process of chromium diethylene complexes, further validating the metallacyclic mechanism. The mechanism for ethylene trimerization is divided into four steps: (i) two ethylene molecules coordinate with Cr; (ii) oxidative coupling to generate metallacyclopentane intermediates; (iii) insert another ethylene molecule to form a metalcycloheptane intermediate; and (iv) H-shift or β-H elimination and reductive elimination to generate 1-hexene. In the process of ethylene tetramerization, the third step of the metal ring mechanism is the insertion of two ethylene molecules into the metallacyclopentane intermediate to form a metalcyclononane intermediate.

This review will discuss the different homogeneous or heterogeneous catalysts reported in ethylene oligomerization with Ni, Fe, Co, and Cr as active centers. Homogeneous catalysts of various metals exhibit high activity, good selectivity, mild reaction conditions, and ligand modifiability in ethylene oligomerization, while heterogeneous catalysts have the characteristics of easy product separation and good thermal stability. In this review, the Ni, Fe, and Co homogeneous catalysts are discussed separately based on different ligand types, while the Cr-based homogeneous catalysts are discussed separately for ethylene trimerization, tetramerization, and non-selective oligomerization. Meanwhile, the Ni, Fe, Co, and Cr heterogeneous catalysts with different supports are discussed. In addition, zinc-, ruthenium-, and gallium-based heterogeneous catalysts are also mentioned. Finally, a summary and outlook on ethylene oligomerization catalysts are provided based on the current research.

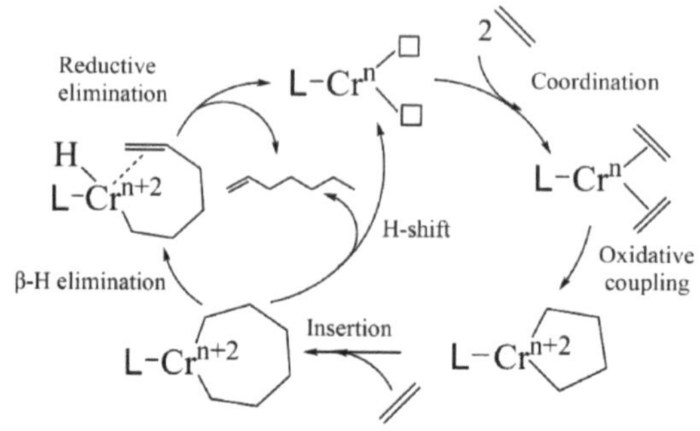

Figure 2. Metallacyclic mechanism.

2. Nickel-Based Catalysts

2.1. Nickel-Based Homogeneous Catalysts

In the 1950s, researchers discovered the "nickel effect" [20], in which ethylene and triethylaluminium were induced to form butenes in the presence of small amounts of nickel salts. The discovery stimulated the investigation of post-transition-metal-catalyzed ethylene oligomerization. In 1966, the concept of "ligand" was introduced, which further improved the development of transition metal catalysts. Shortly thereafter, Keim [21] designed neutral (P, O) ligands for ethylene oligomerization to produce linear α-olefins. In 1990, Keim invented the typical SHOP catalyst (Figure 3), which became a remarkable milestone for the oligomerization reaction in industry. Another important breakthrough was the development of a highly electrophilic cationic nickel complex bearing α-diimine ligands by Brookhart [22] and co-workers in 1995 (Figure 4). When the substituent in the neighboring position of the benzene ring is small (e.g., H), the polymerization of ethylene produces a low-molecular-weight oligomerization product; when the substituent resistance is gradually increased (CH_3 to iPr), the molecular weight of the polyethylene increases as well. This finding suggests that the selectivity can be altered by changing the site resistance of the ligand in nickel complexes. Following this breakthrough, the influence of electronic effects on catalytic performance was investigated by Tuskaev et al. [23,24]. Further investigations of nickel complexes bearing α-diimine ligands have involved the modification of the ligand to alter its spatial site resistance and electronic effects.

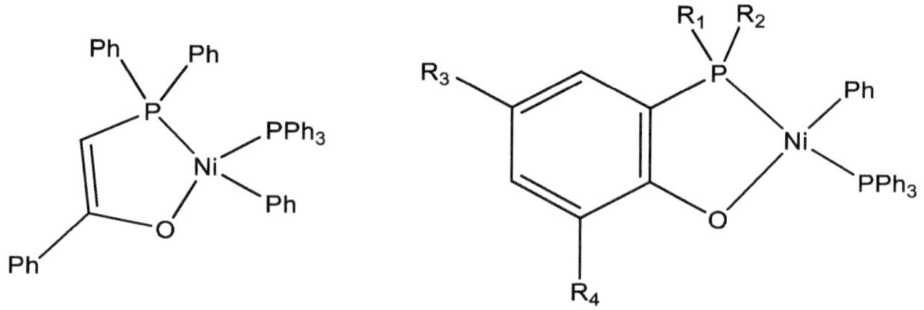

Figure 3. Nickel complex bearing neutral (P, O) ligands.

In addition, other types of bidentate ligands (including (N, O) [25–29], (P, N) [30–32], (P, P) [33–35], (P, S) [36], and (N, S) [37]) as well as tridentate ligands (including (N, N, N) [38–40], (N, N, O) [41,42], (N, O, O) [43,44], etc.) were synthesized. These above ligands have been discussed in detail by Giyjaz E. Bekmukhamedov [45] and H. Olivier-Bourbigou [46] and will not be repeated here. The nickel complexes with tridentate ligands have better selectivity for linear α-olefins than the didentate nickel complexes, but the activity is lower. On the whole, the bidentate ligands are still a hotspot of the current research. The main thrust of the research has been to modify the selectivity and activity by changing the structure of the ligand in the catalyst.

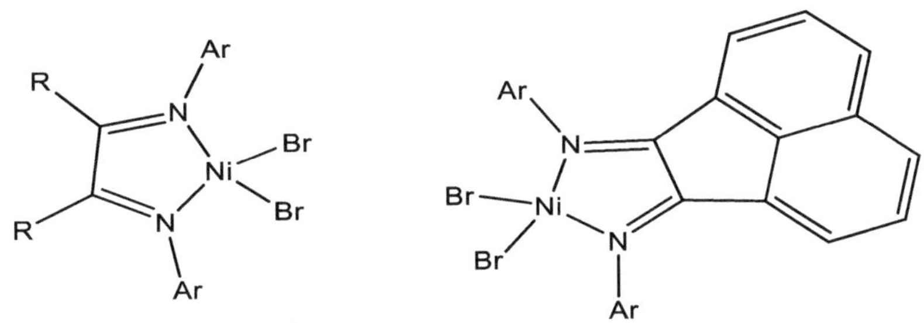

1 R=H; Ar=2,6-$C_6H_3(^{i\text{-}}Pr)_2$
2 R=Me; Ar=2,6-$C_6H_3(^{i\text{-}}Pr)_2$
3 R=H; Ar=2,6-$C_6H_3Me_2$
4 R=Me; Ar=2,6-$C_6H_3Me_2$

5 Ar=2,6-$C_6H_3(^{i\text{-}}Pr)_2$

Figure 4. Nickel complex bearing α-diimine ligands.

2.2. Nickel-Based Heterogeneous Catalysts

Compared with homogeneous catalysts, heterogeneous catalysts have the advantages of easy separation, convenient recovery, and high selectivity. Heterogeneous catalysts have the ability to tailor the synthesized product with controlled chain length and distribution. The use of heterogeneous molecular species allows the active sites to be separated from each other, thus preventing them from aggregating or interacting, which may lead to catalyst deactivation or undesired reactivity [47]. The currently available heterogeneous catalysts for the oligomerization of ethylene include metal–organic frameworks (MOFs), covalent organic frameworks (COFs), molecular sieve materials, and other materials.

2.2.1. Metal–Organic Frameworks (MOFs)

Metal–organic frameworks (MOFs), known as crystalline porous coordination polymers, are crystalline solids assembled by linking metal ions or clusters with ligand molecules [48]. In recent years, MOFs have demonstrated promise as catalysts for olefin oligomerization and polymerization. MOFs possess numerous active sites separated from each other, which offers opportunities for sufficient coordination between the active center and monomer. MOFs not only have unsaturated metal sites themselves (self-assembly) but can also introduce active metal sites through post-synthetic modifications (cation exchange, atomic deposition, or ligand metalation) (Figure 5). In this section, a comprehensive contrast of MOFs for olefin oligomerization in past decades is summarized (Table 1). The highest activity of each catalyst as well as the reaction conditions (including temperature, pressure, and solvent) and selectivity for the product under the highest activity (For calculations of activity, please see Appendix A) were listed. It can be seen from Table 1 that the nickel-based heterogeneous catalysts are dominated by ethylene dimerization, regardless of the method of introducing the nickel active site and structure of the MOFs [49]. One reason may be that the lack of a large steric hindrance group around the nickel active site in the microenvironment of Ni-based MOFs leads to fast β-H elimination [50,51]. Another reason could be the higher nuclear charge number of nickel metals in the microenvironment, which facilitates β-H elimination to produce dimers [52]. As can be seen in Table 1, the current catalyst with the highest ethylene dimerization activity is Ni-ZIF-8 (0.4 wt%), synthesized using the self-assembly method reported by Chen's group [53], and the selectivity of 1-butene can be maintained above 80%. The most promising nickel-based MOFs material capable of catalyzing the trimerization of ethylene is MIL-125(Ti)-NH_2(Ni), obtained by Wang et al. [54]. In the presence of the MAO co-catalyst, the total selectivity to 1-hexene can reach about 80% while achieving the highest activity.

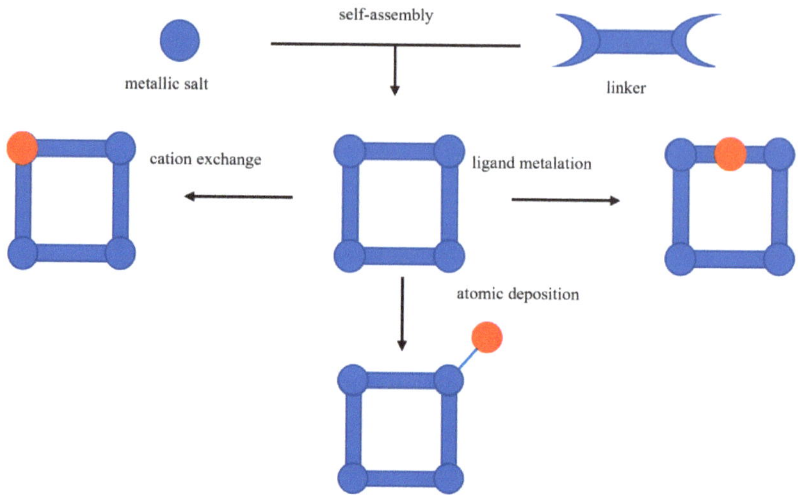

Figure 5. Different synthetic pathways to obtain isolated active sites within MOFs.

All the above-mentioned catalysts in the presence of alkylaluminium co-catalysts form nickel active centers, which undergo coordination insertion with ethylene and β-H elimination to produce linear α-olefins. Moreover, some Ni-MOF materials can directly catalyze ethylene oligomerization without co-catalysts. Bhan [55] and co-workers reported Ni-functionalized UiO-66 via the atomic deposition technique. Ni/UiO-66 that uniquely engenders active sites on stream maintains stable oligomerization rates for over 15 days under gas-phase conditions without any external co-catalyst. The authors demonstrated that the number of active sites involved in the reaction indeed increases with increasing ethylene pressures using in situ NO titration. Steady-state oligomerization kinetics are proposed to follow the Cossee–Arlman mechanism by comparing experimental and computed apparent activation enthalpies. Olsbye [56] and co-workers synthesized a series of UiO-67-bpy MOFs with biphenyl-4,4′-dicarboxylate(bpdc) and different amounts of 2,2′-bipyridine-5,5′-dicarboxylate (bpydc). The post-synthetic metalation of the MOFs was carried out in EtOH with Ni(acetate)$_2$·4H$_2$O as the metal source. Ni-UiO-67-bpy$_{11\%}$ can catalyze the dimerization of flowing ethene (P(ethene) = 26 bar, 250 °C, Time = 240 min) with an activity up to 850 mg butene $g_{cat}^{-1}h^{-1}$ after activation at 300 °C in 10% O$_2$ for 360 min. This catalyst achieves 6% conversion and 99% selectivity for linear butene (1-butene:trans-2-butene:cis-2-butene ratio was 41:26:32), which is attributed to homogeneously distributed active sites in the microenvironment and spatial confinement.

2.2.2. Covalent Organic Frameworks (COFs)

Covalent organic frameworks are a class of organic porous crystalline materials linked with light elements (such as H, B, C, N, O, etc.) by robust covalent bonds via reticular chemistry [57]. COFs have promising applications in the field of heterogeneous catalysis due to their ability to induce catalytically relevant ligands into their structure. Unlike conventional support materials (zeolite, metal oxide, etc.), COFs are structurally predesignable, synthetically controllable, and functionally manageable, suggesting that COFs may be another promising support material for olefin oligomerization analogues to MOFs. Since COFs have no active metal sites, they need to be modified by ligand metallization methods. Gascon et al. [58] synthesized Covalent Triazine Frameworks (CTFs) with micro- and mesoporous structures and a lamellar-structured imine-linked polymer network (IL-PON). The materials were metalated afterwards with nickel(II) bromide ethylene glycol dimethyl ether salt (NiBr$_2$·DME) in THF to afford the three catalysts Ni@meso-CTF, Ni@microCTF, and Ni@IL-PON. Ni can either be chemically coordinated to the nitrogen functional sites of the backbone or remain on the surface and in the pores as uncoordinated complexes. Ni@meso-CTF, Ni@microCTF, and Ni@IL-PON, respectively, achieved an intrinsic catalytic activity of 75.3, 63, and 92.5 $mol_{ethylene\ converted}$ $mol^{-1}{}_{Ni}$ h^{-1}, with respective butene selectivity of 59%, 54%, and 58% in the conditions of 50 °C, 15 bar of ethylene, heptane solvent, and Et_3Al cocatalyst.

Li et al. [59] reported two covalent organic frameworks (COFs), MABD-COF and MAPA-COF, used in ethylene oligomerization. The specific surface area and the pore volume of MAPA-COF with p-phthalaldehyde as a sub-constructing unit were far larger than those of MABD-COF with butanedione as a sub-constructing unit. Ni@MAPA-COF and Ni@MABD-COF were obtained by metallization with nickel dichloride. The Ni-loaded COFs were evaluated for ethylene oligomerization using MAO as a cocatalyst (Al/Ni = 500 to 700), and the best activity was found in toluene as a solvent at 25 °C with a productivity of 8.31×10^4 g $mol^{-1}{}_{Ni}$ h^{-1} and 1-butene selectivity of 59% using Ni@MABD-COF and a productivity of 15.68×10^4 g $mol^{-1}{}_{Ni}$ h^{-1} and 1-butene selectivity of 51% using Ni@MAPA-COF. Recently, another group [60] reported another nickel-coordinated imine-linked covalent organic frameworks material (COF-PD-Ni) for ethylene oligomerization with a productivity of 1.98×10^5 g $mol^{-1}{}_{Ni}$ h^{-1}. In the condition with the highest activity, the selectivity of the product C_4 can reach 82%, and the remaining product is C_6–C_{10}.

2.2.3. Molecular Sieve Materials

Molecular sieves, as a kind of important catalyst carrier, are different from MOFs and COFs. Not only can they introduce active metal sites by ion-exchange or graft modification but they also have their own Brønsted acid sites. Acid sites are responsible not only for isomerization reactions but also for the dimerization of the primary products. Molecular sieve catalysts are able to catalyze ethylene oligomerization at high temperatures without the need for other cocatalysts to initiate the reaction. The activity (ethylene conversion) of these heterogeneous catalysts is determined by the number of accessible nickel sites, whereas the stability of the oligomerization, the carbon number distribution, and the structure of the product are determined by the concentration of Brønsted acid sites and the porosity.

Hulea et al. [61] synthesized the nickel ion-exchanged NiMCM-22 (Si/Al = 14) and NiMCM-36 (Si/Al = 26) for ethylene oligomerization reactions. The NiMCM-36 catalyst with a large mesoporous structure and mild acidity showed good activity (46 g/(g_{cat}·h) of oligomer) and selectivity (100% of olefins with an even number of carbon atoms) for the oligomerization of ethylene, whereas the NiMCM-22 catalyst with a microporous structure and a high concentration of acid sites showed lower catalytic activity and selectivity. The balance between acid and nickel ion sites, as well as the textural properties of catalysts, played a significant role in determining their activity and selectivity. In 2014 [62], Ni-AlSBA-15 was prepared by the post-synthesis alumination reaction of SBA-15 by sodium aluminate followed by exchanging with nickel ions. The catalyst activity of 175 g/(g_{cat}·h) and dimerization selectivity of 77% can be achieved at 150 °C and 35 bar. This catalyst maintains a high and stable conversion rate over eighty hours in a fixed bed reactor at 150 °C, pressure 3.0 MPa, and WHSV10 h^{-1}. This is attributed to the interlinked mesoporous network, which is large enough to allow free diffusion of the product. Shin et al. [63] reported a series of bpy-SBA-15, with different molar amounts of bipyridyl (bpy) sites metalated with $NiCl_2 \cdot H_2O$. Bpy-SBA-15 with 0.21 wt% of Ni achieved 4422 $mol_{oligomers}$·$mol^{-1}{}_{Ni}$·h^{-1} and a selectivity in butene of 77%, with 700 equivalents of Et_2AlCl. The heterogeneous catalyst can be recycled many times without a significant reduction in catalytic activity due to the stability of the SBA-15 support and the appropriate Ni loading density that reduces the possibility of further reaction of the dimer into polymeric oligomers and polymers. Lacarriere et al. [64] prepared Ni-MCM-41 (Si/Al = 9) by the ion exchange of AlMCM-41 precursors. The catalytic activity of Ni-AlMCM-41 reached 180 g/(g_{cat}·h) with a TOF of 16,920 h^{-1}, which was far superior to similar Ni-loaded catalysts and comparable to homogeneous Ni coordination catalysts. Moreover, other molecular sieve catalysts have been used in ethylene oligomerization reactions, like ZSM-5 [65,66], Ni-H-Beta [67,68], and Ni/SIRAL-30 [69], but they did not show desirable linear α-olefins selectivity.

Table 1. A comprehensive contrast of MOFs for olefin oligomerization.

Catalyst	Temperature (°C)	Pressure (bar)	Reaction Time (h)	Co-Catalyst (Al:Ni)	Solvent	Activity or TOF	C_4 (1-C_4)	C_6	C_8	C_{8+}	Ref.
Ni-UMOFNS-190 [1]	25	10	1	Et_2AlCl (500)	toluene	5536 h^{-1}	75.6	0.4	22	2	[70]
CPO-27(Ni) [1]	21	10	1	Et_2AlCl (17)	toluene	1 g/(mol·h)	100				
DUT-8(Ni)-rigid [1]	21	20	1	Et_2AlCl (17)	toluene	182 g/(mol·h)	76	24			[71]
$[Ni(bdc)(dabaco)_{0.5}]n$ [1]	21	10	1	Et_2AlCl (17)	toluene	41 g/(mol·h)	56	44			
$[Ni(bdc)(dabaco)_n]n$ [1]	21	10	1	Et_2AlCl (17)	toluene	41 g/(mol·h)	49	51			
1D-Ni-MIL-77 [1]	30	10	0.5	Et_2AlCl (100)	toluene	5544 h^{-1}	98 (93.3)	0.14		1.84	[72]
3D-Ni-MIL-77 [1]	30	15	0.5	Et_2AlCl (180)	toluene	2226 h^{-1}	99.6 (90)	0.22		0.17	
Ni-ZIF-8 (0.4 wt%) [1]	35	30	0.17	MAO (4640)	toluene	1,116,000 h^{-1}	97 (87.7)				[53]
15Ni-ZIF-L [1]	40	30	0.17	MAO	toluene	342,030 h^{-1}	96.9 (94.7)				[73]
Ni(1%)-MFU-4l [2]	25	50	1	MAO (500)	toluene	41,500 h^{-1}	97.4 (94.5)	2.6			[74]
Ni(7.5%)-CFA-1 [2]	22	50	1	MMAO-12 (2000)	toluene	37,100 h^{-1}	95.5 (91.2)				[75]
20Ni-MOF5 (5.32 wt%) [2]	35	50	0.17	MAO	toluene	352,000 h^{-1}	96.3 (84.2)				[76]
Ni-AIM-NU-1000 [3]	45	2	10	Et_2AlCl		1080 h^{-1}	42	8	46		[77]
Ni-Facac-AIM-NU-1000 [3]	45	2	10	Et_2AlCl		12.6 h^{-1}	100 (78)				[49]
Ni-acac-AIM-NU-1000 [3]	45	2	10	Et_2AlCl		15.84 h^{-1}	100 (82)				
(30)Ni@(Fe)MIL-101 [4]	10	15	1	Et_2AlCl (70)	heptane	3215 h^{-1}	94	5.5	0.5		[78]
MixMOFs-Ni-a [4]	20	20	0.5	Et_2AlCl (100)	toluene	2071 h^{-1}	71.3	6.9	0	21.8	
MixMOFs-Ni-b [4]	40	20	0.5	Et_2AlCl (100)	toluene	16,428 h^{-1}	92.7	6.1	0	1.2	[79]
MixMOFs-Ni-c [4]	20	20	0.5	Et_2AlCl (100)	toluene	2000 h^{-1}	79.5	7.2	0	14.3	
IRMOF-3-Ni-a [4]	20	20	0.5	Et_2AlCl (100)	toluene	2246 h^{-1}	35	9.3	0	55.7	
MIL-125(Ti)-NH_2(Ni) [4]	50	10	0.5	MAO (800)	cyclohexane	6464 h^{-1}	19.6 (87.2)	76.7 (92.5)	1.5	2.2	[54]
NU-1000-bpy-$NiCl_2$ [4]	21	15	1	Et_2AlCl (70)	heptane	1560 h^{-1}	82	18			[80]
[Al]-Ni-bpydc(MOF) [4]	5	15	1	Et_2AlCl (70)	heptane	20 g/(g·h)	89.1 (26.3)	8.9	2.1		[81]
[Ni]-Ni-bpydc(MOF) [4]	5	15	1	Et_2AlCl (70)	heptane	20 g/(g·h)	92.7 (71.3)	7.3			
$Zr_6O_4(OH)_4(bpydc)_{0.84}(bpdc)_{5.16}(NiBr_2)_{0.84}$ [4]	55	59	1	Et_2AlCl (70)	cyclohexane	370 g/(g/h)					[82]
NiCl@DUT-133 [4]	40	35	1	Et_2AlCl (51)	1,2-dichlorobenzene	42 mol/(mol·h)	81	19			[83]

[1] Self-assembly. [2] Cation Exchange. [3] Atomic Deposition. [4] Ligand Metalation. 1-olefin is relative to the total amount of oligomer products of that carbon number.

2.2.4. Other Materials

Additionally, multiwalled carbon nanotubes (MWCNTs) [84], Ionic Liquid Phase (SILP) [85], clays [86], bispyridine-based porous organic polymers [87], and metal oxides [88,89] have all been studied as carriers of ethylene oligomerization catalysts. These provide new perspectives for the development of new heterogeneous catalysts for ethylene oligomerization.

3. Iron-Based Catalysts

3.1. Iron-Based Homogeneous Catalysts

Since M. Brookhart [51,90] and V. C. Gibson [91] discovered that iron- and cobalt-based catalysts with bis(imino)pyridine as the ligand have high activity in ethylene oligomerization reaction in 1998, more and more researchers have been attracted to iron-based catalysts. Usually, the activity of iron catalysts is about one order of magnitude higher than that of cobalt catalysts [92]. At the same time, iron has the advantages of large reserves, cheapness, and environmental friendliness, which makes iron catalysts have broad application prospects [93]. In recent years, research on iron-based catalysts for ethylene oligomerization has mainly focused on ligands [94–96], which could be classified into NNN, NN, NNOO, NNO, and NNNNOO by the element attached to Fe in the ligand. The ligands of iron catalysts are highly modifiable [97–99], and some researchers are devoted to obtaining better catalytic effects by changing the spatial resistances [97,100], substituent groups [101–103], and functional groups [104,105] of the ligands.

3.1.1. NNN Ligand

Newly developed ligands and their complexes with iron are not yet comparable to the bis(imino)pyridine system. The latter is still the most promising iron catalyst for industrialization due to its high catalytic activity, high linear selectivity of products, and mild reaction conditions [106,107]. The catalytic activity of the bis(imino)pyridine iron catalyst is even higher than metallocene catalysts, reaching up to 10^8 g·mol^{-1}·h^{-1}. As shown in Figure 6, the catalyst ligand exhibits a pseudo-tetragonal planar conical geometric configuration, where R_1–R_4 are substituent groups. The two adjacent groups of the benzene ring perpendicular to the metal plane are located, respectively, above and below the plane [108]. The molecular weight of the product can be regulated by modifying the ligand skeleton [109,110] or changing the structure of the benzene ring substituent groups [111].

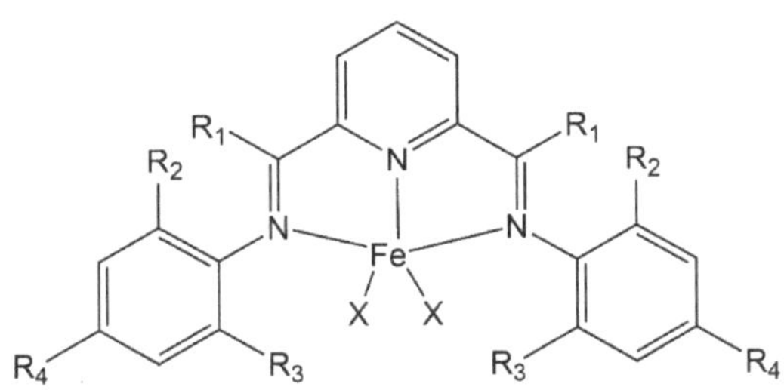

Figure 6. Typical structure of bis(imino)pyridine iron catalysts (complex **1**). Reproduced with permission from Ref. [108]. Copyright 1999, American Chemical Society.

In addition to the symmetric complexes mentioned above, Xie et al. [112] synthesized a series of 2,6-bis(imino)pyridine Fe(II) asymmetric complexes with alkyl and halogen substituents on different aromatic rings. Compared with symmetric alkyl or halogen substituted complexes, due to their spatial and electronic effects, the proportion of C_6-C_{16} in oligomers can significantly increase. Claudio Bianchini et al. [113] changed one of the aryliminos in the main chain of bis(imino)pyridine to alkylimino, demonstrating that the presence of two aryl imines in the complex is not a necessary condition for catalytic reactions. The new complexes are effective and selective for the ethylene oligomerization reaction, and the spatial size of alkyl groups can control the catalyst productivity and Schulz–Flory parameters.

Most NNN ligands have good catalytic activity as well as product selectivity. The spatial site resistance of the benzene ring neighboring substituent groups has an effect on the β-H elimination reaction at the catalyst active center. As the steric hindrance decreases, the β-H elimination reaction intensifies, and the molecular weight of the product will significantly decrease. The product is a series of linear α-olefins when the ortho position of the ligand's aniline group is substituted with a single alkyl group, such as methyl [114].

3.1.2. NN Ligand

For NN ligands, Zhang et al. [115] synthesized a novel star iminopyridyl iron complex using 1.0 G star macromolecule, pyridine-2-carboxaldehyde, and $FeCl_2{\cdot}4H_2O$ as raw materials. After activation by MAO, this complex exhibits high ethylene oligomerization activity (7.5×10^4 g·mol$^{-1}{}_{Fe}$·h^{-1}) while achieving a selectivity of 70.69% for butene. Higher catalytic activity can be obtained by increasing the ethylene pressure and Al/Fe molar ratio.

3.1.3. NNOO Ligand

Zhang et al. [116] synthesized a series of novel hyperbranched salicylaldimine Fe(II) complexes. The catalytic activity and product selectivity of ethylene oligomerization, which uses MAO as co-catalyst and toluene as solvent, can be modified by changing reaction temperature, ethylene pressure, and Al/Fe molar ratio. Under optimal conditions ([Fe] = 7 mmol, 298 K, Al/Fe = 500, 0.5 MPa ethylene), the catalytic activity can reach 6.91×10^4 $g{\cdot}mol^{-1}{}_{Fe}{\cdot}h^{-1}$, and the selectivity for higher-carbon-number olefins (C_{10}+) is 20.48%. Li et al. [117] synthesized iron complexes based on hyperbranched salicylaldimine ligands for ethylene oligomerization reaction. After activation with MAO, the activity of the catalyst reached 13.5×10^4 $g{\cdot}mol^{-1}{}_{Fe}{\cdot}h^{-1}$, and the proportion of C_4 in the product reached 52.1%, while the proportion of C_8 reached 32.63%.

3.1.4. NNO Ligand

George S. Nyamato et al. [118] synthesized (Pyrazolyl)-(phosphinoyl)pyridine iron(II) complex for ethylene oligomerization. Figure 7 shows the molecular structures and synthesis of (pyrazolyl)-(phosphinoyl)pyridine Fe(II), Co(II), and Ni(II) complexes. Compounds **2–7** in Figure 7 were activated with $EtAlCl_2$, MAO, or $AlMe_3$ as co-catalysts, and α-C_4 was obtained as the main product in the ethylene oligomerization reaction using hexane, chlorobenzene, or toluene as solvents. The catalytic activity and product selectivity depend largely on the nature of the alkylaluminium co-catalyst, the nature of the solvent, and the complex structure. Despite the low activity of this catalyst, it has the advantage of high selectivity for α-linear olefins.

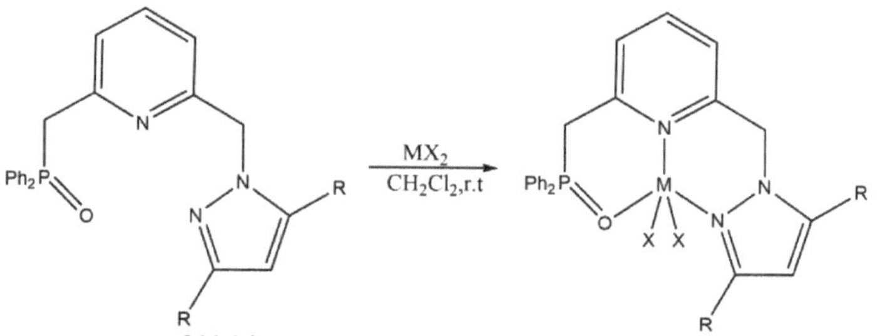

2 M=Ni R=Me X=Cl
3 M=Ni R=Me X=Br
4 M=Ni R=Ph X=Br

5 M=Co R=Ph X=Cl
6 M=Co R=Me X=Cl
7 M=Fe R=Me X=Cl

Figure 7. Molecular structures and synthesis of (pyrazolyl)-(phosphinoyl)pyridine Fe(II), Co(II), and Ni(II) complexes. Reprinted from Journal of Organometallic Chemistry, 783, Nyamato et al., (Pyrazolyl)-(phosphinoyl)pyridine iron(II), cobalt(II) and nickel(II) complexes: Synthesis, characterization and ethylene oligomerization studies, pp. 64–72. Copyright (2015), with permission from Elsevier [118].

3.1.5. NNNNOO Ligand

Makhosonke Ngcobo et al. [93] synthesized Fe(II) complex of bidentate and tridentate (imino)phenol ligands bearing quinoline motifs (Figure 8). Using chlorobenzene as solvent, complex **8** undergoes ethylene oligomerization reaction under the activation of $EtAlCl_2$. Under the optimal reaction conditions, the catalytic activity reached 6.84×10^5 $g{\cdot}mol^{-1}{}_{cat}{\cdot}h^{-1}$. The selectivity of C_4 (α-C_4) and C_6 (α-C_6) in the oligomers is 65% (93%) and 35% (19%), respectively.

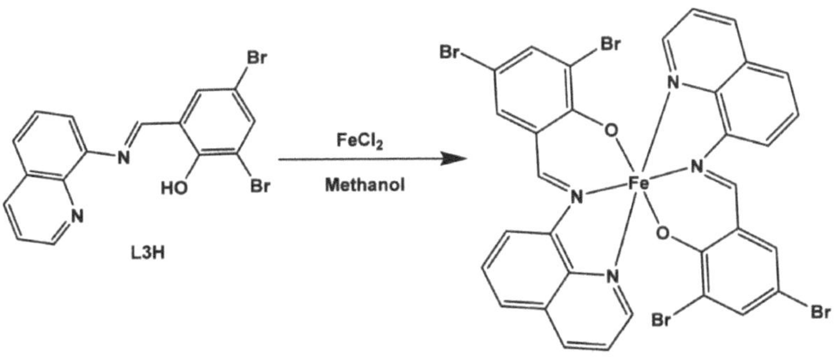

Figure 8. Molecular structures and the synthesis route of Fe(II) complexes of bidentate and tridentate (imino)phenol ligands bearing quinoline motifs (complex **8**).

3.2. Iron-Based Heterogeneous Catalyst

Fe-based homogeneous catalysts exhibit high activity and good selectivity in ethylene oligomerization. Although a better catalytic effect can be achieved by changing the spatial site resistance, substituent group, and functional group of the ligand for the complexes, their industrial application is limited owing to the difficulty of separating the catalyst from the products [119,120] and the large amount of heat released during the reaction process [121,122]. Moreover, the products of most catalytic systems are C_4 or C_6. In recent years, there has been an increasing number of studies on the immobilization of homogeneous catalysts on inorganic supports, such as zeolites [120], silica [123–125], and organic polymers [126,127].

In the study of heterogeneous ethylene oligomerization, the key aspect of effective catalyst design involves the isolated active metal sites within mesoporous support materials. Metal–organic frameworks (MOFs) are materials possessing a large surface area and a highly ordered porous structure, which can be used as effective catalysts for ethylene oligomerization reaction. The products of ethylene oligomerization catalyzed by iron catalysts generally have low carbon numbers. However, Yang Han et al. [128] synthesized MIL-100(Fe) via a hydrothermal method, which can catalyze ethylene tetramerization to produce C_8 liquid products under mild reaction conditions. The proportion of C_8 liquid products is over 70%, but the proportion of α-olefins is very low, only at 3%. Three conditions affect the catalytic activity of MIL-100(Fe): the surface area, the Fe^{2+}/Fe^{3+} ratio, and the vacuum activation temperature. Temperature changes the Fe^{2+}/Fe^{3+} ratio in the catalyst, and the relationship between temperature and catalytic activity is volcanic. Meanwhile, different co-catalysts affect the product distribution. The C_8 product selectivity was 75% when Et_2AlCl or MAO were used as co-catalysts. However, when Et_3Al was used as a co-catalyst, the C_8 product selectivity in the product was only 43.73%.

One way to synthesize loaded catalysts is to immobilize homogeneous catalysts directly onto the support. In 2002, Franz A. R. Kaul et al. [129] immobilized bis(imino)pyridyliron(II) complexes on silica. In recent years, Arumugam Jayamani et al. [125] immobilized Fe(II) catalysts chelated by (phenoxy)imine ligand on MCM-41. The series of catalysts demonstrated high ethylene oligomerization activity with $EtAlCl_2$ as a co-catalyst and toluene as a solvent. The catalytic activity of the homogeneous catalyst was 1.84×10^5 $g \cdot mol^{-1}{}_{cat} \cdot h^{-1}$. Meanwhile, the heterogeneous catalyst using MCM-41 to immobilize the complex presented catalytic activity of 0.97×10^5 $g \cdot mol^{-1}{}_{cat} \cdot h^{-1}$ under the same reaction conditions. The catalytic activity of the catalyst is relatively low after immobilization due to the large spatial site resistance, which limits the binding between ethylene and the metal active centers. In the two cycling experiments, there was almost no decrease in catalytic activity, and the product distribution remained unchanged, so the nature of the active species in the catalyst did not change during the cycling experiments.

Makhosonke Ngcobo et al. [130] synthesized loaded catalysts after the phenol ((triethoxysilyl)propylimino) ligand was immobilized onto SBA-15 or MCM-41 and reacted with $FeCl_2$. The ethylene oligomerization reaction catalyzed by the materials with $EtAlCl_2$ as a co-catalyst provided mainly C_6 oligomers (92% selectivity using Fe-SBA-15 catalyst). Because the ethylene oligomerization reaction occurs in the pores of the support and the pore size of SBA-15 is larger than that of MCM-41, and large pore size reduces the mass transfer resistance, Fe-SBA-15 was more active than Fe-MCM-41 in the reaction. The catalytic activity of Fe-SBA-15 decreased by 30% after two cycling experiments, and the selectivity for the product C_6 was maintained at 92%. The authors attribute the decrease in activity to the loss of active metal atoms and structural changes in the support. It is interesting that this research group previously immobilized homogeneous catalysts onto Fe_3O_4 magnetic nanoparticles to obtain recyclable catalysts [131]. Utilizing the ferromagnetism of the support, the catalyst can be easily separated from the reaction mixture via an external magnet and reused.

Another way to prepare loaded catalysts is to immobilize homogeneous catalysts onto a support that was pretreated by MAO or aluminum alkyls. Guo et al. [107] immobilized bis(imino)pyridine iron complexes on mesoporous molecular sieves (MCM-41 and SBA-15). Prior to this, the mesoporous molecular sieves need to be dried under vacuum at 393 K to constant weight and then refluxed with MAO overnight. During pretreatment, MAO chemically binds to the silanol groups on the inner walls of MCM-41 or SBA-15, so there is no need to add MAO as a co-catalyst during the ethylene oligomerization. The activators and monomer molecules are required to approach metal active sites in the reaction, and pores limit the process. After immobilization, the incidences of chain termination and transfer also increase. So, the loaded catalysts can lead to lower activity and lower carbon number products. Due to the smaller pore size of MCM-41 compared to SBA-15, this phenomenon is more pronounced in the former. Temperature also affects the product distribution of the reaction, and the increase in temperature accelerates the β-H elimination reaction, resulting in a higher proportion of products with low molecular weight. The group also used mesoporous molecular sieves modified with MAO as a support, loaded with iron-based diimine complexes, to catalyze the ethylene oligomerization to produce low-carbon-number α-olefins [124]. Under the same conditions, the catalytic activity of the complex reached 6.06×10^6 $g \cdot mol^{-1}{}_{Fe} \cdot h^{-1}$, while that of the loaded type was only 4.35×10^6 $g \cdot mol^{-1}{}_{Fe} \cdot h^{-1}$.

4. Cobalt-Based Catalysts

4.1. Cobalt-Based Homogeneous Catalysts

Late transition metal cobalt, nickel, and iron are generally used in the oligomerization or dimerization of ethylene. These metals have more d electrons, larger ionic radii, and easier access to ligands, resulting in the formation of catalysts with a strong tendency for β-H elimination. Nickel- and iron-based catalysts are discussed, and then this section will introduce the cobalt-based catalysts employed for ethylene oligomerization over the past decade, which are summarized in terms of the denticity (tri-dentate, bi-dentate, and other types) of the ligands.

4.1.1. Tri-Dentate Cobalt Complex

The tri-dentate-coordinated cobalt catalysts showed good catalytic activity in this reaction, and there has been much progress in this area in recent years.

Sun et al. [132] investigated cobalt complexes **9–12** (Figure 9) bearing 2-(1-(arylimino)ethyl)-7-arylimino-6,6-dimethylcyclopentapyridine for ethylene oligomerization. The molecular structures indicate that 2,7-bis(aryl)cyclopentapyridines is a tridentate ligand; however, one of the Co-N coordination bonds is weak due to the spatial separation of the nitrogen atom. Upon activation by MAO or MMAO, all the cobalt complexes exhibited catalytic activity toward ethylene oligomerization. Under the activation of MMAO, the oligomerization process was highly selective for α-olefins, and the resulting oligomers fully complied with the Schulz–Flory rule. Later, this group successfully synthesized cobalt(II) chloride complexes **13–18** (Figure 10) by a one-pot reaction of 2-benzoimidazolyl-5,6,7-trihydroquinolin-8-one with the corresponding aniline in the presence of cobalt dichloride [133]. Upon activation by MAO, all the complexes were effective in catalyzing the dimerization of ethylene, and especially the activity of compound 16 reached 1.23×10^5 g·mol^{-1}$_{Co}$·h^{-1}, producing products dominated by 1-butene. In contrast, with MMAO as the co-catalyst, the trimerization process of this system was superior, with selectivity for 1-hexene up to 49% (with complexes **15** and **17**). Ortho-benzhydryl-substituted 2-imino-1,10-phenanthroline cobalt complexes **19–26** were also synthesized and characterized by this group (Figure 11) [134]. In the presence of methylaluminoxane (MMAO), all the cobalt complexes showed good ethylene dimerization activity (up to 3.25×10^5 g·mol^{-1}$_{Co}$·h^{-1} at 50 °C), and the selectivity of C_4 was distributed from 92% to 100%, in which the selectivity of 2-butene ranged from 55 to 84.4%. The reason may be that the steric hindrance environment around the cobalt atom is less, which is favorable for the cis–trans conformational rearrangement of C–H.

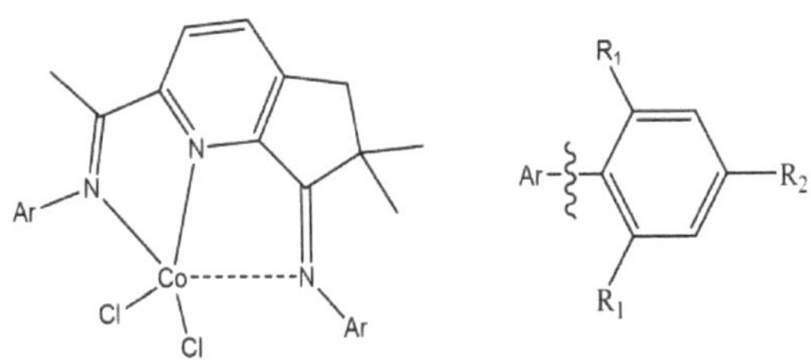

9 R1 = Me R2 = H **11** R1 = Me R2 = Me
10 R1 = Et R2 = H **12** R1 = Et R2 = Me

Figure 9. Structure of cobalt complexes **9–12**.

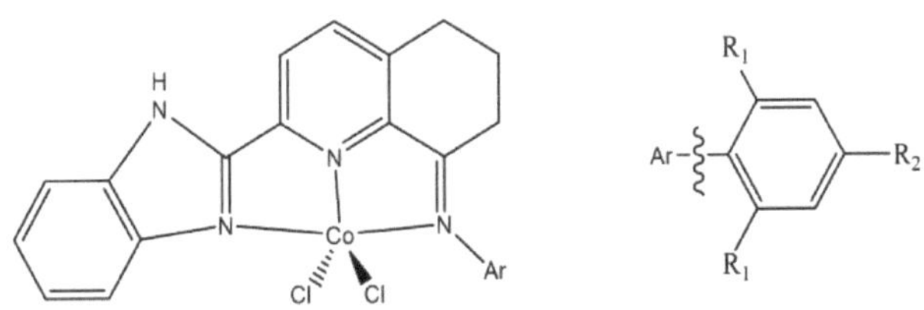

13 R1 = Me R2 = H **16** R1 = Me R2 = Me
14 R1 = Et R2 = H **17** R1 = Et R2 = Me
15 R1 = iPr R2 = H **18** R1 = t-Bu R2 = t-Bu

Figure 10. Structure of cobalt complexes **13–18**.

Konstantin P. Bryliakov and co-workers [135] prepared a series of novel cobalt(II) bis(imino) pyridine complexes bearing one or several electron-withdrawing substituents at the aniline moieties (Figure 12). Activated by MAO, the ethylene oligomerization activity of these complexes reached 1.8×10^7 g$_{products}$·mol^{-1}$_{Co}$·h^{-1}·bar^{-1}. Cobalt complexes **29**, **31**, **32**, and **34** with electron-withdrawing substituents (Cl or CF_3) showed higher catalytic activity (Table 2). Depending on the nature and number of electron-withdrawing groups, the process presents the selectivity ranges from dimerization (up to 98% for 1-butene) to oligomerization (C_4–C_{12+}) and is accompanied by the formation of strictly linear low-molecular-weight polyethylene.

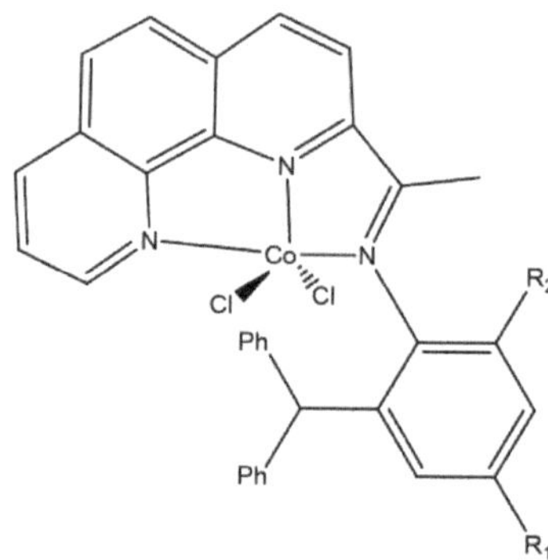

19 R_1 = Me R_2 = $CH(Ph)_2$ **23** R_1 = $CH(Ph)_2$ R_2 = Me
20 R_1 = Et R_2 = $CH(Ph)_2$ **24** R_1 = $CH(Ph)_2$ R_2 = Et
21 R_1 = iPr R_2 = $CH(Ph)_2$ **25** R_1 = $CH(Ph)_2$ R_2 = iPr
22 R_1 = Cl R_2 = $CH(Ph)_2$ **26** R_1 = $CH(Ph)_2$ R_2 = $CH(Ph)_2$

Figure 11. Ortho-benzhydryl-substituted 2-imino-1,10-phenanthroline cobalt complexes **19–26**.

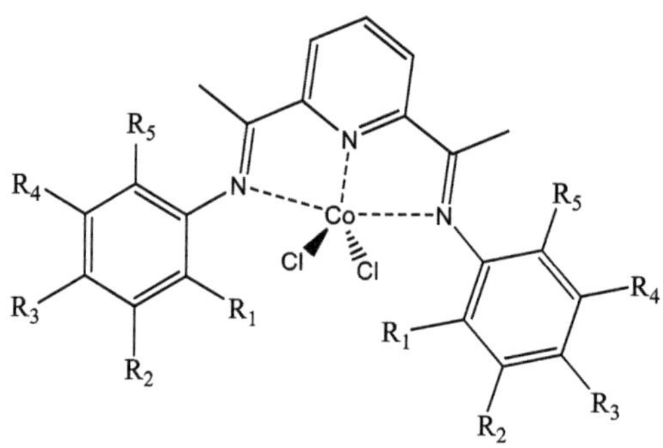

Figure 12. Cobalt(II) bis(imino) pyridine complexes bearing one or several electron-withdrawing substituents at the aniline moieties. Reprinted from Journal of Organometallic Chemistry, 884, Antonov et al., Catalytic ethylene oligomerization on cobalt(II) bis(imino)pyridine complexes bearing electron-withdrawing groups, pp. 55–58. Copyright (2019), with permission from Elsevier [135].

Table 2. The types of R_1–R_5 substituents of Cobalt(II) bis(imino) pyridine complexes in Figure 12.

Complex	R_1	R_2	R_3	R_4	R_5
27	H	H	F	H	H
28	H	H	Br	H	H
29	Cl	H	F	H	H
30	F	H	H	H	H
31	H	Cl	H	Cl	H
32	CF_3	H	H	H	H
33	H	F	H	F	H
34	CF_3	H	F	H	H
35	H	CF_3	F	H	H
36	F	H	F	H	F
37	Cl	H	H	H	Cl
38	H	Cl	H	H	H
39	Br	H	H	H	H
40	Me	H	H	H	Me

4.1.2. Bi-Dentate Cobalt Complex

Bi-dentate ligand cobalt catalysts tend to exhibit relatively low ethylene reactivity compared to tri-dentate ligand cobalt catalysts. Such low activity is probably due to the formation of unstable active species. The metal sites possess greater open spaces and tend to be more prone to coordinate with other species present in the system (leading to deactivation).

N^N bi-dentate ligand cobalt complexes were found to exhibit high activity and selectivity for dimerization. Katia et al. [136] investigated three cobalt-β-diimine complexes **41–43** (Figure 13) for ethylene oligomerization. At 10 bar, 30 °C, using EASC as a co-catalyst and toluene as a solvent, complex **42** exhibited the highest activity (TOF = 34 s^{-1}), while the TOF of complex **43** was only 0.8 s^{-1}. More charge on the center metal leads to higher catalytic activity, so complex **42** with two electron-donating methoxy groups possesses the highest activity. The selectivity of complexes **42** and **43** remained the same, with about 70% selectivity for C_4. Wang et al. [137] synthesized N,N-bidentate iron, nickel, and cobalt complexes bearing (6E,7E)-N^1,N^4-bis((pyridin-2-yl)methylene)benzene-1,4-diamine ligands and investigated their ethylene oligomerization potential. The selectivity toward 1-hexene for cobalt complex **44** was usually higher than that of the iron or nickel complexes at 40 bar pressure. The authors inferred that the N–M–N bite angle had an effect on the catalytic selectivity, and the small N–M–N bite angle produced high-carbon olefins. Another group [115], in 2022, synthesized a novel star iminopyridyl cobalt with 1.0 generation (1.0 G) star macromolecule and pyridine-2-carboxaldehyde as the raw materials. In the presence of methylalumoxane (MAO), cobalt complex **45** demonstrated higher ethylene oligomerization activity (up to 4.28 × 10^5 g·$mol^{-1}{}_{Co}$·h^{-1}), affording C_4 as the major product as well as C_6, C_8, and $C_{10–18}$ oligomers. Sun et al. [138] reported dinuclear pyridyl-imine Co-based complexes **46** and **47** prepared through the one-pot synthesis method. Complexes **46** and **47**, activated by MMAO, were capable of producing ethylene oligomers with moderate activity (up to 5.1 × 10^5 g·$mol^{-1}{}_{Co}$·h^{-1} for complex **47**) in which α-C_8 was the major product.

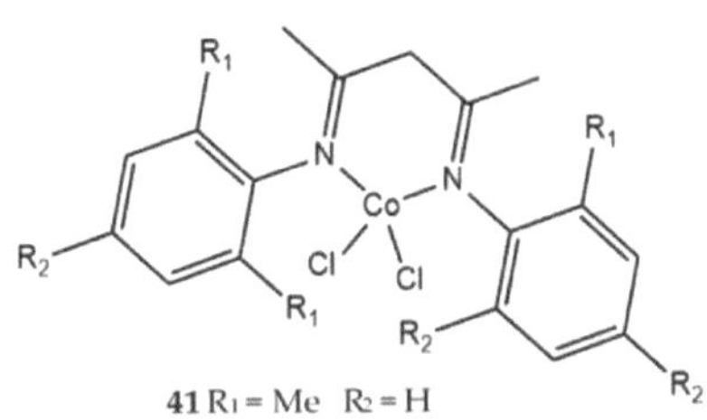

41 R_1 = Me R_2 = H
42 R_1 = Me R_2 = Me
43 R_1 = H R_2 = OMe

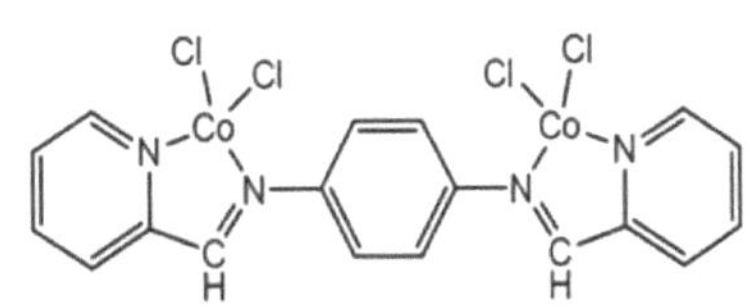

44

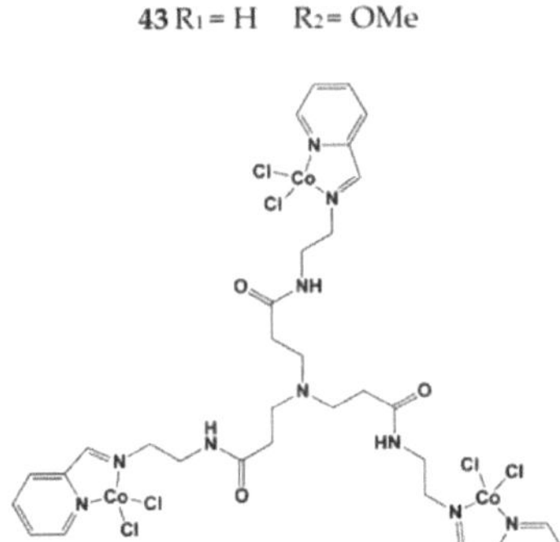

45

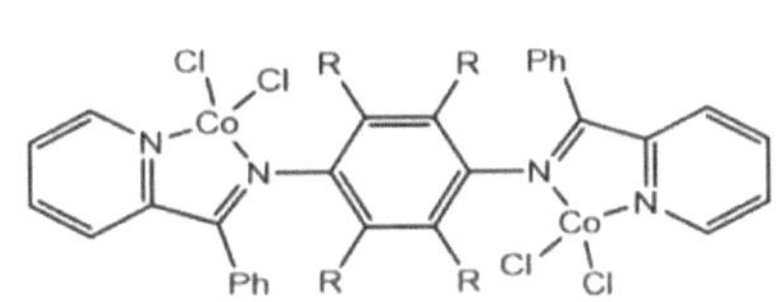

46 R = H
47 R = Me

Figure 13. Structure of cobalt complexes **41–47**.

Jiang and co-workers [139] reported a series of cobalt complexes featuring a monomeric structure with the silicon-bridged diphosphine ligand to the cobalt center (Figure 14). Transition metals supported with diphosphine ligands can exhibit good catalytic performance due to their singly and weakly coordinating moieties that can readily dissociate to create a vacant site for coordination, thereby enhancing the catalytic activity [140,141]. Changing the N-substituent from the larger 2,6-diisopropylphenyl to the smaller cyclopentyl and isopropyl moieties resulted in a decrease in the selectivity of 1-butene from 79.5% to 32.4 and 27.4%, respectively. Under optimal conditions, complex **48** exhibited the highest ethylene dimerization activity (2.3×10^5 g·mol^{-1}$_{Co}$·h^{-1}), providing 100% selectivity toward C_4 and 78.6% selectivity toward 1-butene.

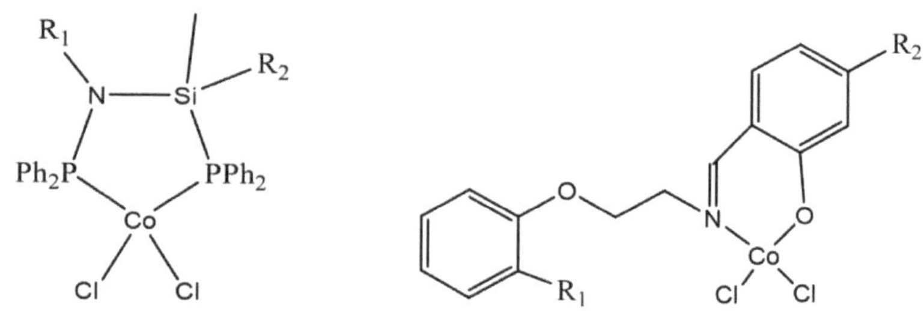

48 R_1 = 2,6-diisopropylphenyl R_2 = Me
49 R_1 = cyclopentyl R_2 = Me
50 R_1 = isopropyl R_2 = Me
51 R_1 = isopropyl R_2 = phenyl

52 R_1 = H R_2 = H
53 R_1 = H R_2 = OCH_3
54 R_1 = OCH_3 R_2 = H
55 R_1 = OCH_3 R_2 = OCH_3

Figure 14. Structure of cobalt complexes **48–55**.

Wang and co-workers [25] explored a series of cobalt complexes as catalyst precursors and reported their performance in the ethylene oligomerization with MAO as an activator (Figure 14). Under the optimum conditions (n (catalyst) = 7 μmol, T = 50 °C, P (ethylene pressure) = 2.0 MPa, n(Al)/n(M) = 300, toluene as solvent), complex **55** produced the selectivity of 34.44% to C_8 and activity of 3.38×10^5 g·mol^{-1}$_{Co}$·h). In this system, the catalytic performance of the cobalt complex is better than that of the nickel complex.

4.1.3. Other Types of Cobalt Complex

Salicylaldimine ligands, a kind of Schiff base, have the ability to coordinate to metals via hard nitrogen and oxygen donor atoms, which results in metal complexes with better anti-reductive stability and unusual thermal stability. Wang et al. [142] synthesized two cobalt complexes based on dendritic PAMAM-bridged salicylaldimine ligands (Figure 15). After EASC activation at 1.0 MPa, 25 °C, and an Al/Co molar ratio of 1500, the highest catalytic activity and selectivity of complex **50** for C_{10}–C_{20} were 3.44×10^6 g·mol^{-1}$_{Co}$·h^{-1} and 76.53%, respectively; and the highest catalytic activity and selectivity of complex **57** for C_{10}–C_{20} were 3.42×10^6 g·mol^{-1}$_{Co}$·h^{-1} and 84.50%. Another group [143], in 2016, explored three cobalt complexes (**58–60**) bearing hyperbranched salicylaldimine ligands with tetradecyl, hexadecyl, or octadecyl as cores (Figure 16). Under the same reaction conditions, it is obvious that the catalytic activity decreases (58 > 59 > 60) with the increase in the length of the alkyl chain in the skeleton, but it has little effect on the selectivity of the products. The main reason for the decrease in activity is that the bulkier ligand hinders the coordination of ethylene to the active metal center, resulting in a lower insertion rate of ethylene. Later, the group synthesized a series of 1.0 G hyperbranched macromolecules bridged salicylaldimine cobalt complexes using 1.0 G hyperbranched macromolecules, salicylaldehyde, and cobalt chloride hexahydrate as raw materials (Figure 17) [144]. To summarize the three types of cobalt complexes of salicylaldimine ligands mentioned above, the backbone structure affects the activity and product selectivity of ethylene oligomerization.

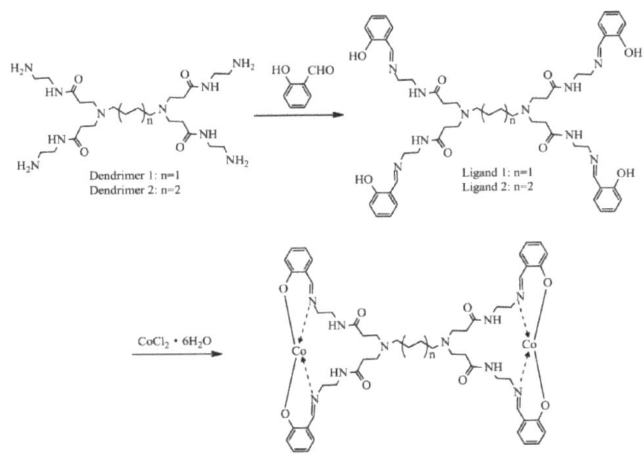

Figure 15. Structure of cobalt complexes **56** (n = 1) and **57** (n = 2). Cobalt complexes based on dendritic PAMAM bridged salicylaldimine ligands: Synthesis, characterization and performance in ethylene oligomerization, [142] Journal of Macromolecular Science, Part A: Pure and Applied Chemistry, 29 September 2016, reprinted by permission of the publisher (Taylor & Francis Ltd., accessed on 12 March 2024, http://www.tandfonline.com).

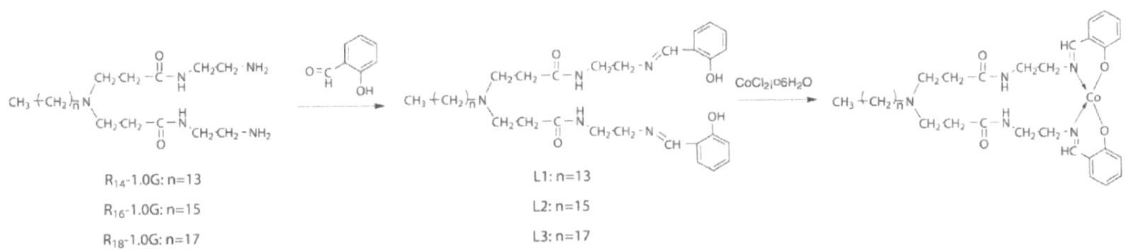

Figure 16. Structure of cobalt complexes **58** (n = 13), **59** (n = 15), and **60** (n = 17) bearing hyperbranched salicylaldimine ligands. Reproduced from Ref. [143] with permission from WILEY.

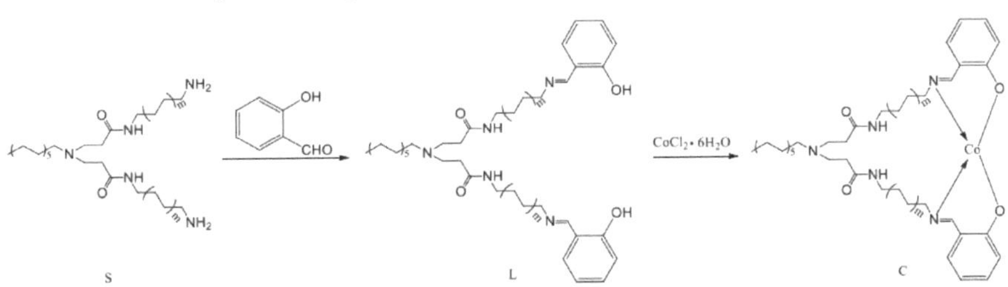

Figure 17. Structure of cobalt complexes **61** (m = 0), 62 (m = 1), **63** (m = 2), and **64** (m = 3).Hyperbranched macromolecules bridged salicylaldimine cobalt complexes: synthesis, characterization and ethylene oligomerization studies, Zhang et al., Chem. Pap, 71:1037–1046, Springer Nature, 2017, reproduced with permission from SNCSC [144].

Stephen O. Ojwach et al. [145] reported Co(II) complexes of 2-[(ethylimino) methyl] phenol ligands as potential catalysts for ethylene oligomerization reactions (Figure 18). The authors established multiple coordination modes of Co(II) complexes that are dependent on the N^O ligand. The ligands of these mono- and dinuclear complexes employ tri/didentate coordination. Under MAO activation, these cobalt complexes produce mainly C_4 (up to 96%), while C_6 is the main product under $EtAlCl_2$ activation. The Co(II) complex of bidentate and tridentate (imino)phenol ligands bearing pyridine and quinoline motifs was also successfully synthesized by this group [93]. The activation of the Co(II) complex by $EtAlCl_2$ co-catalysts makes it active in ethylene oligomerization reactions to produce mainly C_4 and C_6 oligomers.

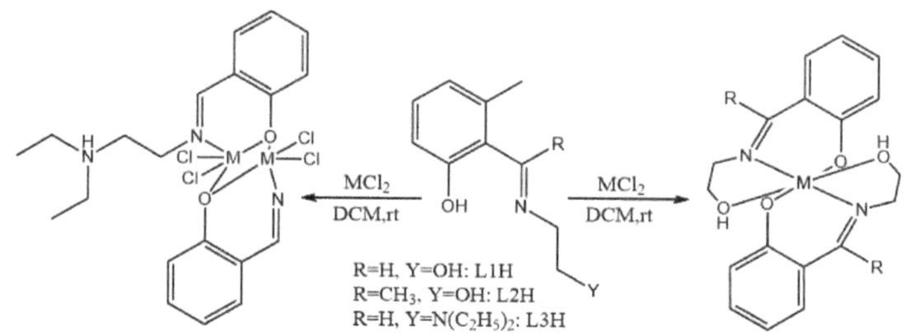

Figure 18. Synthetic protocol of N^O donor (ethylimino)methyl)phenol Co(II) complexes (right: complex **65** R = H and complex **66** R = CH_3; left: complex **67**). Reprinted from Molecular Catalysis, 478, Ngcobo et al., Structural elucidation of N^O (ethylimino-methyl)phenol Fe(II) and Co(II) complexes and their applications in ethylene oligomerization catalysis, p. 110590, Copyright (2019), with permission from Elsevier [145].

Sergey V. Zubkevich et al. [146] designed a novel tetradentate NNNO-heteroscorpionate ligand with an 8-methoxyquinoline pendant arm and synthesized monomeric molecular complex Co(II) halides with it. The complex was inactive under organoaluminum compounds (such as MAO, $EtAl_3$, etc.) activation, while, under Et_2AlCl or $Et_3Al_2Cl_3$ activation, the cobalt complexes have a moderate activity of 60~130 $kg_{olgomer}\cdot mol^{-1}{}_{Co}\cdot h^{-1}\cdot atm^{-1}$ in solvent of toluene and form a mixture of butene and hexene with an α-olefin percentage (1-butene and 1-hexene) of more than 70%. The higher activity (140 to 200 $kg_{oligomer}\cdot mol^{-1}{}_{Co}\cdot h^{-1}\cdot atm^{-1}$) of the catalyst in chlorobenzene is mainly due to the better solubility of the cobalt complex in polar solvents.

4.2. Cobalt-Based Heterogeneous Catalysts

Compared to homogeneous cobalt catalysts, heterogeneous cobalt catalysts have progressed very slowly regarding the ethylene oligomerization reaction. One important reason is that the catalytic activity of heterogeneous cobalt catalysts is lower than that of heterogeneous nickel catalysts.

In recent years, the chemical functionalization of carbon nanotubes (CNTs) with metal complexes has become one of the promising directions for materials development in various fields. Norah Alhokbany [84] reported the preparation of pyridylimine cobalt complex-functionalized multiwalled carbon nanotubes [Pyr-Co(II) MWCNTs] and their catalytic activity for ethylene oligomerization after activation by methylaluminoxane (MAO). The activity and selectivity of this catalyst was found to be sensitive to pressure and the Al/Co ratio. When the amount of co-catalyst is low, it exhibits moderate catalytic activity as well as a predominant selectivity to C_{10}–C_{12}.

Heterogeneous carbon-supported cobalt catalysts have become an important branch of the ethylene oligomerization reaction. Xu et al. [147] demonstrated that (CoO_x/N-C), in a continuous-flow reactor at 80 °C, produced octene with linearity of 77.6% in the ethylene oligomerization reaction, but 1-octene was only 5.2% at 20% conversion. Xu et al. [148] further claimed that incorporation of Cr into the cobalt oxide on a carbon catalyst (Cr-CoO_x/N-C) improved the catalytic activity and stability in the ethylene oligomerization reaction at 80 °C. Alvin Jonathan et al. [149] reported that cobalt oxide on a carbon catalyst is more stable at higher temperatures (~200 °C), likely due to the reduction of Co_3O_4 to CoO, while rapid deactivation is observed at lower temperatures (e.g., 80–140 °C). At a reaction temperature of 200 °C, the catalyst was highly selective for linear olefins, including linear α-olefins. At 48.3% ethylene conversion, the product linearity can reach more than 90%; at low conversion of 20%, the product linearity can reach 60%. Another group [150] demonstrated that increasing the catalyst pretreatment temperature from

230 °C to 560 °C in an argon atmosphere reduces cobalt oxide to cobalt metal and increases activity. However, the Co metal on the carbon catalyst was deactivated owing to the formation of polyethylene on the catalyst (the spent catalyst contained approximately 50 wt% polyethylene after 24 h of reaction). The catalyst support has an effect on the cobalt catalyst for ethylene oligomerization to generate linear olefins [151]. The high-temperature-treated carbon-supported cobalt oxide catalyst (CoO_x/MRX-HTTC) was almost twice as active as the acid-washed carbon-supported cobalt oxide catalyst (CoO_x/MRX-AWC), with more linear alpha olefins. Acid wash treatment oxidizes the surface functional groups and increases the oxygen content of the carbon support, leading to the isomerization of linear α-olefins into linear internal olefins. Zhong et al. [152] investigated the structure-performance evolution of the Co/NAC catalyst for ethylene oligomerization in detail. The initial ethylene conversion was as high as 64%, but it was quickly deactivated and the conversion dropped to 4% after 30 h of reaction. Butene dominated the product, and its selectivity gradually increased in the first 10 h, and then remained at about 74%.

UiO-66-NH_2-grafted pyridinimine ligand [153] was prepared by a post-synthetic modification method using metal–organic skeletons (MOFs) UIO-66-NH_2 and pyridine 2-formaldehyde as raw materials, and then it reacted with $CoCl_2 \cdot 6H_2O$ to obtain the UiO-66-NH_2-grafted pyridinimine cobalt catalyst. When cyclohexane was used as the solvent and methylaluminoxane (MAO) as the co-catalyst, the activity of the catalyst for the oligomerization of ethylene could reach 1.23×10^5 $g \cdot mol^{-1}{}_{Co} \cdot h^{-1}$, and the selectivity for butylene was 88.96% at a pressure of 10 bar using the molar ratio [Al/Co] of 1000:1 at 25 °C.

Stephen O. Ojwach et al. [125] immobilized 2-phenyl-2-((3(triethoxysilyl)propyl) imino)ethanol(L1) or 4-nitro-2-((3(triethoxysilyl)propyl)imino)methyl)phenol(L2) on an MCM-41 support and then reacted with $CoCl_2$. The immobilized catalysts were active regarding the ethylene oligomerization, while their activity was lower than that of the homogeneous compounds. There was no significant loss regarding the catalytic activity or leaching of active substances in three runs.

5. Chromium-Based Catalysts

5.1. Chromium-Based Homogeneous Catalysts

For the catalytic systems mentioned above, it is generally believed that the ethylene oligomerization processes are in compliance with the ethylene insertion/β-H elimination mechanism (Cossee chain growth mechanism). Thus, the reaction products are all mixtures of a range of olefins, consistent with the Schulz–Flory distribution. However, Cr-based catalysts follow the metallacyclic mechanism and mainly catalyze selective ethylene tri-/tetramerization reactions. In recent years, there have been many reports of Cr-based catalysts in the ethylene oligomerization process owing to their high activity [154–156], high selectivity [157,158], and diversity of ligand structures [159–161].

5.1.1. Ethylene Trimerization

As early as 1967, Manyik [162] found the presence of a small amount of 1-hexene in the product of ethylene polymerization catalyzed by chromium(III) 2-ethylhexanoate and partially hydrolyzed triisobutylaluminum. In 1976, Manyik et al. first proposed the metallacyclic mechanism for the selective ethylene trimerization by Cr-based catalysts. Briggs [1] optimized the chromium(III) 2-ethylhexanoate system using dimethoxyethane in 1989 to promote the trimerization of ethylene to form 1-hexene and verified the metallacyclic mechanism that forms a seven-membered ring intermediate from chromium atoms and three ethylene molecules followed by reductive elimination to produce 1-hexene. Chevron Phillips proposed an efficient catalytic ethylene trimerization using an activator consisting of dichloroethylaluminum or triethylaluminum with chromium(III)2-ethylhexanoate combined in situ with 2,5-dimethylpyrrole as the ligand in 1996. This catalytic system can reach an activity value of 3.019 $kg \cdot mol^{-1} \cdot h^{-1}$) and selectivity of 1-hexene up to 96%

at a temperature of 388 K and an ethylene pressure of 100 bar. In 2003, Chevron Phillips achieved industrial production of 1-hexene using this catalytic system.

With the increasing research on the ethylene trimerization reaction catalyzed by Cr catalysts, researchers have found many types of ligands, such as the PNP [163–165], SNS [166–168], NNN [169], PNS [170,171], and PN [172] types.

The PNP ligand is one of the most typical and important ligands for ethylene tri-/tetramerization. In 2002, Wass et al. [156] reported a series of PNP-type Cr complexes that can efficiently catalyze the ethylene trimerization reaction with commercially attractive production rates. The chromium complexes with ligands of the type $Ar_2PN(Me)PAr_2$ (Ar = ortho-substituted aryl group) were synthesized, which catalyze ethylene trimerization after MMAO activation, and the C_6 selectivity in the product was 81%, with 1-hexene accounting for 99.7%. Blann et al. [165] synthesized the PNP-type catalyst, which mainly produced 1-octene. However, when the H in the ortho position of the phenyl group was replaced by an ethyl group (Figure 19), the steric hindrance around the metal center caused by ortho ethyl substituents increased the selectivity of 1-hexene in the product, with the highest C_6 product selectivity reaching 90.7% and 1-hexene accounting for 99.7% in the C_6 product. Similarly, Kuhlmann et al. [173] connected a cyclohexyl group to the N atom of the PNP ligand. When both the ortho positions of the cyclohexane connected to the N atom were completely replaced by methyl groups, the product changed from C_8-dominant to a 1:1 mixture of 1-hexene and 1-octene. McGuinness et al. [174] showed that, when a dicyclohexylphosphine substitution was introduced at the phosphorus atom in the PNP ligand, the product was predominantly polyethylene (PE), and the catalytic activity was reduced. When the dicyclohexylphosphine substituent was replaced with diethylphosphine, the product was dominated by 1-hexene, with a selectivity of 99.2%.

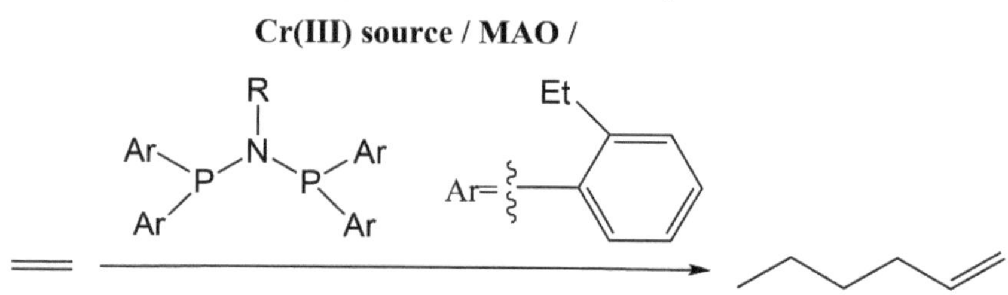

Figure 19. Cr-based catalysts with bulky diphosphinoamine ligands (complex **68**). Reproduced from Ref. [165] with permission from the Royal Society of Chemistry.

Derivative species of PNP-type ligands (including SNS, PCCP, PNCN, etc.) can also be used in ethylene trimerization reactions. McGuinness et al. [174] researched SNS-type ligands as shown in Figure 20 for ethylene trimerization. After activation by the co-catalyst MAO, the SNS ligand also showed excellent ethylene trimerization ability, with TOF up to 263,757 h^{-1} and 1-hexene selectivity up to 97%. Temple et al. [166] synthesized SNS-type ligands with a catalytic activity of 1.6×10^5 $g \cdot g^{-1}{}_{Cr} \cdot h^{-1}$ after MAO activation, for which 98.4% C_6 (including 99.8% 1-hexene) selectivity was achieved.

Boelter et al. [175] synthesized a series of PCCP-type ligands as shown in Figure 21 and found that the size of the phosphorus heterocyclopentane substituent affects the selectivity of the catalysts. With larger substituents, the proportion of 1-hexene in the product increases. Alam et al. [176] researched ethylene trimerization based on chromium(III) silylated diphosphinoamines. When the ligand skeleton in the system was PSiNP, the catalytic activity reached 43,461 $g \cdot g^{-1}{}_{Cr} \cdot h^{-1}$, with a selectivity of 83% for 1-C_8 and only 16% for 1-C_6. When the catalyst ligand skeleton was PSiNSiP, the catalytic activity decreased to 456 $g \cdot g^{-1}{}_{Cr} \cdot h^{-1}$, but the selectivity for 1-C_6 increased to 91%. The authors believe that the steric hindrance of the metal active center and the angle of P–Cr–P affect the catalytic activity and selectivity of the catalyst. Zhang [177] also believes that, as the bond length becomes shorter, the ligand bite angle becomes smaller, and the catalytic activity increases.

Sydora et al. [157] synthesized a class of PNCN ligands that exhibit high catalytic activity for selective ethylene oligomerization. The molar ratio of 1-hexene to 1-octene in the product could be adjusted from 140 to 1.5 by adjusting the steric hindrance. As the steric hindrance increased, the catalyst activity decreased from 54,700 $kg \cdot g^{-1}{}_{Cr} \cdot h^{-1}$ to 35,170 $kg \cdot g^{-1}{}_{Cr} \cdot h^{-1}$, and the selectivity of the C_6 product also decreased from 93.6% to 52.2%, with 33.7% C_8 and 4% PE present in the product. The rate of α-olefin yield in the oligomers also decreased, and the proportion of α-C_6 decreased from 99% to 45%. Radcliffe et al. [178] synthesized a PCN ligand that can selectively catalyze ethylene trimerization. By changing the substituents on the same ligand skeleton, high catalytic activity (Act. = 2799 $kg \cdot g^{-1}{}_{Cr} \cdot h^{-1}$, 1-$C_6$ = 84%) or high product selectivity (Act. = 797 $kg \cdot g^{-1}{}_{Cr} \cdot h^{-1}$, 1-$C_6$ = 97%) can be achieved. The ethylene trimerization using PNPN-type ligands to produce 1-hexene will not elaborated on here; it was discussed in detail by Uwe Rosentha [179] in 2020.

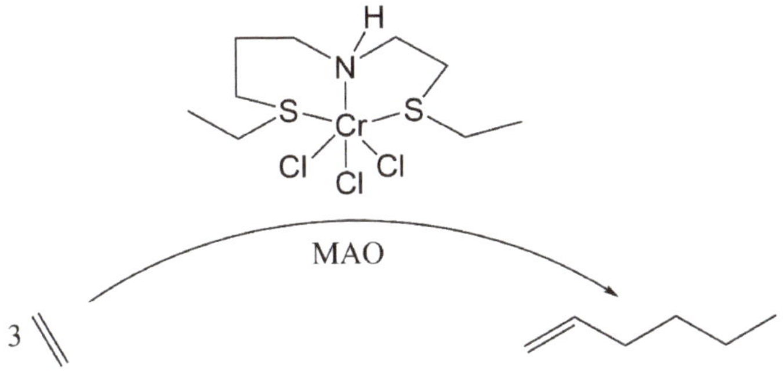

Figure 20. The structure of SNS-type ligand (complex **69**). Reproduced with permission from Ref. [174]. Copyright 2005, American Chemical Society.

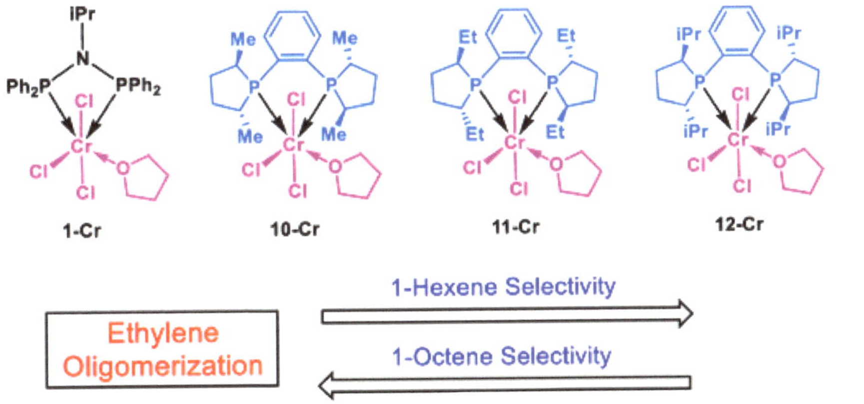

Figure 21. PCCP-type ligand structures with different substituents (from left to right: complexes **70–73**). Reproduced with permission from Ref. [175]. Copyright 2020, American Chemical Society.

In the process of ethylene oligomerization catalyzed by Cr-based catalysts, it is inevitable to produce polyethylene, which will block the pipeline. Many research works suggest that PNP ligands can be induced by alkyl aluminum co-catalysts to form PPN ligands during catalytic reactions [180], which have high activity for ethylene polymerization reaction. Zhao et al. [181] synthesized PCCP-type ligands containing para-alkyl substituents with good results in ethylene trimerization and selectivity to PE below 0.1 wt%. The authors also found that the reactivity will be higher as the σ-donor ability of the ligand increases. The catalytic activity could reach up to 5243 $kg \cdot g^{-1}{}_{Cr} \cdot h^{-1}$, while the selectivity for 1-hexene reached 74.9%.

Temperature and pressure also influence the reaction results. The optimal operating temperature for the catalysts synthesized by McGuinness [174] was in the range of 80–100 °C. At lower temperatures, activity decreases and leads to high polymer content, while, at higher temperatures, catalyst deactivation is observed and oligomers with higher carbon numbers form. Zhang et al. [163] synthesized $PNP(NR_2)_2$-containing catalysts and investigated the effects of temperature and pressure. When the pressure was 1 MPa, the catalytic activity and selectivity became volcanic with the increase in temperature, and the catalytic activity reached the highest at 318 K and the selectivity for C_6 products reached 92.23%, of which the proportion of 1-hexene was 98.87%. An increase in pressure can increase the reaction activity while reducing the selectivity towards C_6. Soheili et al. [167] prepared SNS ligands in which the products varied greatly under different conditions. At a pressure of 15 bar, the reactivity reached 6420 $g_{1\text{-}C6}\cdot g^{-1}{}_{Cr}\cdot h^{-1}$. However, when the pressure increased to 27 bar, the product was predominantly PE and the reactivity increased to 101,282 $g_{PE}\cdot g^{-1}{}_{Cr}\cdot h^{-1}$.

The addition of co-catalysts had a similar effect on the reaction results. Choi et al. [182] synthesized chromium catalysts in situ with trifluoromethyl-containing ligands. When all the other conditions were equal, the proportional mixture of DMAO and TIBA as an activator was more catalytically active, more selective for 1-hexene, and only trace amounts of PE were produced when DMAO was used alone as an activator. The catalytic system achieved an activity value of 1248 $g\cdot mol^{-1}{}_{Cr}\cdot h^{-1}$ with a high percentage (98.5 wt%) of 1-hexene in the product. The catalytic activity of the novel dendritic PNP chromium complexes synthesized by Wang et al. [183] showed a volcanic shape with the amount of co-catalyst MAO. The highest catalyst activity was achieved at Al/Cr = 700. Stennett [184] synthesized a dinuclear PNP-type ligand with higher activity and selectivity using $AlMe_3$ rather than MAO as a co-catalyst. The addition of excess MAO during the catalytic process leads to the production of a large amount of PE in the product.

Currently, there are also many researchers who have been combining computational chemistry and experimental science to study and design Cr-based catalyst ligands [185–187]. Fan et al. [185] combined Density Functional Theory (DFT) and Artificial Neural Networks (ANNs) to aid the design of new PNP ligands and successfully predict their performance. The authors also proposed that the spatial site resistance property affects the performance more significantly than the electronic property. Using DFT theory, Zhong et al. [186] investigated Cr/PCCP-type catalysts and concluded that the semi-instability of methoxy in the PCCP ligand is important for the selective regulation of ethylene trimerization. Similarly, Wang et al. [187], using DFT theory to study PCCP ligands, suggested that the coordination of the oxygen atom of the methoxy to the chromium center may be a key factor in the highly selective formation of 1-hexene.

In summary, different ligand backbones have significant effects on the reaction performance, and changing the backbone structure may lead to catalyst deactivation. The electronic effects and spatial site resistance of different substituents linked to the ligand backbone play a key role in catalyzing the selective oligomerization of ethylene, affecting both the catalytic activity and selectivity. Meanwhile, temperature, pressure, solvent, and co-catalysts are also the main factors affecting catalytic performance. In current research works, more and more researchers are using a combination of computational and experimental methods to study the catalytic reactions, which can help to find suitable catalysts efficiently and predict the catalytic behavior.

5.1.2. Ethylene Tetramerization

At the same time as the development of ethylene trimerization technology, many researchers explored the selective tetramerization of ethylene to produce 1-octene through Cr-based catalysts. Some researchers [188] remain skeptical about this. According to the metallacycle mechanism, obtaining 1-octene products through ethylene oligomerization requires the expansion of metal coordination compound intermediates from a seven-membered to a nine-membered metallacycle. However, in 2004, Bollmann et al. [189] of

SAAOL first reported the aluminum-oxide-activated Cr/$(R2)_2P)_2$NR1 system, where R1 and R2 are different substituents (Me, Ph, i-pr, etc.). The complex produces 1-octene with up to 70% selectivity while also producing small amounts of 1-hexene, methylcyclopentane, and methylene cyclopentane as the main byproducts. The results indicate that changing the steric hindrance and substituent type of ligands can affect the activity and selectivity of the catalyst. The catalyst ligands for ethylene tetramerization can be divided into PNP, PNPO, PNPN, PCP, and other ligands. These above ligands have been discussed in detail by Hao [13] and will not be repeated here.

5.1.3. Unselective Ethylene Oligomerization

The main products, 1-hexene or 1-octene, can be obtained using the above chromium catalysts in the catalytic ethylene oligomerization reaction, and the product distributions can be altered by changing the spatial site resistance of their active center. There are also some catalysts that are not capable of selective catalysis. Most Cr complexes with NNN- [190–192], NNO- [193,194], NO- [195,196], and NN-type [197,198] ligands exhibit high catalytic activity for ethylene oligomerization and provide a wide distribution of oligomer products from C_4 to C_{24}.

Zhang et al. [190] synthesized NNN-type ligands as shown in Figure 22, after activation by MMAO, which exhibit high activity in ethylene oligomerization (up to 7.36×10^6 g·mol$^{-1}{}_{Cr}$·h^{-1}), and the selectivity of C_4, C_6, and $C_8{}^+$ in the product is 19.3%, 32.8%, and 47.9%, respectively. Meanwhile, the catalyst also possesses high ethylene polymerization activity (up to 1.28×10^6 g·mol$^{-1}{}_{Cr}$·h^{-1}). Milani et al. [199] synthesized Cr complexes with bidentate thioether-imine [N, S] ligands to catalyze ethylene oligomerization. The product range is very wide, with the highest selectivity of oligomers reaching 98.8%, and the selectivity of the (C_6 and C_8) products is 35.9%. However, when the reaction temperature decreased, the selectivity of the PE in the product reached 68.1%.

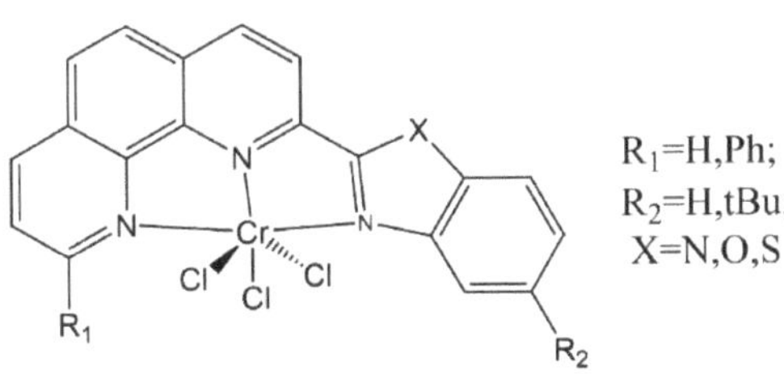

Figure 22. The structure of Cr(III) complexes bearing 2-benzazole-1,10-phenanthrolines (complex **74**). Reproduced from Ref. [190] with permission from the Royal Society of Chemistry.

Generally speaking, as the temperature increases, the activity of ethylene oligomerization increases, but the selectivity of α-olefins in the products decreases due to the accelerated rate of the chain isomerization reaction. However, when the temperature exceeds a certain range, the activity will decrease, which is due to the decomposition of active substances leading to catalyst deactivation or a decrease in the solubility of ethylene in the solvent at high temperatures. Relatively, the proportion of polymerization products decreases because the β-elimination reaction rate accelerates with increasing temperature. As the pressure increases, the catalytic activity increases and the proportion of long-chain products increases. The structures of ligands, co-catalysts [200], substituents [201], and the chemical environment all have a significant impact on the reaction activity.

5.2. Chromium-Based Heterogeneous Catalysts

Although homogeneous catalysts have high activity and good product selectivity, they are prone to deactivation. In contrast, heterogeneous catalysts can not only be separated from the product but also protect the active center of the catalyst. Cr-based heterogeneous catalysts, such as immobilizing homogeneous catalysts onto supports [202,203], dispersing them into ionic liquids [204,205], and Cr-MOFs [206,207], were reported recently.

Lamb et al. [202] immobilized a Cr-based complex on oxides including SiO_2, γ-Al_2O_3, and mixed SiO_2–Al_2O_3 to obtain supported catalysts, which can perform selective ethylene trimerization after treatment via an aluminum activator. When using SiO_2 or SiO_2–Al_2O_3 as the carrier, high reaction activity and moderate trimerization selectivity can be achieved, while, when using γ-Al_2O_3 as the carrier, the reaction activity decreases significantly, and the product is mainly PE.

Gao et al. [205] introduced organ aluminate-based ionic liquids in the PNP/Cr/MAO system and above 98% selectivity for (1-hexene and 1-octene) was obtained. The separation of the catalyst from the product can be achieved by decanting the upper organic phase containing the product, and the catalyst can be reused by adding a new solvent and MAO. The catalytic activity gradually decreased in three cycles because small amounts of impurities were inevitably introduced during the circulation process, but the selectivity for 1-hexene was still above 96%.

Fallahi et al. [203] immobilized SNS ligands onto ionic liquid-functionalized SBA-15 for ethylene trimerization. Under optimal conditions, the catalytic activity can reach 19,394 $g_{1\text{-}C6}\ g^{-1}{}_{Cr}\ h^{-1}$, and the selectivity for 1-hexene can reach 99.7%. Müller et al. [204] dissolved the homogeneous catalyst Cr-PNP in hydrogenated dibenzyltoluene (H18-DBT) together with the co-catalyst MAO and then dispersed the solution onto activated carbon or SiC. This catalyst can be used in continuous gas-phase reactions for up to 220 h. After 40 h of reaction, the selectivity of the product tends to stabilize, with a selectivity of 25.2% for 1-hexene and 44.6% for 1-octene. However, PE, which is continuously produced during the reaction, adheres to the catalyst surface affecting the contact between the ethylene and the active center, leading to deactivation.

Goetjen et al. [206] synthesized Cr-SIM-NU-1000 via solvothermal deposition in MOFs (SIM), and the ethylene conversion rate can reach 20% at 1 bar of ethylene partial pressure and ambient temperature. It is worth noting that the amount of cocatalyst used in the reaction was very low (Al/Cr = 3), and the reaction raw material was a mixture of H_2 and C_2H_4 to suppress the production of polymers. However, the distribution of the products in this catalytic system was very broad, with C_8–C_{18} accounting for 79%. Liu et al. [207] selected MIL-100(Cr) for selective ethylene oligomerization because it can eliminate two water molecules connected to the metal sites after vacuum heating and form coordinatively unsaturated metal sites for reaction. Different pre-treatment temperatures can cause a change in the valence state of Cr, which leads to a transition from ethylene oligomerization to polymerization.

6. Other Metals-Based Heterogeneous Catalysts

In addition to the nickel, iron, cobalt, and chromium metals that have been explored in the heterogeneous ethylene oligomerization reaction, other metals (such as Ru, Zr, and Ga) have been successively applied for this reaction in recent years with promising catalytic performance.

Iker Agirrezabal-Telleria et al. [208] developed a method for tailoring (Ru)HKUST-1 defects via thermal-based protocols. Thermal defect engineering leads to (Ru)HKUST-1, with the defective nodes exhibiting a well-defined active site for ethylene dimerization. In the absence of solvents and co-catalysts, the catalysts can achieve activities up to TOF = 200 h^{-1}, stable catalytic times in excess of 120 h, and selectivity of 99% for 1-butene. At the same time, the authors demonstrated that the behavior of the active Ru-H sites produced by an economical and versatile thermal method is the same as that of the active Ru-H sites prepared by conventional ligand engineering methods.

Nicole J. LiBretto et al. [209] reported that silica-supported single-site catalysts containing fixed main group Zn^{2+} or Ga^{3+} ionic sites catalyze the ethylene oligomerization reaction to reach an equilibrium distribution of linear olefins at rates similar to those of Ni^{2+} (Table 3). By varying the spatial velocity of C_2H_4, at a conversion of 3%, the selectivity of Ga/SiO_2 for butene was 75.9%, for hexene was 16.8%, and for octene was 6.4% at 1 atm and 250 °C. Moreover, Zn/SiO_2 showed 85.5% selectivity for butene, 2% for hexene, and no 1-octene products. Both catalysts were able to catalyze stably for at least more than 25 h under the above conditions. At low conversion (3%), Ga/SiO_2 is more favorable for the generation of high-carbon olefins compared to the Zn/SiO_2 and Ni/SiO_2 catalysts. The product distribution of Ga/SiO_2 does not change significantly, whereas the transfer of the products of Zn/SiO_2 to lower carbon numbers can be achieved at high pressure: 30.6 atm with high conversions. The authors also demonstrated that the catalytic cycle begins with the heterodissociation of the vinyl C–H bond of ethylene, and the reaction intermediates and basic steps were shown to be consistent with the Cossee–Arlman mechanism for Ni^{2+} transition metal catalysts.

Table 3. Product selectivity and conversion for ethylene oligomerization at 250 °C.

Catalyst	Pressure (atm)	Conversion(%) [1]	Selectivity (%)				
			Butenes (C_4=)	Hexenes (C_6=)	Octenes (C_8=)	C_{10}+	Rate (mol C_4H_8 molM^{-1} s^{-1}) [2]
Ni/SiO_2		3	86.1	11.8	0.4	0	15×10^{-4}
Ga/SiO_2	1	3	75.9	16.8	6.4	0	7×10^{-4}
Zn/SiO_2		3	85.5	2.0	0.0	0	1×10^{-4}
Ni/SiO_2		20.7	86.2	11.1	2.9	0	7×10^{-2}
Ga/SiO_2	30.6	20.6	74.2	16.1	4.9	4.2	8×10^{-2}
Zn/SiO_2		15.2	96.0	0.8	0	0	5×10^{-2}

[1] Conversion was varied using different space velocities over 1 g of catalyst in a 3/8 in diameter quartz reactor tube. [2] Rate was calculated by normalizing the mol C_4H_8/s produced by the total mol of metal on the catalyst (rate = (mol C_4H_8)/(mol M ∗ s)).

7. Summary and Outlook

In this article, we summarize the progress of ethylene oligomerization in homogeneous and heterogeneous phases using different metals (Ni, Fe, Co, Cr, etc.) as active sites. In the homogeneous catalytic system, we mainly discuss the effects of the molecular structure, electronic, and coordination states of complexes on their catalytic activity and selectivity. In heterogeneous catalytic systems, we mainly concentrate on the influence of various supports (metal–organic frameworks, covalent organic frameworks, molecular sieves, etc.) and different ways to introduce active centers regarding the activity in ethylene oligomerization.

Due to the discovery of the "nickel effect", nickel, as a post-transition metal, was first used in the ethylene oligomerization reaction. In recent decades, researchers have made significant progress in this field. Firstly, the "SHOP" ethylene oligomerization process derived from the neutral [P, O] ligand catalytic system reached a total olefin production of 1.3 million tons/year in 2018 [2,210]. Secondly, the invention of α-diimine ligands has made nickel-based catalysts designable and enabled the study of the influence of spatial site resistance and electronic effects on catalytic performance [22]. Finally, the development of single-site MOF catalysts bridges the gap between homogeneous and heterogeneous catalysis and provides an ideal platform for elucidating structure–activity relationships using molecular chemistry mechanisms in combination with computational chemistry, transient spectroscopy, and advanced diffraction techniques. For homogeneous systems, the choice of the nickel precursor and the design of ligands naturally play tremendous roles regarding the catalytic effect. In summary, nickel catalysts for the dimerization of ethylene are monocentric asymmetric complexes of Ni^{II} with tridentate ligands containing two or three nitrogen atoms in the conjugated system. The development of new catalytic systems continues to be based on a trial-and-error approach. Although high selectivity can be achieved with homogeneous catalysts, it is poorly recoverable; on the other hand, heterogeneous catalysts are robust but lack selectivity. Among the heterogeneous nickel catalysts mentioned above, metal–organic frameworks are very attractive supports due to their high porosity and site density, unlimited variety of designs, and possible post-synthesis modifications. Ni-ZIF-8 (0.4 wt%), Ni(1%)-MFU-4l, and MIL-125(Ti)-NH_2(Ni) employ

Ni as a catalytic active site and present higher activity and selectivity towards 1-butene. However, some challenges remain for MOFs in olefin oligomerization reactions. On the one hand, secondary reactions such as isomerization and dimerization hinder the generation of α-olefins and linear selectivity, which must be avoided using rationally designed MOF catalysts. On the other hand, the catalyst deactivation due to the accumulation of polymers on the active sites was addressed by rationally selecting and designing the building blocks of MOFs, adjusting the pore size of MOFs, introducing flexible substituents around the active centers, and increasing the percentage of surface active sites.

The discovery of highly active iron and cobalt catalysts is another milestone in the development of olefin oligomerization catalysts using late transition metals. The linearities of the oligomerization products obtained using iron and cobalt catalysts in homogeneous systems are higher compared to nickel-based catalysts. Schiff base ligands exhibit interesting and diverse coordination chemistry and have therefore attracted much attention in the design of catalysts for ethylene oligomerization [211,212]. This ligand and its derivatives can form N–metal bonds via the coordination of nitrogen atoms into monodentate or multidentate. In addition, it can act as a donor and acceptor in hydrogen bonding interactions. Carbocyclic fused N,N,N-iron complexes show real industrial promise with the successful use of an imino-phe nanthroline–iron complex for the production of α-olefins in a 500-ton pilot plant managed by Sinopec in China. In homogeneous catalyzed reactions, iron and cobalt catalysts, although less active than nickel catalysts, have greater selectivity towards higher-carbon products (1-hexene and 1-octene). In heterogeneous catalytic reactions, fewer types of iron and cobalt catalysts have been developed, such as MIL-100(Fe) and carbon-supported cobalt catalysts. Uncontrollable exotherms leading to rapid catalyst deactivation and low activity are problems that need to be solved for the development of heterogeneous iron and cobalt catalysts.

Chromium-based catalysts, as pre-transition metals, have highly electron-deficient properties, are extremely sensitive to water–oxygen, and require large amounts of the expensive co-catalyst MAO for activation, which have led to their limitation in the production of linear α-olefins by ethylene oligomerization. However, these catalysts have become a hot topic in recent years due to their ability to selectively produce high-carbon linear olefins. For chromium-based catalysts, the most commonly used ligands for ethylene trimerization/tetramerization are PNP and the alternative diphosphine ligands PNNP, PCP, PCCP, and PCNCP [213–216]. When the catalyst contains P- and N-electron-donating multidentate ligands, it has a high selectivity potential for the production of 1-hexene and 1-octene [170]. The reactivity of chromium catalysts in heterogeneous reactions is still not comparable to that of post-transition metal catalysts.

In addition, catalysts based on zinc [209], gallium [209], and ruthenium [208] metals are continuously being developed for the heterogeneous catalytic reaction of ethylene oligomerization. The exploration of these catalysts not only extends the types of active metals in the field but also contributes to the understanding of the mechanism of the oligomerization process.

Although there has been extensive research using various metals-based catalysts for ethylene oligomerization, there are still some issues to be solved. First, most catalysts use a large amount of co-catalyst during the catalytic process, but their prices are relatively expensive and their applied conditions are often harsh (without O_2 and H_2O). Finding cheaper and easily operated co-catalysts or modifying the existing co-catalysts has become an important issue. Second, heterogeneous catalysts have the advantages of being easily separated from the products and being recyclable. Compared with homogeneous catalysis, heterogeneous catalysis is more promising for industrial applications, but most of the current heterogeneous catalysts are less active for ethylene oligomerization and less selective for linear α-olefins than homogeneous catalysts. Except SHOP, other industrialized approaches to ethylene oligomerization based on heterogeneous catalysis have not yet been realized. There remains a significant need to optimize heterogeneous catalysts compared to the currently industrialized homogeneous methods, particularly in terms of improving product selectivity. Third, the demand for high-performance polymerization materials prepared using high-carbon α-olefins (1-hexene and 1-octene) is increasing, so the development of catalysts for the efficient production of high-carbon α-olefins is urgent. In addition to changing the reaction conditions (pressure, temperature, amount of co-catalyst, etc.), the development of new metal-based catalysts and new ligands may be a promising solution. Furthermore, the similarities between homogeneous and heterogeneous catalysis may be utilized to gain more insight into the selectivity control mechanism. Finally, despite the general acceptance of the Cossee–Arlman mechanism and metallacyclic mechanism for ethylene oligomerization, there is a lack of in situ real-time evidence. Finding a completely new method or technique might break this bottleneck in the study of these mechanisms. In conclusion, the development of highly selective α-olefin formation processes remains a major challenge for academia and industry.

Author Contributions: A.P. and Z.H. collected the data, writing—original draft. G.L.: writing—review and editing, Supervision. All authors have read and agreed to the published version of the manuscript.

Funding: This research received no external funding.

Data Availability Statement: Data are contained within the article.

Conflicts of Interest: The authors declare no conflicts of interest.

Appendix A

It is necessary to make notes regarding the presentation of the results of the catalytic tests in the review:

(1) There are two types of catalyst activity calculations; one is $[g_{product} \times g_{catalyst}^{-1} \times h^{-1}]$ and the other is $[g_{product} \times mol_{M}^{-1} \times h^{-1}]$, expressed as Equations (A1) and (A2), respectively (M is metal).

$$[\text{Activity}] = \frac{\text{ethylene oligomerization products (g)}}{\text{catalyst(g)} \times \text{reaction time (h)}} = [g_{product} \times g_{catalyst}^{-1} \times h^{-1}] \quad \text{(A1)}$$

$$[\text{Activity}] = \frac{\text{ethylene oligomerization products (g)}}{\text{metal molecular (mol)} \times \text{reaction time (h)}} = \left[g_{product} \times mol_{M}^{-1} \times h^{-1}\right] \quad \text{(A2)}$$

(2) In the experiments, TOF $[mol_{C_2H_4} \times mol_{M}^{-1} \times h^{-1}]$ is calculated using Equation (A3) (M is metal):

$$[\text{TOF}] = \frac{\text{Activity}\left[g \times mol_{M}^{-1} \times h^{-1}\right]}{Mr_{C2H4}\left[g \times mol^{-1}\right]} = [h^{-1}] \quad \text{(A3)}$$

References

1. Speiser, F.; Braunstein, P.; Saussine, L. New Nickel Ethylene Oligomerization Catalysts Bearing Bidentate P,N-Phosphinopyridine Ligands with Different Substituents α to Phosphorus. *Organometallics* **2004**, *23*, 2625–2632. [CrossRef]
2. Breuil, P.A.R.; Magna, L.; Olivier-Bourbigou, H. Role of Homogeneous Catalysis in Oligomerization of Olefins: Focus on Selected Examples Based on Group 4 to Group 10 Transition Metal Complexes. *Catal. Lett.* **2015**, *145*, 173–192. [CrossRef]
3. de Klerk, A. Thermal cracking of Fischer-Tropsch waxes. *Ind. Eng. Chem. Res.* **2007**, *46*, 5516–5521. [CrossRef]
4. Li, C.; Ying, W.; Cao, F.; Zhang, H.; Fang, D. Effects of Impregnation Solvents on Catalytic Performance of Co-Ru-ZrO_2/γ-Al_2O_3 Catalyst for Fischer-Tropsch Synthesis. *Pet. Sci. Technol.* **2008**, *26*, 704–716. [CrossRef]
5. Liu, X.H.; Linghu, W.S.; Li, X.H.; Asami, K.; Fujimoto, K. Effects of solvent on Fischer-Tropsch synthesis. *Appl. Catal. A Gen.* **2006**, *303*, 251–257. [CrossRef]
6. Cheng, Y.; Lin, J.; Wu, T.J.; Wang, H.; Xie, S.H.; Pei, Y.; Yan, S.R.; Qiao, M.H.; Zong, B.N. Mg and K dual-decorated Fe-on-reduced graphene oxide for selective catalyzing CO hydrogenation to light olefins with mitigated CO_2 emission and enhanced activity. *Appl. Catal. B Environ.* **2017**, *204*, 475–485. [CrossRef]
7. Li, A.L.; Tian, D.X.; Zhao, Z.B. DFT studies on the reaction mechanism for the selective oxidative dehydrogenation of light alkanes by BN catalysts. *N. J. Chem.* **2020**, *44*, 11584–11592. [CrossRef]

8. Wang, G.W.; Zhang, S.; Zhu, X.L.; Li, C.Y.; Shan, H.H. Dehydrogenation versus hydrogenolysis in the reaction of light alkanes over Ni-based catalysts. *J. Ind. Eng. Chem.* **2020**, *86*, 1–12. [CrossRef]
9. Yan, B.; Li, W.C.; Lu, A.H. Metal-free silicon boride catalyst for oxidative dehydrogenation of light alkanes to olefins with high selectivity and stability. *J. Catal.* **2019**, *369*, 296–301. [CrossRef]
10. Rajapaksha, R.; Samanta, P.; Quadrelli, E.A.; Canivet, J. Heterogenization of molecular catalysts within porous solids: The case of Ni-catalyzed ethylene oligomerization from zeolites to metal-organic frameworks. *Chem. Soc. Rev.* **2023**, *52*, 8059–8076. [CrossRef]
11. Wang, M.Z.; Wu, W.; Wang, X.; Huang, X.; Nai, Y.N.; Wei, X.Y.; Mao, G.L. Research progress of iron-based catalysts for selective oligomerization of ethylene. *RSC Adv.* **2020**, *10*, 43640–43652. [CrossRef] [PubMed]
12. Wang, Z.; Solan, G.A.; Zhang, W.J.; Sun, W.H. Carbocyclic-fused N,N,N-pincer ligands as ring-strain adjustable supports for iron and cobalt catalysts in ethylene oligo-/polymerization. *Coord. Chem. Rev.* **2018**, *363*, 92–108. [CrossRef]
13. Hao, B.B.; Alam, F.; Jiang, Y.; Wang, L.B.; Fan, H.A.; Ma, J.; Chen, Y.H.; Wang, Y.T.; Jiang, T. Selective ethylene tetramerization: An overview. *Inorg. Chem. Front.* **2023**, *10*, 2860–2902. [CrossRef]
14. Keim, W. Oligomerization of Ethylene to α-Olefins: Discovery and Development of the Shell Higher Olefin Process (SHOP). *Angew. Chem.-Int. Ed.* **2013**, *52*, 12492–12496. [CrossRef] [PubMed]
15. McGuinness, D.S. Olefin Oligomerization via Metallacycles: Dimerization, Trimerization, Tetramerization, and Beyond. *Chem. Rev.* **2011**, *111*, 2321–2341. [CrossRef] [PubMed]
16. Cossee, P. On the mechanism of cis-ligand insertion. *Recl. Trav. Chim. Pays-Bas* **1966**, *85*, 1151–1160. [CrossRef]
17. Arlman, E.J.; Cossee, P. Ziegler-Natta catalysis III. Stereospecific polymerization of propene with the catalyst system $TiCl_3$-$AlEt_3$. *J. Catal.* **1964**, *3*, 99–104. [CrossRef]
18. Briggs, J.R. The selective trimerization of ethylene to hex-1-ene. *J. Chem. Soc.-Chem. Commun.* **1989**, 674–675. [CrossRef]
19. Chabbra, S.; Smith, D.M.; Bell, N.L.; Watson, A.J.B.; Bühl, M.; Cole-Hamilton, D.J.; Bode, B.E. First experimental evidence for a bis-ethene chromium(I) complex forming from an activated ethene oligomerization catalyst. *Sci. Adv.* **2020**, *6*, eabd7057. [CrossRef]
20. Fischer, K.; Jonas, K.; Misbach, P.; Stabba, R.; Wilke, G. The "Nickel Effect" Angew. *Chem. Int. Ed. Engl.* **1973**, *12*, 943–953. [CrossRef]
21. Peuckert, M.; Keim, W. A new nickel complex for the oligomerization of ethylene. *Organometallics* **1983**, *2*, 594–597. [CrossRef]
22. Johnson, L.K.; Killian, C.M.; Brookhart, M. New Pd(II)- and Ni(II)-Based Catalysts for Polymerization of Ethylene and .alpha.-Olefins. *J. Am. Chem. Soc.* **1995**, *117*, 6414–6415. [CrossRef]
23. Tuskaev, V.A.; Gagieva, S.C.; Kurmaev, D.A.; Zvukova, T.M.; Fedyanin, I.V.; Bulychev, B.M. Synthesis, characterization and ethylene oligomerization behavior of new nickel(II) octofluoro-α-diimine complex. *Inorg. Chim. Acta* **2016**, *442*, 167–171. [CrossRef]
24. Rishina, L.A.; Kissin, Y.V.; Lalayan, S.S.; Gagieva, S.C.; Kurmaev, D.A.; Tuskaev, V.A.; Bulychev, B.M. New α-diimine nickel complexes-Synthesis and catalysis of alkene oligomerization reactions. *J. Mol. Catal. A Chem.* **2016**, *423*, 495–502. [CrossRef]
25. Liu, J.; Wang, J.; Chen, L.; Zhang, N.; Lan, T.; Wang, L. Ethylene Oligomerization Study Bearing Nickel (II) and Cobalt (II) Complexes with N, O-Donor Schiff Bases as Ligands. *Chemistryselect* **2019**, *4*, 13959–13963. [CrossRef]
26. Falivene, L.; Wiedemann, T.; Goettker-Schnetmann, I.; Caporaso, L.; Cavallo, L.; Mecking, S. Control of Chain Walking by Weak Neighboring Group Interactions in Unsymmetrical Catalysts. *J. Am. Chem. Soc.* **2018**, *140*, 1305–1312. [CrossRef]
27. Wang, K.; Gao, R.; Hao, X.; Sun, W.-H. Nickel complexes bearing 2-(1H-benzo[d]imidazol-2-yl)-N-benzylidenequinolin-8-amines: Synthesis, structure and catalytic ethylene oligomerization. *Catal. Commun.* **2009**, *10*, 1730–1733. [CrossRef]
28. Zhao, W.; Qian, Y.L.; Huang, J.L.; Duan, J.J. Novel neutral arylnickel(II) phosphine catalysts containing 2-oxazolinylphenolato N-O chelate ligands for ethylene oligomerization and propylene dimerization. *J. Organomet. Chem.* **2004**, *689*, 2614–2623. [CrossRef]
29. Wang, C.; Friedrich, S.; Younkin, T.R.; Li, R.T.; Grubbs, R.H.; Bansleben, D.A.; Day, M.W. Neutral Nickel(II)-Based Catalysts for Ethylene Polymerization. *Organometallics* **1998**, *17*, 3149–3151. [CrossRef]
30. Zhang, S.; Pattacini, R.; Jie, S.; Braunstein, P. A phosphino-oxazoline ligand as a P,N-bridge in palladium/cobalt or P,N-chelate in nickel complexes: Catalytic ethylene oligomerization. *Dalton Trans.* **2012**, *41*, 379–386. [CrossRef]
31. Ortiz de la Tabla, L.; Matas, I.; Palma, P.; Álvarez, E.; Cámpora, J. Nickel and Palladium Complexes with New Phosphinito-Imine Ligands and Their Application as Ethylene Oligomerization Catalysts. *Organometallics* **2012**, *31*, 1006–1016. [CrossRef]
32. Zhang, S.; Pattacini, R.; Braunstein, P. Reactions of a Phosphinoimino-thiazoline-Based Metalloligand with Organic and Inorganic Electrophiles and Metal-Induced Ligand Rearrangements. *Organometallics* **2010**, *29*, 6660–6667. [CrossRef]
33. Boulens, P.; Lutz, M.; Jeanneau, E.; Olivier-Bourbigou, H.; Reek, J.N.H.; Breuil, P.-A.R. Iminobisphosphines to (Non-)Symmetrical Diphosphinoamine Ligands: Metal-Induced Synthesis of Diphosphorus Nickel Complexes and Application in Ethylene Oligomerisation Reactions. *Eur. J. Inorg. Chem.* **2014**, *2014*, 3754–3762. [CrossRef]
34. Song, K.; Gao, H.; Liu, F.; Pan, J.; Guo, L.; Zai, S.; Wu, Q. Syntheses, Structures, and Catalytic Ethylene Oligomerization Behaviors of Bis(phosphanyl)aminenickel(II) Complexes Containing N-Functionalized Pendant Groups. *Eur. J. Inorg. Chem.* **2009**, *2009*, 3016–3024. [CrossRef]
35. Cooley, N.A.; Green, S.M.; Wass, D.F.; Heslop, K.; Orpen, A.G.; Pringle, P.G. Nickel Ethylene Polymerization Catalysts Based on Phosphorus Ligands. *Organometallics* **2001**, *20*, 4769–4771. [CrossRef]

36. Ghisolfi, A.; Fliedel, C.; Rosa, V.; Monakhov, K.Y.; Braunstein, P. Combined Experimental and Theoretical Study of Bis(diphenylphosphino)(N-thioether)amine-Type Ligands in Nickel(II) Complexes for Catalytic Ethylene Oligomerization. *Organometallics* **2014**, *33*, 2523–2534. [CrossRef]
37. Tayade, K.N.; Mane, M.V.; Sen, S.; Murthy, C.N.; Tembe, G.L.; Pillai, S.M.; Vanka, K.; Mukherjee, S. A catalytic and DFT study of selective ethylene oligomerization by nickel(II) oxime-based complexes. *J. Mol. Catal. A Chem.* **2013**, *366*, 238–246. [CrossRef]
38. Xiao, L.; Zhang, M.; Gao, R.; Cao, X.; Sun, W.-H. 2-(1H-2-Benzimidazolyl)-6-(1-(arylimino)ethyl)pyridylnickel Complexes: Synthesis, Characterization, and Ethylene Oligomerization. *Aust. J. Chem.* **2010**, *63*, 109–115. [CrossRef]
39. Yang, Y.; Yang, P.; Zhang, C.; Li, G.; Yang, X.-J.; Wu, B.; Janiak, C. Synthesis, structure, and catalytic ethylene oligomerization of nickel complexes bearing 2-pyrazolyl substituted 1,10-phenanthroline ligands. *J. Mol. Catal. A Chem.* **2008**, *296*, 9–17. [CrossRef]
40. Jie, S.; Zhang, S.; Sun, W.-H. 2-arylimino-9-phenyl-1,10-phenanthrolinyl-iron, -cobalt and -nickel complexes: Synthesis, characterization and ethylene oligomerization behavior. *Eur. J. Inorg. Chem.* **2007**, *2007*, 5584–5598. [CrossRef]
41. Zhang, W.; Wang, Y.; Yu, J.; Redshaw, C.; Hao, X.; Sun, W.-H. 2-(N-Alkylcarboxamide)-6-iminopyridyl palladium and nickel complexes: Coordination chemistry and catalysis. *Dalton Trans.* **2011**, *40*, 12856–12865. [CrossRef]
42. Chen, Y.; Hao, P.; Zuo, W.; Gao, K.; Sun, W.-H. 2-(1-Isopropyl-2-benzimidazolyl)-6-(1-aryliminoethyl)pyridyl transition metal (Fe, Co, and Ni) dichlorides: Syntheses, characterizations and their catalytic behaviors toward ethylene reactivity. *J. Organomet. Chem.* **2008**, *693*, 1829–1840. [CrossRef]
43. Zhang, J.; Liu, S.; Li, A.; Ye, H.; Li, Z. Nickel(II) complexes chelated by 2,6-pyridinedicarboxamide: Syntheses, characterization, and ethylene oligomerization. *N. J. Chem.* **2016**, *40*, 7027–7033. [CrossRef]
44. Ulbrich, A.H.D.P.S.; Bergamo, A.L.; Casagrande, O.D.L., Jr. Oligomerization of ethylene using tridentate nickel catalysts bearing ether-pyrazol ligands with pendant O- and S-donor groups. *Catal. Commun.* **2011**, *16*, 245–249. [CrossRef]
45. Bekmukhamedov, G.E.; Sukhov, A.V.; Kuchkaev, A.M.; Yakhvarov, D.G. Ni-Based Complexes in Selective Ethylene Oligomerization Processes. *Catalysts* **2020**, *10*, 498. [CrossRef]
46. Olivier-Bourbigou, H.; Breuil, P.A.R.; Magna, L.; Michel, T.; Pastor, M.F.E.; Delcroix, D. Nickel Catalyzed Olefin Oligomerization and Dimerization. *Chem. Rev.* **2020**, *120*, 7919–7983. [CrossRef]
47. Crabtree, R.H. Deactivation in Homogeneous Transition Metal Catalysis: Causes, Avoidance, and Cure. *Chem. Rev.* **2015**, *115*, 127–150. [CrossRef]
48. Ockwig, N.W.; Delgado-Friedrichs, O.; O'Keeffe, M.; Yaghi, O.M. Reticular chemistry: Occurrence and taxonomy of nets and grammar for the design of frameworks. *Acc. Chem. Res.* **2005**, *38*, 176–182. [CrossRef]
49. Liu, J.; Ye, J.; Li, Z.; Otake, K.-I.; Liao, Y.; Peters, A.W.; Noh, H.; Truhlar, D.G.; Gagliardi, L.; Cramer, C.J.; et al. Beyond the Active Site: Tuning the Activity and Selectivity of a Metal-Organic Framework-Supported Ni Catalyst for Ethylene Dimerization. *J. Am. Chem. Soc.* **2018**, *140*, 11174–11178. [CrossRef]
50. Britovsek, G.J.P.; Bruce, M.; Gibson, V.C.; Kimberley, B.S.; Maddox, P.J.; Mastroianni, S.; McTavish, S.J.; Redshaw, C.; Solan, G.A.; Strömberg, S.; et al. Iron and cobalt ethylene polymerization catalysts bearing 2,6-bis(imino)pyridyl ligands: Synthesis, structures, and polymerization studies. *J. Am. Chem. Soc.* **1999**, *121*, 8728–8740. [CrossRef]
51. Small, B.L.; Brookhart, M.; Bennett, A.M.A. Highly Active Iron and Cobalt Catalysts for the Polymerization of Ethylene. *J. Am. Chem. Soc.* **1998**, *120*, 4049–4050. [CrossRef]
52. Zhang, C.-L.; Zhou, T.; Li, Y.-Q.; Lu, X.; Guan, Y.-B.; Cao, Y.-C.; Cao, G.-P. Microenvironment Modulation of Metal-Organic Frameworks (MOFs) for Coordination Olefin Oligomerization and (co)Polymerization. *Small* **2023**, *19*, 2205898. [CrossRef] [PubMed]
53. Chen, C.; Alalouni, M.R.; Dong, X.; Cao, Z.; Cheng, Q.; Zheng, L.; Meng, L.; Guan, C.; Liu, L.; Abou-Hamad, E.; et al. Highly Active Heterogeneous Catalyst for Ethylene Dimerization Prepared by Selectively Doping Ni on the Surface of a Zeolitic Imidazolate Framework. *J. Am. Chem. Soc.* **2021**, *143*, 7144–7153. [CrossRef] [PubMed]
54. Chen, L.; Jiang, Y.; Huo, H.; Liu, J.; Li, Y.; Li, C.; Zhang, N.; Wang, J. Metal-organic framework-based composite Ni@MOF as Heterogenous catalyst for ethylene trimerization. *Appl. Catal. A Gen.* **2020**, *594*, 117457. [CrossRef]
55. Yeh, B.; Vicchio, S.P.; Chheda, S.; Zheng, J.; Schmid, J.; Löbbert, L.; Bermejo-Deval, R.; Gutiérrez, O.Y.; Lercher, J.A.; Lu, C.C.; et al. Site Densities, Rates, and Mechanism of Stable Ni/UiO-66 Ethylene Oligomerization Catalysts. *J. Am. Chem. Soc.* **2021**, *143*, 20274–20280. [CrossRef] [PubMed]
56. Komurcu, M.; Lazzarini, A.; Kaur, G.; Borfecchia, E.; Oien-Odegaard, S.; Gianolio, D.; Bordiga, S.; Lillerud, K.P.; Olsbye, U. Co-catalyst free ethene dimerization over Zr-based metal-organic framework (UiO-67) functionalized with Ni and bipyridine. *Catal. Today* **2021**, *369*, 193–202. [CrossRef]
57. Sharma, R.K.; Yadav, P.; Yadav, M.; Gupta, R.; Rana, P.; Srivastava, A.; Zboril, R.; Varma, R.S.; Antonietti, M.; Gawande, M.B. Recent development of covalent organic frameworks (COFs): Synthesis and catalytic (organic-electro-photo) applications. *Mater. Horiz.* **2020**, *7*, 411–454. [CrossRef]
58. Rozhko, E.; Bavykina, A.; Osadchii, D.; Makkee, M.; Gascon, J. Covalent organic frameworks as supports for a molecular Ni based ethylene oligomerization catalyst for the synthesis of long chain olefins. *J. Catal.* **2017**, *345*, 270–280. [CrossRef]
59. Li, D.; Guo, L.; Li, F.; Huang, J.; Li, J.; Li, M.; Li, C. Synthesis and catalytic behavior of nickel heterogenized in covalent organic frameworks as precatalysts in ethylene oligomerization. *Microporous Mesoporous Mater.* **2022**, *338*, 111979. [CrossRef]
60. Guo, L.; Gao, Y.; Huang, J.; Xue, J.; Li, F.; Li, C. Imine-linked covalent organic frameworks coordinated with nickel for ethylene oligomerization. *J. Appl. Polym. Sci.* **2023**, *140*, e53320. [CrossRef]

61. Lallemand, M.; Rusu, O.A.; Dumitriu, E.; Finiels, A.; Fajula, F.; Hulea, V. NiMCM-36 and NiMCM-22 catalysts for the ethylene oligomerization: Effect of zeolite texture and nickel cations/acid sites ratio. *Appl. Catal. A Gen.* **2008**, *338*, 37–43. [CrossRef]
62. Andrei, R.D.; Popa, M.L.; Fajula, F.; Hulea, V. Heterogeneous oligomerization of ethylene over highly active and stable Ni-AlSBA-15 mesoporous catalysts. *J. Catal.* **2015**, *323*, 76–84. [CrossRef]
63. Shin, D.Y.; Yoon, J.H.; Baik, H.; Lee, S.J. A way to avoid polymeric side products during the liquid-phase ethylene oligomerization with SBA-15 supported (bpy)Ni(II)Cl_2 heterogeneous catalyst. *Appl. Catal. A Gen.* **2020**, *590*, 117363. [CrossRef]
64. Lacarriere, A.; Robin, J.; Swierczynski, D.; Finiels, A.; Fajula, F.; Luck, F.; Hulea, V. Distillate-Range Products from Non-Oil-Based Sources by Catalytic Cascade Reactions. *Chemsuschem* **2012**, *5*, 1787–1792. [CrossRef] [PubMed]
65. Mohamed, H.O.; Parsapur, R.K.; Hita, I.; Ramirez, A.; Huang, K.-W.; Gascon, J.; Castano, P. Stable and reusable hierarchical ZSM-5 zeolite with superior performance for olefin oligomerization when partially coked. *Appl. Catal. B Environ.* **2022**, *316*, 121582. [CrossRef]
66. Mohamed, H.O.; Velisoju, V.K.; Hita, I.; Abed, O.; Parsapur, R.K.; Zambrano, N.; Ben Hassine, M.; Morlanes, N.; Emwas, A.-H.; Huang, K.-W.; et al. Highly productive framework bounded Ni^{2+} on hierarchical zeolite for ethylene oligomerization. *Chem. Eng. J.* **2023**, *475*, 146077. [CrossRef]
67. McCaig, J.; Lamb, H.H. Ni-H-Beta Catalysts for Ethylene Oligomerization: Impact of Parent Cation on Ni Loading, Speciation, and Siting. *Catalysts* **2022**, *12*, 824. [CrossRef]
68. Seufitelli, G.V.S.; Park, J.J.W.; Tran, P.N.; Dichiara, A.; Resende, F.L.P.; Gustafson, R. The Role of Nickel and Bronsted Sites on Ethylene Oligomerization with Ni-H-Beta Catalysts. *Catalysts* **2022**, *12*, 565. [CrossRef]
69. Lee, M.; Yoon, J.W.; Kim, Y.; Yoon, J.S.; Chae, H.-J.; Han, Y.-H.; Hwang, D.W. Ni/SIRAL-30 as a heterogeneous catalyst for ethylene oligomerization. *Appl. Catal. A Gen.* **2018**, *562*, 87–93. [CrossRef]
70. Hu, Y.; Zhang, Y.; Han, Y.; Sheng, D.; Shan, D.; Liu, X.; Cheng, A. Ultrathin Nickel-Based Metal–Organic Framework Nanosheets as Reusable Heterogeneous Catalyst for Ethylene Dimerization. *ACS Appl. Nano Mater.* **2019**, *2*, 136–142. [CrossRef]
71. Arrozi, U.S.F.; Bon, V.; Kutzscher, C.; Senkovska, I.; Kaskel, S. Towards highly active and stable nickel-based metal-organic frameworks as ethylene oligomerization catalysts. *Dalton Trans.* **2019**, *48*, 3415–3421. [CrossRef]
72. Wang, C.; Li, G.; Guo, H. Heterogeneous dimerization of ethylene by coordinatively unsaturated metal sites in two forms of Ni-MIL-77. *Mol. Catal.* **2022**, *524*, 112340. [CrossRef]
73. Chen, C.L.; Alalouni, M.R.; Xiao, P.Y.; Li, G.X.; Pan, T.T.; Shen, J.; Cheng, Q.P.; Dong, X.L. Ni-Loaded 2D Zeolitic Imidazolate Framework as a Heterogeneous Catalyst with Highly Activity for Ethylene Dimerization. *Ind. Eng. Chem. Res.* **2022**, *61*, 14374–14381. [CrossRef]
74. Metzger, E.D.; Brozek, C.K.; Comito, R.J.; Dincă, M. Selective Dimerization of Ethylene to 1-Butene with a Porous Catalyst. *ACS Cent. Sci.* **2016**, *2*, 148–153. [CrossRef]
75. Metzger, E.D.; Comito, R.J.; Wu, Z.; Zhang, G.; Dubey, R.C.; Xu, W.; Miller, J.T.; Dinca, M. Highly Selective Heterogeneous Ethylene Dimerization with a Scalable and Chemically Robust MOF Catalyst. *ACS Sustain. Chem. Eng.* **2019**, *7*, 6654–6661. [CrossRef]
76. Chen, C.L.; Meng, L.K.; Alalouni, M.R.; Dong, X.L.; Wu, Z.P.; Zuo, S.W.; Zhang, H.B. Ultra-Highly Active Ni-Doped MOF-5 Heterogeneous Catalysts for Ethylene Dimerization. *Small* **2023**, *19*, e2301235. [CrossRef]
77. Li, Z.; Schweitzer, N.M.; League, A.B.; Bernales, V.; Peters, A.W.; Getsoian, A.B.; Wang, T.C.; Miller, J.T.; Vjunov, A.; Fulton, J.L.; et al. Sintering-Resistant Single-Site Nickel Catalyst Supported by Metal Organic Framework. *J. Am. Chem. Soc.* **2016**, *138*, 1977–1982. [CrossRef]
78. Canivet, J.; Aguado, S.; Schuurman, Y.; Farrusseng, D. MOF-Supported Selective Ethylene Dimerization Single-Site Catalysts through One-Pot Postsynthetic Modification. *J. Am. Chem. Soc.* **2013**, *135*, 4195–4198. [CrossRef]
79. Liu, B.; Jie, S.; Bu, Z.; Li, B.-G. Postsynthetic modification of mixed-linker metal-organic frameworks for ethylene oligomerization. *RSC Adv.* **2014**, *4*, 62343–62346. [CrossRef]
80. Madrahimov, S.T.; Gallagher, J.R.; Zhang, G.; Meinhart, Z.; Garibay, S.J.; Delferro, M.; Miller, J.T.; Farha, O.K.; Hupp, J.T.; Nguyen, S.T. Gas-Phase Dimerization of Ethylene under Mild Conditions Catalyzed by MOF Materials Containing (bpy)Ni^{II} Complexes. *ACS Catal.* **2015**, *5*, 6713–6718. [CrossRef]
81. Kyogoku, K.; Yamada, C.; Suzuki, Y.; Nishiyama, S.; Fukumoto, K.; Yamamoto, H.; Indo, S.; Sano, M.; Miyake, T. Syntheses of Metal-organic Framework Compounds Containing Ni-bipyridyl Complex for Oligomerization of Ethylene. *J. Jpn. Pet. Inst.* **2010**, *53*, 308–312. [CrossRef]
82. Gonzalez, M.I.; Oktawiec, J.; Long, J.R. Ethylene oligomerization in metal-organic frameworks bearing nickel(II) 2,2′-bipyridine complexes. *Faraday Discuss.* **2017**, *201*, 351–367. [CrossRef]
83. Arrozi, U.S.F.; Bon, V.; Krause, S.; Lübken, T.; Weiss, M.S.; Senkovska, I.; Kaskel, S. In Situ Imine-Based Linker Formation for the Synthesis of Zirconium MOFs: A Route to CO_2 Capture Materials and Ethylene Oligomerization Catalysts. *Inorg. Chem.* **2020**, *59*, 350–359. [CrossRef]
84. Alshehri, S.M.; Ahamad, T.; Aldalbahi, A.; Alhokbany, N. Pyridylimine Cobalt(II) and Nickel(II) Complex Functionalized Multiwalled Carbon Nanotubes and Their Catalytic Activities for Ethylene Oligomerization. *Adv. Polym. Technol.* **2016**, *35*, 21528–21538. [CrossRef]

85. Kohler, F.T.U.; Gaertner, K.; Hager, V.; Haumann, M.; Sternberg, M.; Wang, X.; Szesni, N.; Meyer, K.; Wasserscheid, P. Dimerization of ethene in a fluidized bed reactor using Ni-based Supported Ionic Liquid Phase (SILP) catalysts. *Catal. Sci. Technol.* **2014**, *4*, 936–947. [CrossRef]
86. Aid, A.; Andrei, R.D.; Amokrane, S.; Cammarano, C.; Nibou, D.; Hulea, V. Ni-exchanged cationic clays as novel heterogeneous catalysts for selective ethylene oligomerization. *Appl. Clay Sci.* **2017**, *146*, 432–438. [CrossRef]
87. Kim, M.J.; Ahn, S.; Yi, J.; Hupp, J.T.; Notestein, J.M.; Farha, O.K.; Lee, S.J. Ni(II) complex on a bispyridine-based porous organic polymer as a heterogeneous catalyst for ethylene oligomerization. *Catal. Sci. Technol.* **2017**, *7*, 4351–4354. [CrossRef]
88. Yoon, J.S.; Park, M.B.; Kim, Y.; Hwang, D.W.; Chae, H.-J. Effect of Metal Oxide-Support Interactions on Ethylene Oligomerization over Nickel Oxide/Silica-Alumina Catalysts. *Catalysts* **2019**, *9*, 933. [CrossRef]
89. Finiels, A.; Fajula, F.; Hulea, V. Nickel-based solid catalysts for ethylene oligomerization—A review. *Catal. Sci. Technol.* **2014**, *4*, 2412–2426. [CrossRef]
90. Small, B.L.; Brookhart, M. Iron-Based Catalysts with Exceptionally High Activities and Selectivities for Oligomerization of Ethylene to Linear α-Olefins. *J. Am. Chem. Soc.* **1998**, *120*, 7143–7144. [CrossRef]
91. Gibson, V.C.; Redshaw, C.; Solan, G.A. Bis(imino)pyridines: Surprisingly reactive ligands and a gateway to new families of catalysts. *Chem. Rev.* **2007**, *107*, 1745–1776. [CrossRef]
92. Britovsek, G.J.P.; Gibson, V.C.; McTavish, S.J.; Solan, G.A.; White, A.J.P.; Williams, D.J.; Britovsek, G.J.P.; Kimberley, B.S.; Maddox, P.J. Novel olefin polymerization catalysts based on iron and cobalt. *Chem. Commun.* **1998**, 849–850. [CrossRef]
93. Ngcobo, M.; Nose, H.; Jayamani, A.; Ojwach, S.O. Structural and ethylene oligomerization studies of chelating (imino)phenol Fe(II), Co(II) and Ni(II) complexes: An experimental and theoretical approach. *N. J. Chem.* **2022**, *46*, 6219–6229. [CrossRef]
94. Guo, J.; Chen, Q.; Zhang, W.; Liang, T.; Sun, W.-H. The benzhydryl-modified 2-imino-1,10-phenanthrolyliron precatalyst in ethylene oligomerization. *J. Organomet. Chem.* **2021**, *936*, 121713. [CrossRef]
95. Sheng, D.; Zhang, Y.; Wang, Z.; Xu, G.; Song, Q.; Shan, D.; Hou, H. Selective ethylene oligomerization based on Fe-containing metal-organic coordination polymer catalysts. *J. Organomet. Chem.* **2021**, *956*, 122128. [CrossRef]
96. Maksakova, I.S.; Pervova, I.G.; Belov, G.P.; Khasbiullin, I.I.; Ivanova, N.A.; Frolova, E.N.; Lipunov, I.N. Synthesis, structure, and catalytic properties of iron complexes with mono- and bisformazans in sulfide oxidation and ethylene oligomerization reactions. *Pet. Chem.* **2015**, *55*, 217–223. [CrossRef]
97. Appukuttan, V.K.; Liu, Y.; Son, B.C.; Ha, C.-S.; Suh, H.; Kim, I. Iron and Cobalt Complexes of 2,3,7,8-Tetrahydroacridine-4,5(1H,6H)-diimine Sterically Modulated by Substituted Aryl Rings for the Selective Oligomerization to Polymerization of Ethylene. *Organometallics* **2011**, *30*, 2285–2294. [CrossRef]
98. Xiao, T.; Zhang, W.; Lai, J.; Sun, W.-H. Iron-oriented ethylene oligomerization and polymerization: The Iron Age or a flash in the pan. *C. R. Chim.* **2011**, *14*, 851–855. [CrossRef]
99. Tomov, A.K.; Gibson, V.C.; Britovsek, G.J.P.; Long, R.J.; van Meurs, M.; Jones, D.J.; Tellmann, K.P.; Chirinos, J.J. Distinguishing Chain Growth Mechanisms in Metal-catalyzed Olefin Oligomerization and Polymerization Systems: C_2H_4/C_2D_4 Co-oligomerization/Polymerization Experiments Using Chromium, Iron, and Cobalt Catalysts. *Organometallics* **2009**, *28*, 7033–7040. [CrossRef]
100. Görl, C.; Beck, N.; Kleiber, K.; Alt, H.G. Iron(III) complexes with meta-substituted bis(arylimino)pyridine ligands: Catalyst precursors for the selective oligomerization of ethylene. *J. Mol. Catal. A Chem.* **2012**, *352*, 110–127. [CrossRef]
101. Li, G.; Ge, Y.; Xu, G.; Dai, S. The electronic effects on unsymmetrical Bis(imino)pyridyl iron(III) catalyzed ethylene polymerization. *J. Organomet. Chem.* **2020**, *923*, 121457. [CrossRef]
102. Görl, C.; Alt, H.G. Influence of the para-substitution in bis(arylimino)pyridine iron complexes on the catalytic oligomerization and polymerization of ethylene. *J. Organomet. Chem.* **2007**, *692*, 4580–4592. [CrossRef]
103. Kissin, Y.V.; Qian, C.; Xie, G.; Chen, Y. Multi-center nature of ethylene polymerization catalysts based on 2,6-bis(imino)pyridyl complexes of iron and cobalt. *J. Polym. Sci. Part A Polym. Chem.* **2006**, *44*, 6159–6170. [CrossRef]
104. Minaev, B.; Baryshnikova, A.; Sun, W.-H. Spin-dependent effects in ethylene polymerization with bis(imino)pyridine iron(II) complexes. *J. Organomet. Chem.* **2016**, *811*, 48–65. [CrossRef]
105. Cartes, M.Á.; Rodríguez-Delgado, A.; Palma, P.; Sánchez, L.J.; Cámpora, J. Direct evidence for a coordination–insertion mechanism of ethylene oligomerization catalysed by neutral 2,6-bisiminopyridine iron monoalkyl complexes. *Catal. Sci. Technol.* **2014**, *4*, 2504–2507. [CrossRef]
106. Li, H.-y.; Duan, B.-g.; Hu, Y.-l. The influence of substituent electronic effect on ethylene oligomerization activities of bis(imino)pyridyl Fe(II) catalysts: A combined molecular mechanics and charge equilibration method. *Chin. J. Polym. Sci.* **2009**, *27*, 711. [CrossRef]
107. Guo, C.Y.; Xu, H.; Zhang, M.G.; Zhang, X.H.; Yan, F.W.; Yuan, G.Q. Immobilization of bis(imino)pyridine iron complexes onto mesoporous molecular sieves and their catalytic performance in ethylene oligomerization. *Catal. Commun.* **2009**, *10*, 1467–1471. [CrossRef]
108. Ittel, S.D.; Johnson, L.K.; Brookhart, M. Late-Metal Catalysts for Ethylene Homo- and Copolymerization. *Chem. Rev.* **2000**, *100*, 1169–1204. [CrossRef] [PubMed]
109. Bariashir, C.; Zhang, Q.-Y.; Ulambayar, B.; Solan, G.A.; Liang, T.-L.; Sun, W.-H. High Temperature Iron Ethylene Polymerization Catalysts Bearing N, N, N′-2-(1-(2,4-Dibenzhydryl-6-fluorophenylimino)ethyl)-6-(1-(arylphenylimino)ethyl)pyridines. *Chin. J. Polym. Sci.* **2024**, *42*, 188–201. [CrossRef]

110. Zhao, W.; Yue, E.; Wang, X.; Yang, W.; Chen, Y.; Hao, X.; Cao, X.; Sun, W.-H. Activity and stability spontaneously enhanced toward ethylene polymerization by employing 2-(1-(2,4-dibenzhydrylnaphthylimino)ethyl)-6-(1-(arylimino)ethyl) pyridyliron(II) dichlorides. *J. Polym. Sci. Part A Polym. Chem.* **2017**, *55*, 988–996. [CrossRef]
111. Kaul, F.A.R.; Puchta, G.T.; Frey, G.D.; Herdtweck, E.; Herrmann, W.A. Iminopyridine Complexes of 3d Metals for Ethylene Polymerization: Comparative Structural Studies and Ligand Size Controlled Chain Termination. *Organometallics* **2007**, *26*, 988–999. [CrossRef]
112. Xie, G.; Li, T.; Zhang, A. Highly active and selective ethylene oligomerization catalysts: Asymmetric 2,6-bis(imino)pyridyl iron (II) complexes with alkyl and halogen substitutients. *Inorg. Chem. Commun.* **2010**, *13*, 1199–1202. [CrossRef]
113. Bianchini, C.; Mantovani, G.; Meli, A.; Migliacci, F.; Zanobini, F.; Laschi, F.; Sommazzi, A. Oligomerisation of Ethylene to Linear α-Olefins by new C_s- and C_1-Symmetric [2,6-Bis(imino)pyridyl]iron and -cobalt Dichloride Complexes. *Eur. J. Inorg. Chem.* **2003**, *2003*, 1620–1631. [CrossRef]
114. Sun, W.-H.; Xing, Q.; Yu, J.; Novikova, E.; Zhao, W.; Tang, X.; Liang, T.; Redshaw, C. Probing the Characteristics of Mono- or Bimetallic (Iron or Cobalt) Complexes Bearing 2,4-Bis(6-iminopyridin-2-yl)-3H-benzazepines: Synthesis, Characterization, and Ethylene Reactivity. *Organometallics* **2013**, *32*, 2309–2318. [CrossRef]
115. Zhang, N.; Wu, Y.; Li, Y.; Chen, L.; Zhang, M.; Wang, J. Star iminopyridyl iron, cobalt and nickel complexes: Synthesis, molecular structures, and evaluation as ethylene oligomerization catalysts. *Polym. Bull.* **2022**, *79*, 4219–4231. [CrossRef]
116. Zhang, N.; Wang, J.; Huo, H.; Chen, L.; Shi, W.; Li, C.; Wang, J. Iron, cobalt and nickel complexes bearing hyperbranched iminopyridyl ligands: Synthesis, characterization and evaluation as ethylene oligomerization catalysts. *Inorg. Chim. Acta* **2018**, *469*, 209–216. [CrossRef]
117. Li, C.-Q.; Wang, F.-F.; Gao, R.; Sun, P.; Zhang, N.; Wang, J. Bidentate iron complexes based on hyperbranched salicylaldimine ligands and their catalytic behavior toward ethylene oligomerization. *Transit. Met. Chem.* **2017**, *42*, 339–346. [CrossRef]
118. Nyamato, G.S.; Alam, M.G.; Ojwach, S.O.; Akerman, M.P. (Pyrazolyl)-(phosphinoyl)pyridine iron(II), cobalt(II) and nickel(II) complexes: Synthesis, characterization and ethylene oligomerization studies. *J. Organomet. Chem.* **2015**, *783*, 64–72. [CrossRef]
119. Rossetto, E.; Nicola, B.P.; de Souza, R.F.; Bernardo-Gusmão, K.; Pergher, S.B.C. Heterogeneous complexes of nickel MCM-41 with β-diimine ligands: Applications in olefin oligomerization. *J. Catal.* **2015**, *323*, 45–54. [CrossRef]
120. Kim, I.; Han, B.H.; Kim, J.S.; Ha, C.-S. Bis(imino)pyridyl Co(II) and Fe(II) catalysts immobilized on SBA-15 mesoporous material: New highly active supported catalysts for the polymerization of ethylene. *Catal. Lett.* **2005**, *101*, 249–253. [CrossRef]
121. Ionkin, A.S.; Marshall, W.J.; Adelman, D.J.; Bobik Fones, B.; Fish, B.M.; Schiffhauer, M.F.; Spence, R.E.; Xie, T. High-Temperature Catalysts for the Production of α-Olefins Based on Iron(II) and Cobalt(II) Tridentate Bis(imino)pyridine Complexes with a Double Pattern of Substitution: *o*-Methyl plus *o*-Fluorine in the Same Imino Arm. *Organometallics* **2008**, *27*, 1147–1156. [CrossRef]
122. Ionkin, A.S.; Marshall, W.J.; Adelman, D.J.; Fones, B.B.; Fish, B.M.; Schiffhauer, M.F. High-Temperature Catalysts for the Production of α-Olefins Based on Iron(II) and Iron(III) Tridentate Bis(imino)pyridine Complexes with Double Pattern of Substitution: *ortho*-Methyl plus *meta*-Aryl. *Organometallics* **2006**, *25*, 2978–2992. [CrossRef]
123. Zheng, Z.; Liu, J.; Li, Y. Ethylene polymerization with silica-supported bis(imino)pyridyl iron(II) catalysts. *J. Catal.* **2005**, *234*, 101–110. [CrossRef]
124. Guo, C.-Y.; Xu, H.; Zhang, M.; Yang, H.-J.; Yan, F.; Yuan, G. Copolymerization of ethylene and in situ-generated α-olefins to high-performance linear low-density polyethylene with a two-catalyst system supported on mesoporous molecular sieves. *Polym. Int.* **2010**, *59*, 725–732. [CrossRef]
125. Jayamani, A.; Nyamato, G.S.; Ojwach, S.O. Ethylene oligomerization reactions catalyzed by homogeneous and silica immobilized NO Fe(II) and Co(II) complexes. *J. Organomet. Chem.* **2019**, *903*, 120987. [CrossRef]
126. Armspach, D.; Matt, D.; Peruch, F.; Lutz, P. Cyclodextrin-Encapsulated Iron Catalysts for the Polymerization of Ethylene. *Eur. J. Inorg. Chem.* **2003**, *2003*, 805–809. [CrossRef]
127. Bouilhac, C.; Cloutet, E.; Cramail, H.; Deffieux, A.; Taton, D. Functionalized Star-Like Polystyrenes as Organic Supports of a Tridentate Bis(imino)pyridinyliron/Aluminic Derivative Catalytic System for Ethylene Polymerization. *Macromol. Rapid Commun.* **2005**, *26*, 1619–1625. [CrossRef]
128. Han, Y.; Zhang, Y.; Zhang, Y.L.; Cheng, A.C.; Hu, Y.P.; Wang, Z.H. Selective ethylene tetramerization with iron-based metal-organic framework MIL-100 to obtain C_8 alkanes. *Appl. Catal. A Gen.* **2018**, *564*, 183–189. [CrossRef]
129. Kaul, F.A.R.; Puchta, G.T.; Schneider, H.; Bielert, F.; Mihalios, D.; Herrmann, W.A. Immobilization of bis(imino)pyridyliron(II) complexes on silica. *Organometallics* **2002**, *21*, 74–82. [CrossRef]
130. Ngcobo, M.; Ouissa, A.; Kleist, W.; Thiel, W.R.; Ojwach, S.O. Regulating the physical properties of silica immobilized Fe(II), Ni(II) and Co(II) catalysts towards ethylene oligomerization reactions. *Mol. Catal.* **2023**, *549*, 113465. [CrossRef]
131. Ngcobo, M.; Ojwach, S.O. Ethylene oligomerization reactions catalyzed by recyclable Fe(II), Ni(II) and Co(II) complexes immobilized on Fe_3O_4 magnetic nanoparticles. *Mol. Catal.* **2021**, *508*, 111583. [CrossRef]
132. Ba, J.; Du, S.; Yue, E.; Hu, X.; Flisak, Z.; Sun, W.-H. Constrained formation of 2-(1-(arylimino)ethyl)-7-arylimino-6,6-dimethylcyclopentapyridines and their cobalt(II) chloride complexes: Synthesis, characterization and ethylene polymerization. *RSC Adv.* **2015**, *5*, 32720–32729. [CrossRef]
133. Huang, Y.; Zhang, R.; Liang, T.; Hu, X.; Solan, G.A.; Sun, W.-H. Selectivity Effects on N,N,N′-Cobalt Catalyzed Ethylene Dimerization/Trimerization Dictated through Choice of Aluminoxane Cocatalyst. *Organometallics* **2019**, *38*, 1143–1150. [CrossRef]

134. Guo, J.; Zhang, W.; Liang, T.; Sun, W.-H. Revisiting the 2-imino-1,10-phenanthrolylmetal precatalyst in ethylene oligomerization: Benzhydryl-modified cobalt(II) complexes and their dimerization of ethylene. *Polyhedron* **2021**, *193*, 114865. [CrossRef]
135. Antonov, A.A.; Semikolenova, N.V.; Talsi, E.P.; Bryliakov, K.P. Catalytic ethylene oligomerization on cobalt(II) bis(imino)pyridine complexes bearing electron-withdrawing groups. *J. Organomet. Chem.* **2019**, *884*, 55–58. [CrossRef]
136. Caovilla, M.; Thiele, D.; de Souza, R.F.; Gregorio, J.R.; Bernardo-Gusmao, K. Cobalt-β-diimine complexes for ethylene oligomerization. *Catal. Commun.* **2017**, *101*, 85–88. [CrossRef]
137. Chen, L.; Ma, L.; Jiang, Y.; Liu, J.; Li, C.; Zhang, N.; Wang, J. Synthesis and characterization of iron, cobalt and nickel complexes bearing para-phenylene-linked pyridine imine ligand and their catalytic properties for ethylene oligomerization. *Polym. Bull.* **2021**, *78*, 415–432. [CrossRef]
138. Khoshsefat, M.; Ahmadjo, S.; Zohuri, G.H.; Ma, Y.; Sun, W.-H. From tetramerization to oligomerization/polymerization of ethylene by dinuclear pyridyl-imine Co- and Ni-based catalysts. *Appl. Organomet. Chem.* **2023**, *37*, e7226. [CrossRef]
139. Wei, W.; Yu, B.; Alam, F.; Huang, Y.; Cheng, S.; Jiang, T. Cobalt(II)-based ethylene dimerization catalysts with silicon-bridged diphosphine ligands. *Appl. Organomet. Chem.* **2018**, *32*, e4604. [CrossRef]
140. Zhang, L.; Meng, X.J.; Chen, Y.H.; Cao, C.G.; Jiang, T. Chromium-Based Ethylene Tetramerization Catalysts Supported by Silicon-Bridged Diphosphine Ligands: Further Combination of High Activity and Selectivity. *Chemcatchem* **2017**, *9*, 76–79. [CrossRef]
141. Härzschel, S.; Kühn, F.E.; Wöhl, A.; Müller, W.; Al-Hazmi, M.H.; Alqahtani, A.M.; Müller, B.H.; Peulecke, N.; Rosenthal, U. Comparative study of new chromium-based catalysts for the selective tri- and tetramerization of ethylene. *Catal. Sci. Technol.* **2015**, *5*, 1678–1682. [CrossRef]
142. Wang, J.; Li, H.-Y.; Song, L.; Shi, W.-G.; Zhang, N.; Li, C.-Q. Cobalt complexes based on dendritic PAMAM bridged salicylaldimine ligands: Synthesis, characterization and performance in ethylene oligomerization. *J. Macromol. Sci. Part A Pure Appl. Chem.* **2016**, *53*, 709–715. [CrossRef]
143. Li, C.; Wang, F.; Lin, Z.; Zhang, N.; Wang, J. Cobalt complexes based on hyperbranched salicylaldimine ligands as catalyst precursors for ethylene oligomerization. *Appl. Organomet. Chem.* **2017**, *31*, e3756. [CrossRef]
144. Zhang, N.; Wang, S.-h.; Song, L.; Li, C.-q.; Wang, J. Hyperbranched macromolecules bridged salicylaldimine cobalt complexes: Synthesis, characterization and ethylene oligomerization studies. *Chem. Pap.* **2017**, *71*, 1037–1046. [CrossRef]
145. Ngcobo, M.; Nyamato, G.S.; Ojwach, S.O. Structural elucidation of N^O (ethylimino-methyl)phenol Fe(II) and Co(II) complexes and their applications in ethylene oligomerization catalysis. *Mol. Catal.* **2019**, *478*, 110590. [CrossRef]
146. Zubkevich, S.V.; Tuskaev, V.A.; Gagieva, S.C.; Kayda, A.S.; Khrustalev, V.N.; Pavlov, A.A.; Zarubin, D.N.; Bulychev, B.M. NNNO-Heteroscorpionate nickel (II) and cobalt (II) complexes for ethylene oligomerization: The unprecedented formation of odd carbon number olefins. *Appl. Organomet. Chem.* **2020**, *34*, e5873. [CrossRef]
147. Xu, Z.; Zhao, D.; Chada, J.P.; Rosenfeld, D.C.; Rogers, J.L.; Hermans, L.; Huber, G.W. Olefin conversion on nitrogen-doped carbon-supported cobalt catalyst: Effect of feedstock. *J. Catal.* **2017**, *354*, 213–222. [CrossRef]
148. Xu, Z.; Chada, J.P.; Xu, L.; Zhao, D.; Rosenfeld, D.C.; Rogers, J.L.; Hermans, I.; Mavrikakis, M.; Huber, G.W. Ethylene Dimerization and Oligomerization to 1-Butene and Higher Olefins with Chromium-Promoted Cobalt on Carbon Catalyst. *ACS Catal.* **2018**, *8*, 2488–2497. [CrossRef]
149. Jonathan, A.; Eagan, N.M.; Bruns, D.L.; Stahl, S.S.; Lanci, M.P.; Dumesic, J.A.; Huber, G.W. Ethylene oligomerization into linear olefins over cobalt oxide on carbon catalyst. *Catal. Sci. Technol.* **2021**, *11*, 3599–3608. [CrossRef]
150. Jonathan, A.; Tomashek, E.G.; Lanci, M.P.; Dumesic, J.A.; Huber, G.W. Reaction kinetics study of ethylene oligomerization into linear olefins over carbon-supported cobalt catalysts. *J. Catal.* **2021**, *404*, 954–963. [CrossRef]
151. Jonathan, A.; Dastidar, R.G.; Wang, C.; Dumesic, J.A.; Huber, G.W. Effect of catalyst support on cobalt catalysts for ethylene oligomerization into linear olefins. *Catal. Sci. Technol.* **2022**, *12*, 3639–3649. [CrossRef]
152. Zhu, K.; An, Y.; Yu, F.; Liu, L.; Zhong, L. Structure-Performance Evolution of Cobalt-Ammonia Activated Carbon Catalyst for Ethylene Oligomerization. *Catal. Lett.* **2022**, *152*, 2131–2140. [CrossRef]
153. Tang, J.; Zhang, N.; Shi, D.; Zhang, F.; Tang, J. Synthesis of UiO-66-NH_2 Grafted Pyridineimine Cobalt Catalyst and Its Catalytic Performance in Ethylene Oligomerization. *Chin. J. Appl. Chem.* **2022**, *39*, 258–265. [CrossRef]
154. Weng, Z.; Teo, S.; Hor, T.S.A. Chromium(III) catalysed ethylene tetramerization promoted by bis(phosphino)amines with an N-functionalized pendant. *Dalton Trans.* **2007**, 3493–3498. [CrossRef]
155. Mao, G.; Ning, Y.; Hu, W.; Li, S.; Song, X.; Niu, B.; Jiang, T. Synthesis of a novel triple-site diphosphinoamine (PNP) ligand and its applications in ethylene tetramerization. *Chin. Sci. Bull.* **2008**, *53*, 3511–3515. [CrossRef]
156. Carter, A.; Cohen, S.A.; Cooley, N.A.; Murphy, A.; Scutt, J.; Wass, D.F. High activity ethylene trimerisation catalysts based on diphosphine ligands. *Chem. Commun.* **2002**, 858–859. [CrossRef]
157. Sydora, O.L.; Jones, T.C.; Small, B.L.; Nett, A.J.; Fischer, A.A.; Carney, M.J. Selective Ethylene Tri-/Tetramerization Catalysts. *ACS Catal.* **2012**, *2*, 2452–2455. [CrossRef]
158. Zhang, J.; Braunstein, P.; Hor, T.S.A. Highly Selective Chromium(III) Ethylene Trimerization Catalysts with [NON] and [NSN] Heteroscorpionate Ligands. *Organometallics* **2008**, *27*, 4277–4279. [CrossRef]
159. Peng, J.; Zhu, M.; Chen, L.; Liu, Z. Unravelling the chain growth mechanism in Cr/NNN-catalysed ethylene oligomerization. *Catal. Sci. Technol.* **2023**, *13*, 6137–6145. [CrossRef]

160. Ogawa, T.; Lindeperg, F.; Stradiotto, M.; Turculet, L.; Sydora, O.L. Chromium N-phosphinoamidine ethylene tri-/tetramerization catalysts: Designing a step change in 1-octene selectivity. *J. Catal.* **2021**, *394*, 444–450. [CrossRef]
161. Zhou, Y.; Wu, H.; Xu, S.; Zhang, X.; Shi, M.; Zhang, J. Highly active chromium-based selective ethylene tri-/tetramerization catalysts supported by PNPO phosphazane ligands. *Dalton Trans.* **2015**, *44*, 9545–9550. [CrossRef]
162. Manyik, R.M.; Walker, W.E.; Wilson, T.P. A soluble chromium-based catalyst for ethylene trimerization and polymerization. *J. Catal.* **1977**, *47*, 197–209. [CrossRef]
163. Zhang, J.; Alam, F.; Fan, H.; Ma, J.; Jiang, T. Chromium catalysts based on $PNP(NR_2)_2$ ligands for selective ethylene oligomerization. *Appl. Organomet. Chem.* **2022**, *36*, e6454. [CrossRef]
164. Kong, W.; Ma, X.; Zuo, J.; Zhao, X.; Zhang, J. Ethylene Tri-/Tetramerization Catalysts Supported by Phenylene-Bridged Mixed Alkyl/Aryl Diphosphine Ligands. *Organometallics* **2023**, *42*, 651–659. [CrossRef]
165. Blann, K.; Bollmann, A.; Dixon, J.T.; Hess, F.M.; Killian, E.; Maumela, H.; Morgan, D.H.; Neveling, A.; Otto, S.; Overett, M.J. Highly selective chromium-based ethylene trimerisation catalysts with bulky diphosphinoamine ligands. *Chem. Commun.* **2005**, 620–621. [CrossRef]
166. Temple, C.; Jabri, A.; Crewdson, P.; Gambarotta, S.; Korobkov, I.; Duchateau, R. The question of the Cr oxidation state in the {Cr(SNS)} catalyst for selective ethylene trimerization: An unanticipated re-oxidation pathway. *Angew. Chem.-Int. Ed.* **2006**, *45*, 7050–7053. [CrossRef]
167. Soheili, M.; Mohamadnia, Z.; Karimi, B. Switching from Ethylene Trimerization to Ethylene Polymerization by Chromium Catalysts Bearing SNS Tridentate Ligands: Process Optimization Using Response Surface Methodology. *Catal. Lett.* **2018**, *148*, 3685–3700. [CrossRef]
168. McGuinness, D.S.; Brown, D.B.; Tooze, R.P.; Hess, F.M.; Dixon, J.T.; Slawin, A.M.Z. Ethylene trimerization with Cr-PNP and Cr-SNS complexes: Effect of ligand structure, metal oxidation state, and role of activator on catalysis. *Organometallics* **2006**, *25*, 3605–3610. [CrossRef]
169. Zhang, J.; Li, A.; Hor, T.S.A. Crystallographic revelation of the role of $AlMe_3$ (in MAO) in Cr [NNN] pyrazolyl catalyzed ethylene trimerization. *Organometallics* **2009**, *28*, 2935–2937. [CrossRef]
170. Alferov, K.A.; Belov, G.P.; Meng, Y. Chromium catalysts for selective ethylene oligomerization to 1-hexene and 1-octene: Recent results. *Appl. Catal. A Gen.* **2017**, *542*, 71–124. [CrossRef]
171. Alam, F.; Zhang, L.; Zhai, Y.; Wang, J.; Tang, H.; Chen, Y.; Jiang, T. In situ formed Cr(III) based silicon-bridged PNS systems for selective ethylene tri-/tetramerization. *J. Catal.* **2019**, *378*, 312–319. [CrossRef]
172. Zhao, X.; Wang, J.; Liu, D.; Kong, W.; Zhang, J. Chromium Ethylene Tri-/Tetramerization Catalysts Supported by Iminophosphine Ligands. *ACS Omega* **2023**, *8*, 34549–34556. [CrossRef]
173. Kuhlmann, S.; Blann, K.; Bollmann, A.; Dixon, J.T.; Killian, E.; Maumela, M.C.; Maumela, H.; Morgan, D.H.; Prétorius, M.; Taccardi, N.; et al. N-substituted diphosphinoamines: Toward rational ligand design for the efficient tetramerization of ethylene. *J. Catal.* **2007**, *245*, 279–284. [CrossRef]
174. McGuinness, D.S.; Wasserscheid, P.; Morgan, D.H.; Dixon, J.T. Ethylene trimerization with mixed-donor ligand (N,P,S) chromium complexes: Effect of ligand structure on activity and selectivity. *Organometallics* **2005**, *24*, 552–556. [CrossRef]
175. Boelter, S.D.; Davies, D.R.; Margl, P.; Milbrandt, K.A.; Mort, D.; Vanchura, B.A.I.; Wilson, D.R.; Wiltzius, M.; Rosen, M.S.; Klosin, J. Phospholane-Based Ligands for Chromium-Catalyzed Ethylene Tri- and Tetramerization. *Organometallics* **2020**, *39*, 976–987. [CrossRef]
176. Alam, F.; Zhang, L.; Wei, W.; Wang, J.; Chen, Y.; Dong, C.; Jiang, T. Catalytic Systems Based on Chromium(III) Silylated-Diphosphinoamines for Selective Ethylene Tri-/Tetramerization. *ACS Catal.* **2018**, *8*, 10836–10845. [CrossRef]
177. Zhang, C.; Song, L.; Wu, H.; Ji, X.; Jiao, J.; Zhang, J. Ethylene tri-/tetramerization catalysts supported by diphosphinothiophene ligands. *Dalton Trans.* **2017**, *46*, 8399–8404. [CrossRef]
178. Radcliffe, J.E.; Batsanov, A.S.; Smith, D.M.; Scott, J.A.; Dyer, P.W.; Hanton, M.J. Phosphanyl Methanimine (PCN) Ligands for the Selective Trimerization/Tetramerization of Ethylene with Chromium. *ACS Catal.* **2015**, *5*, 7095–7098. [CrossRef]
179. Rosenthal, U. PNPN-H in Comparison to other PNP, PNPN and NPNPN Ligands for the Chromium Catalyzed Selective Ethylene Oligomerization. *ChemCatChem* **2020**, *12*, 41–52. [CrossRef]
180. Lifschitz, A.M.; Hirscher, N.A.; Lee, H.B.; Buss, J.A.; Agapie, T. Ethylene Tetramerization Catalysis: Effects of Aluminum-Induced Isomerization of PNP to PPN Ligands. *Organometallics* **2017**, *36*, 1640–1648. [CrossRef]
181. Zhao, X.; Ma, X.; Kong, W.; Zhang, J. Switchable ethylene tri-/tetramerization with high catalytic performance: Subtle effect presented by P-alkyl substituent of PCCP ligands. *J. Catal.* **2022**, *415*, 95–101. [CrossRef]
182. Choi, E.M.; Park, J.E.; Park, G.S.; Shong, B.; Kang, S.K.; Son, K.s. Selective ethylene oligomerization with in-situ -generated chromium catalysts supported by trifluoromethyl-containing ligands. *J. Polym. Sci. Part A Polym. Chem.* **2018**, *56*, 444–450. [CrossRef]
183. Wang, J.; Gao, R.; Zhang, N.; Tan, X.; Wang, L.; Ma, L.; Shi, W. Novel Dendritic PNP Chromium Complexes: Synthesis, Characterization, and Performance on Ethylene Oligomerization. *Helv. Chim. Acta* **2017**, *100*, e1700162. [CrossRef]
184. Stennett, T.E.; Haddow, M.F.; Wass, D.F. Avoiding MAO: Alternative Activation Methods in Selective Ethylene Oligomerization. *Organometallics* **2012**, *31*, 6960–6965. [CrossRef]
185. Fan, H.; Yang, X.; Ma, J.; Hao, B.; Alam, F.; Huang, X.; Wang, A.; Jiang, T. Computer-assisted design of asymmetric PNP ligands for ethylene tri-/tetramerization: A combined DFT and artificial neural network approach. *J. Catal.* **2023**, *418*, 121–129. [CrossRef]

186. Zhong, X.; Liu, L.; Guo, X.; Sun, L.; Liu, B.; Liu, Z. Cr/PCCP-catalysed selective ethylene oligomerization: Analysis of various conformations and the hemilabile methoxy group. *Catal. Sci. Technol.* **2022**, *12*, 5586–5596. [CrossRef]
187. Wang, Z.; Liu, L.; Ma, X.; Liu, Y.; Mi, P.; Liu, Z.; Zhang, J. Effect of an additional donor on decene formation in ethylene oligomerization catalyzed by a Cr/PCCP system: A combined experimental and DFT study. *Catal. Sci. Technol.* **2021**, *11*, 4596–4604. [CrossRef]
188. Yu, Z.-X.; Houk, K.N. Why Trimerization? Computational Elucidation of the Origin of Selective Trimerization of Ethene Catalyzed by [$TaCl_3(CH_3)_2$] and An Agostic-Assisted Hydride Transfer Mechanism. *Angew. Chem. Int. Ed.* **2003**, *42*, 808–811. [CrossRef]
189. Bollmann, A.; Blann, K.; Dixon, J.T.; Hess, F.M.; Killian, E.; Maumela, H.; McGuinness, D.S.; Morgan, D.H.; Neveling, A.; Otto, S.; et al. Ethylene tetramerization: A new route to produce 1-octene in exceptionally high selectivities. *J. Am. Chem. Soc.* **2004**, *126*, 14712–14713. [CrossRef]
190. Zhang, M.; Wang, K.; Sun, W.-H. Chromium(III) complexes bearing 2-benzazole-1,10-phenanthrolines: Synthesis, molecular structures and ethylene oligomerization and polymerization. *Dalton Trans.* **2009**, 6354–6363. [CrossRef]
191. Hao, Z.; Xu, B.; Gao, W.; Han, Y.; Zeng, G.; Zhang, J.; Li, G.; Mu, Y. Chromium Complexes with N,N,N-Tridentate Quinolinyl Anilido-Imine Ligand: Synthesis, Characterization, and Catalysis in Ethylene Polymerization. *Organometallics* **2015**, *34*, 2783–2790. [CrossRef]
192. Luo, W.; Li, A.; Liu, S.; Ye, H.; Li, Z. 2-Benzimidazol-6-pyrazol-pyridine Chromium(III) Trichlorides: Synthesis, Characterization, and Application for Ethylene Oligomerization and Polymerization. *Organometallics* **2016**, *35*, 3045–3050. [CrossRef]
193. Wang, D.; Liu, S.; Zeng, Y.; Sun, W.-H.; Redshaw, C. 2-Benzimidazolyl-N-phenylquinoline-8-carboxamide Chromium(III) Trichlorides: Synthesis and Application for Ethylene Oligomerization and Polymerization. *Organometallics* **2011**, *30*, 3001–3009. [CrossRef]
194. Junges, F.; Kuhn, M.C.A.; Dos Santos, A.H.D.P.; Rabello, C.R.K.; Thomas, C.M.; Carpentier, J.-F.; Casagrande, O.L., Jr. Chromium catalysts based on tridentate pyrazolyl ligands for ethylene oligomerization. *Organometallics* **2007**, *26*, 4010–4014. [CrossRef]
195. Gibson, V.C.; Newton, C.; Redshaw, C.; Solan, G.A.; White, A.J.P.; Williams, D.J. Chromium complexes bearing pyrrolide-imine N,N-chelate ligands: Synthesis, structures and ethylene polymerisation behaviour. *J. Chem. Soc. Dalton Trans.* **2002**, 4017–4023. [CrossRef]
196. De Oliveira, L.L.; Batista, F.I.; Oliboni, R.S.; Casagrande, A.C.A.; Stieler, R.; Casagrande, O.L. Chromium(III) complexes based on phenoxy-imine ligands with pendant N- and O-donor groups as precatalysts for ethylene oligomerization: Synthesis, characterization, and DFT studies. *J. Organomet. Chem.* **2021**, *936*, 121710. [CrossRef]
197. Wang, J.; Liu, J.; Chen, L.; Lan, T.; Wang, L. Preparation of chromium catalysts bearing bispyridylamine and its performance in ethylene oligomerization. *Transit. Met. Chem.* **2019**, *44*, 681–688. [CrossRef]
198. Klemps, C.; Buchard, A.; Houdard, R.; Auffrant, A.; Mézailles, N.; Le Goff, X.F.; Ricard, L.; Saussine, L.; Magna, L.; Le Floch, P. Chromium (III)-bis(iminophosphoranyl)methanido complexes: Synthesis, X-ray crystal structures and catalytic ethylene oligomerization. *N. J. Chem.* **2009**, *33*, 1748–1752. [CrossRef]
199. Milani, J.L.S.; Biajoli, A.F.P.; Batista, F.I.; Oliboni, R.S.; Casagrande, O.L. Chromium complexes supported by bidentate thioether-imine [N,S] ligands: Synthesis and ethylene oligomerization studies. *N. J. Chem.* **2021**, *45*, 1814–1821. [CrossRef]
200. Grauke, R.; Schepper, R.; Rabeah, J.; Schoch, R.; Bentrup, U.; Bauer, M.; Brückner, A. Impact of Al Activators on Structure and Catalytic Performance of Cr Catalysts in Homogeneous Ethylene Oligomerization—A Multitechnique in situ/operando Study. *ChemCatChem* **2020**, *12*, 1025–1035. [CrossRef]
201. Carney, M.J.; Robertson, N.J.; Halfen, J.A.; Zakharov, L.N.; Rheingold, A.L. Octahedral chromium(III) complexes supported by bis(2-pyridylmethyl)amines: Ligand influence on coordination geometry and ethylene polymerization activity. *Organometallics* **2004**, *23*, 6184–6190. [CrossRef]
202. Lamb, M.J.; Apperley, D.C.; Watson, M.J.; Dyer, P.W. The Role of Catalyst Support, Diluent and Co-Catalyst in Chromium-Mediated Heterogeneous Ethylene Trimerisation. *Top. Catal.* **2018**, *61*, 213–224. [CrossRef]
203. Fallahi, M.; Ahmadi, E.; Ramazani, A.; Mohamadnia, Z. Trimerization of ethylene catalyzed by Cr-based catalyst immobilized on the supported ionic liquid phase. *J. Organomet. Chem.* **2017**, *848*, 149–158. [CrossRef]
204. Müller, T.; Dixon, J.T.; Haumann, M.; Wasserscheid, P. Trimerization and tetramerization of ethylene in continuous gas-phase reaction using a Cr-based supported liquid phase catalyst. *React. Chem. Eng.* **2018**, *4*, 131–140. [CrossRef]
205. Gao, X.; Wang, M.; Chen, Y.; Sun, Y.; Dong, B.; Jiang, T. Biphasic trimerization of ethylene with diphosphinoamine/chromium(III)/methylaluminoxane immobilized in organochloroaluminate ionic liquid. *React. Kinet. Mech. Catal.* **2014**, *113*, 159–167. [CrossRef]
206. Goetjen, T.A.; Zhang, X.; Liu, J.; Hupp, J.T.; Farha, O.K. Metal–Organic Framework Supported Single Site Chromium(III) Catalyst for Ethylene Oligomerization at Low Pressure and Temperature. *ACS Sustain. Chem. Eng.* **2019**, *7*, 2553–2557. [CrossRef]
207. Liu, S.Y.; Zhang, Y.; Han, Y.; Feng, G.L.; Gao, F.; Wang, H.; Qiu, P. Selective Ethylene Oligomerization with Chromium-Based Metal Organic Framework MIL-100 Evacuated under Different Temperatures. *Organometallics* **2017**, *36*, 632–638. [CrossRef]
208. Agirrezabal-Telleria, I.; Luz, I.; Ortuno, M.A.; Oregui-Bengoechea, M.; Gandarias, I.; Lopez, N.; Lail, M.A.; Soukri, M. Gas reactions under intrapore condensation regime within tailored metal-organic framework catalysts. *Nat. Commun.* **2019**, *10*, 2076. [CrossRef]
209. LiBretto, N.J.; Xu, Y.; Quigley, A.; Edwards, E.; Nargund, R.; Vega-Vila, J.C.; Caulkins, R.; Saxena, A.; Gounder, R.; Greeley, J.; et al. Olefin oligomerization by main group Ga^{3+} and Zn^{2+} single site catalysts on SiO_2. *Nat. Commun.* **2021**, *12*, 2322. [CrossRef]

210. Keim, W. Nickel—An element with application in industrial homogeneous catalysis. *Angew. Chem.-Int. Ed. Engl.* **1990**, *29*, 235–244. [CrossRef]
211. Trofymchuk, O.S.; Gutsulyak, D.V.; Quintero, C.; Parvez, M.; Daniliuc, C.G.; Piers, W.E.; Rojas, R.S. N-Arylcyano-β-diketiminate Methallyl Nickel Complexes: Synthesis, Adduct Formation, and Reactivity toward Ethylene. *Organometallics* **2013**, *32*, 7323–7333. [CrossRef]
212. Wang, S.D.; Zhang, W.J.; Du, S.Z.; Asuha, S.; Flisak, Z.; Sun, W.H. Propyl substituted 4-arylimino-1,2,3-trihydroacridylnickel complexes: Their synthesis, characterization and catalytic behavior toward ethylene. *J. Organomet. Chem.* **2015**, *798*, 408–413. [CrossRef]
213. Fan, H.A.; Zhang, Y.; Alam, F.; Ma, J.; Hao, B.B.; Chen, Y.H.; Wang, Y.T.; Huang, J.M.; Jiang, T. Highly efficient PCCP/Cr(III) ethylene tetramerization catalyst designed using artificial neural network approach. *J. Catal.* **2024**, *429*, 115237. [CrossRef]
214. Fan, H.N.; Hao, B.B.; Jiang, Y.; Yang, X.D.; Wang, Y.T.; Jiang, T. Selective ethylene tri-/tetramerization chromium catalysts based on binuclear PCCP ligands. *Mol. Catal.* **2023**, *549*, 113545. [CrossRef]
215. Boelter, S.D.; Davies, D.R.; Milbrandt, K.A.; Wilson, D.R.; Wiltzius, M.; Rosen, M.S.; Klosin, J. Evaluation of Bis(phosphine) Ligands for Ethylene Oligomerization: Discovery of Alkyl Phosphines as Effective Ligands for Ethylene Tri- and Tetramerization. *Organometallics* **2020**, *39*, 967–975. [CrossRef]
216. Ji, X.Y.; Song, L.B.; Zhang, C.Y.; Jiao, J.J.; Zhang, J. Highly active chromium-based selective ethylene tri-/tetramerization catalysts supported by N,N-diphospholylamines. *Inorg. Chim. Acta* **2017**, *466*, 117–121. [CrossRef]

catalysts

MDPI

Article

Visible-Light-Photocatalyzed C5-H Nitration of 8-Aminoquinoline Amides

Pugen Liu, Huijie Qiao, Xiaoxue Su, Peirong Bai and Fan Yang *

Henan Key Laboratory of Chemical Biology and Organic Chemistry, Key Laboratory of Applied Chemistry of Henan Universities, Green Catalysis Center, College of Chemistry, Zhengzhou University, Zhengzhou 450052, China; liupugen@163.com (P.L.)
* Correspondence: yangf@zzu.edu.cn

Abstract: A mild and efficient protocol for visible-light-photocatalyzed C5 nitration of 8-aminoquinoline derivatives was developed utilizing $Cu(NO_3)_2{\cdot}3H_2O$ as a nitro source. The reaction proceeded smoothly under very mild conditions, employing Acid Red 94 and a commercial household light bulb as an organic photosensitizer and a light source, respectively, making this synthetic procedure green and easy to operate. Furthermore, most products could be obtained through recrystallization, which enhanced the operational simplicity of this method.

Keywords: photocatalysis; nitration; aminoquinoline

1. Introduction

Nitroarenes play an important role in the synthesis of dyes, pharmaceuticals, plastics, explosives, and perfumes [1,2]. The classical methodologies for nitration are based on a mixed acid H_2SO_4/HNO_3 protocol, but usually generate the complex mixtures of regioisomers (*o-/m-/p-* = 19:80:1 at 30 °C), thereby exhibiting from regioselectivity and functional group compatibility issues [3,4]. In order to overcome these problems, efforts have been directed towards the development of new nitration protocols. For example, nitroarenes can be readily accessed via the nitration of aryl boronic acids, aryl halides, pseudo halides, and aryl carboxylic acids, as well as the oxidation of aromatic primary amines or azides [5–11]. However, all these protocols require prefunctionalized arenes as the starting materials.

The auxiliary-assisted C–H functionalization has become a reliable and robust tool in modern organic synthesis in recent years [12–22]. Specifically, the direct functionalization of quinoline ring at the C5 position using the amide unit as a directing group has also been explored in recent years [23–34], and the nitration of C5 quinoline ring has especially received considerable attention [35–46]. Recently, the Ribas group described a mild methodology for C–H nitration of 8-aminoquinoline amides utilizing $Co(NO_3)_2{\cdot}6H_2O$ as the catalyst and *t*BuONO as the nitrating agent [38]. However, it tended to suffer from regioselectivity for quinoline rings (ratio of C5:C7 = 3:1). Subsequently, the Zhang group developed the copper-catalyzed nitration of 8-aminoquinolines at C5 position using $Cu(NO_3)_2/NaNO_2/PhI(TFA)_2$ as the nitration system (Scheme 1, eq 1) [39]. Recently, Fan also reported an alternative transformation with $Fe(NO_3)_3{\cdot}9H_2O$ as the nitro source [40], albeit this protocol should be performed at 100 °C (Scheme 1, eq 2). Despite these achievements attained in this area, these synthetic routes still require expensive oxidants, expensive metals, or a high reaction temperature. Therefore, the development of a mild and environmentally benign route to synthesize nitro compounds is highly desirable.

The recent developments on visible light photoredox catalysis provides a new possibility for modern organic synthesis due to their catalysis process with the mild, environmentally benign, operationally simple advantages and commercially available photocatalysts [47]. The research interest of our group is focused on regioselective C–H functionalization, especially that takes place at an unusual C–H site [28,29,48]. For example,

Citation: Liu, P.; Qiao, H.; Su, X.; Bai, P.; Yang, F. Visible-Light-Photocatalyzed C5-H Nitration of 8-Aminoquinoline Amides. *Catalysts* **2024**, *14*, 263. https://doi.org/10.3390/catal14040263

Academic Editors: Leonarda Liotta, Narendra Kumar and Konstantin Ivanov Hadjiivanov

Received: 18 March 2024
Revised: 30 March 2024
Accepted: 9 April 2024
Published: 15 April 2024

we demonstrated the C5–H sulfonylation [28] and phosphonation [29] of 8-aminoquinoline amides and developed the palladium-catalyzed picolinamide-directed C8–H amination of 1-naphthylamine derivatives with simple secondary aliphatic amines [47]. Inspired by these previous and our own reports, we envisioned achieving a mild and efficient protocol for visible-light-photocatalyzed C5 nitration on the quinoline ring (Scheme 1, eq 3).

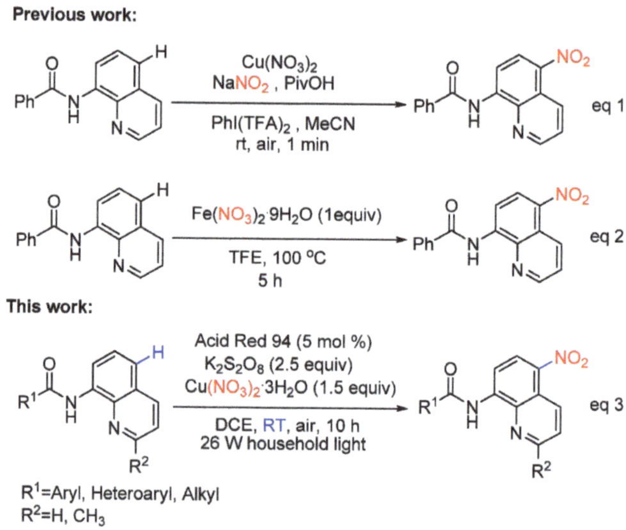

Scheme 1. The direct nitration of C5 quinoline ring.

2. Results and Discussion

Initially, the reaction of N-(quinolin-8-yl)benzamide (**1a**) was performed as the model reaction to optimize the reaction parameters at room temperature under air as Scheme 2, and the results are displayed in Table 1. After various catalysts, solvents, and oxidants were investigated (see Supporting Information), the desired product **2a** was obtained in 68% yield in DCE using $K_2S_2O_8$, $Cu(NO_3)_2{\cdot}3H_2O$, and Eosin B as the oxidant, the nitrating agent, and the photoredox catalyst, respectively (Table 1, entry 1) [49]. Thereafter, various photocatalysts were checked, and Acid Red 94 could produce the product in a good yield of 82% (Table 1, entry 7 vs. entries 2–6). However, reducing the amount of Acid Red 94 from 5 mol % to 4 mol % resulted in a lower yield of 68% (Table 1, entry 8 vs. entry 7). Additionally, performing reaction under a nitrogen or an O_2 atmosphere did not afford better results (Table 1, entries 9 and 10 vs. entry 7). When the reaction time decreased to 8 h, a lower yield of 73% was observed (Table 1, entry 11 vs. entry 7). And the conditions in entry 7, Table 1 was selected as the optimized reaction conditions.

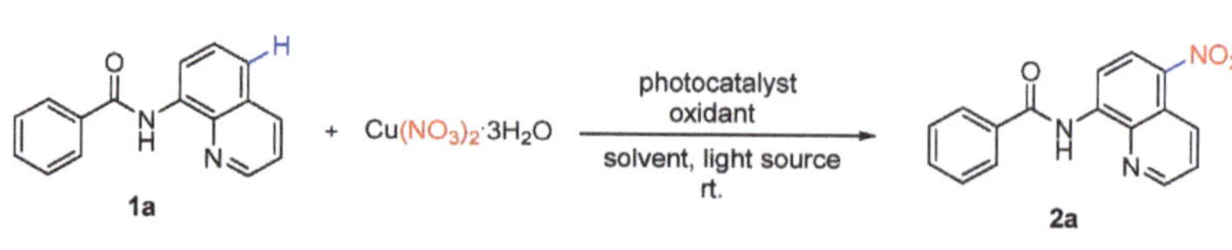

Scheme 2. The nitration of 1-naphthylamine derivatives.

Table 1. Screening optimal conditions [a,b].

Entry	Catalyst	Oxidant	Light Source	Yield (%) [b]
1	EosinB	$K_2S_2O_8$	household light	68
2	EosinY	$K_2S_2O_8$	household light	65
3	$Ru(bpy)_3(PF_6)_2$	$K_2S_2O_8$	household light	69
4	$Ru(bpy)_3Cl_2$	$K_2S_2O_8$	household light	75
5	$Ru(bpy)_3Cl_2{\cdot}6H_2O$	$K_2S_2O_8$	household light	76
6	Alizarin Red S	$K_2S_2O_8$	household light	72
7	**Acid Red 94**	$\mathbf{K_2S_2O_8}$	**household light**	**82**
8 [c]	Acid Red 94	$K_2S_2O_8$	household light	69
9 [d]	Acid Red 94	$K_2S_2O_8$	household light	76
10 [e]	Acid Red 94	$K_2S_2O_8$	household light	68
11 [f]	Acid Red 94	$K_2S_2O_8$	household light	73
12	Acid Red 94	$K_2S_2O_8$	blue	66
13	Acid Red 94	$K_2S_2O_8$	green	62
14	Acid Red 94	$K_2S_2O_8$	red	43
15	Acid Red 94	$K_2S_2O_8$	dark	23
16	—	$K_2S_2O_8$	household light	25
17 [g]	Acid Red 94	$K_2S_2O_8$	household light	51
18 [h]	Acid Red 94	$K_2S_2O_8$	household light	NR

[a] reaction conditions: **1a** (0.20 mmol), $Cu(NO_3)_2{\cdot}3H_2O$ (1.5 equiv), photocatalyst (5 mol %), and $K_2S_2O_8$ (2.5 equiv) in DCE (1.5 mL) at room temperature under air for 10 h. [b] isolated yield of **2a**. [c] photocatalyst (4 mol %). [d] under N_2. [e] under O_2. [f] for 8 h. [g] $Cu(NO_3)_2{\cdot}3H_2O$ (1.0 equiv). [h] $AgNO_3$ (2.0 equiv) instead of $Cu(NO_3)_2{\cdot}3H_2O$ (1.5 equiv). Bold represents optimized conditions.

Finally, some other light sources were investigated (e.g., blue LED, green LED, and red LED), but they could not match the effect of household light (Table 1, entries 12–14 vs. entry 7); the yield of the target product dropped drastically in the absence of household light irradiation or photocatalysts (Table 1, entries 15 and 16). The amount of nitro source $Cu(NO_3)_2{\cdot}3H_2O$ when decreased to 1.0 equiv resulted in a lower yield of 51% (Table 1, entry 17). However, when $AgNO_3$ was utilized as the nitro source, the product was not obtained (Table 1, entry 18).

With the optimized conditions in hand, the scope of this reaction was investigated, and the results were summarized in Scheme 3. The results showed that the electronic effect is not obvious on the benzene ring of benzamides, and the desired products could be obtained in good yields. When the benzamides possessed electron-donating groups (e.g., Me and MeO), the desired products were obtained in a yield range of 76–82% (**2b**–**2g**); substrates bearing electron-withdrawing groups (e.g., F, Cl, Br, and CF_3) provided the corresponding products in yields of 72–78% (**2h**–**2n**). However, 3-nitro-*N*-(quinolin-8-yl)benzamide did not produce the corresponding nitration product (**2o**). In addition, the heterocyclic amide could produce the nitration product in a good yield of 69% (**2q**), and to our delight, aliphatic amide could also afford the corresponding product in 73% yield (**2r**).

The scope of 2-methyl-8-aminoquinoline amides was also investigated (Scheme 4). Both electron-rich and electron-poor groups on the benzene ring of benzamides were successfully tolerated in the reaction, and comparatively, substrates containing electron-donating groups could result in slightly higher yields of the coupling products than those of electron-withdrawing groups (**2a′**–**2m′**). The substrate bearing Me or MeO group on the phenyl ring could afford the nitrated products in 75–79% yield (**2b′**–**2e′**). Moreover, aromatic amides, possessing F, Cl, Br and CF_3, could afford the target compounds **2f′**–**2m′** in moderate yields (59–71%). Notably, the method was also applicable to five-membered heterocycles such as thiophene, providing the corresponding products in 72% yield (**2n′**).

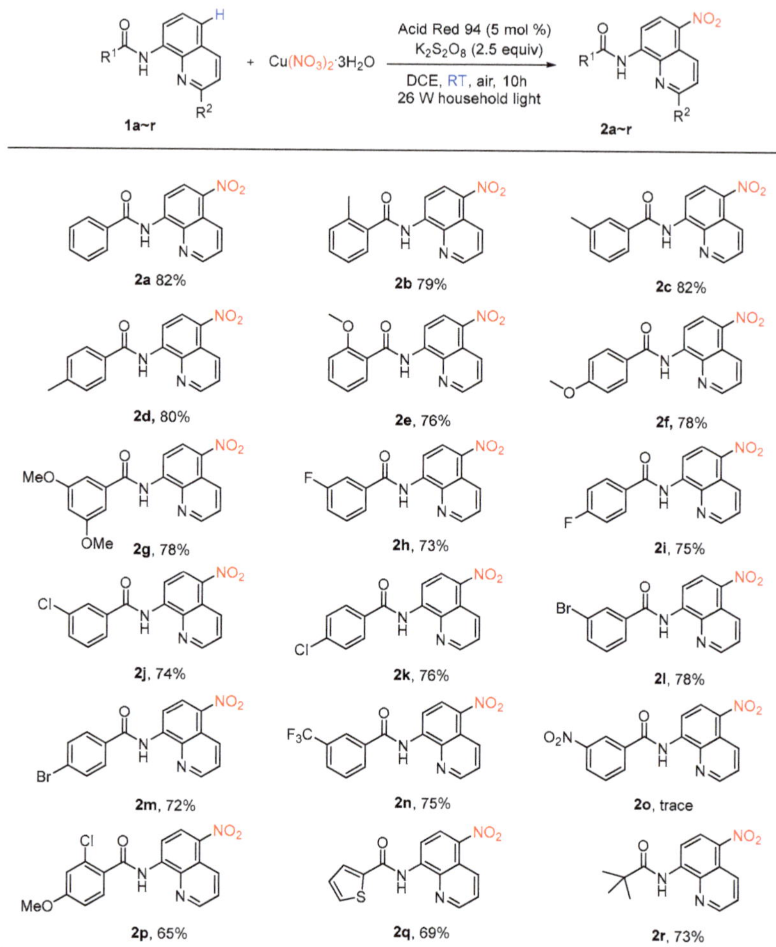

[a] reaction conditions: **1a** (0.20 mmol), $Cu(NO_3)_2{\cdot}3H_2O$ (1.5 equiv), Acid Red 94 (5 mol %), and $K_2S_2O_8$ (2.5 equiv) in DCE (1.5 mL) at room temperature under air for 10 h. [b] isolated yield.

Scheme 3. Substrate scope of 8-aminoquinoline amides [a,b].

In order to prove the synthetic value of this method, we performed a gram-scale reaction (Scheme 5), and the nitrated product **2a** was isolated in 74% yield. Furthermore, the obtained product **2a** could be easily transformed into 5-nitro-8-aminoquinoline **3** in a yield of 93% via a hydrolysis process.

In order to obtain the mechanistic information about this transformation, control experiments were carried out (Scheme 6). No desired product was observed when the benzamides (**4** and **5**) and benzoate **6** participated in this reaction under the standard conditions. These results showed that the formation of a chelated complex between 8-aminoquinoline and the copper salt was crucial to the remote C–H nitration [50]. The addition of TEMPO to the reaction mixture resulted in the inhibition of the corresponding nitration products, suggesting that radical steps might be involved in this reaction. In addition, our previous report has pointed out that the C5 site on the quinoline ring has the largest p_z orbital occupancies and may be the electrophilic reactive site [28].

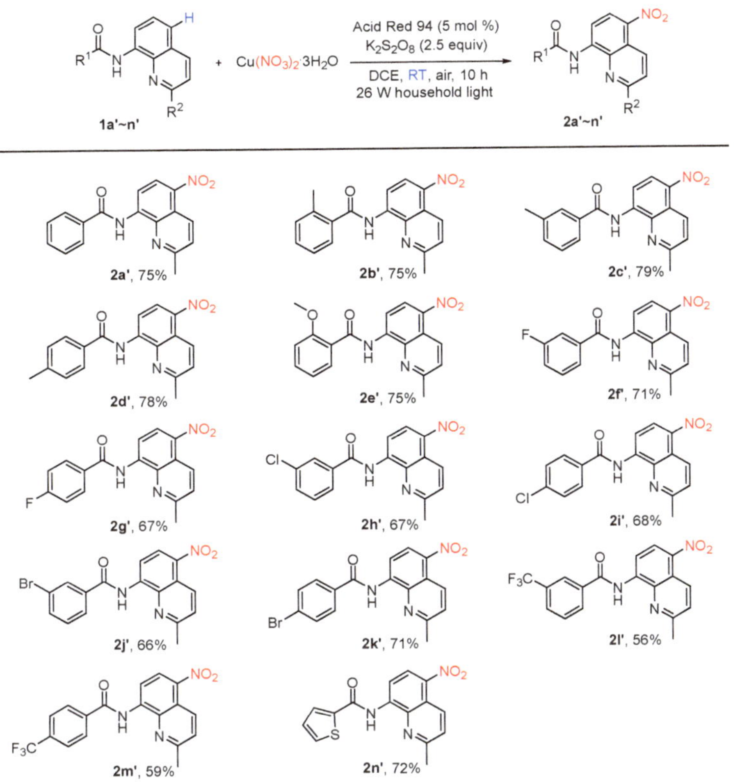

[a] reaction conditions: **1a** (0.20 mmol), $Cu(NO_3)_2{\cdot}3H_2O$ (1.5 equiv), Acid Red 94 (5 mol %), and $K_2S_2O_8$ (2.5 equiv) in DCE (1.5 mL) at room temperature under air for 10 h. [b] isolated yield.

Scheme 4. Substrate scope of 2-methyl-8-aminoquinoline amides [a,b].

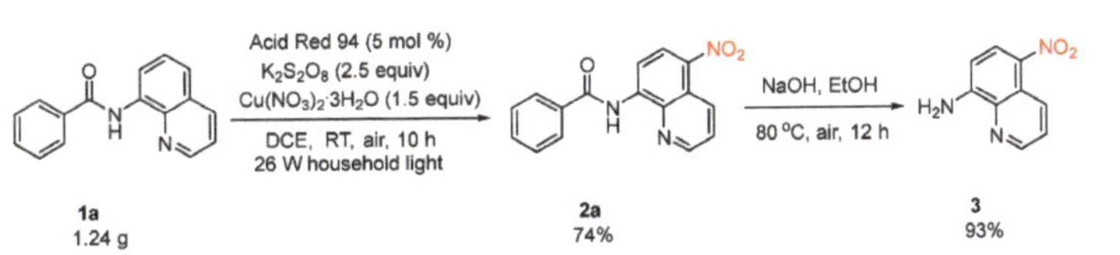

Scheme 5. A gram-scale experiment.

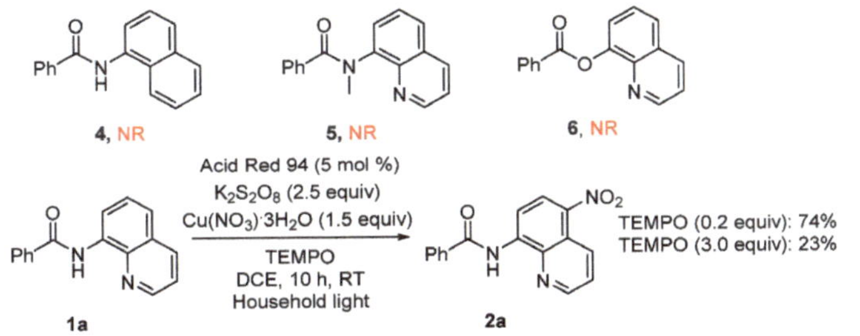

Scheme 6. Control experiments.

On the basis of the above experiments and previous reports, a plausible mechanism for this visible-light-promoted remote C–H nitration is proposed in Scheme 7. First, Acid Red 94 is excited by the household light, generating the excited state Acid Red 94*. Then, Acid Red 94* undergoes a reductive quenching process with $Cu(NO_3)_2$, yielding a $NO_2\bullet$ and Acid Red 94 radical anion. The Acid Red 94 radical anion could be oxidized by the $K_2S_2O_8$, regenerating the ground state Acid Red 94 to complete the photocatalytic cycle. On the other hand, the coordination of **1a** with $Cu(NO_3)_2$ affords the chelated intermediate **A**. Afterwards, the deprotonation of the amide group leads to the formation of the complex **B**. Then, an intermediate radical **C** is formed by the electronic attack of the nitro radical ($NO_2\bullet$) at the C5 position of intermediate **B**. After that, the cationic intermediate **D** is generated through the oxidation of the intermediate radical **C** by the sulfate radical anion. Finally, a metal dissociation process of **D** occurs after the proton transfer process, providing the desired product **2a**, along with the regeneration of $Cu(NO_3)_2$ to fulfill the catalytic cycle.

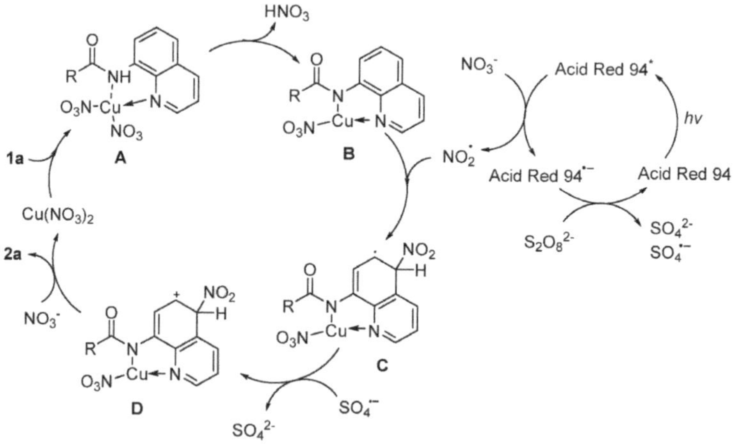

Scheme 7. A plausible reaction mechanism.

3. Conclusions

We have developed a mild and green protocol for the C5 nitration on the quinoline ring of 8-aminoquinoline amides, producing the corresponding product in moderate-to-good yields. This reaction proceeded smoothly at room temperature in air under visible light photoredox catalysis. Moreover, this transformation also showed good functional group tolerance and provided a practical synthetic route for 5-nitro-8-aminoquinoline derivatives.

4. Experimental

4.1. General Information

The ^{1}H and ^{13}C NMR spectra were recorded on a Bruker DPX-400 spectrometer (Bruker BioSpin, Faellanden, Switzerland) with $CDCl_3$ as the solvent and TMS (Alfa Aesar, Shanghai, China) as an internal standard. Melting points were measured using a X-5 microscopic (Gongyi Yuhua Yiqi, Zhengzhou, China) apparatus and are uncorrected. High-resolution mass spectra were obtained with Agilent Technologies 1290-6540 UHPLC/Accurate-Mass Quadrupole Time-of-Flight LC/MS (Agilent Technologies, Wilmington, NC, USA). All solvents were used directly without further purification. Dichloromethane, ethyl acetate, and hexane were used for column chromatography. The commercials were obtained from commercial sources and used as-received without further purification unless otherwise noted. For the detailed experimental information, please see the Supplementary Materials.

4.2. Typical Procedure for Synthesizing the Catalytic C5-H Nitration of 8-Aminoquinoline Amides

(2a-2p, 2a′-2m′): To a 10 mL reaction tube, the mixture of amide (0.2 mmol), $Cu(NO_3)_2 \cdot 3H_2O$ (0.3 mmol, 1.5 equiv), Acid Red 94 (0.01 mmol, 5 mol %), $K_2S_2O_8$ (0.5 mmol, 2.5 equiv), and DCE (1.5 mL) was added. The resulting mixture was stirred under the irradiation of 26 W household light under air at room temperature for 10 h. Upon completion, the mixture was filtered through a celite pad and washed with CH_2Cl_2, the solvent was removed under reduced pressure and then followed by recrystallization to produce the corresponding product using hexane–CH_2Cl_2 as a solvent.

(2q, 2r and 2n′): To a 10 mL reaction tube, the mixture of amide (0.2 mmol), $Cu(NO_3)_2 \cdot 3H_2O$ (0.3 mmol, 1.5 equiv), Acid Red 94 (0.01 mmol, 5 mol %), $K_2S_2O_8$ (0.5 mmol, 2.5 equiv), and DCE (1.5 mL) was added. Upon completion, the solvent was removed under reduced pressure, and the residue was purified by silica gel column chromatography to produce the corresponding product.

***N*-(5-nitroquinolin-8-yl)benzamide (2a) [36]:** Yellow solid in 82% yield; mp 215–216 °C; ^{1}H NMR (400 MHz, $CDCl_3$) δ 11.07 (s, 1H), 9.30 (d, J = 8.6 Hz, 1H), 9.00 (d, J = 8.8 Hz, 1H), 8.95 (d, J = 3.3 Hz, 1H), 8.59 (d, J = 8.8 Hz, 1H), 8.08 (d, J = 7.28 Hz, 2H), 7.75 (dd, J = 8.8, 4.1 Hz, 1H), 7.66–7.63 (m, 1H), 7.60–7.56 (m, 2H); ^{13}C NMR (100 MHz, $CDCl_3$) δ 165.7, 149.1, 140.9, 138.7, 137.8, 134.1, 133.4, 132.7, 129.0, 127.9, 127.5, 124.7, 121.8, 113.7.

2-methyl-*N*-(5-nitroquinolin-8-yl)benzamide(2b) [36]: Yellow solid in 79% yield; mp 192–193 °C; ^{1}H NMR (400 MHz, $CDCl_3$) δ 10.58 (s, 1H), 9.30 (d, J = 8.8 Hz, 1H), 9.01 (d, J = 8.8 Hz, 1H), 8.88 (d, J = 2.8 Hz, 1H), 8.61 (d, J = 8.8 Hz, 1H), 7.74–7.69 (m, 2H), 7.74–7.69 (m, 1H), 7.38–7.34 (m, 2H), 2.62 (s, 3H); ^{13}C NMR (100 MHz, $CDCl_3$) δ 168.3, 149.1, 141.0, 138.8, 137.6, 137.3, 135.5, 133.4, 131.7, 131.1, 127.9, 127.3 126.2, 124.7, 121.8, 113.7, 20.3.

3-methyl-*N*-(5-nitroquinolin-8-yl)benzamide (2c): Yellow solid in 82% yield; mp 201–202 °C; ^{1}H NMR (400 MHz, $CDCl_3$) δ 11.04 (s, 1H), 9.30 (d, J = 8.5 Hz, 1H), 9.01 (d, J = 8.9 Hz, 1H), 8.96 (d, J = 2.8 Hz, 1H), 8.60 (d, J = 8.8Hz, 1H), 7.88 (d, J = 11.9 Hz, 2H), 7.75 (dd, J = 9.1, 4.1 Hz, 1H), 7.49–7.45 (m, 2H), 2.49 (s, 3H); ^{13}C NMR (100 MHz, $CDCl_3$) δ 165.94, 149.1, 141.0, 139.0, 138.6, 137.8, 134.2, 133.4, 128.9, 128.2, 127.9, 124.7, 124.4, 121.8, 113.7, 21.5; HRMS (ESI): calculated for $C_{17}H_{13}N_3O_3$: $[M + H]^+$ requires 308.1030, found 308.1031.

4-methyl-*N*-(5-nitroquinolin-8-yl)benzamide (2d) [37]: Yellow solid in 80% yield; mp 210–211 °C; ^{1}H NMR (400 MHz, $CDCl_3$) δ 11.05 (s, 1H), 9.31 (d, J = 8.7 Hz, 1H), 9.02 (d, J = 8.8 Hz, 1H), 8.96 (d, J = 3.4 Hz, 1H), 8.61 (d, J = 8.8 Hz, 1H), 7.99 (d, J = 7.8 Hz, 2H), 7.74 (dd, J = 8.9, 4.1 Hz, 1H), 7.38 (d, J = 7.9 Hz, 2H), 2.48 (s, 3H); ^{13}C NMR (100 MHz, $CDCl_3$) δ 165.7, 149.0, 143.4, 141.0, 138.6, 137.8, 133.5, 131.4, 129.7, 128.0, 127.5, 124.7, 121.9, 113.7, 21.6; HRMS (ESI): calculated for $C_{17}H_{13}N_3O_3$: $[M + H]^+$ requires 308.1030, found 308.1033.

2-methoxy-*N*-(5-nitroquinolin-8-yl)benzamide (2e): Yellow solid in 75% yield; mp 208–209 °C; ^{1}H NMR (400 MHz, $CDCl_3$) δ 12.68 (s, 1H), 9.26 (d, J = 7.9 Hz, 1H), 9.06 (d, J = 8.9 Hz, 1H), 8.93 (d, J = 2.8 Hz, 1H), 8.56 (d, J = 8.8Hz, 1H), 8.31 (d, J = 6.7, 1H), 7.69 (dd, J = 8.8, 4.1 Hz, 1H), 7.57–7.54 (m, 1H), 7.17–7.14 (m, 1H), 7.08 (d, J = 8.2 Hz, 1H), 4.20 (s, 3H); ^{13}C NMR (100 MHz, $CDCl_3$) δ 164.2, 157.9, 148.9, 142.2, 138.4, 138.2, 134.0, 133.1, 132.6, 128.1, 124.4, 121.9, 121.5, 121.4, 114.4, 111.7, 56.2; HRMS (ESI): calculated for $C_{17}H_{13}N_3O_4$: $[M + H]^+$ requires 324.0979, found 324.0977.

4-methoxy-*N*-(5-nitroquinolin-8-yl)benzamide (2f) [35]: Yellow solid in 78% yield; mp 260–261 °C; ^{1}H NMR (400 MHz, $CDCl_3$) δ 11.03 (s, 1H), 9.34 (d, J = 8.6 Hz, 1H), 9.01 (d, J = 8.9 Hz, 1H), 8.96 (d, J = 3.6 Hz, 1H), 8.63 (d, J = 8.9 Hz, 1H), 8.07 (d, J = 8.7 Hz, 2H), 7.76 (dd, J = 8.9, 4.1 Hz, 1H), 7.07 (d, J = 8.6 Hz, 2H), 3.92 (s, 3H); ^{13}C NMR (100 MHz, $CDCl_3$) δ 165.2, 163.2, 149.0, 141.2, 138.5, 137.8, 133.5, 129.5, 128.1, 126.4, 124.7, 121.9, 114.3, 113.5, 55.6.

3,5-dimethoxy-*N*-(5-nitroquinolin-8-yl)benzamide (2g): Yellow solid in 78% yield; mp 200–201 °C; ^{1}H NMR (400 MHz, $CDCl_3$) δ 11.00 (s, 1H), 9.30 (d, J = 8.8 Hz, 1H), 8.99–8.95 (m, 2H), 8.61 (d, J = 8.8 Hz, 1H), 7.76 (dd, J = 8.9, 4.1 Hz, 1H), 7.20 (s, 2H), 6.71 (s, 2H), 3.91 (s, 6H); ^{13}C NMR (100 MHz, $CDCl_3$) δ 165.6, 161.2, 149.2, 140.8, 138.8, 137.8, 136.3, 133.4, 127.9, 124.8, 121.8, 113.8, 105.5, 104.4, 55.7; HRMS (ESI): calculated for $C_{18}H_{15}N_3O_5$: $[M + H]^+$ requires 354.1084, found 354.1087.

3-fluoro-*N*-(5-nitroquinolin-8-yl)benzamide (2h): Yellow solid in 73% yield; mp 236–237 °C; ^{1}H NMR (400 MHz, $CDCl_3$) δ 11.06 (s, 1H), 9.31 (d, *J* = 8.7 Hz, 1H), 9.01–8.98 (m, 2H), 8.61 (d, *J* = 8.8 Hz, 1H), 7.88 (d, *J* = 7.6 Hz, 1H), 7.80–7.76 (m, 2H), 7.60–7.55 (m, 1H),7.36–7.33 (m, 1H); ^{13}C NMR (100 MHz, $CDCl_3$) δ 164.4, 163.0, 149.2, 140.5, 139.0, 137.8, 136.4, 133.5, 130.8, 127.8, 124.8, 122.9, 121.8, 119.7, 114.9, 113.9; HRMS (ESI): calculated for $C_{16}H_{10}FN_3O_3$: [M + H]$^+$ requires 312.0779, found 312.0781.

4-fluoro-*N*-(5-nitroquinolin-8-yl)benzamide (2i): Yellow solid in 75% yield; mp 218–219 °C; ^{1}H NMR (400 MHz, $CDCl_3$) δ 11.03 (s, 1H), 9.31 (d, *J* = 8.6 Hz, 1H), 9.01–8.96 (m, 2H), 8.60 (d, *J* = 8.8 Hz, 1H), 8.13–8.09 (m, 2H), 7.77 (dd, *J* = 9.2, 4.1 Hz, 1H), 7.29–7.24 (m, 2H); ^{13}C NMR (100 MHz, $CDCl_3$) δ 165.5, 164.6, 149.1, 140.7, 138.8, 137.7, 133.5, 130.4, 130.0, 127.9, 124.8, 121.8, 116.2, 113.7; HRMS (ESI): calculated for $C_{16}H_{10}FN_3O_3$: [M + H]$^+$ requires 312.0779, found 312.0778.

3-chloro-*N*-(5-nitroquinolin-8-yl)benzamide (2j): Yellow solid in 74% yield; mp 234–235 °C; ^{1}H NMR (400 MHz, $CDCl_3$) δ 11.04 (s, 1H), 9.31 (d, *J* = 8.7 Hz, 1H), 8.99 (d, *J* = 8.8Hz, 2H), 8.62 (d, *J* = 8.8 Hz, 1H),8.07 (s, 1H), 7.96 (d, *J* = 7.5 Hz, 1H), 7.78 (dd, *J* = 9.1, 4.1 Hz, 1H), 7.62 (d, *J* = 7.8 Hz, 1H), 7.52 (t, *J* = 7.8Hz, 1H); ^{13}C NMR (100 MHz, $CDCl_3$) δ 164.4, 149.2, 140.5, 139.0, 137.8, 135.9, 135.3, 133.5, 132.7, 130.3, 127.9, 127.8, 125.4, 124.8, 121.8, 114.0; HRMS (ESI): calculated for $C_{16}H_{10}ClN_3O_3$: [M + H]$^+$ requires 328.0483, found 328.0483.

4-chloro-*N*-(5-nitroquinolin-8-yl)benzamide (2k) [37]: Yellow solid in 76% yield; mp 210–212 °C; ^{1}H NMR (400 MHz, $CDCl_3$) δ 11.05 (s, 1H), 9.32 (d, *J* = 8.1 Hz, 1H), 9.01–8.97 (m, 2H), 8.62 (d, *J* = 8.8 Hz, 1H), 8.04 (d, *J* = 8.5 Hz, 2H), 7.77 (dd, *J* = 9.1, 4.1 Hz, 1H), 7.56 (d, *J* = 8.4 Hz, 2H); ^{13}C NMR (100 MHz, $CDCl_3$) δ 164.7, 149.2, 140.6, 139.1, 138.9, 137.8, 133.5, 132.5, 129.4, 128.9, 127.8, 124.8, 121.8, 113.9; HRMS (ESI): calculated for $C_{16}H_{10}ClN_3O_3$: [M + H]$^+$ requires 328.0483, found 328.0484.

3-bromo-*N*-(5-nitroquinolin-8-yl)benzamide (2l): Yellow solid in 78% yield; mp 213–214 °C; ^{1}H NMR (400 MHz, $CDCl_3$) δ 11.00 (s, 1H), 9.36 (d, *J* = 8.8 Hz, 1H), 8.96 (d, *J* = 8.9 Hz, 2H), 8.61 (d, *J* = 8.8 Hz, 1H), 8.23 (s, 1H), 8.01 (d, *J* = 7.7 Hz, 1H), 7.78–7.75 (m, 2H), 7.49 (t, *J* = 7.8 Hz, 1H); ^{13}C NMR (100 MHz, $CDCl_3$) δ 164.2, 149.2, 140.4, 139.0, 137.8, 136.1, 135.6, 133.5, 130.8, 130.5, 127.8, 125.9, 124.8, 123.3, 121.8, 114.0; HRMS (ESI): calculated for $C_{16}H_{10}BrN_3O_3$: [M + H]$^+$ requires 371.9978, found 371.9979.

4-bromo-*N*-(5-nitroquinolin-8-yl)benzamide (2m): Yellow solid in 72% yield; mp 279–280 °C; ^{1}H NMR (400 MHz, $CDCl_3$) δ 11.06 (s, 1H), 9.32 (d, *J* = 8.2 Hz, 1H), 9.01–8.97 (m, 2H), 8.62 (d, *J* = 8.9 Hz, 1H),7.96 (d, *J* = 8.3Hz, 2H), 7.78 (dd, *J* = 8.9, 4.3 Hz, 1H), 7.72 (d, *J* = 8.5 Hz, 2H); ^{13}C NMR (100 MHz, $CDCl_3$) δ 164.8, 149.2, 140.6, 138.9, 137.8, 133.5, 133.0, 132.3, 129.0, 127.8, 127.6, 124.8, 121.8, 113.9; HRMS (ESI): calculated for $C_{16}H_{10}BrN_3O_3$: [M + H]$^+$ requires 371.9978, found 371.9977.

***N*-(5-nitroquinolin-8-yl)-3-(trifluoromethyl)benzamide (2n):** Yellow solid in 75% yield; mp 181–182 °C; ^{1}H NMR (400 MHz, $CDCl_3$) δ 11.08 (s, 1H), 9.30 (d, *J* = 8.6 Hz, 1H), 8.99–8.97 (m, 2H), 8.61 (d, *J* = 8.8 Hz, 1H), 8.36 (s, 1H), 8.25 (d, *J* = 7.6 Hz, 1H), 7.90 (d, *J* = 7.6 Hz, 1H), 7.79–7.72 (m, 2H); ^{13}C NMR (100 MHz, $CDCl_3$) δ 164.2, 149.3, 140.3, 139.1, 137.7, 135.0, 133.7, 133.5, 130.4, 129.7, 129.2, 127.7, 124.9, 124.7, 122.1, 114.0; HRMS (ESI): calculated for $C_{17}H_{10}F_3N_3O_3$: [M + H]$^+$ requires 362.0747, found 362.0748.

2-chloro-4-methoxy-*N*-(5-nitroquinolin-8-yl)benzamide (2p): Yellow solid in 65% yield; mp 262–263 °C; ^{1}H NMR (400 MHz, $CDCl_3$) δ 10.99 (s, 1H), 9.33 (d, *J* = 8.8 Hz, 1H), 8.99–8.97 (m, 2H), 8.62 (d, *J* = 8.8 Hz, 1H), 8.14 (d, *J* = 1.2 Hz, 1H), 8.02 (d, *J* = 8.6 Hz, 1H), 7.77 (dd, *J* = 8.9, 4.1 Hz, 1H), 7.10 (d, *J* = 8.5 Hz, 1H), 4.03 (s, 3H); ^{13}C NMR (100 MHz, $CDCl_3$) δ 164.1, 158.5, 149.1, 140.8, 138.7, 137.8, 133.5, 129.7, 127.9, 127.7, 127.2, 124.8, 123.3, 121.9, 113.7, 111.8, 56.5; HRMS (ESI): calculated for $C_{17}H_{12}ClN_3O_4$: [M + H]$^+$ requires 358.0589, found 358.0591.

***N*-(5-nitroquinolin-8-yl)thiophene-2-carboxamide (2q) [35]:** Yellow solid in 69% yield; mp 220–221 °C; ^{1}H NMR (400 MHz, $CDCl_3$) δ 10.90 (s, 1H), 9.30 (d, *J* = 8.8 Hz, 1H), 8.96–8.95 (m, 1H), 8.91 (d, *J* = 8.8 Hz, 1H), 8.59 (d, *J* = 8.8 Hz, 1H), 7.88 (d, *J* = 3.1 Hz, 1H), 7.76 (dd,

J = 8.9, 4.1 Hz, 1H), 7.65 (d, J = 4.8 Hz, 1H), 7.24–7.21 (m, 1H); ^{13}C NMR (100 MHz, $CDCl_3$) δ 160.2, 149.1, 140.6, 139.0, 138.7, 137.5, 133.5, 132.2, 129.4, 128.2, 127.9, 124.8, 121.9, 113.7.

***N*-(5-nitroquinolin-8-yl)pivalamide (2r) [37]:** Yellow solid in 73% yield; mp 172–173 °C; ^{1}H NMR (400 MHz, $CDCl_3$) δ 10.59 (s, 1H), 9.27–9.25 (m, 1H), 8.92 (d, J = 3.0 Hz, 1H), 8.84 (d, J = 8.8 Hz, 1H), 8.54 (d, J = 8.8 Hz, 1H), 7.71 (dd, J = 8.8, 4.1 Hz, 1H), 1.45 (s, 9H); ^{13}C NMR (100 MHz, $CDCl_3$) δ 177.8, 149.0, 141.0, 138.3, 137.7, 133.3, 127.9, 124.6, 121.7, 113.4, 40.7, 27.6.

***N*-(2-methyl-5-nitroquinolin-8-yl)benzamide (2a′) [37]:** Yellow solid in 75% yield; mp 202–203 °C; ^{1}H NMR (400 MHz, $CDCl_3$) δ 11.08 (s, 1H), 9.12 (d, J = 8.9 Hz, 1H), 8.91 (d, J = 8.8, 1H), 8.49 (d, J = 8.8 Hz, 1H), 8.05 (d, J = 7.3 Hz, 2H), 7.66–7.57 (m, 4H), 2.81 (s, 3H); ^{13}C NMR (100 MHz, $CDCl_3$) δ 165.5, 158.5, 140.1, 138.7, 137.3, 134.3, 133.3, 132.6, 129.1, 127.4, 126.7, 125.6, 119.9, 113.7, 25.2; HRMS (ESI): calculated for $C_{17}H_{13}N_3O_3$: $[M + H]^+$ requires 308.1030, found 308.1035.

2-methyl-*N*-(2-methyl-5-nitroquinolin-8-yl)benzamide (2b′): Yellow solid in 75% yield; mp 209–210 °C; ^{1}H NMR (400 MHz, $CDCl_3$) δ 10.62 (s, 1H), 9.15 (d, J = 8.9 Hz, 1H), 8.96 (d, J = 8.8, 1H), 8.51 (d, J = 8.8 Hz, 1H), 7.72 (d, J = 7.4 Hz, 1H), 7.58 (d, J = 8.9 Hz, 1H), 7.48–7.44 (m, 1H), 7.39–7.35 (m, 2H), 2.75 (s, 3H), 2.64 (s, 3H); ^{13}C NMR (100 MHz, $CDCl_3$) δ 168.2, 158.6, 140.3, 138.9, 137.2, 137.2, 135.6, 133.3, 131.8, 131.1, 127.5, 126.7, 126.3, 125.6, 120.0, 113.7, 25.1, 20.4; HRMS (ESI): calculated for $C_{18}H_{15}N_3O_3$: $[M + H]^+$ requires 322.1186, found 322.1190.

3-methyl-*N*-(2-methyl-5-nitroquinolin-8-yl)benzamide (2c′): Yellow solid in 79% yield; mp 184–185 °C; ^{1}H NMR (400 MHz, $CDCl_3$) δ 11.04 (s, 1H), 9.15 (d, J = 8.9 Hz, 1H), 8.93 (d, J = 8.8 Hz, 1H), 8.51 (d, J = 8.8 Hz, 1H), 7.91 (s, 1H), 7.81 (d, J = 6.9 Hz, 1H), 7.59 (d, J = 8.9 Hz, 1H), 7.49–7.43 (m, 2H), 2.83 (s, 3H), 2.47 (s, 3H); ^{13}C NMR (100 MHz, $CDCl_3$) δ 165.7, 158.4, 140.2, 139.0, 138.7, 137.3, 134.3, 133.4, 133.3, 128.9, 128.3, 126.8, 125.6, 124.2, 120.0, 113.7, 25.2, 21.5; HRMS (ESI): calculated for $C_{18}H_{15}N_3O_3$: $[M + H]^+$ requires 322.1186, found 322.1190.

4-methyl-*N*-(2-methyl-5-nitroquinolin-8-yl)benzamide (2d′): Yellow solid in 78% yield; mp 206–207 °C; ^{1}H NMR (400 MHz, $CDCl_3$) δ 10.99 (s, 1H), 9.09 (d, J = 8.9 Hz, 1H), 8.86 (d, J = 8.8 Hz, 1H), 8.44 (d, J = 8.8 Hz, 1H), 7.92 (d, J = 7.9 Hz, 2H), 7.55 (d, J = 8.9 Hz, 1H), 7.37 (d, J = 7.8 Hz, 2H), 2.81 (s, 3H), 2.45 (s, 3H); ^{13}C NMR (100 MHz, $CDCl_3$) δ 165.4, 158.4, 143.3, 140.2, 138.5, 137.2, 133.2, 131.3, 129.7, 127.4, 126.8, 125.6, 120.0, 113.5, 25.2, 21.6; HRMS (ESI): calculated for $C_{18}H_{15}N_3O_3$: $[M + H]^+$ requires 322.1186, found 322.1189.

2-methoxy-*N*-(2-methyl-5-nitroquinolin-8-yl)benzamide (2e′): Yellow solid in 75% yield; mp 258–259 °C; ^{1}H NMR (400 MHz, $CDCl_3$) δ 12.46 (s, 1H), 9.17 (d, J = 8.9 Hz, 1H), 9.10 (d, J = 8.9 Hz, 1H), 8.51 (d, J = 8.9 Hz, 1H), 8.33 (d, J = 6.6 Hz, 1H), 7.60–7.55 (m, 2H), 7.18 (t, J = 7.6 Hz, 1H), 7.11 (d, J = 8.3 Hz, 1H), 4.23 (s, 3H), 2.86 (s, 3H); ^{13}C NMR (100 MHz, $CDCl_3$) δ 164.3, 157.9, 157.8, 141.6, 137.8, 134.0, 133.2, 132.8, 127.0, 125.3, 121.7, 121.6, 120.1, 114.7, 111.6, 56.2, 25.3; HRMS (ESI): calculated for $C_{18}H_{15}N_3O_4$: $[M + H]^+$ requires 338.1135, found 338.1139.

3-fluoro-*N*-(2-methyl-5-nitroquinolin-8-yl)benzamide (2f′): Yellow solid in 71% yield; mp 251–252 °C; ^{1}H NMR (400 MHz, $CDCl_3$) δ 11.10 (s, 1H), 9.17 (d, J = 8.9 Hz, 1H), 8.94 (d, J = 8.8 Hz, 1H), 8.53 (d, J = 8.8 Hz, 1H), 7.85 (d, J = 7.7 Hz, 1H), 7.78 (d, J = 9.2 Hz, 1H), 7.63–7.55 (m, 2H), 7.36–7.32 (m, 1H), 2.84 (s, 3H);^{13}C NMR (100 MHz, $CDCl_3$) δ 164.3, 161.8, 158.7, 139.8, 139.1, 137.4, 136.6, 133.5, 130.8, 126.7, 125.7, 122.8, 120.0, 119.6, 114.9, 113.9, 25.3; HRMS (ESI): calculated for $C_{17}H_{12}FN_3O_3$: $[M + H]^+$ requires 326.0935, found 326.0938.

4-fluoro-*N*-(2-methyl-5-nitroquinolin-8-yl)benzamide (2g′): Yellow solid in 66% yield; mp 218–219 °C; ^{1}H NMR (400 MHz, $CDCl_3$) δ 11.07 (s, 1H), 9.16 (d, J = 8.9 Hz, 1H), 8.93 (d, J = 8.8, 1H), 8.52 (d, J = 8.8 Hz, 1H), 8.11–8.08 (m, 2H), 7.62 (d, J = 8.9 Hz, 1H), 7.29–7.25 (m, 2H), 2.83 (s, 3H);^{13}C NMR (100 MHz, $CDCl_3$) δ 166.7, 164.3, 158.5, 140.0, 138.9, 137.3, 133.5, 130.5, 129.9, 126.7, 125.7, 120.0, 116.2, 113.8, 25.3; HRMS (ESI): calculated for $C_{17}H_{12}FN_3O_3$: $[M + H]^+$ requires 326.0935, found 326.0938.

3-chloro-*N*-(2-methyl-5-nitroquinolin-8-yl)benzamide (2h′): Yellow solid in 67% yield; mp 215–216 °C; ^{1}H NMR (400 MHz, $CDCl_3$) δ 11.08 (s, 1H), 9.15 (d, J = 8.9 Hz, 1H),

8.91 (d, J = 8.8 Hz, 1H), 8.51 (d, J = 8.8 Hz, 1H), 8.07 (s, 1H), 7.94 (d, J = 7.6 Hz, 1H), 7.62 (d, J = 9.2 Hz, 2H), 7.53 (t, J = 7.8 Hz, 1H), 2.83 (s, 3H); ^{13}C NMR (100 MHz, $CDCl_3$) δ 164.1, 158.7, 139.7, 139.1, 137.3, 136.1, 135.4, 133.4, 132.6, 130.3, 127.9, 126.6, 125.7, 125.2, 120.0, 113.9, 25.3; HRMS (ESI): calculated for $C_{17}H_{12}ClN_3O_3$: [M + H]$^+$ requires 342.0640, found 342.0642.

4-chloro-*N*-(2-methyl-5-nitroquinolin-8-yl)benzamide (2i′) [37]: Yellow solid in 68% yield; mp 221–222 °C; ^1H NMR (400 MHz, $CDCl_3$) δ 11.08 (s, 1H), 9.16 (d, J = 8.9 Hz, 1H), 8.93 (d, J = 8.8 Hz, 1H), 8.52 (d, J = 8.8 Hz, 1H), 8.02 (d, J = 8.2 Hz, 2H), 7.62 (d, J = 8.9 Hz, 1H), 7.56 (d, J = 8.2 Hz, 2H), 2.84 (s, 3H); ^{13}C NMR (100 MHz, $CDCl_3$) δ 164.5, 158.6, 139.9, 140.0, 139.0, 137.3, 133.5, 132.7, 129.4, 128.8, 126.7, 125.7, 120.0, 113.9, 25.3; HRMS (ESI): calculated for $C_{17}H_{12}ClN_3O_3$: [M + H]$^+$ requires 342.0640, found 342.0641.

3-bromo-*N*-(2-methyl-5-nitroquinolin-8-yl)benzamide (2j′): Yellow solid in 66% yield; mp 203–204 °C; ^1H NMR (400 MHz, $CDCl_3$) δ 11.09 (s, 1H), 9.16 (d, J = 8.9 Hz, 1H), 8.91 (d, J = 8.8 Hz, 1H), 8.51 (d, J = 8.8 Hz, 1H), 8.23 (s, 1H), 7.99 (d, J = 7.7 Hz, 1H), 7.76 (d, J = 7.9 Hz, 1H), 7.62 (d, J = 8.9 Hz, 1H), 7.47 (t, J = 7.8 Hz, 1H), 2.88 (s, 3H); ^{13}C NMR (100 MHz, $CDCl_3$) δ 164.0, 158.7, 139.7, 139.1, 137.3, 136.2, 135.5, 133.4, 130.9, 130.5, 126.6, 125.7, 125.7, 123.3, 119.9, 113.9, 25.3; HRMS (ESI): calculated for $C_{17}H_{12}BrN_3O_3$: [M + H]$^+$ requires 386.0135, found 386.0136.

4-bromo-*N*-(2-methyl-5-nitroquinolin-8-yl)benzamide (2k′): Yellow solid in 71% yield; mp 249–250 °C; ^1H NMR (400 MHz, $CDCl_3$) δ 11.09 (s, 1H), 9.15 (d, J = 8.9 Hz, 1H), 9.17 (d, J = 9.0, 1H), 8.93 (d, J = 8.8 Hz, 1H), 8.52 (d, J = 8.8 Hz, 1H), 7.94(d, J = 8.4 Hz, 2H), 7.44 (d, J = 8.4 Hz, 2H), 7.61 (d, J = 8.9 Hz, 1H), 2.83 (s, 3H); ^{13}C NMR (100 MHz, $CDCl_3$) δ 164.7, 158.6, 139.8, 139.0, 137.3, 133.5, 133.2, 132.4, 128.9, 127.5, 126.7, 125.7, 120.0, 113.9, 25.3; HRMS (ESI): calculated for $C_{17}H_{12}BrN_3O_3$: [M + H]$^+$ requires 386.0135, found 386.0136.

***N*-(2-methyl-5-nitroquinolin-8-yl)-3-(trifluoromethyl)benzamide (2l′):** Yellow solid in 56% yield; mp 191–192 °C; ^1H NMR (400 MHz, $CDCl_3$) δ 11.19 (s, 1H), 9.15 (d, J = 8.9 Hz, 1H), 8.92 (d, J = 8.8 Hz, 1H), 8.53 (d, J = 8.8 Hz, 1H), 8.34 (s, 1H), 8.26 (d, J = 7.7 Hz, 1H), 7.90 (d, J = 7.7 Hz, 1H), 7.74 (t, J = 7.7 Hz, 1H), 7.63 (d, J = 8.9 Hz, 1H), 2.84 (s, 3H); ^{13}C NMR (100 MHz, $CDCl_3$) δ 164.0, 158.7, 139.6, 139.2, 137.3, 135.1, 133.5, 131.7, 130.5, 129.8, 129.1, 126.6, 125.8, 124.5, 123.6, 119.9, 113.9, 25.2; HRMS (ESI): calculated for $C_{18}H_{12}F_3N_3O_3$: [M + H]$^+$ requires 376.0904, found 376.0907.

***N*-(2-methyl-5-nitroquinolin-8-yl)-4-(trifluoromethyl)benzamide (2m′):** Yellow solid in 59% yield; mp 208–209 °C; ^1H NMR (400 MHz, $CDCl_3$) δ 11.16 (s, 1H), 9.18 (d, J = 8.9 Hz, 1H), 8.96 (d, J = 8.8 Hz, 1H), 8.54 (d, J = 8.8 Hz, 1H), 8.20 (d, J = 8.0 Hz, 2H), 7.87 (d, J = 8.1 Hz, 2H), 7.64 (d, J = 8.9 Hz, 1H), 2.84 (s, 3H); ^{13}C NMR (100 MHz, $CDCl_3$) δ 164.3, 158.7, 139.6, 139.3, 137.6, 137.4, 134.1, 133.5, 127.9, 126.6, 126.2, 125.8, 123.6, 120.0, 114.1, 25.3; HRMS (ESI): calculated for $C_{18}H_{12}F_3N_3O_3$: [M + H]$^+$ requires 376.0904, found 376.0905.

***N*-(2-methyl-5-nitroquinolin-8-yl)thiophene-2-carboxamide (2n′):** Yellow solid in 72% yield; mp 257–258 °C; ^1H NMR (400 MHz, $CDCl_3$) δ 10.98 (s, 1H), 9.16 (d, J = 8.8 Hz, 1H), 8.85 (d, J = 8.8 Hz, 1H), 8.51 (d, J = 8.7 Hz, 1H), 7.85 (d, J = 2.4 Hz, 1H), 7.67 (d, J = 4.5 Hz, 1H), 7.61 (d, J = 8.9 Hz, 1H), 7.23 (d, J = 3.9 Hz, 1H), 2.84 (s, 3H); ^{13}C NMR (100 MHz, $CDCl_3$) δ 160.1, 158.5, 139.9, 139.2, 138.8, 137.1, 133.4, 132.0, 129.3, 128.2, 126.8, 125.7, 120.0, 113.7, 25.2; HRMS (ESI): calculated for $C_{15}H_{11}N_3O_3S$: [M + H]$^+$ requires 314.0594, found 314.0596.

Supplementary Materials: The following are available online at https://www.mdpi.com/article/10.3390/catal14040263/s1, Table S1: Screening of reaction conditions. Copies of 1H and 13C NMR spectra for the products.

Author Contributions: F.Y.: review and editing and supervision. P.L. and X.S.: the main part of the experimental and data collection. P.L., H.Q. and P.B.: writing—original draft preparation. All authors have read and agreed to the published version of the manuscript.

Funding: We are grateful to the National Natural Science Foundation of China (no. 21102134) for its financial support.

Data Availability Statement: The Supporting Information is available free of charge on the website and contains experimental details.

Conflicts of Interest: The authors declare no conflicts of interest.

References

1. Feuer, H.; Nielsen, A.T. *Nitro Compounds: Recent Advances in Synthesis and Chemistry*; Wiley-VCH: Weinheim, Germany, 1990.
2. Ono, N. *The Nitro Group in Organic Synthesis*; Wiley-VCH: Weinheim, Germany, 2001.
3. Feldman, M.; Wheeler, J.W.J. Nitration of benzoic acid: Determination of isomer distribution by the isotope dilution technique. *Chem. Educ.* **1967**, *44*, 464. [CrossRef]
4. Olah, G.A.; Malhotra, R.; Narang, S.C. *Nitration: Methods and Mechanisms*; VCH: New York, NY, USA, 1989.
5. Manna, S.; Maity, S.; Rana, S.; Agasti, S.; Maiti, D. ipso-Nitration of Arylboronic Acids with Bismuth Nitrate and Perdisulfate. *Org. Lett.* **2012**, *14*, 1736–1739. [CrossRef] [PubMed]
6. Prakash, G.K.S.; Panja, C.; Mathew, T.; Surampudi, V.; Petasis, N.A.; Olah, G.A. ipso-Nitration of Arylboronic Acids with Chlorotrimethylsilane-Nitrate Salts. *Org. Lett.* **2004**, *6*, 2205–2207. [CrossRef] [PubMed]
7. Saito, S.; Koizumi, Y. Copper-catalyzed coupling of aryl halides and nitrite salts: A mild Ullmann-type synthesis of aromatic nitro compounds. *Tetrahedron Lett.* **2005**, *46*, 4715–4717. [CrossRef]
8. Fors, B.P.; Buchwald, S.L.J. Pd-Catalyzed Conversion of Aryl Chlorides, Triflates, and Nonaflates to Nitroaromatics. *Am. Chem. Soc.* **2009**, *131*, 12898–12899. [CrossRef]
9. Natarajan, P.; Chaudhary, R.; Venugopalan, P.J. Silver(I)-Promoted ipso-Nitration of Carboxylic Acids by Nitronium Tetrafluoroborate. *Org. Chem.* **2015**, *80*, 10498–10504. [CrossRef] [PubMed]
10. Rozen, S.; Carmeli, M.J. From azides to nitro compounds in a few seconds using HOF·CH_3CN. *Am. Chem. Soc.* **2003**, *125*, 8118–8119. [CrossRef]
11. Reddy, K.R.; Maheswari, C.U.; Venkateshwar, M. Selective Oxidation of Aromatic Amines to Nitro Derivatives using Potassium Iodide-tert-Butyl Hydroperoxide Catalytic System. *Adv. Synth. Catal.* **2009**, *351*, 93–96. [CrossRef]
12. Mkhalid, I.A.I.; Barnard, J.H.; Marder, T.B.; Murphy, J.M.; Hartwig, J.F. C–H Activation for the Construction of C–B Bonds. *Chem. Rev.* **2010**, *110*, 890–931. [CrossRef] [PubMed]
13. Colby, D.A.; Bergman, R.G.; Ellman, J.A. Rhodium-Catalyzed C–C Bond Formation via Heteroatom-Directed C–H Bond Activation. *Chem. Rev.* **2010**, *110*, 624–655. [CrossRef] [PubMed]
14. Sehnal, P.; Taylor, R.J.K.; Fairlamb, I.J.S. Emergence of Palladium(IV) Chemistry in Synthesis and Catalysis. *Chem. Rev.* **2010**, *110*, 824–889. [CrossRef] [PubMed]
15. Xu, L.-M.; Li, B.-J.; Yang, Z.; Shi, Z.-J. Organopalladium(IV) chemistry. *Chem. Soc. Rev.* **2010**, *39*, 712–733. [CrossRef] [PubMed]
16. Shi, W.; Liu, C.; Lei, A. Transition-metal catalyzed oxidative cross-coupling reactions to form C–C bonds involving organometallic reagents as nucleophiles. *Chem. Soc. Rev.* **2011**, *40*, 2761–2776. [CrossRef] [PubMed]
17. Song, G.; Wang, F.; Li, X. C–C, C–O and C–N bond formation via rhodium(III)-catalyzed oxidative C–H activation. *Chem. Soc. Rev.* **2012**, *41*, 3651–3678. [CrossRef] [PubMed]
18. Arockiam, P.B.; Bruneau, C.; Dixneuf, P.H. Ruthenium(II)-Catalyzed C–H Bond Activation and Functionalization. *Chem. Rev.* **2012**, *112*, 5879–5918. [CrossRef] [PubMed]
19. Ackermann, L. Carboxylate-Assisted Ruthenium-Catalyzed Alkyne Annulations by C–H/Het–H Bond Functionalizations. *Acc. Chem. Res.* **2014**, *47*, 281–295. [CrossRef] [PubMed]
20. Yi, J.; Yang, L.; Xia, C.-G.; Li, F.-W. Nickel-Catalyzed Alkynylation of a C(sp^2)–H Bond Directed by an 8-Aminoquinoline Moiety. *J. Org. Chem.* **2015**, *80*, 6213–6221. [CrossRef] [PubMed]
21. Xia, C.-C.; Wei, Z.-J.; Shen, C.; Xu, J.; Yang, Y.; Su, W.-K.; Zhang, P.-F. Palladium-catalyzed direct ortho-sulfonylation of azobenzenes with arylsulfonyl chlorides via C–H activation. *RSC Adv.* **2015**, *5*, 52588–52594. [CrossRef]
22. Pawar, G.G.; Brahmanandan, A.; Kapur, M. Palladium(II)-Catalyzed, Heteroatom-Directed, Regioselective C–H Nitration of Anilines Using Pyrimidine as a Removable Directing Group. *Org. Lett.* **2016**, *18*, 448–451. [CrossRef]
23. Suess, A.M.; Ertem, M.Z.; Cramer, C.J.; Stahl, S.S.J. Divergence between Organometallic and Single-Electron-Transfer Mechanisms in Copper(II)-Mediated Aerobic C–H Oxidation. *Am. Chem. Soc.* **2013**, *135*, 9797–9804. [CrossRef]
24. Wang, Z.; Kuninobu, Y.; Kanai, M. Copper-Mediated Direct C(sp^3)–H and C(sp^2)–H Acetoxylation. *Org. Lett.* **2014**, *16*, 4790–4793. [CrossRef] [PubMed]
25. Cong, X.; Zeng, X. Iron-Catalyzed, Chelation-Induced Remote C–H Allylation of Quinolines via 8-Amido Assistance. *Org. Lett.* **2014**, *16*, 3716–3719. [CrossRef] [PubMed]
26. Liang, H.-W.; Jiang, K.; Ding, W.; Yuan, Y.; Shuai, L.; Chen, Y.-C.; Wei, Y. Selective remote C–H sulfonylation of aminoquinolines with arylsulfonyl chlorides via copper catalysis. *Chem. Commun.* **2015**, *51*, 16928–16931. [CrossRef] [PubMed]
27. Zhu, L.-Z.; Qiu, R.-H.; Cao, X.; Xiao, S.; Xu, X.-H.; Au, C.-T.; Yin, S.-F. Copper-Mediated Remote C–H Bond Chalcogenation of Quinolines on the C5 Position. *Org. Lett.* **2015**, *17*, 5528–5531. [CrossRef] [PubMed]
28. Qiao, H.-J.; Sun, S.-Y.; Yang, F.; Zhu, Y.; Zhu, W.-G.; Dong, Y.-X.; Wu, Y.-S.; Kong, X.-T.; Jiang, L.; Wu, Y.-J. Copper(I)-Catalyzed Sulfonylation of 8-Aminoquinoline Amides with Sulfonyl Chlorides in Air. *Org. Lett.* **2015**, *17*, 6086–6089. [CrossRef] [PubMed]
29. Sun, M.-M.; Sun, S.-Y.; Qiao, H.-J.; Yang, F.; Zhu, Y.; Kang, J.-X.; Wu, Y.-S.; Wu, Y.-J. Silver (i)-promoted C5–H phosphonation of 8-aminoquinoline amides with H-phosphonates. *Org. Chem. Front.* **2016**, *3*, 1646–1650. [CrossRef]
30. Wei, J.; Jiang, J.; Xiao, X.; Lin, D.; Deng, Y.; Ke, Z.; Jiang, H.; Zeng, W.J. Copper-Catalyzed Regioselective C–H Sulfonylation of 8-Aminoquinolines. *Org. Chem.* **2016**, *81*, 946–955. [CrossRef] [PubMed]

31. Chen, H.; Li, P.; Wang, M.; Wang, L. Nickel-Catalyzed Site-Selective C–H Bond Difluoroalkylation of 8-Aminoquinolines on the C5-Position. *Org. Lett.* **2016**, *18*, 4794–4797. [CrossRef] [PubMed]
32. Sahoo, H.; Reddy, M.K.; Ramakrishna, I.; Baidya, M. Copper-Catalyzed 8-Amido Chelation-Induced Remote C–H Amination of Quinolines. *Chem. Eur. J.* **2016**, *22*, 1592–1596. [CrossRef] [PubMed]
33. Xu, J.; Zhu, X.-L.; Zhou, G.-B.; Ying, B.-B.; Ye, P.-P.; Su, L.-Y.; Shen, C.; Zhang, P.-F. Copper(II)-catalyzed C5 and C7 halogenation of quinolines using sodium halides under mild conditions. *Org. Biomol. Chem.* **2016**, *14*, 3016–3021. [CrossRef] [PubMed]
34. Xia, C.-C.; Wang, K.; Xu, J.; Wei, Z.-J.; Shen, C.; Duan, G.-Y.; Zhu, Q.; Zhang, P.-F. Copper(II)-catalyzed remote sulfonylation of aminoquinolines with sodium sulfinates via radical coupling. *RSC Adv.* **2016**, *6*, 37173–37179. [CrossRef]
35. Whiteoak, C.J.; Planas, O.; Company, A.; Ribas, X. A First Example of Cobalt-Catalyzed Remote C–H Functionalization of 8-Aminoquinolines Operating through a Single Electron Transfer Mechanism. *Adv. Synth. Catal.* **2016**, *358*, 1679–1688. [CrossRef]
36. Zhu, X.; Qiao, L.; Ye, P.; Ying, B.; Xu, J.; Shen, C.; Zhang, P. Copper-catalyzed rapid C–H nitration of 8-aminoquinolines by using sodium nitrite as the nitro source under mild conditions. *RSC Adv.* **2016**, *6*, 89979–89983. [CrossRef]
37. He, Y.; Zhao, N.-N.; Qiu, L.-Q.; Zhang, X.-Y.; Fan, X.-S. Regio-and chemoselective mono-and bisnitration of 8-amino quinoline amides with Fe $(NO_3)_3{\cdot}9H_2O$ as promoter and nitro source. *Org. Lett.* **2016**, *18*, 6054–6057. [CrossRef] [PubMed]
38. Nicewicz, G.A.; MacMillan, D.W.C. Merging photoredox catalysis with organocatalysis: The direct asymmetric alkylation of aldehydes. *Science* **2008**, *322*, 77–80. [CrossRef] [PubMed]
39. Narayanam, J.M.R.; Stephenson, C.R.J. Visible light photoredox catalysis: Applications in organic synthesis. *Chem. Soc. Rev.* **2011**, *40*, 102–113. [CrossRef] [PubMed]
40. Shi, L.; Xia, W. Photoredox functionalization of C–H bonds adjacent to a nitrogen atom. *Chem. Soc. Rev.* **2012**, *41*, 7687–7697. [CrossRef] [PubMed]
41. Prier, C.K.; Rankic, D.A.; MacMillan, D.W.C. Visible light photoredox catalysis with transition metal complexes: Applications in organic synthesis. *Chem. Rev.* **2013**, *113*, 5322–5363. [CrossRef] [PubMed]
42. Beatty, J.W.; Stephenson, C.R.J. Amine Functionalization via Oxidative Photoredox Catalysis: Methodology Development and Complex Molecule Synthesis. *Acc. Chem. Res.* **2015**, *48*, 1474–1484. [CrossRef] [PubMed]
43. Le, C.C.; MacMillan, D.W.C.J. Fragment Couplings via CO_2 Extrusion–Recombination: Expansion of a Classic Bond-Forming Strategy via Metallaphotoredox. *Am. Chem. Soc.* **2015**, *137*, 11938–11941. [CrossRef] [PubMed]
44. Schultz, D.M.; Yoon, T.P. Solar synthesis: Prospects in visible light photocatalysis. *Science* **2014**, *343*, 1239176. [CrossRef] [PubMed]
45. Yang, F.; Koeller, J.; Ackermann, L. Photoinduced Copper-Catalyzed C–H Arylation at Room Temperature. *Angew. Chem. Int. Ed.* **2016**, *55*, 4759–4762. [CrossRef] [PubMed]
46. Li, L.; Mu, X.; Liu, W.; Wang, Y.; Mi, Z.; Li, C.-J. Simple and Clean Photoinduced Aromatic Trifluoromethylation Reaction. *J. Am. Chem. Soc.* **2016**, *138*, 5809–5812. [CrossRef] [PubMed]
47. Li, Z.-X.; Sun, S.-Y.; Qiao, H.-J.; Yang, F.; Zhu, Y.; Kang, J.-X.; Wu, Y.-S.; Wu, Y.-J. Palladium-Catalyzed Regioselective C8–H Amination of 1-Naphthylamine Derivatives with Aliphatic Amines. *Org. Lett.* **2016**, *18*, 4594–4597. [CrossRef] [PubMed]
48. Clark, A.M.; Rickard, C.E.F.; Roper, W.R.; Wright, L.J. Electrophilic Substitution Reactions at the Phenyl Ring of the Chelated 2-(2′-Pyridyl)phenyl Ligand Bound to Ruthenium(II) or Osmium(II). *Organometallics* **1999**, *18*, 2813–2820. [CrossRef]
49. Tang, H.; Zhou, B.; Huang, X.-R.; Wang, C.; Yao, J.; Chen, H. Origins of Selective $C(sp^2)$–H Activation Using Transition Metal Complexes with N,N-Bidentate Directing Groups: A Combined Theoretical–Experimental Study. *ACS Catal.* **2014**, *4*, 649–656. [CrossRef]
50. Aihara, Y.; Chatani, N.J. Nickel-Catalyzed Direct Arylation of $C(sp^3)$–H Bonds in Aliphatic Amides via Bidentate-Chelation Assistance. *Am. Chem. Soc.* **2014**, *136*, 898–901. [CrossRef] [PubMed]

catalysts

Article

Versatile Polyoxometalates of Different Structural Dimensionalities for Liquid Phase Catalytic Oxidation

Patrícia Neves [1], Guilherme Simões [1], Bogna D. Napruszewska [2], Katarzyna Pamin [2], Paweł Serda [3], Wieslaw Łasocha [2,3,*] and Anabela A. Valente [1,*]

[1] Department of Chemistry, CICECO-Aveiro Institute of Materials, University of Aveiro, 3810-193 Aveiro, Portugal; pneves@ua.pt (P.N.); gui.simoes@ua.pt (G.S.)

[2] Jerzy Haber Institute of Catalysis and Surface Chemistry, Polish Academy of Sciences, Niezapominajek 8, 30-239 Kraków, Poland; ncnaprus@cyf-kr.edu.pl (B.D.N.); ncpamin@cyf-kr.edu.pl (K.P.)

[3] Faculty of Chemistry, Jagiellonian University, Gronostajowa 2, 30-387 Kraków, Poland; serda@chemia.uj.edu.pl

* Correspondence: lasocha@chemia.uj.edu.pl (W.Ł.); atav@ua.pt (A.A.V.)

Abstract: Ionic polymolybdate compounds (IPOM) possessing the anions $[Mo_8O_{26}]^{4-}$ and $[Mo_3O_{10}]^{2-}$, and cyclohexylammonium (Cy6N) or anilinium (Anil) as organic cations, namely cyclohexylammonium β-octamolybdate dihydrate (**1**), cyclohexylammonium trimolybdate hydrate (**2**), anilinium β-octamolybdate dihydrate (**3**), anilinium trimolybdate tetrahydrate (**4**) and anilinium trimolybdate dihydrate (**5**), were synthesized via simple, eco-friendly one-pot routes. New crystal structures of **1**, **2** and **5** were discovered. IPOM compounds with different structural dimensionality, density and ratio of the number of terminal oxo groups/molybdenum atoms (n(oxo)/Mo) were developed. The IPOM compounds promoted the epoxidation of biobased olefins such as the fatty acid methyl esters methyl oleate and methyl linoleate with *tert*-butylhydroperoxide as oxidant, leading to conversions of at least 81% at 4 h, 70 °C and the corresponding epoxides. The reaction scope of applications for the IPOM catalysts covered cyclooctane oxidation.

Keywords: ionic; polymolybdate; epoxidation; fatty acid methyl esters; anilinium; cyclohexylammonium

Citation: Neves, P.; Simões, G.; Napruszewska, B.D.; Pamin, K.; Serda, P.; Łasocha, W.; Valente, A.A. Versatile Polyoxometalates of Different Structural Dimensionalities for Liquid Phase Catalytic Oxidation. *Catalysts* **2024**, *14*, 251. https://doi.org/10.3390/catal14040251

Academic Editors: Leonarda Liotta, Narendra Kumar, Konstantin Ivanov Hadjiivanov and Vincenzo Vaiano

Received: 5 January 2024
Revised: 15 March 2024
Accepted: 29 March 2024
Published: 10 April 2024

1. Introduction

In the transition to a bio-based economy, the epoxidation of olefins such as fatty acid methyl esters (FAMEs) has attracted much attention because the epoxidized FAMEs (EFAMEs) have broad application profiles, e.g., for plasticizers, lubricants, polyurethanes, painting, coatings and adhesives, and improving the oxidation stability of biodiesel [1–6].

Moreover, EFAMEs can be sustainably produced from waste and non-edible sources such as waste cooking oil and agricultural or industrial residues [7]. For example, crude tall oil is a by-product of the wood pulp industry with an average production of 30–50 kg per 1000 kg of processed wood. This oil contains *ca.* 30–50 wt.% of free fatty acids, which are mainly oleic acid and linoleic acid, and has high iodine value, i.e., many -C=C- double bonds for epoxidation. Prominent players in the EFAMEs market include DOW Chemical Company and KH Chemicals [8]. The industrial production of epoxidized vegetable oils based on the Prilezhaev peracid process involves percarboxylic acids, e.g., peracetic acid or performic acid, which may be formed via the reaction of the corresponding carboxylic acid with hydrogen peroxide [2,9–11]. These routes suffer from several drawbacks such as low epoxide selectivity, significant waste generation, safety, and corrosion issues associated with the strongly acidic media. This has led to research efforts in the fields of catalysis to develop efficient and less-hazardous epoxidation processes.

Molybdenum-based homogeneous catalysts reached industrial implementation in the 20th century for large-scale liquid phase epoxidation using *tert*-butylhydroperoxide (TBHP) as an oxidant [9,12]. The recognized effectiveness of molybdenum-based catalysts for liquid phase epoxidation has led to the development of Mo catalysts with different architectures to meet superior performances in diverse reactions systems [13–15]. The ionic polymolybdates family of compounds has impressive structural diversity. They may be designed with different structural dimensionalities, types of anion and cations, etc. The primary units of the anions are $\{MoO_n\}$ coordination polyhedra (e.g., distorted $\{MoO_6\}$ octahedra or $[MoO_4\}$ tetrahedra), which assemble in different fashions, leading to different nuclearities and types of metal and oxygen sites (e.g., internal or terminal sites, low high coordinate sites, different metal–oxygen distances and angles) [16]. On the other hand, the type of cations may influence the structural and chemical features such as interlayer distances and solubility properties of the ionic polymolybdates [17,18]. The different physicochemical properties of ionic polymolybdates may determine the catalytic performances [19,20].

Although there are relatively few literature studies on the topic, ionic polymolybdates with organic cations (IPOM) may be effective catalysts for upgrading FAMEs [19,21–23], and they may be designed to meet superior catalytic performances. For example, a 2-D bis(anilinium) pentamolybdate, which possessed the simplest anilinium (Anil) cation, was a stable catalyst, although it possessed inferior activity to other 0-D and 1-D compounds, partly due to the presence of fewer accessible active sites in the 2-D structure [19]. Hence, the development of lower dimensionality Anil IPOM catalysts may lead to improved performances. On the other hand, IPOMs possessing the aliphatic counterpart of (aromatic) Anil, namely cyclohexylammonium (Cy6N), have not been reported to date, to the best of our knowledge. Cy6N may be less acidic than Anil [24] and influence the catalytic properties.

In this work, 0-D and 1-D IPOM compounds possessing the anions $[Mo_8O_{26}]^{4-}$ and $[Mo_3O_{10}]^{2-}$, and Cy6N or Anil as cations, were synthesized, characterised, and studied for liquid phase epoxidation systems, including FAMEs conversion (Scheme 1). The IPOMs were cyclohexylammonium β-octamolybdate dihydrate (**1**), cyclohexylammonium trimolybdate hydrate (**2**), anilinium β-octamolybdate dihydrate (**3**), anilinium trimolybdate tetrahydrate (**4**) and anilinium trimolybdate dihydrate (**5**). The crystal structures of the new compounds **1**, **2** and **5** were identified. The selected compounds possessed identical anions and different cations (e.g., **1**/**3**, or **2**/(**4**, **5**)) or vice versa (**1-2**, or **3-6**), and, on the other hand, include two pseudo-polymorphs (different hydrates) of anilinium trimolybdate (**4**, **5**). The previously reported (stable) 2-D compound anilinium pentamolybdate (**6**) was included for comparative studies. The prepared compounds promoted olefin epoxidation with TBHP. To the best of our knowledge, **1** and **2** are the first molybdenum compounds with a cyclohexylamine-type organic component reported for catalytic epoxidation. The reaction scope of applications of the IPOM catalysts included cyclooctane oxidation.

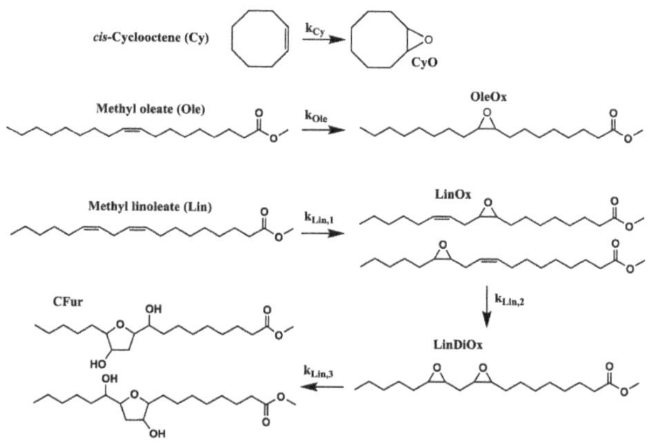

Scheme 1. Olefin epoxidation reaction systems studied for the IPOM catalysts **1–6**, using TBHP as oxidant, at 70 °C, and the respective kinetic constants (k_i) per step.

2. Results and Discussion

2.1. Structures of the IPOM Catalysts

Table 1 summarises the structural dimensionality and chemical formulae of the IPOM compounds **1–5**, and, for comparison, **6**. The most important crystallographic data for the new compounds **1**, **2** and **5**, based on single crystal methods, are indicated in Table 2, and further details are given in Tables S1–S3.

Table 1. Structural dimensionality and chemical formulae of compounds **1-6**.

	Chemical Name	Chemical Formula	Structural Features	Dim [a]	Dx/ DC [b]	n(oxo)/Mo [c]
1	Cyclohexylammonium β-octamolybdate dihydrate	$(C_6H_{14}N)_4[Mo_8O_{26}]\cdot 2H_2O$	isolated clusters	0-D	2.264/ 4	1.75
2	Cyclohexylammonium trimolybdate hydrate	$(C_6H_{14}N)_2[Mo_3O_{10}]\cdot H_2O$	polymeric	1-D	2.056/ 3	2
3	Anilinium β-octamolybdate dihydrate	$(C_6H_8N)_4[Mo_8O_{26}]\cdot 2H_2O$	isolated clusters	0-D	2.520/ 4	1.75
4	Anilinium trimolybdate tetrahydrate	$(C_6H_8N)_2[Mo_3O_{10}]\cdot 4H_2O$	polymeric	1-D	2.180/ 3	2
5	Anilinium trimolybdate dihydrate	$(C_6H_8N)_2[Mo_3O_{10}]\cdot 2H_2O$	polymeric	1-D	2.399/ 3	2
6	Anilinium pentamolybdate	$(C_6H_8N)_2[Mo_5O_{16}]$	layered	2-D	2.850/ 5	0.8

[a] Dim. = structural dimensionality. [b] Dx = calculated density (by X-ray diffraction technique) and DC = degree of condensation, which, for isopolymolybdates, equals (number of Mo atoms)/(0.5 × (cluster charge)); DC represents the number of Mo atoms in the polymolybdate formula written in the form $M(I)_2O.xMoO_3$, where M(I) represents a monovalent metal cation. [c] Number (n(oxo)) of oxo groups (terminal oxygen atoms) per Mo atoms of the anion.

A search into the crystallographic database for **1** indicated that it is a new polymorphic modification of cyclohexylamine β-octamolybdate. In comparison to a previously reported monoclinic C-centred structure [25], a much smaller triclinic structure was obtained for **1** (i.e., it is not a simple subgroup of the previously reported monoclinic compound, as may be confirmed by the Checkcif reports). An essential building unit (with labelling scheme) and packing of the molecules are presented in Figure 1a and b, respectively, and a simplified representation of how the structure is built is presented in Figure 1c. The β-octamolybdate anion can be summarized as a polyoxometalate ion (IPOM) formed by 8 Mo atoms and 26 O atoms, among which there exist 14 terminal O atoms (Mo=O with ~1.707(8) Å length), 6 O atoms of μ_2 type, 4 O atoms of μ_3 type and 2 O atoms of μ_5 type.

Table 2. Summary of crystal data of the new compounds **1**, **2** and **5**.

Compound (XRD Technique)	**1 (Single Crystals)**	**2 (Powder Data)**	**5 (Powder Data)**
Chemical formula	$C_{24}H_{60}Mo_8N_4O_{28}$	$C_{12}H_{30}Mo_3N_2O_{11}$	$C_{12}H_{20}Mo_3N_2O_{12}$
MW (g/mol)	1620.27	666.20	672.12
T(K)	100	293(2)	293(2)
Wavelength, [Å]	MoKα: 0.71073	CuKα: 1.54187	CuKα: 1.54187
Crystal system, space group	triclinic, P-1	Orthorhombic, P nma	Monoclinic, P 2_1/c
Cell parameters			
a [Å]	8.7255(2)	8.763(3)	14.711
b [Å]	10.5278(2)	7.645(3)	7.578
c [Å]	14.4778(4)	32.114(4)	17.852
α [°]	75.019(2)	90	90
β [°]	76.273(2)	90	110.81
γ [°]	69.550(2)	90	90
V (Å^3)	1187.79(5)	2151.4(4)	1860.31
Z, calculated density (g/cm^3)	1, 2.264	4, 2.056	4, 2.399
Absorption coefficient (mm^{-1})	12.775	14.627	16.959 na
F(000)	792	1200	996
2Theta range	5.04–31.9363.86	3.5–80	3.5–70
Limiting indices	$-12 \leq h \leq 12; -15 \leq k \leq 15; -20 \leq l \leq 21$	7; 6; 26	10; 5; 13
Reflections collected	7841	524	526

Table 2. *Cont.*

Compound (XRD Technique)	1 (Single Crystals)	2 (Powder Data)	5 (Powder Data)
Completeness to theta(max)	95.5%	100% (xrpd)	100% (xrpd)
Absorption correction	Multi-scan	const. (Bragg-Brentano geom.)	const. (Bragg-Brentano geom.)
Maximum and minimum transmission	n.a. (powder sample)		
Refinement method	wR2	Rietveld	Rietveld
Data/restraints/parameters	7841/0/321	3824/36/25	3325/30/20
Goodness of fit on F2	1.23	8	11
Final R indices wR2 ($I > 2\sigma$) or R_p/R_{wp}	8.70	22.8/26.0	15.1/20.9
R_F (all data)	3.15	14.29	19.11
Largest difference peak and hole (eA^{-3})	1.612/−1.447	PD	PD
CCDC	2,320,511	PD−1	PD−1

PD Usually not reported in XRPD studies. PD−1 Submitted to ICDD to supplement deposited data.

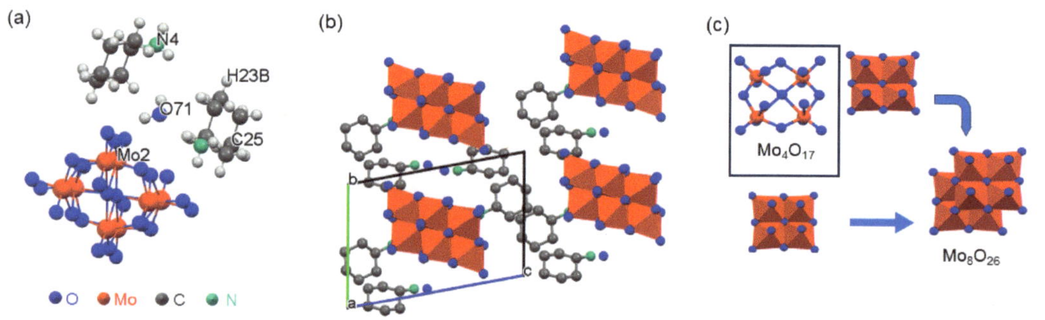

Figure 1. (**a**) Essential building unit (with labelling scheme), (**b**) packing of the molecules and (**c**) simplified representation how the structure is built for **1**.

Compound **2** is a new trimolybdate of cyclohexylammonium, which is crystallized (fibrillar crystals) in an orthorhombic system. Polymeric anionic chains of Mo_3O_{10} stoichiometry (monomers) are built of distorted $\{MoO_6\}$ octahedra, linked together by edges and vertices. By sharing one edge, the two octahedra form Mo_2O_{10} groups, and by sharing the vertices, these groups form infinite double chains of Mo_2O_8 units. This ribbon is joined by alternate MoO_6 octahedra (on both sides of it). The resultant composition is Mo_3O_{10}, and the length of the identity period along the polymer axis is equal to two heights of the MoO_6 octahedron ($2 \times 3.8 = 7.6$ Å). In each octahedron, there exist two terminal O atoms, and four O of μ_3 type. Between chains running along the [010] direction, there exist molecules of protonated amines, and water. The essential building unit (with labelling scheme) and packing of the molecules for **2** are presented in Figures 2a and 2b, respectively, and a simplified representation of how the structure is built is presented in Figure 2c.

For the new 1-D compound **5**, a similar arrangement of inorganic polymeric chains to that of compound **2** was obtained; the molecular packing is given in Figure 3. However, the monoclinic structure of **5** was slightly denser than the orthorhombic structures of **2** and **4** (Table 1). The calculated density (Dx) equals 2.399 for **5** versus 2.180 and 2.056 for **4** and **2**, respectively (Table 1). Based on a comparison for all IPOM compounds, the Dx was highest for the 2-D **6**, which is a layered material with infinite anionic layers $[(Mo_5O_{16})^{2-}]_n$ formed by distorted MoO_6 octahedra [26].

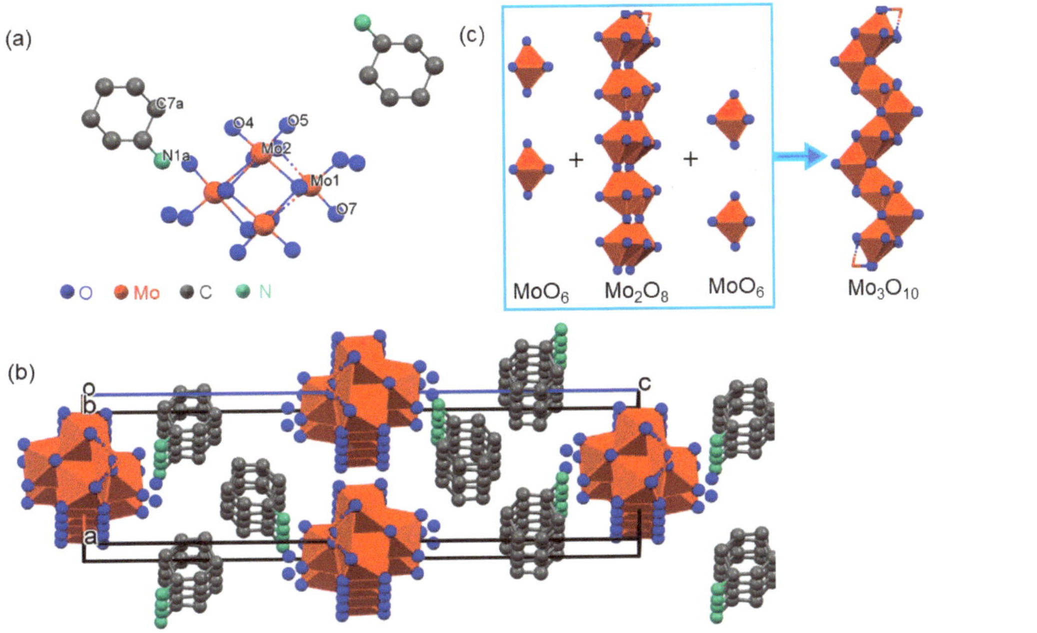

Figure 2. (**a**) Essential building unit (with labelling scheme), (**b**) packing of the molecules and (**c**) simplified representation how the structure is built for **2**.

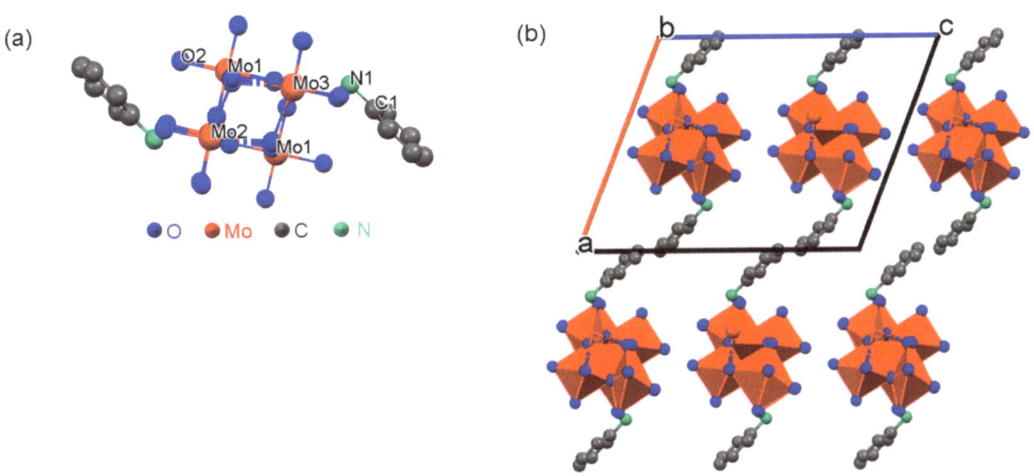

Figure 3. (**a**) Essential building unit (with labelling scheme), and (**b**) packing of the molecules for **5**.

The variable temperature X-ray powder diffraction (XRPD/temp) and thermal analyses (TG/DSC) results for **1**, **2** and **5** are given in the Supporting information (Figures S1–S3 and S4–S6). The corresponding data for the remaining compounds was previously reported by some of us [26]. In general, octamolybdates (0-D) seem slightly more thermally stable than the trimolybdates (1-D), and layered (2-D) pentamolybdates appeared to be the most stable. Through prolonged heating, the anilinium trimolybdate tetrahydrate **4** (which was synthesized in aqueous conditions) converted completely to a solid with same crystal structure as that of **5**. Importantly, compounds

1–6 are thermally stable in the temperature range of the catalytic studies reported in this work (up to 120 °C).

The structural dimensionality seems to be essentially related to the type of anion: 0-D for $[Mo_8O_{26}]^{4-}$, 1-D for $[Mo_3O_{10}]^{2-}$ and 2-D for $[Mo_5O_{16}]^{2}$. On the other hand, depending on the type of anion, the IPOMs possess different amounts of terminal oxygen atoms (Mo=O) per total amount of molybdenum atoms (n(oxo)/Mo), which equals 2 for trimolybdates (1-D), 1.75 for octamolybdates (0-D) and 0.8 for pentamolybdates (2-D) (Table 1). For each family of IPOMs, the n(oxo)/Mo ratio increased with decreasing density Dx (Figure 4).

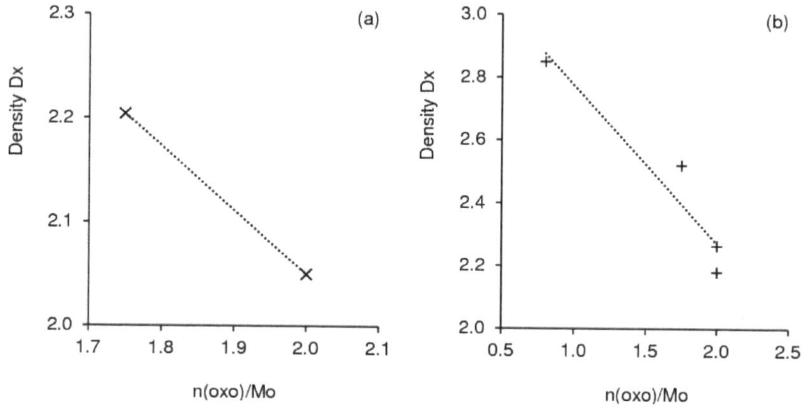

Figure 4. Relationship between Dx and the ratio n(oxo)/Mo for the Cy6N (**a**) and Anil (**b**) families of IPOM compounds.

2.2. Catalytic Studies

2.2.1. Olefin Epoxidation

Impact of IPOM Features

The IPOM catalysts were firstly compared based on the model reaction of *cis*-cyclooctene (Cy) with *tert*-butyl hydroperoxide (TBHP) at 70 °C. The initial molar ratios of Mo:olefin and Mo:TBHP were kept constant in all catalytic tests, i.e., the catalytic performances were compared on the basis of the same amount of molybdenum in the reactors. Cyclooctene oxide (CyO) was the only product formed for all catalytic tests (Scheme 1). Blank reactions without catalyst and/or without TBHP led to negligible Cy conversion at 24 h, indicating the need for both a molybdenum catalyst and TBHP for the epoxidation reaction to occur. It is consensual from the literature that molybdenum-catalysed epoxidation of olefins occurs via a heterolytic mechanism which may be initialised via the coordination of the hydroperoxide oxidant (TBHP) to the Lewis acid metal centre, leading to active oxidising species. The latter species are responsible for the oxygen atom transfer to the olefin, giving the corresponding epoxide product plus *tert*-butanol, which is the coproduct of TBHP [27–30]. The types of active oxidising species may possess a {Mo(OH)(OOtBu)} (*t*Bu = *tert*-butyl) moiety, and its formation involves a oxomolybdenum (Mo=O) group [27–30].

Figure 5 compares the catalytic results for the IPOM compounds based on Cy conversions at different reaction times. The catalytic results were similar for the pseudo-polymorphs (different hydrates) of anilinium trimolybdate (**4** and **5**); 100% Cy conversion at 1 h. These results were correlated with the identical anions and cations (independently of the extent of hydration) of **4** and **5**. On the other hand, a comparative study of the IPOMs with the same type of anion (and structural dimensionality) but different cations suggested that the anilinium (Anil) family (**3**, **4**, **5**) was more active than the cyclohexylamine (Cy6N) one (**1**, **2**). Specifically, the Anil IPOMs led to 100% Cy conversion within 1 h, whereas for the Cy6N ones (**1**, **2**), the reaction was complete only after 4 h (**1** vs. **3**, and **2** vs. **4** and **5**).

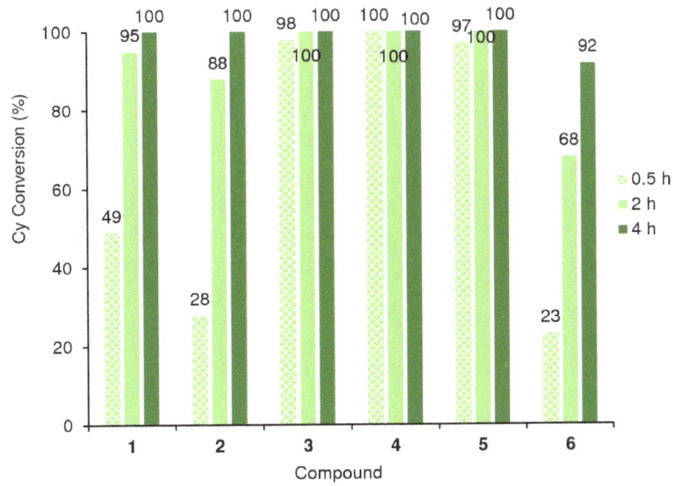

Figure 5. *cis*-Cyclooctene (Cy) epoxidation with TBHP, in the presence of **1–6**, at 70 °C. Epoxide selectivity was always 100%.

Catalysts **3**, **4** and **5** remained very active at the lower reaction temperature of 55 °C; 63, 87 and 100% Cy conversion was achieved, respectively, at 1 h (Figure S14). Catalyst **5** led to superior results (based on Cy conversion at 1 h, 55 °C) to literature data for IPOMs possessing different anilinium type cations, namely 1-D $(C_8H_{12}N)_2[Mo_3O_{10}]$ (91%) and 1-D $(C_7H_{10}N)_2[Mo_3O_{10}]$ (71%) (Table S6) [19].

The reaction mixtures were biphasic solid–liquid. High Cy conversions were reached in the initial stages (38–100% within 30 min reaction, in the presence of **1–5**). The recovered solids **1–5** were reused and led to roughly comparable catalytic results to the respective original catalysts (Figure 6). The characterisation studies of the used catalysts suggested that the chemistry (ATR FT-IR spectroscopy) and crystalline structures (XRPD) seemed to be preserved (similar characterisation results were obtained for the fresh and used catalysts, Figure S16). This parallels that reported for compound **6** [19]. Hence, the constituent ions seem to withstand the catalytic reaction conditions. Nevertheless, one cannot exclude contributions of dissolved metal species. Contact tests were carried out for **1–6**, which suggested that the catalytic reactions were essentially homogeneous (details in experimental section; Figure S15). Attempts to precipitate dissolved species (please see details in the experimental section, for **1–5**) did not lead to measurable (or visible) amounts, suggesting that the dissolved fractions were very soluble.

The catalytic performances may depend on the interplay of several factors, such as the type of organic cation, ratio (n(oxo)/Mo), and/or catalyst solubility. Based on the above mechanistic considerations, more available Mo=O groups may favour the epoxidation reaction kinetics, since the reaction begins with the coordination of the oxidant to a metal centre (of the anion) with the concomitant protonation of an oxomolybdenum group (Mo=O) [27–30]. On the other hand, one cannot exclude the possibility of the structural properties of the IPOMs (e.g., Dx) influencing the kinetics of dissolution of metal species, and consequently the overall epoxidation reaction kinetics (which is essentially homogeneous). In this sense, Figure S17 shows the kinetic constants (Table 3; Supplementary Materials for kinetic modelling details) as a function of Dx or n(oxo)/Mo ratio. For the Anil family, **4** and **5**, which possessed lower Dx (and higher n(oxo)/Mo, Table 1) than **3**, led to faster reaction kinetics (k_{Cy} = 11.5946 and 10.9802 h^{-1} for **4** and **5**, respectively, and k_{Cy} = 6.3279 for **3**, Table 3). Compound **6**, with the highest Dx (and lowest n(oxo)/Mo, Table 1), led to the slowest reaction kinetics (k_{Cy} = 0.5565 h^{-1} for **6** versus 1.0774–11.5946 h^{-1} for the remaining catalysts). However, the opposite was verified for the Cy6N family.

Specifically, **1**, with higher Dx (and lower n(oxo)/Mo), led to faster reaction kinetics than **2** (k_{Cy} = 1.3989 and 1.0774 h^{-1}, respectively). Hence, Dx does not solely explain the observed differences of reaction kinetics. Another possible factor is the different structural dimensionalities of the IPOMs, since a lower structural dimensionality may facilitate dissolution of metal species, and consequently the overall reaction. This could partly explain the catalytic differences between **1** (0-D) and **2** (1-D) but does not explain the differences observed for all compounds. The above results suggest that the (yet unknown) soluble active metal species may be structurally different for the different IPOMs, contributing with different intrinsic activities to the overall reaction.

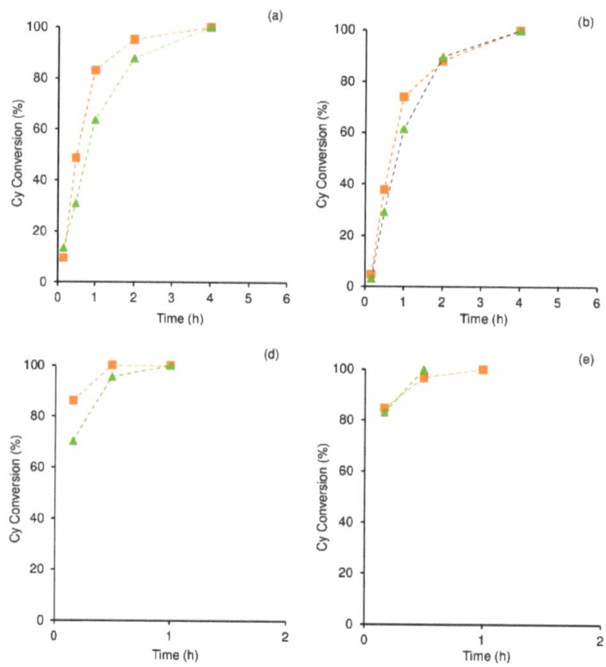

Figure 6. First (■) and second (▲) catalytic runs of *cis*-cyclooctene (Cy) epoxidation with TBHP, in the presence of **1** (**a**), **2** (**b**), **3** (**c**), **4** (**d**) and **5** (**e**), at 70 °C. Epoxide selectivity was always 100%.

Table 3. Kinetic constants (k_i) and objective function (F_{obj}) obtained from the kinetic models for Cy, Ole and Lin reactions, in the presence of the IPOM catalysts, at 70 °C.

k_i (h^{-1})	Reaction	1	2	3	4	5	6
k_{Cy} F_{obj}	Cy-CyO	1.3989 4.05×10^{-2}	1.0774 4.70×10^{-2}	6.3279 1.20×10^{-3}	11.5946 2.00×10^{-5}	10.9802 1.60×10^{-3}	0.5565 1.88×10^{-2}
k_{Ole} F_{obj}	Ole-OleOx	0.5356 5.61×10^{-2}	0.3002 2.27×10^{-2}	2.4035 4.26×10^{-4}	3.1423 3.33×10^{-6}	2.4031 6.24×10^{-4}	0.3149 2.76×10^{-2}
$k_{Lin,1}$ $k_{Lin,2}$ $k_{Lin,3}$ F_{obj}	Lin-LinOx LinOx-LinDiOx LinDiOx-LinFur	0.5827 0.1262 0.0594 5.20×10^{-2}	0.4767 0.1087 0.0013 7.20×10^{-2}	0.6392 0.1888 0.2527 1.36×10^{-2}	0.7294 0.2248 0.2811 7.48×10^{-2}	0.7607 0.2191 0.2594 6.70×10^{-2}	0.3957 0.0893 0.0647 7.00×10^{-2}

Table 4 compares the catalytic results for **1–6** to literature data for different IPOMs possessing anilinium derivates as organic components. To the best of our knowledge, there are no literature studies on IPOM epoxidation catalysts possessing cyclohexylammonium derivatives as organic components. A comparative study for the anilinium family (entries

4, 5, 7 and 8), suggested that the 1-D catalysts are highly active, and, on the other hand, the 2-D catalysts (entries 6 and 9–11) seemed generally less active than the 1-D ones.

Table 4. Comparison of the catalytic results for **1–6** to literature data for IPOM compounds possessing anilinium derivatives as organic components, tested for the Cy/TBHP reaction at 70 °C [a].

#	Catalyst	Dim. [b]	t (h) [c]	Conv., CyO yield (%) [d]	Ref.
1	**1**	0-D	1/2/4	83/95/100	-
2	**2**	1-D	1/2/4	74/88/100	-
3	**3**	0-D	0.5/1	98/100	-
4	**4**	1-D	0.5	100	-
5	**5**	1-D	0.5/1	97/100	-
6	**6**	2-D	1/4	48/86	[19]
7	$(C_8H_{12}N)_2[Mo_3O_{10}]$	1-D	1/6	97/100	[19]
8	$(C_7H_{10}N)_2[Mo_3O_{10}]$	1-D	1	100	[19]
9	$(C_8H_{12}N)_2[Mo_5O_{16}]$	2-D	1/6	54/98	[19]
10	$(C_7H_{10}N)_2[Mo_5O_{16}]$	2-D	1/6	40/94	[19]
11	$(C_8H_{12}N)_2[Mo_5O_{16}]$	2-D	1/6	44/92	[19]

[a] Initial molar ratio Mo:olefin:oxidante = 1:100:153. [b] Dim. = Structural dimensionality. [c] t = Reaction time. [d] Conv. = Cy conversion, CyO yield = cyclooctene oxide yield (epoxide selectivity was 100%).

Biobased Olefin Epoxidation

The synthesized IPOMs were further explored for the epoxidation of FAMEs, namely monounsaturated methyl oleate (Ole) and double unsaturated methyl linoleate (Lin), with TBHP, at 70 °C. All catalysts led to 100% Ole conversion within 24 h, and the epoxide 9,10-epoxyoctadecanoate (OleOx) was formed with 100% selectivity. Without catalyst, the reaction was sluggish (12% Ole conversion at 24 h).

The reaction kinetics depended on the type of IPOM catalysts. The Anil family of compounds **3–5** was more active than the Cy6N one (Figure 7), in parallel to the results verified with Cy as the substrate (Figures 5 and S17). For the Cy6N family, **1** led to a faster reaction than **2** (k_{Ole} = 0.5356 and 0.3002 h^{-1}, respectively, Table 3; kinetic curves in Figure S12). On the other hand, for the Anil family, the reaction kinetics was faster for **3–5** (k_{Ole} = 2.4031–3.1423 h^{-1}) than **6** (k_{Ole} = 0.3149 h^{-1}). Ole conversion at 2 h was 95–100% for **3–5** and only 43% for **6**. After 2 h reaction, the differences in kinetic profiles were less pronounced (Figure S12), e.g., at 4 h and 24 h, **1–5** led to 81–84% and 96–100% Ole conversion, respectively.

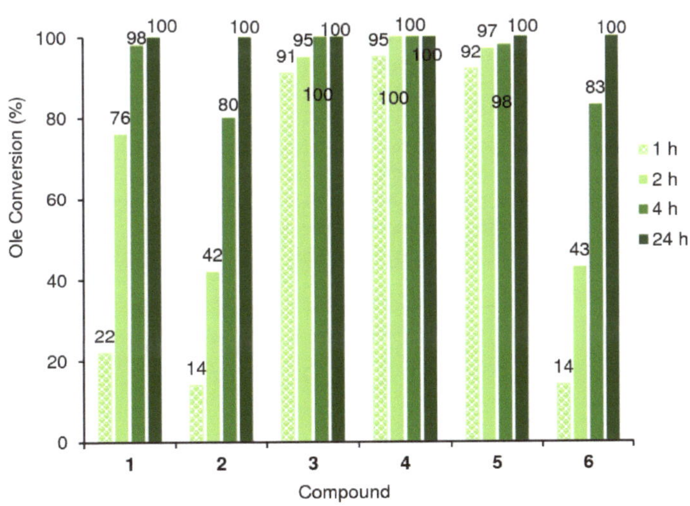

Figure 7. Methyl oleate (Ole) epoxidation with TBHP, in the presence of **1–6**, at 70 °C. Epoxide selectivity was always 100%.

Based on a literature survey for IPOM catalysts tested for the target reaction (Ole/TBHP), it seems that, in general, **3–5** performed superiorly to other catalysts (Table 5). The 1-D compound bis(3,5-dimethylanilinium) trimolybdate ($(C_8H_{12}N)_2[Mo_3O_{10}]$) led to 84% OleOx yield at 89% Ole conversion at 6 h (70 °C), which is inferior to the 1-D **3–5** IPOMs (91–95% Ole conversion at 1 h; entry 7 *vs.* entries 3–5, Table 5) [19]. The introduction of alkyl substituent groups in the anilinium cation did not seem to improve the catalytic performance, which may be partly due to differences in catalyst solubility and/or acid–base properties of the anilinium derivatives [31]. The 0-D pyridinium-tetrazole-based polyoxometallates $(Hptz)_4[SiMo_{12}O_{40}]\cdot nH_2O$ and $(t\text{Bu-ptz})_2[Mo_6O_{19}]$ performed differently under somewhat comparable reaction conditions, with the latter being less active (88% Ole conversion at 6 h) than the former (97% Ole conversion at 3 h) (entries 8 and 9) [32,33]. Both compounds were less active than **3–5**. At a higher temperature (75 °C), the clusters $(py)_4[Mo_8O_{26}]$ and $(t\text{Bu-py})_4[Mo_8O_{26}]$ (py = pyridine, *t*Bu-py = 4-*tert*-butylpyridine) led to *ca.* 97%/99% OleOx yield after 2 h/6 h reaction (entries 10 and 11) [23].

Table 5. Comparison of the catalytic results for **1–6** to literature data for IPOM compounds tested for the Ole/TBHP reaction.

#	Catalyst [a]	Dim. [b]	Reaction Conditions [c]				Conv. [d] (%)	OleOx Yield (%)	Ref
			Solv.	T (°C)	Mo:S:ox	t (h)			
1	**1**	0-D	TFT	70	1:100:300	½/4	22/76/98	22/74/96	-
2	**2**	1-D	TFT	70	1:100:300	½/4	14/42/80	14/42/80	-
3	**3**	0-D	TFT	70	1:100:300	½/4	91/97/100	91/97/100	-
4	**4**	1-D	TFT	70	1:100:300	½	95/100	95/100	-
5	**5**	1-D	TFT	70	1:100:300	½/4	92/97/98	92/97/98	-
6	**6**	2-D	TFT	70	1:100:300	½/4/24	14/43/83/100	14/43/83/96	-
7	$(C_8H_{12}N)_2[Mo_3O_{10}]$	1-D	TFT	70	1:100:153	1/6/24	70/89/99	66/84/92	[19]
8	$(Hptz)_4[SiMo_{12}O_{40}]\cdot nH_2O$	0-D	TFT	70	1:100:250	1/6/24	56/88/100	56/88/100	[32]
9	$(t\text{Bu-ptz})_2[Mo_6O_{19}]$	0-D	TFT	70	1:100:210	0.5/3	73/97	73/97	[33]
10	$(py)_4[Mo_8O_{26}]$	0-D	DCE	75	1:100:152	2/6	~97/100	~97/99	[23]
11	$(t\text{Bu-py})_4[Mo_8O_{26}]$	0-D	DCE	75	1:100:152	2/6	~97/100	~97/99	[23]

[a] ptz = 5-(2-pyridyl)tetrazole; *t*Bu-ptz = 2-*tert*-butyl-5-(2-pyridyl)−2H-tetrazole; py = pyridine; *t*Bu-py = 4-*tert*-butylpyridine. [b] Dim. = structural dimensionality. [c] Solv. = solvent; TFT = α,α,α-trifluorotoluene, DCE = 1,2-dichloroethane; Mo:S:ox = initial molar ratio of molybdenum:substrate:oxidant; T = reaction temperature; t = time. [d] Conv. = Ole conversion.

IPOMs **1–6** promoted the epoxidation of methyl linoleate (Lin) with TBHP, leading to 98–100% Lin conversion within 24 h (Figure 8); the kinetic constants were greater for **3-5** than **1-2** (Table 3; calculated kinetic curves in Figure S13). The products were monoepoxides, diepoxides and furan-type compounds (Scheme 1, Figure 9. The monoepoxides methyl-9,10-epoxy-12-octadecenoate and 12,13-epoxy-9-octadecenoate (LinOx) were formed initially. LinOx selectivity decreased with increasing Lin conversion, which was accompanied by the formation of the diepoxides methyl 9,10-12,13-diepoxyoctadecanoate (LinDiOx) and the furanic products methyl 10,13-epoxy-9,12-dihydroxyoctadecanoate and methyl 9,12-epoxy-10,13-dihydroxyoctadecanoate (LinFur) (Figure 9). The LinFur products may be formed via epoxide ring opening of LinDiOx and intramolecular cyclization [34–37]. According to the literature, due to the proximity of the two epoxide groups (separated by a -CH_2 group), the ring opening of one of these groups may give unstable intermediates which rapidly undergo cyclization to LinFur [34–37].

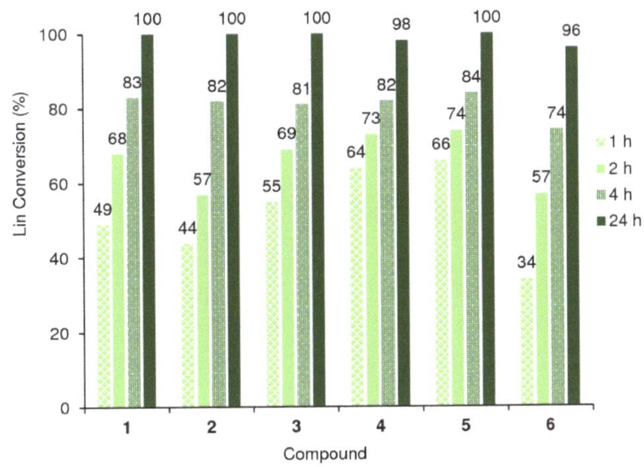

Figure 8. Methyl linoleate (Lin) epoxidation with TBHP, in the presence of **1–6**, at 70 °C.

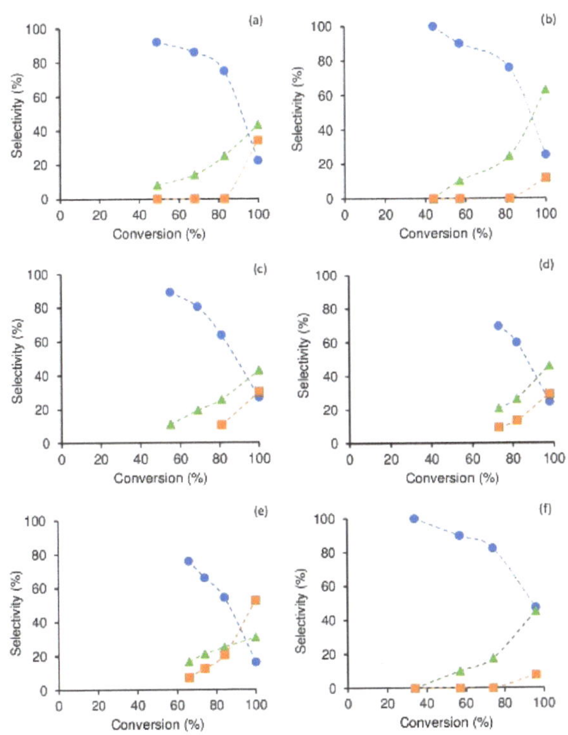

Figure 9. Dependency of products selectivity on methyl linoleate (Lin) conversion (LinOx (●), LinDiOx (▲) and LinFur (■)), in the presence of **1** (**a**), **2** (**b**), **3** (**c**), **4** (**d**), **5** (**e**) and **6** (**f**), at 70 °C.

Table 6 compares the catalytic results for **1–6** to literature data for IPOM type catalysts tested for the Lin reaction. The 1-D bis(3,5-dimethylanilinium) trimolybdate $(C_8H_{12}N)_2[Mo_3O_{10}]$ led to 39% LinOx yield, 39% LinDiOx yield and 16% yield LinFur at 95% Lin conversion (70 °C, 24 h, molar ratio Mo:LinOle:TBHP = 1:100:250) [19]. Under somewhat similar reaction conditions to those used for **1–6**, the catalyst $(Hptz)_4[SiMo_{12}O_{40}]\cdot nH_2O$ (ptz = 5-(2-pyridyl)tetrazole) led to 60%/28% LinOx selectivity, 31%/33% LinDiOx selectivity and 9%/40% LinFur selectivity at

87%/100% Lin conversion (6 h/24 h) [32]. On the other hand, only epoxides were formed for a related ionic compound with a substituted ptz organic fragment (tBu-Hptz)$_2$[Mo_6O_{19}], which led to 83% LinOx and 17% LinDiOx selectivity at 85% Lin conversion, 24 h/70 °C [33]. For this catalyst, the selectivity towards epoxides remained high at 90 °C (63% and 37% of LinOx and LinDiOx selectivity, respectively, at 100% Lin conversion, 2 h), and no LinFur products were formed [33].

Table 6. Comparison of the catalytic results for **1–6** to literature data for IPOM compounds tested for the Lin/TBHP reaction.

Catalyst [a]	Reaction Conditions [b]			Conv. (%) [c]	Selectivity (%)			Ref.
	Solv.; T (°C)	Mo:S:ox	t (h)		LinOx	LinDiOx	LinFur	
1	TFT; 70	1:100:300	1/4/24	49/83/100	92/75/23	8/25/43	0/0/34	-
2	TFT; 70	1:100:300	1/4/24	44/82/100	100/76/25	0/24/63	0/0/12	-
3	TFT; 70	1:100:300	1/4/24	55/81/100	89/64/27	11/25/43	0/11/30	-
4	TFT; 70	1:100:300	1/4/24	64/82/98	77/60/25	18/27/46	5/13/29	-
5	TFT; 70	1:100:300	1/4/24	66/84/100	76/54/16	17/25/31	7/21/53	-
6	TFT; 70	1:100:300	1/4/24	34/74/96	100/82/47	0/18/45	0/0/8	-
($C_8H_{12}N$)$_2$[Mo_3O_{10}]	TFT; 70	1:100:250	24	95	41	41	17	[19]
[tBu-Hptz]$_2$[Mo_6O_{19}]	TFT; 70	1:100:210	0.5/24	36/85	100/83	0/17	0	[33]
[tBu-Hptz]$_2$[Mo_6O_{19}]	TFT; 90	1:100:210	0.5/2	80/100	81/63	19/37	0	[33]
(Hptz)$_4$[$SiMo_{12}O_{40}$]·$n$$H_2O$	TFT; 70	1:100:250	1/6/24	45/87/100	100/60/28	0/31/33	0/9/40	[32]

[a] tBu-ptz = 2-*tert*-butyl-5-(2-pyridyl)−2H-tetrazole; ptz = 5-(2-pyridyl)tetrazole. [b] Solv. = solvent; TFT = α,α,α-trifluorotoluene, Mo:S:ox = initial molar ratio of molybdenum:substrate:oxidant, T = reaction temperature, t = time. [c] Conv. = Lin conversion.

In summary, there have been few ionic molybdenum compounds tested for Lin reaction, and the results for **3–5** seem promising. Outstandingly, **5** led to the highest LinFur yield (Table 6). The conversion of FAMEs to higher polarity products such as LinFur may further broaden the applications profile of FAMEs [34], e.g., there is an increasing interest in substituted tetrahydrofurans as potential plasticizers, antibiotics and antitumour, pesticidal, antiprotozoal and antimicrobial agents [38–41].

2.2.2. Cyclooctane Oxidation

The prepared IPOMs were further explored for the oxidation of cyclooctane in air at 10 atm and 120 °C (Table 7, Scheme 2). The main products were cyclooctanone (A) and cyclooctanol (B), and to a lower extent, suberic acid (C), which were formed in a similar molar ratio (A + B)/(A + B + C) of 0.80–0.86, at 6 h. According to the literature, cyclooctane oxidation in the presence of IPOMs may occur via homolytic mechanism [26].

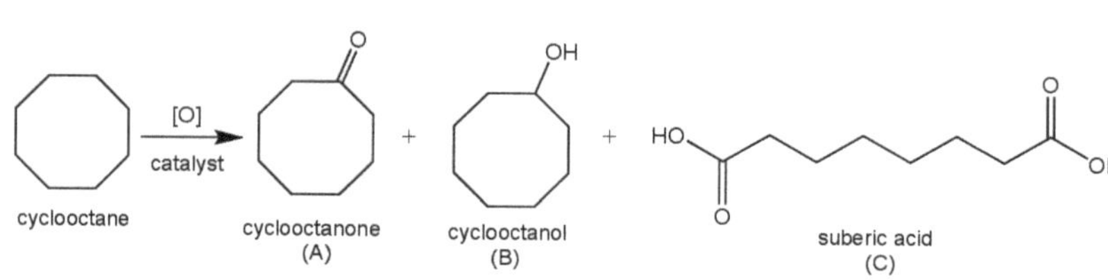

Scheme 2. Cyclooctane oxidation, in the presence of the IPOMs **1–6**, at 120 °C.

Table 7. Comparison of the catalytic results for **1–6** to literature data for IPOMs possessing anilinium derivatives as organic components, tested for the cyclooctane/O_2 reaction at 120 °C [a].

Catalyst	Dim	Ketone Yield (A) (%)	Alcohol Yield (B) (%)	Suberic Acid Yield (C) (%)	Conv. [b] (%)	A/B	(A + B)/(A + B + C)	Ref.
1	0-D	35	13	8	56	2.29	0.86	-
2	1-D	41	13	12	66	3.15	0.82	-
3	0-D	37	14	13	64	2.64	0.80	-
4	1-D	41	15	11	67	2.73	0.83	-
5	1-D	39	14	11	64	2.79	0.83	-
6	2-D	27	11	8	46	2.45	0.83	-
$(C_7H_{10}N)_4[Mo_8O_{26}]\cdot 2H_2O$	0-D	11	9	0	20	1.14	1.00	[26]
$(C_8H_{12}N)_2[Mo_3O_{10}]$	1-D	30	16	0	46	1.91	1.00	[26]
$(C_7H_{10}N)_2[Mo_3O_{10}]$	1-D	26	16	0	42	1.61	1.00	[26]
$(C_7H_{10}N)_2[Mo_5O_{16}]$	2-D	34	11	~5	~50	3.16	~0.90	[26]
$(C_6H_7NBr)_2[Mo_5O_{16}]$	2-D	34	11	~4	~49	3.21	~0.92	[26]
$(C_8H_{12}N)_2[Mo_5O_{16}]$	2-D	23	15	~2	~41	1.53	~0.95	[26]
$(C_9H_{14}N)_2[Mo_5O_{16}]$	2-D	29	15	5	49	1.97	0.91	[42]
$(C_9H_{14}N)_4[Mo_8O_{26}]\cdot H_2O$	0-D	30	18	0	48	1.69	1.00	[42]
$(C_9H_{14}N)_6[Mo_8O_{26}(C_2H_3O_2)_2]$	0-D	28	17	0	45	1.66	1.00	[42]

[a] Reaction conditions: mole ratio cyclooctane:oxygen = 6.5, air atmosphere (10 bar), 6 h. [b] Conversion (A + B + C) on the basis of oxygen quantity in the reaction mixture.

The molar ratio A/B was slightly higher for **2** (3.15 versus 2.29–2.79 for the remaining IPOMs). The cyclooctane conversions were in the range 46–67%, being lowest for compound **6**, somewhat in parallel to the results verified for olefin epoxidation systems. However, for cyclooctane oxidation, the catalytic results for the remaining IPOMs **1–5** were not considerably different (56–67% cyclooctane conversion at 6 h). More pronounced differences in catalytic results were observed for the epoxidation systems. A comparative study with the literature indicated that **3–5** led to higher cyclooctane conversions than other 0-D, 1-D and 2-D anilinium-type IPOMs tested under similar cyclooctane reaction conditions (Table 7) [26,42]. For all cyclooctane reaction systems in the presence of IPOMs, the ketone product was predominant. The compounds reported in the literature led to 41–50% cyclooctane conversion at 6 h, which is comparable to the results for 2-D **6** (Table 7). The IPOM tetrakis(4-methylanilinium) octamolybdate dihydrate ($(C_7H_{10}N)_4[Mo_8O_{26}]\cdot H_2O$) was the least active, leading to only 20% conversion [26,42].

3. Materials and Methods

3.1. Materials

For the synthesis of the compounds, MoO_3, acetic acid, HCl and molybdic acid ($H_2MoO_4.H_2O$ (HMO)) were acquired from AVANTOR (Gliwice, Polska) (acs purity). Aniline and anilinium salts (99%) were acquired from Sigma/Aldrich and used as received. For catalytic studies: *cis*-cyclooctene (Cy, 95%, Aldrich, St. Louis, MO, USA), methyl oleate (Ole, 99%, Aldrich), methyl linoleate (LinOle, 95%, Alfa Aesar, Haverhill, MA, USA), α,α,α-trifluorotoluene (TFT, anhydrous, 99.8%, Aldrich)), *tert*-butylhydroperoxide (TBHP, 5.5 M in decane, Aldrich), pentane (95%, Carlo Erba, Emmendingen, Germany), 1-butyl-3-methylimidazolium bis(trifluoromethylsulfonyl)imide ([bmim][NtF_2], 99%, Iolitec, Heilbronn, Germany) and acetonitrile (99.9%, Panreac, Heilbronn, Germany) were acquired from commercial sources and used as received.

The syntheses of the new compounds **1**, **2** and **5** are described, and the remaining compounds were synthesized in a similar fashion.

3.2. Synthesis of the IPOM Compounds

Compound 1. Molybdic acid (HMO; 0.01 mol 1.79 g) was added to 150 mL of hot water under reflux. Then, a mixture of 0.01 mol of cyclohexylamine with 50 mL of acetic acid was added and boiled for 0.5 h. After filtering, the filtrate was allowed to crystallize (room temperature). Yield: 85%. Transparent colourless crystals.

Compound 2. Molybdenum oxide (0.01 mol) was added to 15 mL of water and 0.75 mmol of cyclohexylamine. This mixture was kept in an oven for 7 days at 75 °C. Every day, the solution was thoroughly mixed by shaking. The resultant (white coloured) precipitate was separated by filtration, washed with a water/2-propanol mixture and finally dried in air at room temperature. Yield: *ca.* 100%, white-coloured, fibrillar crystals.

Compound 3. First, 0.01 mol HMO was dissolved in boiling water (150 mL) and then added to a boiling solution mixture of 50 mL acetic acid and 0.01 mol aniline. HCl was added and heated (reflux conditions) for *ca.* 1 h. The solution was left for crystallization. Yield: 92%. The structure of anilinium β-octamolybdate dihydrate was confirmed by XRPD: PDF-4+ card 50-2366.

Compound 4. To a boiling solution of HMO in water (0.01 mol, 150 mL), aniline was added (0.02 mol). This mixture (which became transparent after adding aniline) was heated (reflux conditions) for ca. 2 h, and then left for crystallization under ambient conditions. After ca. 2–3 months, white-coloured fibrillar crystals were obtained. Yield: *ca.* 60%. Crystallization can be accelerated by nucleation using previously obtained crystals (or even crystallites of compound **5**). The structure of anilinium trimolybdate tetrahydrate was confirmed by XRPD: PDF-4+ 49-2407.

Compound 5. To a boiling solution of HMO in water (0.01 mol, 150 mL H_2O), aniline was added (0.02 mol) and kept under reflux for ca. 2 h. Then, the solution was cooled down slowly and then kept at ca. 80 °C overnight. After ca. 12 h, the reaction mixture turned into a white gelatinous product. Microscope observation indicated that fibrillar crystals were formed. Yield: 66%. The structure of anilinium trimolybdate dihydrate was confirmed by XRPD: PDF-4+ card 50-2402.

Compound 6. Molybdic acid (0.01 mol, 1.79 g) was dissolved in 150 mL water under reflux. Then, a solution of 0.0075 mol of anilinium sulphate dissolved in 25 mL water was added and heated (reflux conditions) for 24 h. The obtained precipitate was separated by filtration and dried in air at room temperature. Yield: *ca.* 100%. Very fine, white-greyish-coloured powder. The structure of anilinium pentamolybdate was confirmed by XRPD: PDF-4+ 50-2221.

3.3. Characterisation of the Compounds

X-ray powder/single crystal diffraction. The X-ray powder diffraction (XRPD) investigations were performed for **2** and **5** and single crystal X-ray diffraction (XRD) for **1**. The temperature during measurements was 293(2) K in case of XRPD and 100 K for the single crystal investigations.

The XRPD data were collected using PANalytical X'Pert Pro MPD diffractometer, and single-crystal data using a SuperNova (Oxford Diffraction). High Score software and the PDF-4+ [ICDD, 2022] database were used for the analysis of the XRPD data. Powder diffraction studies were performed using the Expo2014 program [43] to locate Mo and O atoms, and FOX [44] to find the position and orientation of the organic moieties. Finally, restrained Rietveld refinement was performed with the use of Jana2006 [45]. Figures S7 and S8) present Rietveld refinement plots for **2** and **3**, respectively.

Structure solution and refinement of **1** was performed using SHELXS-97 and SHELXL-2013 programs [46]. To find the location of the hydrogen atoms, difference Fourier maps were used. For all non-hydrogen atoms, anisotropic refinement was used.

The Diamond [47] and Mercury [48] programs were used for visualization of the crystal structures of the obtained compounds.

The density Dx each compound was calculated as the ratio of mass/volume (determined by X-ray). For isopolymolybdates, the degree of condensation (DC) equals the ratio of (number of Mo atoms)/(0.5 × (cluster charge)).

X-ray thermal decomposition. X-ray thermal decomposition patterns were measured using Philips X'Pert Pro MPD. X-ray data for compounds **1, 2** and **5** was collected at the following temperatures: 25, 50, 75, 100, 125, 150, 175, 200, 225, 250, 300, 350, 400, 450, 500, 550, 600 °C and again at 25 °C. The 2θ range was 5–70°. The heating rate was 5 °C min^{-1} and the cooling rate was 10 °C min^{-1}.

3.4. Catalytic Tests

3.4.1. Olefin Epoxidation

In this work, the substrates tested for epoxidation reactions were *cis*-cyclooctene (Cy), methyl oleate (Ole) and methyl linoleate (Lin). The catalytic reactions were carried out in 10 mL borosilicate reactors equipped with a Teflon valve (for sampling), at 55 or 70 °C with magnetic stirring, using an amount of catalyst equivalent to 18 μmol of Mo, 1.8 mmol of olefin and 1 mL of α,α,α-trifluorotoluene (TFT). The mixture of catalyst/solvent/olefin was pre-heated at the reaction temperature. Then, the oxidant TBHP (also preheated) was added (2.75 mmol (for Cy) or 5.5 mmol (for Ole and Lin)) to the reaction mixture, and this was taken as the initial instant of the reaction.

The reactions were monitored by gas chromatography, using an Agilent 7820A GC equipped with a DB-5 capillary column (30 m × 0.25 mm × 0.25 μm) and an FID detector with H_2 as the carrier gas (for supportive chromatograms please see Figure S18). Quantifications were based on calibrations using undecane (for Cy reactions) or methyl decanoate (for Ole and Lin reactions) as internal standards. The reaction products were identified by GC-MS (Trace GC 2000 Series Thermo Quest CE Instruments GC; Thermo-Scientific DSQ II), using He as the carrier gas. The product identifications were based on commercial mass spectrometry databases (Wiley229, NIST147, NIST107, NIST27, NISTChemistry WebBook, MAINLIB) and mass spectra similarities.

The reaction systems consisted of a biphasic solid–liquid mixture. The homogeneous or heterogeneous nature of the catalytic reaction Cy/TFT/TBHP was studied by carrying out contact tests (ct) as follows: the mixture of catalyst/TFT/TBHP was stirred (1000 rpm) for 1 h at 70 °C, and then cooled to room temperature and passed through a PTFE membrane (0.2 μm pore size) filter. The filtrate was transferred immediately to a separate (empty) reactor, preheated at 70 °C, and subsequently Cy (also preheated) was added in an amount equivalent to that typically used (initial molar ratio Cy:TBHP = 1:1.5). The evolution of the homogeneous phase reaction was monitored by GC as described above. The attempts to precipitate the species from the reaction mixtures involved the addition of solvents such as *n*-pentane, *n*-hexane and acetone, and keeping the solutions at 4 °C, which unfortunately was not sufficient to precipitate measurable amounts of solid.

At the end of Cy reaction, the solids were separated from the reaction mixture by centrifugation (5000 rpm) and thoroughly washed with pentane. The solids were dried overnight under atmospheric conditions, followed by a vacuum drying for 1 h at 60 °C. The recovered solids were used in a second batch run at 70 °C, keeping constant the initial mass ratio of catalyst/olefin/oxidant used in run 1.

3.4.2. Kinetic Modelling

Kinetic models were developed considering perfectly stirred, isothermal batch reactors, and corresponding material balance according to Equation (1):

$$\frac{dC_i}{dt} = r_i \tag{1}$$

where C_i is the molar concentration of species i (M), t is the reaction time (h) and r_i is the reaction rate of species i. The material balances closed in 100% for the substrates Cy (considering CyO), Ole (considering OleOx) and Lin (considering monoepoxides, diepoxides and LinFur). Details of the kinetic models are described in the Supplementary Materials. Three different kinetic models were firstly developed (for Cy epoxidation). The simplest model considered irreversible, first-order reactions for all steps; the second model considered second-order reactions with the oxidant (these two models considered that the epoxidation reaction kinetics was not governed by the rate of formation of active oxidizing species); and the third model considered second-order reactions and the kinetics of formation of active oxidizing species (Mo+TBHP). The three models fitted the experimental data well. Hence, the discussed kinetic models for the Cy, Ole and Lin reactions were based on the assumptions for the simplest model, e.g., for Cy, the equations 1, 2 and 3 with the rate constant k_{Cy} (h^{-1}):

$$\frac{dC_{Cy}}{dt} = -k_{Cy}\, C_{Cy} \tag{2}$$

$$\frac{dC_{CyO}}{dt} = k_{Cy}\, C_{Cy} \tag{3}$$

3.4.3. Cyclooctane Oxidation

The oxidation of cyclooctane was experimentally performed as previously described [26,49,50]. The experiment was carried out in a stainless-steel batch reactor system at 120 °C and under an air pressure of 10 atm. The cyclooctane-to-oxygen molar ratio was 6.5. The reactor was cooled after 6 h, followed by sampling and analytic analysis. The oxidation products cyclooctanone (A) and cyclooctanol (B) were analysed by an Agilent 6890 N Gas Chromatograph (FID detector) equipped with an Innowax (30 m × 0.53 mm × 1.0 μm) column, using an internal standard (chlorobenzene). The amount of suberic acid was determined by the back titration method (the sample is mixed with excess NaOH/KOH and stirred, followed by titration using HCl).

4. Conclusions

The ionic polymolybdate compounds (IPOM) cyclohexylammonium β-octamolybdate dihydrate (**1**), cyclohexylammonium trimolybdate hydrate (**2**), anilinium β-octamolybdate dihydrate (**3**), anilinium trimolybdate tetrahydrate (**4**) and anilinium trimolybdate dihydrate (**5**) were synthesized via simple, eco-friendly one-pot routes and studied as epoxidation catalysts of olefins (*cis*-cyclooctene and FAMEs) using *tert*-butylhydroperoxide as oxidant. Depending on the IPOM synthesis conditions and the type of molybdenum precursor, IPOM compounds with different structural dimensionality, density Dx and ratio of terminal oxo groups/molybdenum atoms (n(oxo)/Mo) were formed. The different compounds led to different olefin epoxidation reaction kinetics, and no direct structure–activity correlations could be established for all IPOMs. The catalytic studies indicated that the reactions were essentially homogeneous in nature. Accordingly, the (yet unknown) types of soluble metal species may be structurally different, contributing with different intrinsic activities to the overall reaction kinetics.

Nevertheless, the IPOM compounds promoted the epoxidation of biobased fatty acid methyl esters, namely methyl oleate and methyl linoleate, leading to conversions of at least 81% at 4 h, 70 °C, and the corresponding epoxides, as well as furan-type products in the case of the Lin reaction. They also promoted cyclooctane oxidation, although the difference in the catalytic performances of the catalysts seemed less pronounced than for the epoxidation systems. Overall, for the studied substrate scope and reaction scope, **1–5** seemed more active than **6**. Insights into the factors influencing the catalytic performances may call for in situ characterization of dissolved metal species involved in the catalytic processes, which is challenging partly due to the complexity of the liquid–solid, multicomponent systems.

Supplementary Materials: The following supporting information can be downloaded at: https://www.mdpi.com/article/10.3390/catal14040251/s1, Figure S1: TG/DSC study of cyclohexylammonium β-octamolybdate dihydrate (**1**). Figure S2: TG/DSC study of cyclohexylammonium trimolybdate hydrate (**2**). Figure S3: TG/DSC study of anilinium trimolybdate dihydrate (**5**). Figure S4: XRPD/temp. study of cyclohexylammonium β-octamolybdate dihydrate (**1**). Figure S5: XRPD/temp study of cyclohexylammonium trimolybdate hydrate (**2**). Figure S6: XRPD/temp study of anilinium trimolybdate dihydrate (**5**). Figure S7: Cyclohexylammonium trimolybdate hydrate—Rietveld refinement plots (**2**). Figure S8: Anilinium trimolybdate dihydrate—Rietveld refinement plots (**5**). Table S1: Cyclohexylammonium β-octamolybdate dihydrate—interatomic distances (**1**). Table S2: Cyclohexylammonium trimolybdate hydrate—interatomic distances (**2**). Table S3: Anilinium trimolybdate dihydrate—interatomic distances (**5**). Figure S9: Experimental data (markers) and calculated kinetic curves (lines) for Cy (o) and epoxide CyO (Δ) concentration, based on the model A, for catalyst **1** (a), **2** (b), **3** (c), **4** (d), **5** (e) and **6** (f), at 70 °C. Figure S10: Experimental data (markers) and calculated kinetic curves (lines) for Cy (o) and epoxide CyO (Δ) concentration, based on the model B, for catalyst **1** (a), **2** (b), **3** (c), **4** (d), **5** (e) and **6** (f), at 70 °C. Figure S11: Experimental data (markers) and calculated kinetic curves (lines) for Cy (o) and epoxide CyO (Δ) concentration, based on the model C, for catalyst 1 **1** (a), **2** (b), **3** (c), **4** (d), **5** (e) and **6** (f), at 70 °C. Table S4: Calculated data based on kinetic models A and B, for the Cy reaction, at 70 °C. Table S5: Calculated data based on kinetic model C, for the Cy reaction, at 70 °C. Scheme S2: Mechanistic proposal for the reaction of the olefins, considering irreversible, first-order reactions (same as Scheme 1 of the main text, serves as guide). Figure S12: Experimental data (markers) and calculated kinetic curves (lines) for Ole (o) and OleOx (Δ) concentration for catalyst **1** (a), **2** (b), **3** (c), **4** (d), **5** (e) and **6** (f), at 70 °C. Figure S13: Experimental data (markers) and calculated kinetic curves (lines) for Lin (o), LinOx (Δ), LimDiOx (×) and LinFur (+) concentration for catalyst **1** (a), **2** (b), **3** (c), **4** (d), **5** (e) and **6** (f), at 70 °C. Figure S14: cis-Cyclooctene epoxidation with TBHP, in the presence of the IPOM catalysts **3–5**, at 55 °C. Epoxide selectivity was always 100 %. Table S6: Comparison of the catalytic results for **3–5** to literature data for IPOM compounds possessing anilinium derivatives as organic components, tested for the Cy/TBHP reaction, at 55 °C. Figure S15: Typical epoxidation catalytic test (with catalyst) (○) and contact tests (×) for cis-cyclooctene epoxidation with TBHP, in the presence of the IPOM catalysts **1-6**, at 70 °C. Figure S16: ATR FT-IR spectra (a) and powder XRD patterns (b) of the original and recovered solids **1**, **2**, **3** and **5**. The experimental details of the characterisation techniques are indicated. Figure S17: Kinetic constants as a function of density Dx (left) and n(oxo)/Mo ratio (right), for the Cy6N (a, b) and Anil (c, d) families of IPOM catalysts with different substrates (Cy (circles), Ole (squares) and Lin (triangles); the dotted lines are trendlines). Figure S18: Examples of chromatograms for Cy, Ole and Lin reactions.

Author Contributions: Conceptualization, W.Ł. and A.A.V.; validation, P.N. (epoxidation), G.S. (kinetic modelling), B.D.N. (crystalline phase analysis, TG/DSC analyses), K.P. (cyclooctane oxidation), P.S. (single crystal), A.A.V. (epoxidation) and W.Ł. (XRPD); investigation, P.N., G.S., B.D.N., K.P. and P.S.; resources, W.Ł. and A.A.V.; writing—original draft preparation, P.N.; writing—review and editing, A.A.V. and W.Ł.; visualization, P.N., W.Ł. and A.A.V.; supervision, P.N., W.Ł. and A.A.V.; project administration, W.Ł. and A.A.V.; funding acquisition, W.Ł. and A.A.V. All authors have read and agreed to the published version of the manuscript.

Funding: This work was developed within the scope of the project CICECO-Aveiro Institute of Materials, UIDB/50011/2020, UIDP/50011/2020, and LA/P/0006/2020, financed by national funds through the FCT/MCTES (PIDDAC). The study (at Jagiellonian University) was carried out using research infrastructure purchased with the funds of the European Union in the framework of the Smart Growth Operational Programme, Measure 4.2; Grant No. POIR.04.02.00-00-D001/20, ATOMIN 2.0—ATOMic scale science for the INnovative economy.

Data Availability Statement: Data is essentially contained within the article or Supplementary Materials.

Acknowledgments: The authors are grateful to the mentioned funders.

Conflicts of Interest: The authors declare no conflicts of interest. The funders had no role in the design of the study; in the collection, analyses, or interpretation of data; in the writing of the manuscript; or in the decision to publish the results.

References

1. Lligadas, G.; Ronda, J.C.; Galià, M.; Biermann, U.; Metzger, J.O. Synthesis and Characterization of Polyurethanes from Epoxidized Methyl Oleate Based Polyether Polyols as Renewable Resources. *J. Polym. Sci. A Polym. Chem.* **2006**, *44*, 634–645. [CrossRef]
2. Musik, M.; Janus, E.; Pełech, R.; Sałaciński, Ł. Effective Epoxidation of Fatty Acid Methyl Esters with Hydrogen Peroxide by the Catalytic System $H_3PW_{12}O_{40}$/Quaternary Phosphonium Salts. *Catalysts* **2021**, *11*, 1058. [CrossRef]
3. Hájek, M.; Hájek, T.; Kocián, D.; Frolich, K.; Peller, A. Epoxidation of Methyl Esters as Valuable Biomolecules: Monitoring of Reaction. *Molecules* **2023**, *28*, 2819. [CrossRef]
4. Sharma, B.K.; Doll, M.; Erhan, S.Z. Oxidation, Friction Reducing, and Low Temperature Properties of Epoxy Fatty Acid Methyl Esters. *Green Chem.* **2007**, *9*, 469–474. [CrossRef]
5. Wadumesthrige, K.; Salley, S.O.; Ng, K.Y.S. Effects of Partial Hydrogenation, Epoxidation, and Hydroxylation on the Fuel Properties of Fatty Acid Methyl Esters. *Eur. J. Lipid Sci. Technol.* **2009**, *90*, 1292–1299. [CrossRef]
6. Available online: https://polymer-additives.specialchem.com/selectors/tr-vikoflex (accessed on 21 December 2023).
7. Turco, R.; Tesser, R.; Russo, V.; Vitiello, R.; Fagnano, M.; Di Serio, M. Comparison of Different Possible Technologies for Epoxidation of Cynara Cardunculus Seed Oil. *Eur. J. Lipid Sci. Technol.* **2020**, *122*, 1900100. [CrossRef]

8. Available online: https://www.fortunebusinessinsights.com/epoxidized-soybean-oil-market-104343 (accessed on 21 December 2023).
9. Meng, Y.; Taddeo, F.; Aguilera, A.F.; Cai, X.; Russo, V.; Tolvanen, P.; Leveneur, S. The Lord of the Chemical Rings: Catalytic Synthesis of Important Industrial Epoxide Compounds. *Catalysts* **2021**, *11*, 765. [CrossRef]
10. Yan, W.; Wang, Z.; Luo, C.; Xia, X.; Liu, Z.; Zhao, Y.; Du, F.; Jin, X. Opportunities and Emerging Challenges of the Heterogeneous Metal-Based Catalysts for Vegetable Oil Epoxidation. *ACS Sustain. Chem. Eng.* **2022**, *10*, 7426–7446. [CrossRef]
11. Cucciolito, M.E.; Di Serio, M.; Esposito, R.; Melchiorre, M.; Ruffo, F.; Russo, V.; Tesser, R.; Turco, R.; Vitiello, R. Catalysis for Oleochemical Platforms. *Eur. J. Inorg. Chem.* **2023**, *26*, e202200783. [CrossRef]
12. Pinaeva, L.G.; Noskov, A.S. Prospects for the Development of Catalysts for the Oxidation Processes of Advanced Propylene Processing. *Catal. Ind.* **2020**, *12*, 176–200. [CrossRef]
13. Shen, Y.; Jiang, P.; Wai, P.T.; Gu, Q.; Zhang, W. Recent Progress in Application of Molybdenum-Based Catalysts for Epoxidation of Alkenes. *Catalysts* **2019**, *9*, 31. [CrossRef]
14. Brégeault, J.M. Transition-Metal Complexes for Liquid-Phase Catalytic Oxidation: Some Aspects of Industrial Reactions and of Emerging Technologies. *J. Chem. Soc. Dalton Trans.* **2003**, *3*, 3289–3302. [CrossRef]
15. Pisk, J.; Agustin, D. Molybdenum, Vanadium, and Tungsten-Based Catalysts for Sustainable (Ep)Oxidation. *Molecules* **2022**, *27*, 6011. [CrossRef]
16. Bridgeman, A.J.; Cavigliasso, G. Electronic Structure of the α and β Isomers of $[Mo_8O_{26}]^{4-}$. *Inorg. Chem.* **2002**, *41*, 3500–3507. [CrossRef]
17. Available online: https://colapret.cm.utexas.edu/courses/320M%20Syllabus_Sp2019.pdf (accessed on 21 December 2023).
18. Available online: https://openstax.org/books/organic-chemistry/pages/24-4-basicity-of-arylamines?query=cyclohexylammonium (accessed on 21 December 2023).
19. Bożek, B.; Neves, P.; Valente, A.A. Ionic Ammonium and Anilinium Based Polymolybdate Hybrid Catalysts for Olefin Epoxidation. *Appl. Catal. A Gen.* **2018**, *564*, 13–25. [CrossRef]
20. Zhou, M.-D.; Liu, M.-J.; Huang, L.-L.; Zhang, J.; Wang, J.-Y.; Li, X.-B.; Kühn, F.E.; Zang, S.-L. Olefin Epoxidation with Hydrogen Peroxide Using Octamolybdate-Based Self-Separating Catalysts. *Green Chem.* **2015**, *17*, 1186–1193. [CrossRef]
21. Kumar, A.; Gupta, A.K.; Devi, M.; Gonsalves, K.E.; Pradeep, C.P. Engineering Multifunctionality in Hybrid Polyoxometalates: Aromatic Sulfonium Octamolybdates as Excellent Photochromic Materials and Self-Separating Catalysts for Epoxidation. *Inorg. Chem.* **2017**, *56*, 10325–10336. [CrossRef]
22. Neves, P.; Amarante, T.R.; Valente, A.A.; Pillinger, M.; Gonçalves, I.S. Catalytic Application of an Octamolybdate Salt (H3biim)4[β-Mo8O26] in Olefin Epoxidation (H2biim = 2,2′-Biimidazole). *Catal. Lett.* **2016**, *146*, 841–850. [CrossRef]
23. Gamelas, C.A.; Neves, P.; Gomes, A.C.; Valente, A.A.; Romão, C.C.; Gonçalves, I.S.; Pillinger, M. Molybdenum(II) Diiodo-Tricarbonyl Complexes Containing Nitrogen Donor Ligands as Catalyst Precursors for the Epoxidation of Methyl Oleate. *Catal. Lett.* **2012**, *142*, 1218–1224. [CrossRef]
24. Available online: https://www.masterorganicchemistry.com/2017/04/18/basicity-of-amines-and-pkah/ (accessed on 21 December 2023).
25. Kama, A.B.; Dessapt, R.; Serier-Brault, H.; Sidibe, M.; Diop, C.A.K.; Gautier, R. Stabilization of β-Octamolybdate with Large Counterions. *J. Mol. Struct.* **2017**, *1141*, 698–702. [CrossRef]
26. Szyman, A.; Pamin, K.; Połtowicz, J. Molybdenum Complexes as Catalysts for the Oxidation of Cycloalkanes with Molecular Oxygen. *Catal. Lett.* **2016**, *146*, 998–1010. [CrossRef]
27. Veiros, L.F.; Prazeres, A.; Costa, P.J.; Romão, C.C.; Kühn, F.E.; José Calhorda, M. Olefin Epoxidation with Tert-Butyl Hydroperoxide Catalyzed by MoO_2X_2L Complexes: A DFT Mechanistic Study. *Dalton Trans.* **2006**, *11*, 1383–1389. [CrossRef] [PubMed]
28. Comas-Vives, A.; Lledós, A.; Poli, R. A Computational Study of the Olefin Epoxidation Mechanism Catalyzed by Cyclopentadienyloxidomolybdenum(VI) Complexes. *Chem. Eur. J.* **2010**, *16*, 2147–2158. [CrossRef] [PubMed]
29. Al-Ajlouni, A.; Valerie, A.A.; Nunes, C.D.; Pillinger, M.; Santos, A.M.; Zhao, J.; Romão, C.C.; Gonçalves, I.S.; Kühn, F.E. Kinetics of Cyclooctene Epoxidation with *tert*-Butyl Hydroperoxide in the Presence of $[MoO_2X_2L]$-Type Catalysts (L = Bidentate Lewis Base). *Eur. J. Inorg. Chem.* **2005**, *2005*, 1716–1723. [CrossRef]
30. Prazeres, A.; Santos, A.M.; Calhorda, M.J.; Roma, C.C.; Gonçalves, I.S. Octahedral Bipyridine and Bipyrimidine Dioxomolybdenum (VI) Complexes. *Chem. Eur. J.* **2002**, *8*, 2370–2383. [CrossRef]
31. Garrido, G.; Rosés, M.; Ràfols, C.; Bosch, E. Acidity of Several Anilinium Derivatives in Pure Tetrahydrofuran. *J. Solut. Chem.* **2008**, *37*, 689–700. [CrossRef]
32. Neves, P.; Gomes, A.C.; Cunha-Silva, L.; Valente, A.A.; Pillinger, M.; Gonçalves, I.S. Silicododecamolybdate/Pyridinium-Tetrazole Hybrid Molecular Salt as a Catalyst for the Epoxidation of Bio-Derived Olefins. *Inorg. Chim. Acta A* **2021**, *516*, 120129. [CrossRef]
33. Nunes, M.S.; Gomes, A.C.; Neves, P.; Mendes, R.F.; Almeida Paz, F.A.; Lopes, A.D.; Pillinger, M.; Gonçalves, I.S.; Valente, A.A. Molybdenum(VI) Complexes with Ligands Derived from 5-(2-Pyridyl)-2H-Tetrazole as Catalysts for the Epoxidation of Olefins. *Catal. Today* **2023**, *423*, 114273. [CrossRef]
34. Herman, K.; Foodtechn, J. Catalytic Epoxidation of Methyl Linoleate-Cyclisation Products of Epoxyacid Esters. *Fat. Sci. Technol.* **1995**, *97*, 269–273.
35. Dorado, V.; Herrerías, C.I.; Fraile, J.M. Catalytic Hydrolysis of Epoxyfatty Esters with Solid Sulfonic Acids. *Mol. Catal.* **2023**, *547*, 113282. [CrossRef]

36. Piazza, G.J.; Nuñz, A.; Foglia, T.A. Hydrolysis of Mono- and Diepoxyoctadecanoates by Alumina. *J. Am. Oil Chem. Soc.* **2003**, *80*, 901–904. [CrossRef]
37. Bantchev, G.B.; Doll, K.M.; Biresaw, G.; Vermillion, K.E. Formation of Furan Fatty Alkyl Esters from Their Bis-Epoxide Fatty Esters. *J. Am. Oil Chem. Soc.* **2014**, *91*, 2117–2123. [CrossRef]
38. Stolp, L.J.; Joseph, E.; Kodali, D.R. Synthesis and Evaluation of Soy Fatty Acid Ester Estolides as Bioplasticizers in Poly(Vinyl Chloride). *J. Am. Oil Chem. Soc.* **2019**, *96*, 1291–1302. [CrossRef]
39. Wolfe, J.P.; Hay, M.B. Recent Advances in the Stereoselective Synthesis of Tetrahydrofurans. *Tetrahedron* **2007**, *63*, 261–290. [CrossRef] [PubMed]
40. Capon, R.J.; Barrow, R.A. Acid-Mediated Conversion of Methylene-Interrupted Bisepoxides to Tetrahydrofurans: A Biomimetic Transformation. *J. Org. Chem.* **1998**, *63*, 75–83. [CrossRef] [PubMed]
41. Araji, N.; Chatel, G.; Moores, A.; Jérôme, F.; De Oliveira Vigier, K. Oxidative Cyclization of Linoleic Acid in the Presence of Hydrogen Peroxide and Phosphotungstic Acid. *Mol. Catal.* **2020**, *493*, 111084. [CrossRef]
42. Oszajca, M.; Nitek, W.; Rafalska-Łasocha, A.; Pamin, K.; Połtowicz, J.; Łasocha, W. Synthesis, Crystal Structure and Selected Properties of Three New 4-Propylanilinium Polyoxomolybdates. *J. Mol. Struct.* **2023**, *1273*, 134292. [CrossRef]
43. Altomare, A.; Cuocci, C.; Giacovazzo, C.; Moliterni, A.; Rizzi, R.; Corriero, N.; Falcicchio, A. EXPO2013: A Kit of Tools for Phasing Crystal Structures from Powder Data. *J. Appl. Crystallogr.* **2013**, *46*, 1231–1235. [CrossRef]
44. Favre-Nicolin, V.; Cerný, R. FOX, 'free Objects for Crystallography': A Modular Approach to Ab Initio Structure Determination from Powder Diffraction. *J. Appl. Crystallogr.* **2002**, *35*, 734–743. [CrossRef]
45. Petricek, V.; Dusek, M.; Palatinus, L. *The Crystallographic Computing System*; Institute of Physics: Jana, Czech Republic, 2006.
46. Sheldrick, G.M. A Short History of SHELX. *Acta Crystallogr. A* **2008**, *64*, 112–122. [CrossRef]
47. Brandenburg, K. Diamond Version 3.2g, Crystal Impact GbR, Bonn, Germany, 1997–2001. Available online: http://www.crystalimpact.com/diamond (accessed on 4 January 2023).
48. Macrae, C.F.; Edgington, P.R.; McCabe, P.; Pidcock, E.; Shields, G.P.; Taylor, R.; Towler, M.; Van De Streek, J. Mercury: Visualization and Analysis of Crystal Structures. *J. Appl. Crystallogr.* **2006**, *39*, 453–457. [CrossRef]
49. Bożek, B.; Neves, P.; Oszajca, M.; Valente, A.A.; Połtowicz, J.; Pamin, K.; Łasocha, W. Simple Hybrids Based on Mo or W Oxides and Diamines: Structure Determination and Catalytic Properties. *Catal. Lett.* **2020**, *150*, 713–727. [CrossRef]
50. Pamin, K.; Jachimska, B.; Onik, K.; Połtowicz, J.; Grabowski, R. Electrostatic Self-Assembly of Polyoxometalates on Chitosan as Catalysts of Oxidation of Cyclic Hydrocarbons. *Catal. Lett.* **2009**, *127*, 167–174. [CrossRef]

catalysts

MDPI

Article

Bimetallic Pt-IrO_x/g-C_3N_4 Photocatalysts for the Highly Efficient Overall Water Splitting under Visible Light

Nikolay D. Sidorenko [1], Polina A. Topchiyan [2], Andrey A. Saraev [1], Evgeny Yu. Gerasimov [1], Angelina V. Zhurenok [1], Danila B. Vasilchenko [2] and Ekaterina A. Kozlova [1,*]

[1] Federal Research Center Boreskov Institute of Catalysis, 630090 Novosibirsk, Russia; gerasimov@catalysis.ru (E.Y.G.); angelinazhurenok@gmail.com (A.V.Z.)

[2] Nikolaev Institute of Inorganic Chemistry, Siberian Branch of the Russian Academy of Science, 630090 Novosibirsk, Russia; vasilchenko@niic.nsc.ru (D.B.V.)

* Correspondence: kozlova@catalysis.ru

Abstract: Two series of bimetallic photocatalysts (0.5% Pt/0.01–0.5% IrO_x/g-C_3N_4 and 0.1% Pt/0.01–0.1% IrO_x/g-C_3N_4) were synthesized by the thermolysis of melamine cyanurate and a successive deposition of platinum and iridium labile complexes $(Me_4N)_2[Pt_2(\mu\text{-}OH)_2(NO_3)_8]$ and *fac*-$[Ir(H_2O)_3(NO_2)_3$. The synthesized photocatalysts were studied by a set of physicochemical analysis techniques. Platinum exists in two states, with up to 60% in metallic form and the rest in the Pt^{2+} state, while iridium is primarily oxidized to the Ir^{3+} state, which was determined by X-ray photoelectron spectroscopy (XPS). The specific surface area (S_{BET}), which is determined by low-temperature nitrogen adsorption, ranges from 80 to 100 $m^2\ g^{-1}$ and the band gap energy (E_g) value is in the range of 2.75–2.80 eV as found by diffuse reflectance spectroscopy (DRS). The activity of the photocatalysts was tested in the photocatalytic production of hydrogen from ultrapure water under visible light (λ = 400 nm). It was found that the splitting of water occurs with the formation of the stochiometric amount of H_2O_2 as an oxidation product. Two photocatalysts 0.5% Pt/0.01% IrO_x/g-C_3N_4 and 0.1% Pt/0.01% IrO_x/g-C_3N_4 showed the highest activity at 100 $\mu mol\ h^{-1}\ g_{cat}^{-1}$, which is among the highest in H_2 production published for such systems.

Keywords: g-C_3N_4; photocatalytic water splitting; hydrogen peroxide; visible light

Citation: Sidorenko, N.D.; Topchiyan, P.A.; Saraev, A.A.; Gerasimov, E.Y.; Zhurenok, A.V.; Vasilchenko, D.B.; Kozlova, E.A. Bimetallic Pt-IrO_x/g-C_3N_4 Photocatalysts for the Highly Efficient Overall Water Splitting under Visible Light. *Catalysts* **2024**, *14*, 225. https://doi.org/10.3390/catal14040225

Academic Editors: Narendra Kumar, Leonarda Liotta and Konstantin Ivanov Hadjiivanov

Received: 26 February 2024
Revised: 24 March 2024
Accepted: 26 March 2024
Published: 28 March 2024

1. Introduction

The transition to a carbon-free and low-carbon economy is now being seriously considered, and therefore there is a huge interest in hydrogen as an ideal energy carrier [1,2]. Photocatalytic water splitting with hydrogen production is widely regarded as the most promising strategy for the conversion of solar energy into chemical energy [3–7].

Unfortunately, the quantum efficiency of this process is usually low due to the back reaction between the evolved hydrogen and oxygen and the recombination of electron–hole pairs [8]. Therefore, most investigations are devoted to photocatalytic hydrogen production not by water splitting, but by its reduction through the use of available organic and inorganic electron donors in aqueous solutions [9]. The addition of so-called "sacrificial agents" reduces the recombination of electron–hole pairs and thus increases the quantum efficiency of hydrogen production [10]. The maximum rate of hydrogen production in systems with electron donors is several orders of magnitude higher than the rate of hydrogen production during water splitting [11]. However, the use of "sacrificial agents" has a fundamental disadvantage—the need for the constant introduction of an electron donor into the system. Moreover, it is necessary to select a photocatalyst that is active in the presence of a specific "sacrificial agent". Therefore, recent research has tended to move from systems containing electron donors to photocatalytic splitting of pure water [12–15].

For the complete water splitting under normal conditions, the standard Gibbs free energy is very high and amounts to $\Delta_r G°_{298}$ = +237.2 $kJ \cdot mol^{-1}$. In addition, the oxidation of water usually requires a much higher overvoltage than the reduction in protons [9]. It is known that the four-electron oxidation of water is a complex process that occurs with great kinetic constraints [16]. In addition, when obtaining a mixture of hydrogen and oxygen, safety issues arise, and efforts must be made to separate the components of the mixture [17].

Photocatalytic water splitting is known to follow two different mechanisms: a thermodynamically favored (E = 1.23 eV vs. NHE) four-electron pathway [18]:

$$2H_2O \rightarrow 2H_2 + O_2, \tag{1}$$

and a kinetically favored (E = 1.78 eV vs. NHE) two-electron pathway:

$$2H_2O \rightarrow H_2O_2 + H_2. \tag{2}$$

It is believed that under ambient conditions, process (2) is more probable [16]. Note that this process is also considered favorable because hydrogen peroxide is a valuable product itself [18–20].

The most frequently studied photocatalysts for water splitting are graphene-based structures [15,21–23] and polymeric graphite-like carbon nitride g-C_3N_4 [18,19,24]. Unlike photocatalysts based on transition metal sulfides, g-C_3N_4 is an environmentally friendly material. g-C_3N_4 is a semiconductor with a band gap of 2.7 eV ($\lambda = 460$ nm) and with valence band levels at +1.6 V vs. NHE and the conduction band at −1.1 V vs. NHE [25,26]. The latter is one of the most negative values for known semiconductor photocatalysts, which favors photocatalytic reduction. In addition, g-C_3N_4 is able to catalyze the two-electron process of formation of H_2O_2 [27,28]. Graphitic carbon nitride is usually obtained by thermal polycondensation of inexpensive nitrogen-containing precursors, such as cyanamide and dicyandiamide, melamine, urea, and thiourea [29]. However, the resulting g-C_3N_4 has a small specific surface area. In order to increase the surface area and obtain materials with higher catalytic activity, various methods are proposed. For example, in our group, a new synthetic technique, based on thermolysis of a melamine cyanurate complex, has been proposed for the formation of g-C_3N_4 with a high surface area and photocatalytic activity [30–32].

For the splitting of water by reaction (1) and reaction (2), the surface of the photocatalyst must contain additional cocatalysts for the reduction and oxidation of water, respectively [16,33]. Suitable cocatalysts can reduce energy barriers for reduction/oxidation by reducing the activation energy [33]. In order to reduce the energy barriers to reduction/oxidation, appropriate cocatalysts are applied. For example, platinum group metals such as Pt, Rh, Ru, and Ir, as well as Ni, can be used as a cocatalyst for hydrogen evolution [34,35]. Metal oxides such as Co, Fe, Ni, Mn, Ru, and Ir oxides are commonly used as oxygen production cocatalysts [34,36]. For the complete water splitting, systems based on graphitic carbon nitride with cocatalysts such as Pt-CoP/g-C_3N_4 [37], Pt-PtO_x-CoOx/g-C_3N_4 [38], Rh-RhO_x/g-C_3N_4 [39], Pt-IrO_2/g-C_3N_4 [6], Pt/(K^+-doped)g-C_3N_4 [20], or Pt-CoP/g-C_3N_4 [8] have been proposed. These systems showed high activity in complete water splitting; however, the content of noble metals used in them was quite high—from 1 to 5 wt.%. The possibility of synthesizing photocatalysts with small amounts of noble metals has not been studied, although the high content of these components leads to a significant increase in the cost of the material.

This research was aimed at the synthesis of g-C_3N_4-based photocatalysts containing platinum and iridium oxide cocatalysts with low content of noble metals (<0.5 wt.%). In our previous works, we have developed novel approaches for the synthesis of active photocatalysts for two processes—hydrogen evolution from aqueous solutions of TEOA and oxygen evolution from aqueous solutions of $NaIO_4$. For the first process, 0.01–0.5% Pt/g-C_3N_4 photocatalysts were prepared by thermal treatment of melamine–cyanuric acid supramolecular complex and further sorption of platinum nitrato complexes as a precursor of Pt cocatalyst [30]. For the second process, 0.005–0.5% IrO_x/g-C_3N_4 photocatalysts were prepared via deposition of Ir aqua/nitro complex onto the surface of g-C_3N_4 with further thermal treatment in air [32]. It was shown that at high activity in the production of oxygen from an electron acceptor solution, $NaIO_4$, the $IrO_x/g\text{-}C_3N_4$ photocatalysts show very low activity in the production of hydrogen, i.e., they catalyze only the process of oxygen formation [32]. Thus, we assumed that a system $IrO_x/Pt/g\text{-}C_3N_4$ with two supported cocatalysts ("reductive" Pt and "oxidative" IrO_x) would be active in the complete decomposition of water. The main task was to provide the synthesis of photocatalysts in such a way that platinum would be reduced to the metallic state, while iridium would remain in an oxide state. Based on this idea, we synthesized two series of samples $IrO_x/Pt/g\text{-}C_3N_4$ with the Pt weight content of 0.5 and 0.1% and the Ir content of 0.01–0.5 wt.%. The catalysts provided photocatalytic splitting of water without the addition of donors or acceptors of electrons under the action of visible light.

Thus, the systematic analysis carried out to understand the mutual influence of platinum and iridium particles in bimetallic $IrO_x/Pt/g\text{-}C_3N_4$ photocatalysts on the overall activity of such materials in the reaction of water splitting into H_2 and H_2O_2 is the main novelty of this work. By using an original synthetic approach to deposit cocatalysts based on our previous studies of $Pt/g\text{-}C_3N_4$ photocatalyst [30], several different combinations of Ir and Pt states of cocatalysts with different Pt/Ir ratios were realized on $g\text{-}C_3N_4$ surface. The activities of the resulting ensembles were analyzed along with the data about their structure revealed by the series of methods.

2. Results

2.1. Preparation of the Pt-Ir/C_3N_4 Photocatalysts

Earlier it was shown that the best cocatalysts for the production of hydrogen and oxygen are platinum in the metallic state and ionic forms of iridium, respectively [29–32]. The objective of this study was to select a combination of cocatalysts that would produce hydrogen from water without the addition of organic substances—electron donors.

Our main synthesis strategy was Method 1. This procedure, with the metal deposition steps separated from each other, leads to an independent distribution of Ir and Pt particles (see below). However, to check the influence of a synthesis method on the structure and activity of catalysts, two modified approaches (Methods 2 and 3) were also tested.

2.2. Photocatalyst Characterization

The synthesized photocatalysts were characterized by XRD. It was proved that all three synthesis methods provide the formation of graphite-like carbon nitride $g\text{-}C_3N_4$: in the X-ray patterns, the most intense reflections (at 13° and 27°) correspond to the layered structure of graphite-like carbon nitride. The interplanar spacing is about 3.26 Å, which agrees with the data for materials based on $g\text{-}C_3N_4$. Unfortunately, XRD turned out to be of little use for analysis of deposited metals in such systems: it could only identify metallic platinum with a mass fraction of at least 0.5% (Figure 1A). Iridium compounds could not be identified even at this high mass fraction.

To illustrate how metal concentration affects catalytic activity, two series of catalysts were prepared using Method 1. The series contained Pt with concertation fixed at either 0.5 or 0.1 wt.%, while iridium concentration was varied in rows 0.01, 0.05, 0.1, and 0.5 wt.%. In the case of metal cocatalysts, the actual metal content is a very important factor. Therefore, it was experimentally determined using ICP-AES. The chemical analysis showed that the actual atomic ratios Pt:Ir were in a good linear correlation with the introduced quantities of precursors (Figure S1). The noticeable overconcentration of Pt in the samples can be attributed to the partial decomposition of the $g\text{-}C_3N_4$ support during the synthesis. Indeed, the overconcentration correlates with the mass loss previously detected for $Pt/g\text{-}C_3N_4$ materials during their calcination in H_2 [30].

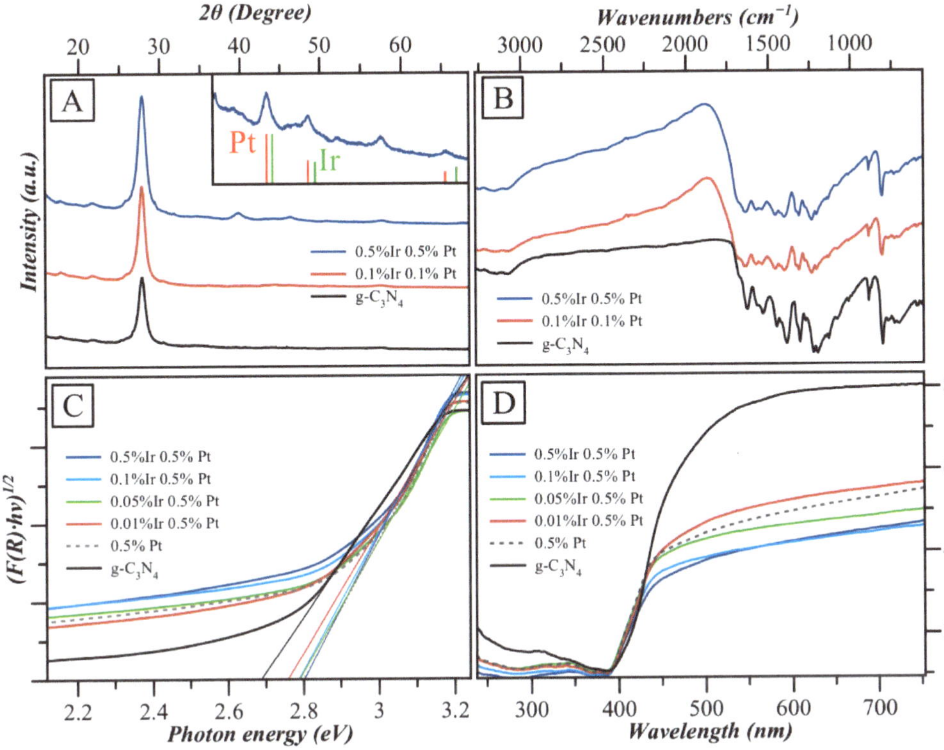

Figure 1. (**A**) XRD patterns of the $Pt_{0.5}Ir_{0.5}/C_3N_4(1)$ and $Pt_{0.1}Ir_{0.1}/C_3N_4(1)$ catalysts and g-C_3N_4; (**B**) FTIR spectra of the $Pt_{0.5}Ir_{0.5}/C_3N_4(1)$, $Pt_{0.1}Ir_{0.1}/C_3N_4(1)$, and g-$C_3N_4$; (**C**) Tauc plots performed for the $Pt_{0.5}Ir_y/C_3N_4(1)$ series of catalysts, $Pt_{0.5}/C_3N_4(1)$, and g-C_3N_4; and (**D**) DRS of the $Pt_{0.5}Ir_y/C_3N_4(1)$ series of catalysts, $Pt_{0.5}/C_3N_4(1)$ and g-C_3N_4.

The FTIR spectra of the materials demonstrate the series of signals in the range 1800–900 cm^{-1} which are assigned to the C–N and C–C stretching vibration in agreement with the IR fingerprints of the heptazine-based g-C_3N_4 materials reported previously [30]. According to the IR spectroscopy data, metal deposition does not lead to any significant changes in the structure of the parent g-C_3N_4 material (Figure 1B). This result can be ascribed to the low loadings of metals, which makes their influence on the structure of g-C_3N_4 negligible.

On the other hand, the specific surface area of the samples prepared via Method 1 shows a clear correlation with the amount of Pt: a higher percentage of Pt leads to a higher BET area of a sample. Indeed, the surface area of pristine g-C_3N_4 is about 50 m^2·g^{-1}, whereas for the samples with 0.1% and 0.5% Pt, the surface area achieves 80 m^2·g^{-1} and 100 m^2·g^{-1}, respectively (Table 1). This result agrees well with our data on monometallic Pt/C_3N_4 catalysts prepared by reductive (H_2) thermolysis of PtO_x/C_3N_4 precursors [30]. In that case, an increase in the specific surface area was a result of the following process. In a hydrogen atmosphere, Pt nanoparticles serve as catalytic centers for the hydrogenation of the C_3N_4 framework, resulting in its partial destruction and noticeable porosity growth. Obviously, this process is more pronounced for the samples with higher Pt loadings. The deposition of Pt in the case of Pt-Ir/g-C_3N_4 catalysts proceeded in the identical manner as for monometallic Pt/g-C_3N_4, and therefore the tendency of the BET surface in both cases is similar.

Table 1. Textural properties of the Pt-Ir/g-C_3N_4 catalysts.

Sample (Preparation Method)	S_{BET}, $m^2 \cdot g^{-1}$	V, $cm^3 \cdot g^{-1}$
g-C_3N_4	47.5	0.24
$Pt_{0.5}Ir_{0.1}/C_3N_4(1)$	97.2	0.37
$Pt_{0.5}Ir_{0.05}/C_3N_4(1)$	99.3	0.28
$Pt_{0.5}Ir_{0.01}/C_3N_4(1)$	101	0.36
$Pt_{0.1}Ir_{0.1}/C_3N_4(1)$	76.8	0.41
$Pt_{0.1}Ir_{0.5}/C_3N_4(1)$	86.4	0.45
$Pt_{0.5}Ir_{0.5}/C_3N_4(2)$	120	0.40

In contrast, the fraction of loaded Ir almost does not affect the specific surface area of the samples (Table 1). At the same time, a comparison with the mentioned monometallic Pt/C_3N_4 catalysts shows that the BET surface of the Pt-Ir/g-C_3N_4 catalysts is noticeably lower. For example, the surface area of monometallic $Pt_{0.5}$/g-C_3N_4 achieves 300 m^2/g, whereas for the bimetallic systems, it does not exceed 120 m^2/g. This can be due to the partial collapse and blocking of pores after deposition of Ir via impregnation. Note that the highest surface area (120 $m^2 \cdot g^{-1}$) is achieved for the sample $Pt_{0.5}Ir_{0.5}/C_3N_4(2)$, prepared by calcination in hydrogen at the final stage of the synthesis.

The BET surface of Pt-Ir/g-C_3N_4 catalysts can be particularly enhanced by their calcination in hydrogen after Ir deposition, but (as it has been shown previously and will be clear from the results below) metallic Ir formed in a reductive atmosphere does not possess catalytic activity toward water oxidation.

Diffuse reflectance spectra were obtained for the supported bimetallic photocatalysts with various compositions. Compared to unmodified g-C_3N_4, they show a characteristic increase in absorption in the region of 500–700 nm, due to the presence of deposited metal nanoparticles (Figure 1D). This effect is barely visible for catalysts with low metal content but clearly observed for the series $Pt_{0.5}Ir_y/C_3N_4(1)$. That said, even for the $Pt_{0.5}Ir_{0.5}/C_3N_4(1)$ catalyst, changes in the position of the band gap are negligible with respect to that for pristine g-C_3N_4 material, calculated from corresponding Tauc plots (Figure 1C). For bimetallic photocatalysts, the E_g value varies from 2.75 to 2.80 eV, whereas for pristine g-C_3N_4 this value amounts to 2.7 eV. This fact agrees with the IR data, and it emphasizes the intactness of the g-C_3N_4 material after all treatments involved in the preparation of catalysts. In Figure 1D, absorption in the range of 290–400 nm was observed, caused by the surface plasmon resonance effect of platinum and iridium nanoparticles. Most likely, the increase in the energy gap can be associated with the surface plasmon resonance effect inherent in platinum group metals. Previously, the authors [40] also wrote about an increase in the band gap when adding silver nanoparticles.

The surface properties of synthesized Pt-Ir/g-C_3N_4 photocatalysts were studied by XPS (Figure 2). Peaks corresponding to carbon bonded to nitrogen atoms in the g-C_3N_4 structure (288.1 eV) were identified in the C1s spectra (Figure S2). The N1s spectrum exhibits four peaks with binding energies at 398.6, 400.0, 401.0, and 404.5 eV (Figure S2). These peaks can be attributed to nitrogen atoms forming, respectively, a CN=C bond, an N-(C)3 bond, an N-H terminal group, and an excited π-bond. The main task was to find the charge state of metals in the bimetallic systems depending on the deposition method. Table 2 shows the surface concentration and oxidation state of the metals, calculated from the XPS data. As seen, the photocatalysts synthesized by Method 1 contain both metals coexisting in the form of metallic (Pt^0, Ir^0) and ionic (Pt^{2+} and Ir^{3+}) species.

The fraction of ionic platinum gradually grows for both series (with 0.1 and 0.5% Pt) with an increase in the iridium loading. This indicates that iridium species additionally promote the oxidation of platinum nanoparticles. For the series with 0.5% Pt, the percentage of metallic platinum varies from 50 to 68%, while for the series with 0.1% Pt, it varies from 24 to 60%. Interestingly, the calcination of the parent monometallic Pt/C_3N_4 material in air at 400 °C leads to the transformation of only a small part of metallic platinum into the 2+ charge state: in the starting material, there is 95% of Pt^0, while after calcination in air, the fraction of Pt^0 decreases to 87% (Figure S3). Therefore, it can be concluded that iridium species can promote the partial oxidation of Pt nanoparticles or isolated Pt atoms.

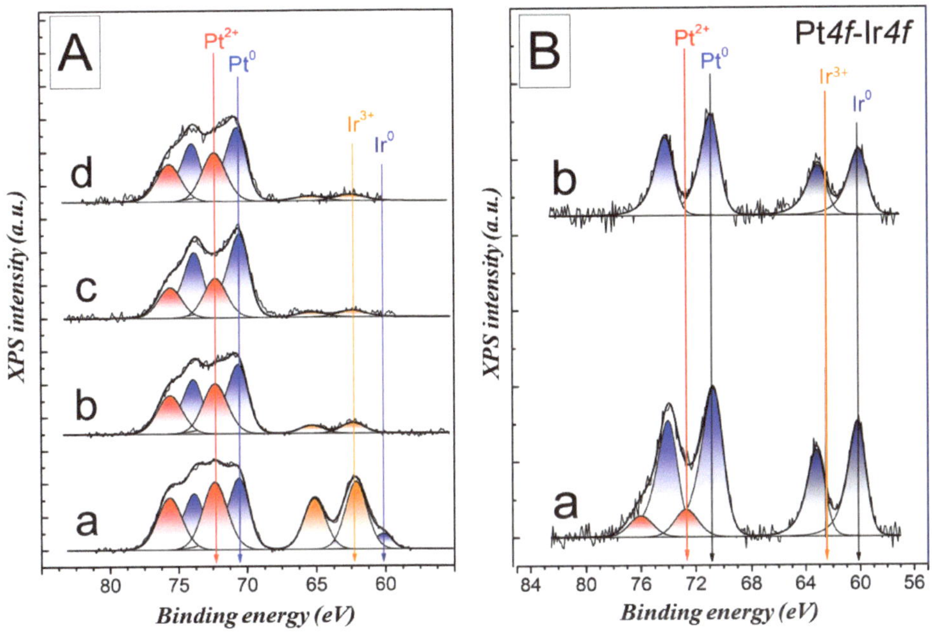

Figure 2. (**A**) Pt4f and Ir4f regions of the XPS spectra of the $Pt_{0.5}Ir_y/C_3N_4(1)$ series of catalysts (y = 0.5, 0.1, 0.05, and 0.01 for curves a, b, c, and d correspondingly); and (**B**) XPS spectra in the same region recorded for the catalysts $Pt_{0.5}Ir_{0.5}/C_3N_4(2)$ (a) and $Pt_{0.5}Ir_{0.5}/C_3N_4(3)$ (b).

Table 2. Ir and Pt concentrations and ratios in the catalyst as found by XPS.

#	Sample	[Pt]/[C]	%, Pt^0	[Ir]/[C]	%, Ir^0	[Pt]/[Ir]	$[Pt^0]/[Ir^{3+}]$
			Method 1				
1	$Pt_{0.1}Ir_{0.5}/C_3N_4(1)$	0.0003	24	0.0015	14	0.2	0.06
2	$Pt_{0.1}Ir_{0.1}/C_3N_4(1)$	0.0004	39	0.0003	13	1.5	0.6
3	$Pt_{0.1}Ir_{0.05}/C_3N_4(1)$	0.0003	53	0.0002	33	1.8	1.7
4	$Pt_{0.1}Ir_{0.01}/C_3N_4(1)$	0.0003	60	-	-	-	-
5	$Pt_{0.5}Ir_{0.5}/C_3N_4(1)$	0.0021	50	0.0012	17	1.7	1.1
6	$Pt_{0.5}Ir_{0.1}/C_3N_4(1)$	0.0018	57	0.0002	0	10.5	5.9
7	$Pt_{0.5}Ir_{0.05}/C_3N_4(1)$	0.0018	68	0.0001	0	14.2	9.7
8	$Pt_{0.5}Ir_{0.01}/C_3N_4(1)$	0.0018	60	-	-	-	-
			Method 2				
9	$Pt_{0.5}Ir_{0.5}/C_3N_4(2)$	0.0006	100	0.0005	100	1.23	-
			Method 3				
10	$Pt_{0.5}Ir_{0.5}/C_3N_4(3)$	0.0011	84	0.0006	100	1.16	-

In all samples prepared by Method 1, iridium is represented mainly by ionic species (Ir^{3+}) that can be ascribed to a highly dispersed oxide IrO_x. It should be noted that for the samples with 0.01 wt.% Ir, the metal is found at the noise level and its oxidation state cannot be identified. For the samples with 0.1% Pt, the surface [Pt]/[Ir] ratio agrees well with its loaded content, whereas in the case of 0.5% Pt, the surface [Pt]/[Ir] content is at least 1.5–2 times higher than the loaded ratio. It can be explained by the effect described earlier—the loss of mass of g-C_3N_4 during Pt deposition.

For the sample $Pt_{0.5}Ir_{0.5}/C_3N_4$(2), prepared by Method 2, which includes the final treatment in hydrogen at 400 °C, both platinum and iridium exist in the metallic state. Under such treatment, the alloying and combined agglomeration of Pt and Ir become more pronounced. For Method 3 (with the "reverse" deposition), the percentage of metallic Pt exceeds 80%, while iridium is reduced to the metallic state completely.

Figure 3 and Figure S4 represent HR-TEM and EDS elemental mapping images of the photocatalyst $Pt_{0.5}Ir_{0.5}/C_3N_4$(1). The data presented in Figure S4 and Table S1 confirm the presence of C and N elements with the ratio ca. 3:4 in the $Pt_{0.5}Ir_{0.5}/C_3N_4$(1) photocatalyst. The HR-TEM micrographs of the as-prepared catalysts show that platinum tends to aggregate into relatively large nanoparticles on the surface of g-C_3N_4. In contrast, ionic iridium species IrO_x are uniformly distributed over the surface of g-C_3N_4 and do not form distinct particles. This is consistent with the XRD data revealing that at high metal loadings, the reflexes of Pt become clear, while phases of iridium remain undetectable. On the other hand, the elemental EDS mapping and HR-TEM micrographs (Figure 3B–D) obtained at lower magnification reveal that the overall distribution of both metals over g-C_3N_4 particles is quite homogeneous. The interplanar distances (1.95 and 2.26 Å) (Figure 3A and Figure S5) observed for metallic nanoparticles are clearly due to the (200) and (111) indexes of the fcc phase of metallic Pt (1.962 and 2.265 Å distances are reported in the PDF-2 database (card #00-004-0802)) [41]. As seen from Figure S5, isolated (not aggregated) particles show a clear spherical geometry. Due to the proximity of the metallic Pt and Ir cell parameters, it is quite difficult to distinguish their alloys from individual particles. However, even for the $Ir_{0.5}Pt_{0.5}/C_3N_4$(1) sample, the EDS mapping shows that there is no noticeable agglomeration of iridium in the area of the observed metallic nanoparticles compared to the average Ir distribution. Nevertheless, partial alloying of Pt with Ir cannot be excluded.

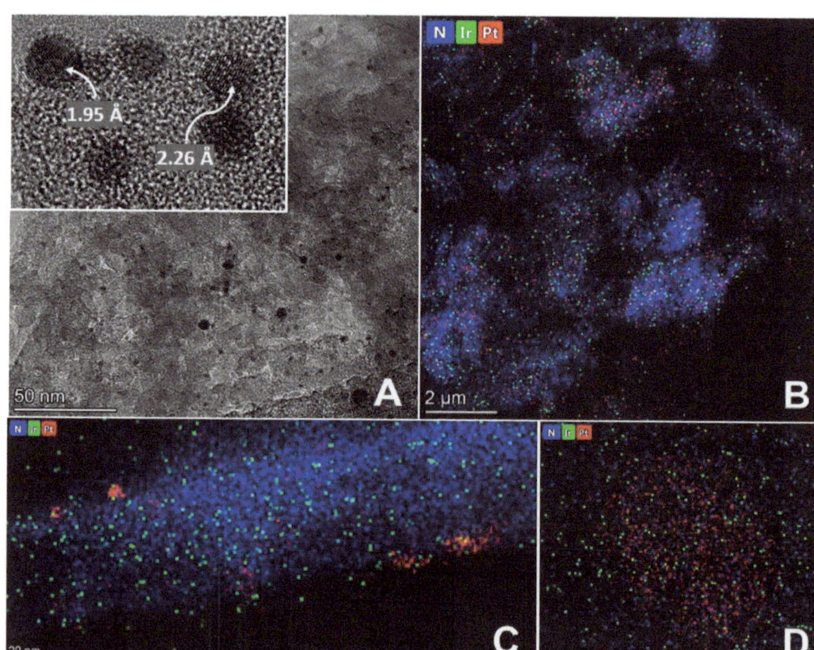

Figure 3. HR-TEM (**A**) and EDS (**B–D**) elemental mapping data depicting the metallic nanoparticles observed in the $Ir_{0.5}Pt_{0.5}$/g-C_3N_4(1) photocatalyst.

For comparison, Figure 4 shows TEM images with elemental mapping of the photocatalyst $Ir_{0.5}Pt_{0.5}/g\text{-}C_3N_4(2)$, which has the same metal content, but prepared by Method 2. For this sample, elemental mapping shows the formation of large bimetallic Pt-Ir particles up to 10 nm in size. This confirms the alloying of Pt and Ir under hydrogen treatment at 400 °C used in the synthesis by Method 2.

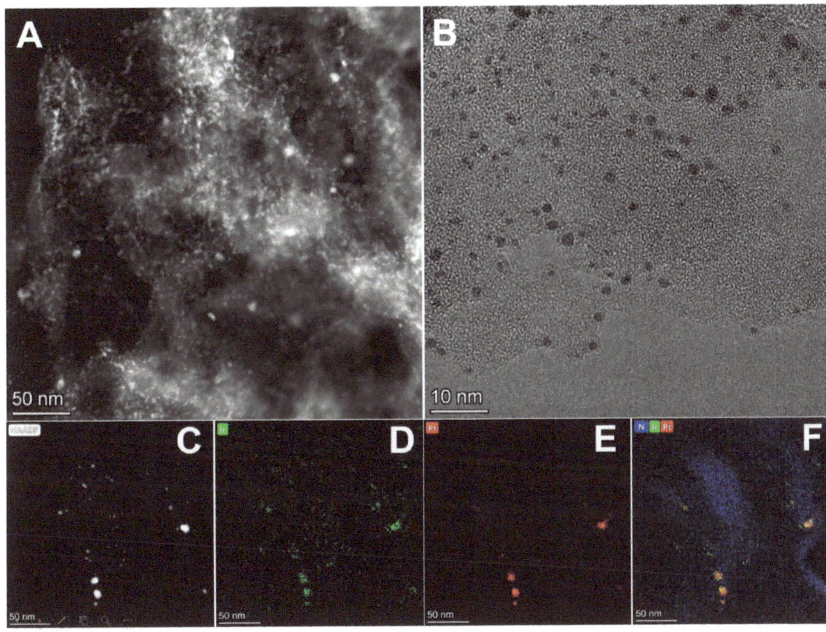

Figure 4. (**A**,**C**) HAADF, (**B**) HR-TEM images, and (**D**–**F**) EDS elemental mapping representing the metals distribution in $Ir_{0.5}Pt_{0.5}/g\text{-}C_3N_4(2)$ photocatalyst.

Based on the results above, the following conclusions can be drawn. In the catalysts prepared by Method 1, platinum exists in two states—metallic (up to 60%) and Pt^{2+}, while iridium is oxidized predominantly to the Ir^{3+} state. However, after Steps 1 and 2 of synthesis, platinum is completely reduced to the metallic state (this was shown earlier for $Pt/g\text{-}C_3N_4$ samples). This means that after Steps 3 and 4, platinum becomes partially oxidized. Moreover, it tends to aggregate into relatively large nanoparticles, while ionic iridium species IrO_x are evenly distributed over the surface of $g\text{-}C_3N_4$ and do not form distinct particles. In the case of Method 2, with treatment in hydrogen at Step 4, both metals are in the metallic state. In this case, the formation of large alloys of bimetallic Ir-Pt nanoparticles on the surface of $g\text{-}C_3N_4$ is observed. For Method 3, with the "reverse" deposition, most of the platinum (over 80%) is in the metallic state, while iridium is reduced to the metallic state completely.

2.3. Photocatalytic Tests

The results of kinetic experiments (Table 3) show that the formation of hydrogen from pure water occurs only with the use of the bimetallic photocatalysts prepared by Method 1. In contrast, the photocatalysts $Pt_{0.5}Ir_{0.5}/C_3N_4(2)$ and $Pt_{0.5}Ir_{0.5}/C_3N_4(3)$, as well as the reference samples $Pt_{0.5}/C_3N_4(1)$ and $Ir_{0.5}/C_3N_4(1)$, possess zero activity. The activity in hydrogen production from pure water was compared with the activity with the use of a TEOA solution as a substrate. It should be noted that when hydrogen is evolved from aqueous solutions of TEOA, another dependence is observed—for bimetallic photocatalysts $Pt_{0.5}Ir_{0.5}/C_3N_4(2)$ and $Pt_{0.5}Ir_{0.5}/C_3N_4(3)$, the activity is much higher than that of the sample

$Pt_{0.5}Ir_{0.5}/C_3N_4(1)$. Indeed, the activity in H_2 evolution from TEOA solutions is higher than for pure water by an order of magnitude.

Table 3. Activity of Pt_xIr_y/C_3N_4 photocatalysts in hydrogen production from ultrapure water and triethanolamine aqueous solution (10 vol.%). Conditions: V_{susp} = 50 mL (H_2O) or 100 mL (TEOA solution), m_{cat} = 25 mg, λ_{LED} = 400 nm.

#	Method	Metal Loading	Ultrapure H_2O			10% TEOA Solution	
			W(H_2), µmol min^{-1}	Activity, µmol H_2 h^{-1} $g_{cat}{}^{-1}$	Activity, µmol H_2O_2 h^{-1} $g_{cat}{}^{-1}$	W(H_2), µmol min^{-1}	Activity, µmol H_2 h^{-1} $g_{cat}{}^{-1}$
1		$Pt_{0.5}/C_3N_4(1)$ *	-	-	-	6.28	7540
2		$Pt_{0.5}/C_3N_4(1)$	-	-	-	4.85	5820
3		$Ir_{0.5}/C_3N_4(1)$	-	-	-	-	-
4		$Pt_{0.5}Ir_{0.5}/C_3N_4(1)$	0.023	55.2	29.4	1.06	1270
5		$Pt_{0.5}Ir_{0.1}/C_3N_4(1)$	0.023	55.2	44.1	0.46	552
6	1	$Pt_{0.5}Ir_{0.05}/C_3N_4(1)$	0.026	62.4	62.7	0.36	432
7		$Pt_{0.5}Ir_{0.01}/C_3N_4(1)$	0.043	103	16.8	0.46	552
8		$Pt_{0.1}Ir_{0.5}/C_3N_4(1)$	0.016	38.4	27.3	0.13	156
9		$Pt_{0.1}Ir_{0.1}/C_3N_4(1)$	0.019	45.6	35.7	0.06	72
10		$Pt_{0.1}Ir_{0.05}/C_3N_4(1)$	0.026	62.4	39.9	0.37	444
11		$Pt_{0.1}Ir_{0.01}/C_3N_4(1)$	0.036	86.4	71.4	0.45	540
12		$Pt_{0.1}/C_3N_4(1)$	-	-	-	2.79	3350
13	2	$Pt_{0.5}Ir_{0.5}/C_3N_4(2)$	-	-	-	7.18	8620
14	3	$Pt_{0.5}Ir_{0.5}/C_3N_4(3)$	-	-	-	2.74	3290

* without air treatment at 350 °C.

For comparison, the activity of two monometallic photocatalysts $Pt_{0.5}/C_3N_4(1)$ was studied. The first of the reference samples was prepared via Steps 1 and 2, while the second one was additionally treated in air at 350 °C, which essentially imitated the preparation of bimetallic samples by Method 1. It was shown that the activity of the first reference photocatalyst ($Pt_{0.5}/C_3N_4(1)$ *, 100% Pt in the metallic state) is higher than the activity of the second one ($Pt_{0.5}/C_3N_4(1)$; 11% Pt in the metallic state and 89% Pt in the Pt^{2+} state) in the hydrogen production from an aqueous solution of TEOA, 7.5 and 5.8 mmol h^{-1} $g_{cat}{}^{-1}$, respectively. In addition, their activity in the production of hydrogen from aqueous solutions of TEOA significantly exceeds the activity of all bimetallic samples prepared by Method 1. Surprisingly, the highest activity in this reaction, 8.6 mmol h^{-1} $g_{cat}{}^{-1}$, was shown by the bimetallic photocatalyst $Pt_{0.5}Ir_{0.5}/C_3N_4(2)$, in which both platinum and iridium are reduced into the metallic state, which can be explained by the high content of metallic cocatalysts in the necessary oxidation state.

Thus, photocatalysts with platinum in the metallic state exhibit the highest activity in the production of hydrogen from aqueous solutions of an organic electron donor. The transition of platinum to the Pt^{2+} state or the deposition of iridium oxides suppresses the photocatalytic activity in this reaction. Regarding the photocatalytic evolution of hydrogen from pure water (without the addition of electron donors), the catalytic effect of platinum is provided by two charge states—Pt^0 and Pt^{2+}, but it is necessary that most of the iridium be in the oxidized form.

The next step of the work was to study the effect of the content of iridium and platinum in the samples prepared by Method 1 on the activity in the evolution of hydrogen from pure water. Typical kinetics of hydrogen evolution over bimetallic photocatalysts prepared by Method 1 are shown in Figure 5A,C. The net activities are represented in Table 3 and Figure 5B,D. One can see that the kinetics of hydrogen formation for all the catalysts are linear over 2 h. At that, among photocatalysts $Pt_{0.5}Ir_y/C_3N_4(1)$ and $Pt_{0.1}Ir_y/C_3N_4(1)$, the maximum rate of hydrogen evolution is about the same, which certainly makes photocatalysts with 0.1% Pt more cost-efficient.

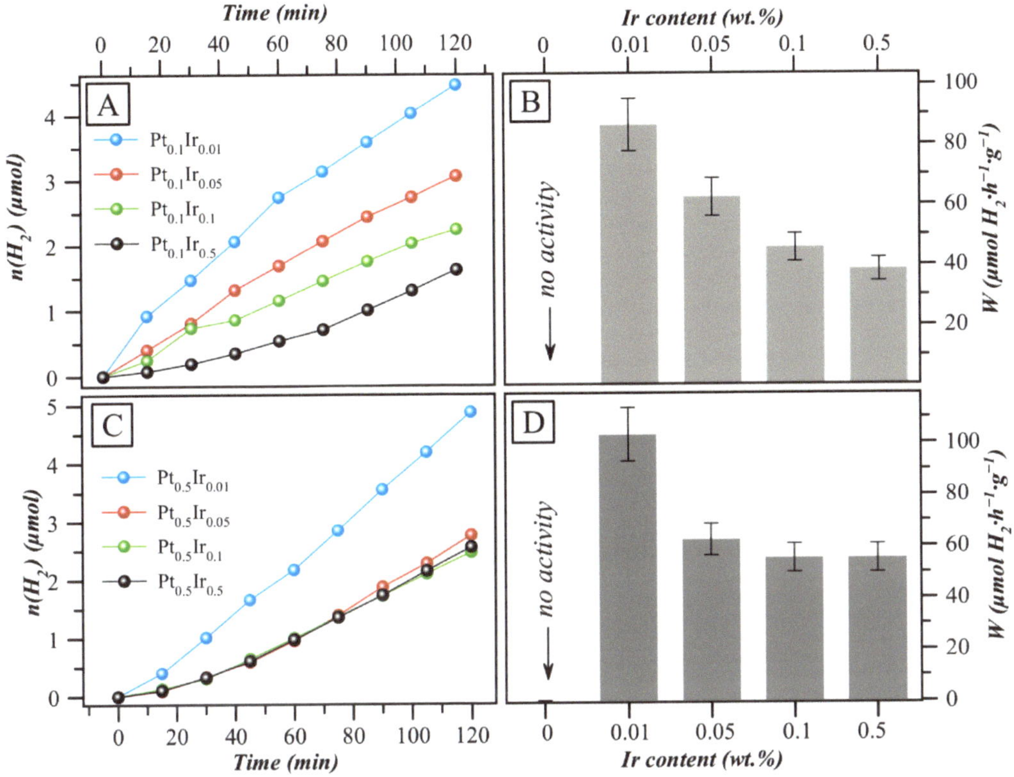

Figure 5. (**A**,**C**) Kinetic traces for the H_2 generation from ultrapure water by $Pt_{0.5}Ir_y/C_3N_4(1)$ and $Pt_{0.1}Ir_y/C_3N_4(1)$ photocatalysts (y = 0.01, 0.05, 0.1 and 0.5) correspondingly under irradiation with 400 nm LED; and (**B**,**D**) represent net H_2 evolution rates calculated from the kinetic traces (**A**,**B**).

In addition, it is interesting that at a fixed content of platinum, the activity increases with a decrease in the fraction of iridium. Thus, the photocatalysts $Pt_{0.5}Ir_{0.01}/C_3N_4(1)$ and $Pt_{0.1}Ir_{0.01}/C_3N_4(1)$ show the highest activity. In other words, for the splitting of water, it is necessary that platinum content be at least 20 times greater than iridium.

According to our previous results, the rate of hydrogen production in the presence of TEOA over the best 0.5% Pt/g-C_3N_4 photocatalysts is at the level of 10 mmol h^{-1} $g_{cat}{}^{-1}$, while the rate of oxygen production from $NaIO_4$ solutions for 0.5% IrO_x/g-C_3N_4 photocatalysts achieves 80 mmol h^{-1} $g_{cat}{}^{-1}$ [30,32]. In addition, as seen from Table 2, the proportion of metallic platinum [Pt^0]/[Pt] is less than the proportion of oxidized iridium [IrO_x]/[Ir], i.e., the [Pt^0]/[Ir^{3+}] surface ratio is less than the ratio [Pt]/[Ir]. Accordingly, it can be assumed that iridium oxide catalyzes the formation of oxygen more intensely than metallic platinum catalyzes the formation of hydrogen, which is why active photocatalysts require a large loading of platinum compared to iridium. To the best of our knowledge, this effect has not been published before [6,42].

It should be noted that oxygen was not detected during the reaction, which suggests the presence of other products of water oxidation. It seems most logical that the product of water oxidation is hydrogen peroxide, as has been noted by many researchers [16,20,24,43]. Therefore, we analyzed the reaction product filtrate for the presence of hydrogen peroxide by tests with horseradish peroxidase. Table 3 represents the activities in H_2O_2 production for $Pt_{0.5}Ir_y/C_3N_4(1)$ and $Pt_{0.1}Ir_y/C_3N_4(1)$ series. For the $Pt_{0.1}Ir_y/C_3N_4(1)$ series, the activity and therefore the amount of evolved H_2O_2 is at the level of 65–80% of the stoichiometric amount according to Equation (2). For the samples ranging from $Pt_{0.5}Ir_{0.05}/C_3N_4(1)$ to $Pt_{0.5}Ir_{0.5}/C_3N_4(1)$, approximately the same relationship is observed. Only the point on the concentration of hydrogen peroxide for the photocatalyst $Pt_{0.5}Ir_{0.01}/C_3N_4(1)$ is an outlier. Incomplete compliance with stoichiometry can be explained by the decomposition of hydrogen peroxide during sample preparation, in particular, the separation of the photocatalyst from the suspension.

An analysis of published data (Table S2 [3,6,8,13,44–47] and Table S3 [18,20,24,48–55]) shows that the obtained activities, ca. 100 μmol h^{-1} $g_{cat}{}^{-1}$, are among the highest for such systems. For example, a photocatalyst 1% Pt/3% Ir/g-C_3N_4 shows activity at the same level of 100 μmol h^{-1} $g_{cat}{}^{-1}$, but at a much higher mass fraction of both noble metals [6]. In general, according to Table S2, for the water splitting with the formation of H_2 and H_2O_2, photocatalysts based on single-phase g-C_3N_4 doped with metal cocatalysts are used quite rarely. Moreover, in this case, a discrepancy between the stoichiometry of hydrogen and peroxide is often observed [20,24]. In addition to activity, the stability of photocatalysts is one of the most important aspects. The stability of our photocatalysts was tested in four successive 1.5 h runs of hydrogen production from ultrapure water (Figure 6B). After each run, the reaction suspension was purged with argon to remove gaseous reaction products. One can see that the hydrogen evolution rate remains constant during the runs. The XPS study of the photocatalysts before and after photocatalytic tests (Figure S6) shows that both the surface [Pt]/[C] ratio and the percentage of Pt^0 also remain almost unchanged. These results are fundamentally different from the data obtained for Pt/g-C_3N_4 photocatalysts in TEOA solutions, since in that case, the photocatalyst undergoes noticeable deactivation due to the adsorption of oxidation intermediates on its surface [30]. Thus, hydrogen production from pure water seems more promising not only because it is unnecessary to add electron donors but also because of the higher stability of photocatalysts.

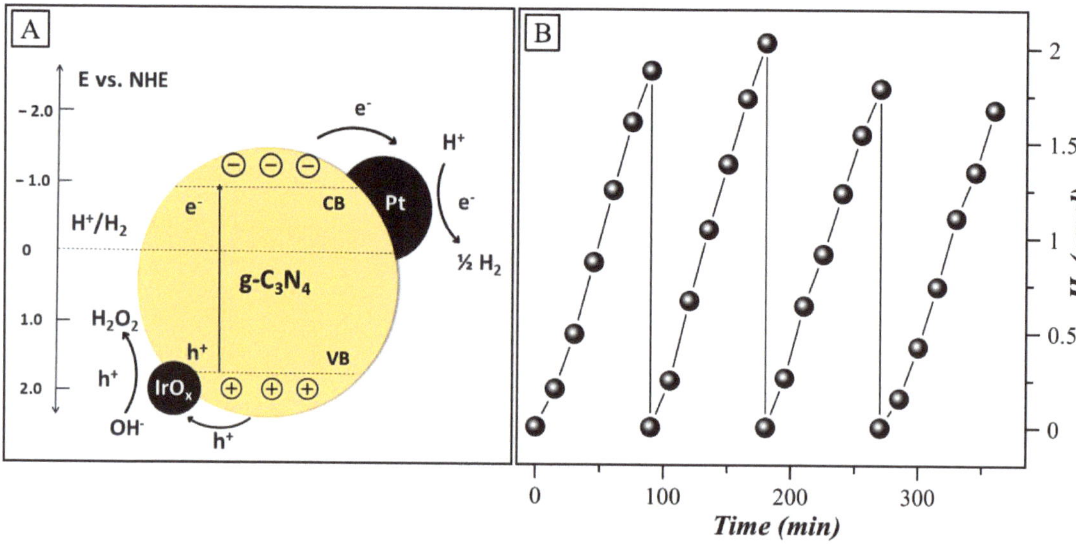

Figure 6. (**A**) Diagram for the proposed mechanism of the H_2 and H_2O_2 generation from water by a visible-light irradiated Pt_xIr_y/C_3N_4 photocatalysts; and (**B**) the cyclic tests of the H_2 generation from the ultrapure water with the $Pt_{0.1}Ir_y/C_3N_4$ photocatalyst.

A probable mechanism of water splitting over a bimetallic photocatalyst is represented in Figure 6A. For efficient water splitting into H_2 and H_2O_2, the conduction band (CB) level should be more negative than the reduction potential of H^+ to H_2 (0 vs. NHE). In addition, the valence band (VB) should be more positive than the oxidation potential of H_2O to H_2O_2

(1.78 V) [9,56]. Therefore, g-C_3N_4, with the bottom of its CB at -0.8 V and the top of its VB at 1.9 V, meets the requirements for a water splitting photocatalyst [56]. Irradiation at 400 nm excites an electron from the VB of g-C_3N_4 to its CB. Most likely, photogenerated electrons and holes migrate to particles of metallic platinum and iridium oxide, respectively, where the formation of molecular hydrogen and hydrogen peroxide takes place.

3. Materials and Methods

3.1. *Starting Reagents and Physical Measurements*

All reagents were purchased from commercial suppliers and were used as received. $(NH_4)_3[Ir(NO_2)_6]$ was synthesized from hydrated iridium trichloride according to the procedure presented in [32]. The obtained $(NH_4)_3[Ir(NO_2)_6]$ was then characterized by XRD. $(Me_4N)_2[Pt_2(\mu\text{-}OH)_2(NO_3)_8]$ was obtained from $[Pt(H_2O)_2(OH)_4]$ according to the procedure described in [57]. The purity of $(Me_4N)_2[Pt_2(\mu\text{-}OH)_2(NO_3)_8]$ was confirmed by XRD and NMR ^{195}Pt. Graphitic g-C_3N_4 was obtained by thermal condensation of a supramolecular complex consisting of melamine and cyanuric acid, according to the method presented in [30]. A solution of *fac*-$[Ir(H_2O)_3(NO_2)_3]$ (0.15M) was prepared from $(NH_4)_3[Ir(NO_2)_6]$ by its hydrothermal treatment in a 0.3M solution of HNO_3 [32]. Ultrapure water was produced using a Milli-Q system (total org. carbon < 5 ppb (μg/L)).

3.2. *Preparation of Pt-Ir/g-C_3N_4 Catalysts*

The catalysts were synthesized using three methods.

Method 1 (Pt-Ir/C_3N_4(1)) consisted of the following steps.

1. Deposition of Pt on g-C_3N_4. To prepare catalysts with 0.5 or 0.1 wt.% of Pt, an appropriate aliquot (18.0 or 3.60 mL, respectively) of a $(Me_4N)_2[Pt_2(\mu\text{-}OH)_2(NO_3)_8]$ acetone solution (1.82 mM) was added to a suspension of *g*-C_3N_4 (2500 mg) in acetone (40 mL). The resulting suspension was stirred for 12 h at room temperature in a closed vial. The completion of sorption of platinum was checked by the absence of light absorption at 380 nm. The solid was collected by filtration, washed with a copious amount of acetone, and dried in an airflow at room temperature for 20 min.
2. Calcination in H_2. The resulting material was calcined in hydrogen (400 °C, 10 °C/min ramping, 1 h exposure time) to give a Pt(X)/g-C_3N_4 precursor. Here, X is the loading of Pt (0.5 or 0.1).
3. Deposition of Ir. The precursor (500.0 mg) was dispersed in 6 mL of acetone, and then an appropriate aliquot (Table 4) of a *fac*-$[Ir(H_2O)_3(NO_2)_3]$ solution ([Ir] = 4.00 mM) was added. The prepared suspension was thoroughly mixed using ultrasonication (10 min) and then dried in an airflow (200 °C) until complete removal of the solvent.
4. Calcination in air. The obtained powder was calcined in an airflow at 350 °C for 1 h.

Table 4. Volumes of *fac*-$[Ir(H_2O)_3(NO_2)_3]$ solution used for the preparation of catalysts.

Ir wt.% in a Final Catalyst	0.5	0.1	0.05	0.01
V_{Ir} (μL)	3266	653	327	65

The resulting materials were designated as $Pt_xIr_y/C_3N_4(1)$. Here and below, X and Y are nominal loadings of Pt and Ir, respectively. The actual loadings were determined by ICP-AES.

For comparison reasons, monometallic $Pt_x/C_3N_4(1)$ and $Ir_y/C_3N_4(1)$ catalysts were also synthesized. These materials were prepared from the pristine g-C_3N_4 using Method 1, excluding Step 3 (for the preparation of $Pt_x/C_3N_4(1)$) or Steps 1–2 (for the preparation of $Ir_y/C_3N_4(1)$), respectively.

Method 2 (Pt-Ir/C_3N_4(2)). Here, the same protocol was used, except that Step 4 was calcination in a hydrogen flow at 400 °C. The materials obtained by this method were designated as $Pt_xIr_y/C_3N_4(2)$.

Method 3 (Pt-Ir/C_3N_4(3)). In this protocol, the components were deposited in the reverse order: deposition of Ir on g-C_3N_4, calcination in air, deposition of Pt, and calcination in H_2. The materials obtained in this case were designated as Pt_xIr_y/C_3N_4(3).

3.3. Apparatus

XRD. X-ray powder diffraction analysis of the polycrystalline samples was carried out on a DRON-RM4 diffractometer (Bourevestnik JSC, Saint-Petersburg, Russia, Cu Kα radiation, graphite monochromator in the reflected beam, scintillation detector with amplitude discrimination). The samples were prepared by deposition of a suspension in hexane on the polished side of a cell made of fused quartz. A sample of polycrystalline silicon (a = 5.4309 Å), prepared similarly, was used as an external standard. The lattice parameters were refined by the full-profile technique using the TOPAS-Academic program, Version 6 [58].

HR-TEM. The structure and microstructure of the photocatalysts were studied by HRTEM using a ThemisZ electron microscope (Thermo Fisher Scientific, Waltham, MA, USA) operated at an accelerating voltage of 200 kV. The microscope was equipped with a corrector of spherical aberrations, which provided a maximum lattice resolution of 0.06 nm, and a SuperX spectrometer (TFS, Plano, TX, USA). Images were recorded using a Ceta 16 CCD sensor (Thermo Fisher Scientific, USA). For electron microscopy studies, samples were deposited on perforated carbon substrates attached to aluminum grids using an ultrasonic dispersant.

UV–vis and DRS. Electron absorbance spectra of the solutions were analyzed with a PG Instruments T60 UV–vis single-beam spectrophotometer. Diffuse reflectance spectroscopy was carried out with a Shimadzu UV–VIS–NIR spectrometer UV-3101 PC (Kyoto, Japan) using $BaSO_4$ as a reference. The reflectance spectra were recorded in the range of 240–800 nm and presented using the Kubelka–Munk function (K-M).

IR. Infrared spectra were recorded in the range of 400–4000 cm^{-1} on a Scimitar FTS 2000 (Varian, Palo Alto, CA, USA) apparatus using tablets of KBr.

XPS. The chemical composition of the samples was studied by the XPS technique on electronic spectrometers SPECS Surface NanoAnalysis GmbH and FLEXPS (Berlin, Germany). The spectrometers were equipped with a hemispherical Phoibos 150 MCD-9 analyzer and Phoibos 150 analyzer with a DLD electron detector, respectively. The spectra were recorded using non-monochromatized AlKα radiation ($h\nu$ = 1486.61 eV). The background was subtracted from the spectra using the Shirley method [59].

ICP-AES. Elemental analysis was carried out for several samples with the ICP-AES technique with the use of an iCAP-6500 high-resolution spectrometer (Thermo Scientific, Waltham, MA, USA, (ICP-AES)) with a cyclone-type spray chamber and a SeaSpray nebulizer. Samples were dissolved completely in a mixture of sulfuric, nitric, and hydrochloric acid prior to analysis.

3.4. Photocatalytic Activity Tests

The synthesized photocatalysts were tested in water splitting. Ordinary distilled water usually contains dissolved organic impurities that can act as electron donors. Therefore, to avoid their influence, we used ultrapure water. The experiments were carried out using a photocatalytic reactor (Figure S7) described in detail elsewhere [32]. First, 25.0 mg of the photocatalyst was suspended in 50 mL of water. After that, the suspension was then purged with argon for 15 min to remove oxygen from the reactor. The reactor was then irradiated with an LED with a wavelength of 400 nm (Figure S7) with parameters (30 V, 1 A; 19 $mW \cdot cm^{-2}$). The hydrogen concentration was recorded on a gas chromatograph "Khromos GC-1000" (manufactured by "Khromos", Dzerzhinsk, Russia), using a zeolite column and a thermal conductivity detector. Argon was used as a carrier gas. The power density of the LED light source was measured using a ThorLabs Optical Power Monitor (PM16-401).

The amount of hydrogen peroxide formed during the photocatalytic process was measured by UV–vis spectroscopy (UV-250 spectrophotometer (Shimadzu, Kyoto, Japan)) using the reaction of H_2O_2 with horseradish peroxidase (HRP) (CAS 9003-99-0, "Dia-m", Moscow, Russia) and 3,3′,5,5′-tetramethylbenzidine (TMB) (CAS 54827-17-7, NIOPIK, Moscow, Russia) [60]. An acetate–acetic buffer solution with pH 4 was prepared to create the appropriate environment for hydrogen peroxide determination while preventing the overoxidation of TMB. After each water-splitting experiment, a 30 mL aliquot of the suspension was centrifuged for 10 min. The volume of the decanted solution was determined in a measuring cylinder. Subsequently, the solution was transferred to a chemical flask, and 30 mL of buffer solution was added to achieve suitable pH. Additionally, 0.3 mL of a TMB solution in dimethyl sulfoxide (DMSO) and 20 µL of an HRP solution were added. The resulting blue-colored solution was subjected to spectrophotometric analysis using a 1 cm cuvette. The maximum absorption was observed at 652 nm, at which the charge transfer complex, which is formed during partial oxidation with TMB peroxide, has a maximum emission. The concentration of hydrogen peroxide was determined using the Buger–Lambert–Beer equation.

In a separate set of experiments, we studied the activity of synthesized photocatalysts in the production of hydrogen from aqueous solutions of triethanolamine (TEOA, 10 vol.%).

4. Conclusions

Thus, this work proposes a new strategy for the synthesis of bimetallic photocatalysts for water splitting into H_2 and H_2O_2: the synthesis of $Pt/IrO_x/g\text{-}C_3N_4$ photocatalysts with low content of both noble metals. In such systems, platinum acts as a cocatalyst for the reduction of water with the formation of hydrogen, and iridium oxide acts as a cocatalyst for oxidation with the formation of peroxide. A distinctive feature of the proposed synthesis method is the combination of two elements: the preparation of $g\text{-}C_3N_4$ from melamine cyanurate and the successive deposition of platinum and iridium oxides from their labile complexes. By regulating the order of precursor deposition and the conditions of subsequent thermal treatment, one can directly manage the state of both metals in the resulting materials. From our experiments, we have concluded that the catalytic activity of platinum in the generation of H_2 is provided by two charge states—Pt^0 and Pt^{2+}. However, for the efficient production of H_2O_2, most of the iridium must be oxidized— photocatalysts with iridium in the metallic state are completely inactive in the decomposition of H_2O. To synthesize such materials, platinum was first deposited from a solution of $(Me_4N)_2[Pt_2(OH)_2(NO_3)_8]$ and then treated with H_2 at 400 °C. After that, the resulting Pt/C_3N_4 material was impregnated with an acetone solution of ($[Ir(H_2O)_3(NO_2)_3]$) and calcined in air at 350 °C.

It was found that two tested series of photocatalysts—$Pt_{0.5}Ir_y/C_3N_4$ and $Pt_{0.1}Ir_y/C_3N_4$—show approximately the same maximum rate of hydrogen evolution, which certainly makes photocatalysts with 0.1% Pt more promising. At a fixed amount of platinum, the catalyst activity increases with a decrease in the amount of iridium, and for the splitting of water, the platinum content must exceed the iridium content at least 20-fold. The literature data show that the obtained activities ca. 100 µmol h^{-1} $g_{cat}{}^{-1}$ are among the highest achieved for H_2 production without the use of electron donors over photocatalysts based on $g\text{-}C_3N_4$. The proposed synthetic approaches look very promising: they require very low contents of noble metals (0.1 wt.% Pt; 0.01 wt.% Ir), and the "support" they use is available and non-toxic graphite-like carbon nitride without any modifications (doping with non-metals, creating heterostructures, etc.).

Supplementary Materials: The following supporting information can be downloaded at: https://www.mdpi.com/article/10.3390/catal14040225/s1, Figure S1: Correlation between loaded Ir:Pt ratios and the same value determined with ICP AES chemical analysis in the as-prepared $Ir_xPt_{0.1}/g\text{-}C_3N_4(1)$ catalysts (x = 0.5, 0.1, 0.01); Figure S2: C1s and N1s core-level spectra of photocatalysts. The N1s spectra are normalized to the integral intensity of the C1s peaks corresponding to the spectrum of $g\text{-}C_3N_4$; Figure S3: XPS spectra of Pt4f of the photocatalysts $Pt_{0.5}/g\text{-}C_3N_4$ after treatment and hydrogen and consecutive treatment in air. Spectra are normalized to the integral peak intensity C1s corresponding to the carrier spectrum (g-C_3N_4); Figure S4: HR-TEM of the $Ir_{0.5}Pt_{0.5}$/g-C_3N_4(1) photocatalyst, EDS mapping of the C and N elements in the $Ir_{0.5}Pt_{0.5}$/g-C_3N_4(1) photocatalyst and spectrum from EDS mapping; Figure S5: Analysis of the interplanar distances observed in HR-TEM micrographs of Pt nanoparticles in the $Ir_{0.5}Pt_{0.5}$/g-C_3N_4(1) photocatalyst; Figure S6: XPS spectra of Pt4f and Ir4f of the spent and fresh $Ir_{0.01}Pt_{0.1}$/g-C_3N_4(1) photocatalyst. Spectra are normalized to the integral peak intensity C1s corresponding to the carrier spectrum (g-C_3N_4); Figure S7: (a) The scheme of photocatalytic reactor utilized in this work and (b) the spectrum of 400 nm LED used for irradiation of catalysts suspensions; Table S1: The percentage composition of all elements calculated from spectrum from EDS mapping; Table S2: Data on the photocatalytic hydrogen and oxygen evolution from water without the addition of electron donors; Table S3: Data on the photocatalytic hydrogen and peroxide evolution from pure water presented in the literature.

Author Contributions: N.D.S.: conceptualization, validation, methodology, investigation; P.A.T.: investigation, visualization; A.A.S.: validation, formal analysis, investigation; E.Y.G.: data curation, visualization; A.V.Z.: formal analysis, investigation and data curation; D.B.V.: conceptualization, investigation, visualization and writing—original draft preparation; E.A.K.: conceptualization, visualization, resources, writing—original draft preparation, supervision, project administration, funding acquisition. All authors have read and agreed to the published version of the manuscript.

Funding: This work was funded by the Russian Science Foundation (Grant #21-13-00314; https://rscf.ru/en/project/21-13-00314/).

Data Availability Statement: The data presented in this study are available on request from the corresponding author.

Acknowledgments: The TEM and XPS study was carried out using the facility of the shared research center "National center of investigation of catalysts" at Boreskov Institute of Catalysis. The authors deeply thank the XRD Facility of NIIC SB RAS for the FTIR data collection, P.E. Plyusnin for thermogravimetric analysis, and S.V. Cherepanova for XRD analysis.

Conflicts of Interest: The authors declare that they have no known competing financial interests or personal relationships that could have appeared to influence the work reported in this paper.

References

1. Tian, N.; Zhang, Y.; Li, X.; Xiao, K.; Du, X.; Dong, F.; Waterhouse, G.I.N.; Zhang, T.; Huang, H. Precursor-Reforming Protocol to 3D Mesoporous g-C_3N_4 Established by Ultrathin Self-Doped Nanosheets for Superior Hydrogen Evolution. *Nano Energy* **2017**, *38*, 72–81. [CrossRef]
2. Zhang, G.; Lan, Z.-A.; Wang, X. Conjugated Polymers: Catalysts for Photocatalytic Hydrogen Evolution. *Angew. Chem. Int. Ed.* **2016**, *55*, 15712–15727. [CrossRef] [PubMed]
3. Mo, Z.; Xu, H.; Chen, Z.; She, X.; Song, Y.; Lian, J.; Zhu, X.; Yan, P.; Lei, Y.; Yuan, S.; et al. Construction of MnO_2/Monolayer g-C_3N_4 with Mn Vacancies for Z-Scheme Overall Water Splitting. *Appl. Catal. B Environ.* **2019**, *241*, 452–460. [CrossRef]
4. Zheng, D.; Cao, X.-N.; Wang, X. Precise Formation of a Hollow Carbon Nitride Structure with a Janus Surface to Promote Water Splitting by Photoredox Catalysis. *Angew. Chem. Int. Ed.* **2016**, *55*, 11512–11516. [CrossRef] [PubMed]
5. Fujishima, A.; Honda, K. Electrochemical Photolysis of Water at a Semiconductor Electrode One and Two-Dimensional Structure of Alpha-Helix and Beta-Sheet Forms of Poly (L-Alanine) Shown by Specific Heat Measurements at Low Temperatures (1.5–20 K). *Nature* **1972**, *238*, 37–38. [CrossRef] [PubMed]
6. Chen, X.; Shi, R.; Chen, Q.; Zhang, Z.; Jiang, W.; Zhu, Y.; Zhang, T. Three-Dimensional Porous g-C_3N_4 for Highly Efficient Photocatalytic Overall Water Splitting. *Nano Energy* **2019**, *59*, 644–650. [CrossRef]
7. Navakoteswara Rao, V.; Lakshmana Reddy, N.; Preethi, V.; Karthik, M.; Yu, Y.T.; Yang, J.M.; Mamatha Kumari, M.; Shankar, M.V. A Critical Review on Core/Shell-Based Nanostructured Photocatalysts for Improved Hydrogen Generation. *Int. J. Hydrogen Energy* **2023**, *48*, 11754–11774. [CrossRef]
8. Liu, Y.; Li, X.; He, H.; Yang, S.; Jia, G.; Liu, S. CoP Imbedded G-C_3N_4 Heterojunctions for Highly Efficient Photo, Electro and Photoelectrochemical Water Splitting. *J. Colloid Interface Sci.* **2021**, *599*, 23–33. [CrossRef] [PubMed]
9. Kozlova, E.A.; Parmon, V.N. Heterogeneous Semiconductor Photocatalysts for Hydrogen Production from Aqueous Solutions of Electron Donors. *Russ. Chem. Rev.* **2017**, *86*, 870–906. [CrossRef]
10. Bharagav, U.; Ramesh Reddy, N.; Nava Koteswara Rao, V.; Ravi, P.; Sathish, M.; Rangappa, D.; Prathap, K.; Shilpa Chakra, C.; Shankar, M.V.; Appels, L.; et al. Bifunctional G-C_3N_4/Carbon Nanotubes/WO_3 Ternary Nanohybrids for Photocatalytic Energy and Environmental Applications. *Chemosphere* **2023**, *311*, 137030. [CrossRef]
11. Hainer, A.S.; Hodgins, J.S.; Sandre, V.; Vallieres, M.; Lanterna, A.E.; Scaiano, J.C. Photocatalytic Hydrogen Generation Using Metal-Decorated TiO_2: Sacrificial Donors vs. True Water Splitting. *ACS Energy Lett.* **2018**, *3*, 542–545. [CrossRef]

12. Solakidou, M.; Giannakas, A.; Georgiou, Y.; Boukos, N.; Louloudi, M.; Deligiannakis, Y. Efficient Photocatalytic Water-Splitting Performance by Ternary CdS/Pt-N-TiO_2 and CdS/Pt-N,F-TiO_2: Interplay between CdS Photo Corrosion and TiO_2-Dopping. *Appl. Catal. B Environ.* **2019**, *254*, 194–205. [CrossRef]
13. He, H.; Cao, J.; Guo, M.; Lin, H.; Zhang, J.; Chen, Y.; Chen, S. Distinctive Ternary CdS/Ni_2P/g-C_3N_4 Composite for Overall Water Splitting: Ni_2P Accelerating Separation of Photocarriers. *Appl. Catal. B Environ.* **2019**, *249*, 246–256. [CrossRef]
14. Yan, J.; Wu, H.; Chen, H.; Zhang, Y.; Zhang, F.; Liu, S.F. Fabrication of TiO_2/C_3N_4 Heterostructure for Enhanced Photocatalytic Z-Scheme Overall Water Splitting. *Appl. Catal. B Environ.* **2016**, *191*, 130–137. [CrossRef]
15. Yu, J.; Wu, S.; Zhao, X.; Li, Z.; Yang, X.; Shen, Q.; Lu, M.; Xie, X.; Zhan, D.; Yan, J. Progress on Two-Dimensional Transitional Metal Dichalcogenides Alloy Materials: Growth, Characterisation, and Optoelectronic Applications. *Nanomaterials* **2023**, *13*, 2843. [CrossRef] [PubMed]
16. Xue, F.; Si, Y.; Wang, M.; Liu, M.; Guo, L. Toward Efficient Photocatalytic Pure Water Splitting for Simultaneous H_2 and H_2O_2 Production. *Nano Energy* **2019**, *62*, 823–831. [CrossRef]
17. Wang, Z.; Inoue, Y.; Hisatomi, T.; Ishikawa, R.; Wang, Q.; Takata, T.; Chen, S.; Shibata, N.; Ikuhara, Y.; Domen, K. Overall Water Splitting by Ta_3N_5 Nanorod Single Crystals Grown on the Edges of $KTaO_3$ Particles. *Nat. Catal.* **2018**, *1*, 756–763. [CrossRef]
18. Zhu, M.; Zhu, C.; Wu, D.; Wang, X.; Wang, H.; Gao, J.; Huang, H.; Shi, C.; Liu, Y.; Kang, Z. Efficient Photocatalytic Water Splitting through Titanium Silicalite Stabilized CoO Nanodots. *Nanoscale* **2019**, *11*, 15984–15990. [CrossRef]
19. Xiang, Q.; Yu, J.; Jaroniec, M. Preparation and Enhanced Visible-Light Photocatalytic H 2 -Production Activity of Graphene/C_3N_4 Composites. *J. Phys. Chem. C* **2011**, *115*, 7355–7363. [CrossRef]
20. Hu, S.; Sun, X.; Zhao, Y.; Li, W.; Wang, H.; Wu, G. The Effective Photocatalytic Water Splitting to Simultaneously Produce H_2 and H_2O_2 over Pt Loaded K-g-C_3N_4 Catalyst. *J. Taiwan Inst. Chem. Eng.* **2020**, *107*, 129–138. [CrossRef]
21. Lu, W.; Yi, Z.; Zhang, J.; Xu, X.; Tang, B.; Li, G.; Zeng, L.; Chen, J.; Sun, T. A Tunable Broadband Absorber in the Terahertz Band Based on the Proportional Structure of a Single Layer of Graphene. *Diam. Relat. Mater.* **2023**, *140*, 110481. [CrossRef]
22. Zhou, S.; Bi, K.; Li, Q.; Mei, L.; Niu, Y.; Fu, W.; Han, S.; Zhang, S.; Mu, J.; Tan, L.; et al. Patterned Graphene-Based Metamaterials for Terahertz Wave Absorption. *Coatings* **2023**, *13*, 59. [CrossRef]
23. Lai, R.; Chen, H.; Zhou, Z.; Yi, Z.; Tang, B.; Chen, J.; Yi, Y.; Tang, C.; Zhang, J.; Sun, T. Design of a Penta-Band Graphene-Based Terahertz Metamaterial Absorber with Fine Sensing Performance. *Micromachines* **2023**, *14*, 1802. [CrossRef] [PubMed]
24. Liang, F.; Sun, X.; Hu, S.; Ma, H.; Wang, F.; Wu, G. Photocatalytic Water Splitting to Simultaneously Produce H_2 and H_2O_2 by Two-Electron Reduction Process over Pt Loaded Na+ Introduced g-C_3N_4 Catalyst. *Diam. Relat. Mater.* **2020**, *108*, 107971. [CrossRef]
25. Wang, X.; Maeda, K.; Thomas, A.; Takanabe, K.; Xin, G.; Carlsson, J.M.; Domen, K.; Antonietti, M. A Metal-Free Polymeric Photocatalyst for Hydrogen Production from Water under Visible Light. *Nat. Mater.* **2009**, *8*, 76–80. [CrossRef] [PubMed]
26. Ye, S.; Wang, R.; Wu, M.-Z.; Yuan, Y.-P. A Review on G-C_3N_4 for Photocatalytic Water Splitting and CO_2 Reduction. *Appl. Surf. Sci.* **2015**, *358*, 15–27. [CrossRef]
27. Liu, J.; Liu, Y.; Liu, N.; Han, Y.; Zhang, X.; Huang, H.; Lifshitz, Y.; Lee, S.-T.; Zhong, J.; Kang, Z. Metal-Free Efficient Photocatalyst for Stable Visible Water Splitting via a Two-Electron Pathway. *Science* **2015**, *347*, 970–974. [CrossRef] [PubMed]
28. Wu, X.; Zhu, C.; Wang, L.; Guo, S.; Zhang, Y.; Li, H.; Huang, H.; Liu, Y.; Tang, J.; Kang, Z. Control Strategy on Two-/Four-Electron Pathway of Water Splitting by Multidoped Carbon Based Catalysts. *ACS Catal.* **2017**, *7*, 1637–1645. [CrossRef]
29. Zhurenok, A.V.; Vasilchenko, D.B.; Kozlova, E.A. Comprehensive Review on G-C3N4-Based Photocatalysts for the Photocatalytic Hydrogen Production under Visible Light. *Int. J. Mol. Sci.* **2023**, *24*, 346. [CrossRef]
30. Vasilchenko, D.; Zhurenok, A.; Saraev, A.; Gerasimov, E.; Cherepanova, S.; Tkachev, S.; Plusnin, P.; Kozlova, E. Highly Efficient Hydrogen Production under Visible Light over G-C_3N_4-Based Photocatalysts with Low Platinum Content. *Chem. Eng. J.* **2022**, *445*, 136721. [CrossRef]
31. Vasilchenko, D.; Tkachenko, P.; Tkachev, S.; Popovetskiy, P.; Komarov, V.; Asanova, T.; Asanov, I.; Filatov, E.; Maximovskiy, E.; Gerasimov, E.; et al. Sulfuric Acid Solutions of [$Pt(OH)_4(H_2O)_2$]: A Platinum Speciation Survey and Hydrated Pt(IV) Oxide Formation for Practical Use. *Inorg. Chem.* **2022**, *61*, 9667–9684. [CrossRef] [PubMed]
32. Topchiyan, P.; Vasilchenko, D.; Tkachev, S.; Sheven, D.; Eltsov, I.; Asanov, I.; Sidorenko, N.; Saraev, A.; Gerasimov, E.; Kurenkova, A.; et al. Highly Active Visible Light-Promoted Ir/g-C_3N_4 Photocatalysts for the Water Oxidation Reaction Prepared from a Halogen-Free Iridium Precursor. *ACS Appl. Mater. Interfaces* **2022**, *14*, 35600–35612. [CrossRef] [PubMed]
33. Li, R.; Li, C. Photocatalytic Water Splitting on Semiconductor-Based Photocatalysts. *Adv. Catal.* **2017**, *60*, 1–57. [CrossRef]
34. Wang, Z.; Li, C.; Domen, K. Recent Developments in Heterogeneous Photocatalysts for Solar-Driven Overall Water Splitting. *Chem. Soc. Rev.* **2019**, *48*, 2109–2125. [CrossRef] [PubMed]
35. Maheskumar, V.; Saravanakumar, K.; Yea, Y.; Yoon, Y.; Park, C.M. Construction of Heterostructure Interface with $FeNi_2S_4$ and CoFe Nanowires as an Efficient Bifunctional Electrocatalyst for Overall Water Splitting and Urea Electrolysis. *Int. J. Hydrog. Energy* **2023**, *48*, 5080–5094. [CrossRef]
36. Maheskumar, V.; Saravanakumar, K.; Govindan, J.; Park, C.M. Rational Design of Double Perovskite $La_2Ni_{0.5}Co_{0.5}MnO_6$ Decorated Polyaniline Array on MoO3 Nanobelts with Strong Heterointerface Boosting Oxygen Evolution Reaction and Urea Oxidation. *Appl. Surf. Sci.* **2023**, *612*, 155737. [CrossRef]
37. Pan, Z.; Zheng, Y.; Guo, F.; Niu, P.; Wang, X. Decorating CoP and Pt Nanoparticles on Graphitic Carbon Nitride Nanosheets to Promote Overall Water Splitting by Conjugated Polymers. *ChemSusChem* **2017**, *10*, 87–90. [CrossRef] [PubMed]

38. Zhang, G.; Lan, Z.-A.; Lin, L.; Lin, S.; Wang, X. Overall Water Splitting by Pt/g-C_3N_4 Photocatalysts without Using Sacrificial Agents. *Chem. Sci.* **2016**, *7*, 3062–3066. [CrossRef]
39. Pan, Z.; Wang, S.; Niu, P.; Liu, M.; Wang, X. Photocatalytic Overall Water Splitting by Spatially-Separated Rh and RhOx Cocatalysts on Polymeric Carbon Nitride Nanosheets. *J. Catal.* **2019**, *379*, 129–137. [CrossRef]
40. Mosiori, C.O.; Njoroge, W.K.; Ochoo, L.O. Optical Analysis of Ag-NPs Containing Methyl Ammonium Lead Tri-Iodide Thin Films. *Path Sci.* **2017**, *3*, 2007–2015. [CrossRef]
41. Navakoteswara Rao, V.; Ravi, P.; Sathish, M.; Cheralathan, K.K.; Neppolian, B.; Mamatha Kumari, M.; Shankar, M.V. Manifestation of Enhanced and Durable Photocatalytic H_2 Production Using Hierarchically Structured Pt@Co_3O_4/TiO_2 Ternary Nanocomposite. *Ceram. Int.* **2021**, *47*, 10226–10235. [CrossRef]
42. Moon, H.S.; Hsiao, K.; Wu, M.; Yun, Y.; Hsu, Y.; Yong, K. Spatial Separation of Cocatalysts on Z-Scheme Organic/Inorganic Heterostructure Hollow Spheres for Enhanced Photocatalytic H_2 Evolution and In-Depth Analysis of the Charge-Transfer Mechanism. *Adv. Mater.* **2023**, *35*, 2200172. [CrossRef]
43. Li, Z.; Kong, C.; Lu, G. Visible Photocatalytic Water Splitting and Photocatalytic Two-Electron Oxygen Formation over Cu- and Fe-Doped g-C_3N_4. *J. Phys. Chem. C* **2016**, *120*, 56–63. [CrossRef]
44. Xie, H.; Zhao, Y.; Li, H.; Xu, Y.; Chen, X. 2D $BiVO_4$/g-C_3N_4 Z-Scheme Photocatalyst for Enhanced Overall Water Splitting. *J. Mater. Sci.* **2019**, *54*, 10836–10845. [CrossRef]
45. Dai, D.; Wang, P.; Bao, X.; Xu, Y.; Wang, Z.; Guo, Y.H.; Wang, Z.; Zheng, Z.; Liu, Y.; Cheng, H.; et al. G-C_3N_4/ITO/Co-$BiVO_4$ Z-Scheme Composite for Solar Overall Water Splitting. *Chem. Eng. J.* **2022**, *433*, 134476. [CrossRef]
46. Xu, B.B.; Fu, X.B.; You, X.M.; Zhao, E.; Li, F.F.; Chen, Z.; Li, Y.X.; Wang, X.L.; Yao, Y.F. Synergistic Promotion of Single-Atom Co Surrounding a PtCo Alloy Based on a g-C_3N_4 Nanosheet for Overall Water Splitting. *ACS Catal.* **2022**, *12*, 6958–6967. [CrossRef]
47. Gong, Q.; Zhou, Y.; Wang, R.; Jiao, W. Enhanced Photocatalytic Pure Water Splitting of Porous G-C3N4/CdS Composite by the Bimetallic Phosphide. *J. Environ. Chem. Eng.* **2022**, *10*, 108046. [CrossRef]
48. Xue, F.; Si, Y.; Cheng, C.; Fu, W.; Chen, X.; Shen, S.; Wang, L.; Liu, M. Electron Transfer via Homogeneous Phosphorus Bridges Enabling Boosted Photocatalytic Generation of H_2 and H_2O_2 from Pure Water with Stoichiometric Ratio. *Nano Energy* **2022**, *103*, 107799. [CrossRef]
49. Liu, Y.; Zhao, Y.; Wu, Q.; Wang, X.; Nie, H.; Zhou, Y.; Huang, H.; Shao, M.; Liu, Y.; Kang, Z. Charge Storage of Carbon Dot Enhances Photo-Production of H_2 and H_2O_2 over Ni_2P/Carbon Dot Catalyst under Normal Pressure. *Chem. Eng. J.* **2021**, *409*, 128184. [CrossRef]
50. Ai, Z.; Zhang, K.; Xu, L.; Huang, M.; Shi, D.; Shao, Y.; Shen, J.; Wu, Y.; Hao, X. In Situ Configuration of Dual S-Scheme BP/(Ti3C2Tx@TiO_2) Heterojunction for Broadband Spectrum Solar-Driven Photocatalytic H_2 Evolution in Pure Water. *J. Colloid Interface Sci.* **2022**, *610*, 13–23. [CrossRef]
51. Zhang, L.; Dong, W.; Zhang, Y.; Song, X.; Jiang, H. Precisely Anchoring Ni-Doped Cobalt Phosphide Nanoparticles on Phosphatized Carbon Nitride for Efficient Photocatalytic Water Splitting. *Chem. Eng. J.* **2023**, *472*, 144898. [CrossRef]
52. Liu, Y.; Zhao, Y.; Sun, Y.; Cao, J.; Wang, H.; Wang, X.; Huang, H.; Shao, M.; Liu, Y.; Kang, Z. A $4e^-$–$2e^-$ Cascaded Pathway for Highly Efficient Production of H_2 and H_2O_2 from Water Photo-Splitting at Normal Pressure. *Appl. Catal. B Environ.* **2020**, *270*, 118875. [CrossRef]
53. Dou, Y.; Zhu, C.; Zhu, M.; Fu, Y.; Wang, H.; Shi, C.; Huang, H.; Liu, Y.; Kang, Z. Highly Mesoporous Carbon Nitride Photocatalysts for Efficient and Stable Overall Water Splitting. *Appl. Surf. Sci.* **2020**, *509*, 144706. [CrossRef]
54. Zhang, E.; Zhu, Q.; Huang, J.; Liu, J.; Tan, G.; Sun, C.; Li, T.; Liu, S.; Li, Y.; Wang, H.; et al. Visually Resolving the Direct Z-Scheme Heterojunction in CdS@$ZnIn_2S_4$ Hollow Cubes for Photocatalytic Evolution of H_2 and H_2O_2 from Pure Water. *Appl. Catal. B Environ.* **2021**, *293*, 120213. [CrossRef]
55. Fang, W.L.; Wang, L.; Li, C.H. Preparation of Au-OVs-BiOBr-P25 Z-Scheme Photocatalyst and Its Photocatalytic Performance in Overall Water Splitting. *Ranliao Huaxue Xuebao/J. Fuel Chem. Technol.* **2022**, *50*, 446–455. [CrossRef]
56. Cao, J.; Wang, H.; Zhao, Y.; Liu, Y.; Wu, Q.; Huang, H.; Shao, M.; Liu, Y.; Kang, Z. Phosphorus-Doped Porous Carbon Nitride for Efficient Sole Production of Hydrogen Peroxide via Photocatalytic Water Splitting with a Two-Channel Pathway. *J. Mater. Chem. A* **2020**, *8*, 3701–3707. [CrossRef]
57. Vasilchenko, D.; Topchiyan, P.; Berdyugin, S.; Filatov, E.; Tkachev, S.; Baidina, I.; Komarov, V.; Slavinskaya, E.; Stadnichenko, A.; Gerasimov, E. Tetraalkylammonium Salts of Platinum Nitrato Complexes: Isolation, Structure, and Relevance to the Preparation of PtOx/CeO_2 Catalysts for Low-Temperature CO Oxidation. *Inorg. Chem* **2019**, *58*, 6075–6087. [CrossRef] [PubMed]
58. Coelho, A.A. *TOPAS-Academic, Version 6 (Computer Software)*; Coelho Softwafe: Brisbane, Australia, 2007.
59. Shirley, D.A. High-Resolution X-ray Photoemission Spectrum of the Valence Bands of Gold. *Phys. Rev. B* **1972**, *5*, 4709–4714. [CrossRef]
60. Bally, R.W.; Gribnau, T.C.J. Some Aspects of the Chromogen 3,3′,5,5′-Tetramethylbenzidine as Hydrogen Donor in a Horseradish Peroxidase Assay. *Clin. Chem. Lab. Med.* **1989**, *27*, 791–796. [CrossRef]

catalysts

MDPI

Article

Naphthalene Dehydrogenation on Ni(111) in the Presence of Chemisorbed Oxygen and Nickel Oxide

Kess Marks [1,2], Axel Erbing [2], Lea Hohmann [3], Tzu-En Chien [3], Milad Ghadami Yazdi [1], Matthias Muntwiler [4], Tony Hansson [2], Klas Engvall [3], Dan J. Harding [3], Henrik Öström [2], Michael Odelius [2] and Mats Göthelid [1,*]

1 Materials and Nano Physics, School of Engineering Sciences, KTH Royal Institute of Technology, SE-11419 Stockholm, Sweden; kessm@kth.se (K.M.)
2 Department of Physics, AlbaNova University Center, Stockholm University, SE-10691 Stockholm, Sweden; tony.hansson@su.se (T.H.); ostrom@fysik.su.se (H.Ö.); odelius@fysik.su.se (M.O.)
3 Department of Chemical Engineering, KTH Royal Institute of Technology, SE-10044 Stockholm, Sweden; lhohmann@kth.se (L.H.); chie@kth.se (T.-E.C.); kengvall@kth.se (K.E.); djha@kth.se (D.J.H.)
4 Photon Science Division, Paul Scherrer Institut, 5232 Villigen PSI, Switzerland
* Correspondence: gothelid@kth.se

Abstract: Catalyst passivation through carbon poisoning is a common and costly problem as it reduces the lifetime and performance of the catalyst. Adding oxygen to the feed stream could reduce poisoning but may also affect the activity negatively. We have studied the dehydrogenation, decomposition, and desorption of naphthalene co-adsorbed with oxygen on Ni(111) by combining temperature-programmed desorption (TPD), sum frequency generation spectroscopy (SFG), photoelectron spectroscopy (PES), and density functional theory (DFT). Chemisorbed oxygen reduces the sticking of naphthalene and shifts H_2 production and desorption to higher temperatures by blocking active Ni sites. Oxygen increases the production of CO and reduces carbon residues on the surface. Chemisorbed oxygen is readily removed when naphthalene is decomposed. Oxide passivates the surface and reduces the sticking coefficient. But it also increases the production of CO dramatically and reduces the carbon residues. Ni_2O_3 is more active than NiO.

Keywords: dehydrogenation; decomposition; naphthalene; nickel; oxygen; nickel oxide

Citation: Marks, K.; Erbing, A.; Hohmann, L.; Chien, T.-E.; Yazdi, M.G.; Muntwiler, M.; Hansson, T.; Engvall, K.; Harding, D.J.; Öström, H.; et al. Naphthalene Dehydrogenation on Ni(111) in the Presence of Chemisorbed Oxygen and Nickel Oxide. *Catalysts* **2024**, *14*, 124. https://doi.org/10.3390/catal14020124

Academic Editor: Zhong-Wen Liu

Received: 18 December 2023
Revised: 11 January 2024
Accepted: 23 January 2024
Published: 5 February 2024

1. Introduction

Naphthalene is a prominent component in tar, an unwanted by-product of biomass gasification consisting of a mixture of lighter to heavy hydrocarbons, including a number of polycyclic aromatic hydrocarbons (PAHs). Catalytic conversion of PAHs is a promising method to reduce tar and mitigate their negative effects on gas cleaning and upgrading processes downstream from the biomass gasification process [1–3]. Aside from the destruction of the unwanted tars, converting them into useful permanent gases such as CO and H_2 via steam reforming will also increase the total syngas yield of the process. A typical reforming catalyst used under industrial conditions is nickel [4]. Nickel-based catalysts are effective catalysts for the decomposition and dehydrogenation of hydrocarbons but are also prone to poisoning via graphite formation due to carbon formation on the surface [5,6]. The level of carbon poisoning, as well as the ratio of product gases, such as CO and CO_2, is heavily influenced by the content of oxygen-containing compounds, such as H_2O, in the gas mixture.

Gas-phase water provides surface oxygen in the form of OH and O. Higher content of oxygen-containing compounds therefore leads to more efficient hydrocarbon conversion and oxidation of the carbon deposits [6]. However, a too-high content may lead to oxidation of the nickel catalyst and a subsequent lowered catalytic activity [7].

To study this catalytic system, we use the simplified model catalytic reaction of naphthalene over a Ni(111) single-crystal surface, as naphthalene is one of the most stable compounds in tar and a frequently used model compound. [8]. In a previous study, we have shown that naphthalene adsorbed on a Ni(111) single-crystal surface dehydrogenates, producing H_2 gas, as well as graphitic and carbidic surface carbon, rendering the surface inert after several dehydrogenation cycles [9]. In the present paper, we use the same model system to study the effect of oxygen on this decomposition reaction of naphthalene. Whereas in industrial catalytic steam reforming, oxygen is provided via steam in the raw producer gas feed gas [4], in this study, we use varying doses of surface oxygen to simulate the catalytic surface under different oxidative atmospheres from chemisorbed oxygen to oxidation of the outer nickel layer. We use temperature-programmed desorption (TPD), sum frequency generation (SFG), and X-ray photoelectron spectroscopy (XPS) to investigate the surface experimentally. Additionally, density functional theory (DFT) calculations are used to find adsorption geometries and electron-binding energies.

2. Results and Discussion

Different starting positions of the naphthalene were tested to probe the set of possible adsorption geometries. The energetically most favorable geometries for naphthalene on Ni(111), on Ni(111)-O (2 × 2), and on NiO are shown in Figure 1a–c. In all three cases, the molecule lies down but with a small tilt.

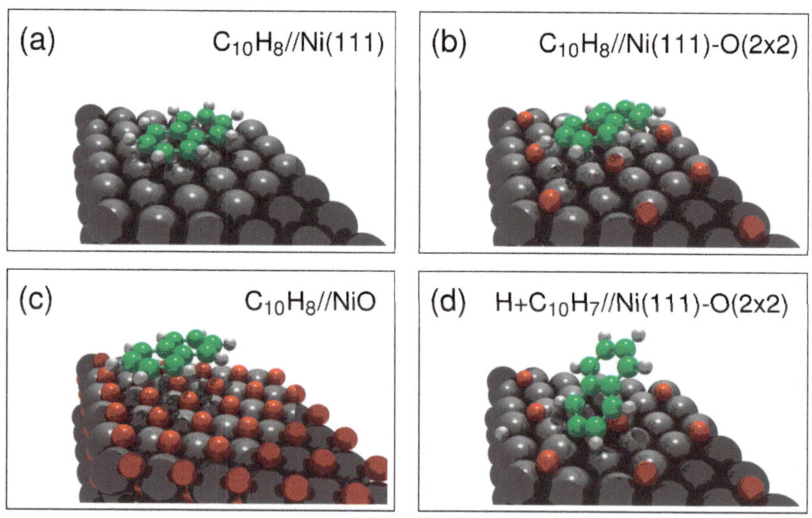

Figure 1. The energetically preferred geometries for naphthalene on (**a**) Ni(111), (**b**) Ni(111)-O (2 × 2), (**c**) NiO and (**d**) the first dehydrogenation step for naphthalene on Ni(111)-O (2 × 2).

The adsorption geometry of naphthalene on Ni(111) has been discussed previously [9,10]. The molecule is chemisorbed on the surface, which is indicated by the relatively small distance to the surface [11], approximately 2 Å, close to the sum of the covalent radii of C and Ni [12], as well as by direct inspection of the highest occupied molecular orbitals, which display covalent bonds between the molecule and the surface. Additional details about the adsorption geometries are presented in the Supplementary Material.

In addition, the first dehydrogenation step on the Ni(111)-O (2 × 2) surface is shown in Figure 1d. Previous results from naphthalene dehydrogenation on Ni(111) showed a preference for dehydrogenation of the α-hydrogens (side) [9]. In the case of Ni(111)-O, it is instead the β-hydrogen (end) that is dissociated first. The calculated difference in adsorption energy between α and β dissociation was only 7 kcal/mol [10], and the presence of surface oxygen changes the balance and instead favors the bonding of the short end to the Ni.

Figure 2 shows TPD spectra from 110 to 700 K of the mass-to-charge ratios 2, 18, 28, 44, and 128 recorded for naphthalene on Ni(111) (a), on (2 × 2) (b), and on NiO (c).

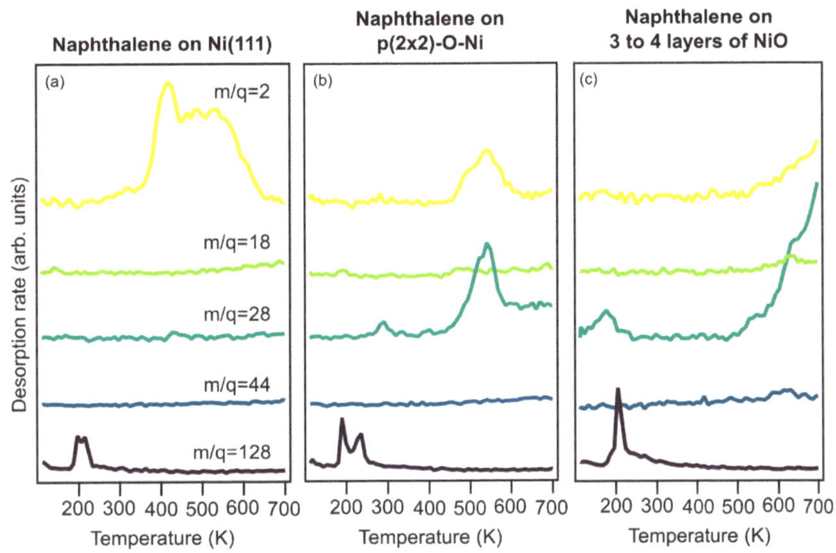

Figure 2. TPD spectra for naphthalene on Ni(111) (**a**), on Ni(111)-O (2 × 2) (**b**), and on NiO (**c**). The traces are shown in order of increasing m/q, starting at the top with 2 (H_2) followed by 18 (H_2O), 28 (CO), 44 (CO_2), and 128 (naphthalene).

Without oxygen, naphthalene dehydrogenation causes a double-peak structure for H_2 desorption starting at 380 K and ending at 650 K, as described by Yazdi et al. [9]. The first peak represents the desorption of two α-hydrogen atoms from the side of the molecule, where the C-H bonds are replaced by C-Ni bonds together with a tilting of the molecule. Desorption at higher temperatures includes the remaining hydrogens at the same time as a graphene layer is formed on the surface [9].

The only other desorption product from the oxygen-free surface is molecular, physisorbed naphthalene desorbing around 220 K. For all preparations with oxygen, a similar desorption peak of molecular naphthalene (mass 128) is observed, indicating that all preparations produce physisorbed naphthalene. Desorption from the (2 × 2) surface (Figure 2b) exhibits a H_2 desorption peak at 540 K, thus a substantial shift to a higher temperature. A similar decrease in H_2 production was previously observed for benzene co-adsorbed with oxygen on Ni(110) [13] and (100) [14]. In addition to H_2 desorption at 540 K, CO desorption is observed at the same temperature, indicating that some of the dehydrogenated carbon atoms react with the chemisorbed oxygen and are removed from the surface. This can potentially have a positive effect on the poisoning of the catalyst.

Desorption from NiO decreases hydrogen production even more and moves CO desorption to higher temperatures; the activation energies for naphthalene decomposition are higher on NiO. Although studies of benzene co-adsorbed with oxygen on Pt(111) [15] and Pd(111) [16] show that H_2O and CO_2 are the main reaction products, this is not the case for naphthalene co-adsorbed with oxygen on Ni(111). No significant amounts of H_2O (mass 18) and CO_2 (mass 44) are observed from any of the surfaces in this study. Only for NiO are small amounts of CO_2 and H_2O desorption observed around 600 K. But the main combustion product is CO. This is similar to findings by Viste et al. [17], who found that CO production dominated over CO_2 production with a CO:CO_2 ratio above 10, for benzene and oxygen on Rh(111).

Results from the SFG experiments, measuring the C-H stretching region, are presented in Figures 3 and 4. Figure 3 shows spectra at selected temperatures from naphthalene on clean Ni(111), on (2 × 2), and on NiO. All three preparations show a resonance below 220 K at similar wavenumbers. Numerical fits according to Equation (1) place the resonances at 3052 cm^{-1} for the (2 × 2) surface and 3057 cm^{-1} for the NiO surface. This matches with results by Marks et al. [10] that show a resonance at 3057 cm^{-1} for naphthalene on clean Ni(111), which was attributed to tilted physisorbed multilayers of naphthalene. Marks et al. observed desorption of the physisorbed naphthalene around 220 K, which seems to be insensitive to oxygen on the surface.

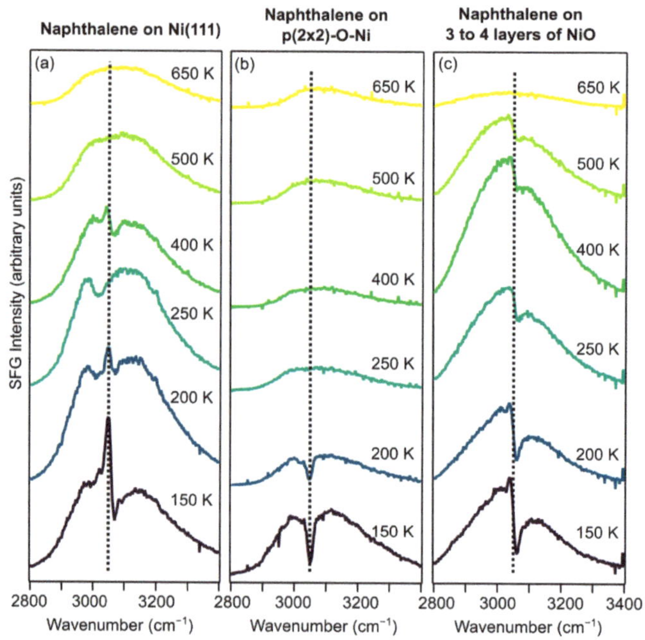

Figure 3. Selected SFG spectra at relevant temperatures of the temperature-dependent SFG spectra for the different preparations. The black dotted vertical line is used in each panel to indicate the position of the resonance at 3052 cm^{-1} (**b**) and 3057 cm^{-1} (**a**,**c**).

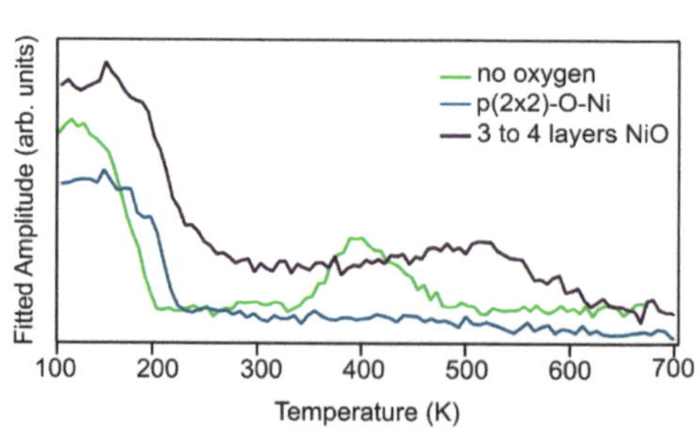

Figure 4. Fitted absolute amplitudes of the C-H resonance around 3050 cm^{-1} as a function of temperature, for different surfaces.

The temperature evolution of the SFG amplitudes is plotted in Figure 4. The amplitudes were acquired by fitting each line in the 2D datasets of Figure S1 in the Supplementary Materials, assuming that the resonance positions (3052 cm^{-1} and 3057 cm^{-1}) as well as the other fit parameters are constant, only letting the amplitude vary freely. Around 220 K, the amplitudes decrease for all three surfaces, and for the (2 × 2) surface, the resonance disappears completely. In TPD, we observed desorption of molecular naphthalene (m/q = 128) around 220 K for each preparation, implying that the decrease in the SFG signal around this temperature is due to the desorption of physisorbed naphthalene, similar to what was observed for naphthalene on clean Ni(111) [9].

The SFG signal disappears after the desorption of physisorbed naphthalene from the (2 × 2) surface, as seen in Figure 3. This absence of any resonant signal could imply that no naphthalene molecules remain on the surface. However, since we do observe H_2 and CO desorption for these preparations at higher temperatures, it is more likely that chemisorbed naphthalene remains on the surface but is invisible to SFG. There are several reasons why an adsorbate vibration would not be visible in SFG spectroscopy, such as disordered layers and low coverages. Additionally, in SFG, only vibrations that are both Raman and IR active will be observable [18], and according to the metal surface selection rule, only vibrations with a dynamical dipole moment along the surface normal will be active [19]. Thus, a stretching vibration can be invisible if its dynamic dipole moment is parallel to the surface. For naphthalene on Ni(111), C-H stretches are visible in SFG because the nickel repels the hydrogens such that the C-H bonds are tilted with respect to the surface, and thus have a component of their dynamic dipole moment perpendicular to the surface. Chemisorbed oxygen increases the distance and reduces the repulsion of the hydrogens, leaving the C-H bonds essentially parallel to the surface.

For the NiO surface, the resonance at 3057 cm^{-1} decreases but does not completely disappear upon desorption of physisorbed naphthalene. In fact, between 300 and 520 K, the amplitude of this resonance, depicted in red in Figure 4, increases slightly before it decreases and disappears at 620 K. This behavior is different from naphthalene on clean Ni(111), where the chemisorbed naphthalene molecules are strongly de-aromatized[5] and initially give rise to resonance at 2980–3000 cm^{-1} as can be seen in Figure 3a (200–400 K) and is discussed in detail in the paper by Marks et al. [10]. When dehydrogenation starts at 380 K, the partially dehydrogenated naphthalene molecules tilt with respect to the surface and regain aromaticity, which in turn gives rise to a new resonance at 3057 cm^{-1} [10]. The SFG spectra for naphthalene on NiO only have a resonance at 3057 cm^{-1}, which is visible from 110 K to 620 K. This implies that naphthalene retains its aromaticity on NiO and adopts a tilted geometry.

For cyclic hydrocarbons, aromaticity is a stabilizing factor and the aromatic resonance persists more than 100 degrees higher on NiO than on Ni(111), implying that naphthalene is more stable on NiO. This reasoning can be turned around with the claim that Ni(111) is a better catalyst for naphthalene decomposition than NiO since the profound de-aromatization upon adsorption weakens the internal bonds of naphthalene, facilitating C-C cleavage at lower temperatures than on NiO. Although co-adsorption of oxygen decreases hydrogen production, it might be beneficial in the long run to add small amounts of oxygen to the Ni(111) catalyst to reduce carbon residues. If the amounts of oxygen are small enough, naphthalene decomposition and hydrogen production can still occur.

Figure 5 shows a C1s spectrum recorded from a monolayer of naphthalene on Ni(111) at room temperature. The numerical fit includes two peaks, one at 284.6 eV and one at 284.1 eV, both with a width of ~0.6 eV. The peak at 284.1 eV represents the eight carbon atoms in the naphthalene molecule with a C-H bond and the peak at 284.6 eV the two carbon atoms with only C-C bonds at the center of the molecule. This assignment is confirmed by DFT calculations, which predict the presence of two peaks 0.5 eV apart. The atomic origins are indicated in the figure. We used the energetically most favorable adsorption geometry, di-bridge- [10], for the C1s calculations. In the experimental spectrum, the ratio is not 1:4 but rather 1:1.4. The additional intensity of the high BE peak is attributed to vibrational excitations. A photoelectron may lose kinetic energy by exciting a C-H vibration on its way from the surface, thus appearing at an apparently higher binding energy in the spectrum. From SFG, we know that the C-H vibrational energy is around 3000 and

3050 cm^{-1}, which corresponds to approximately 0.38 eV. Thus, the energy loss resulting from the C-H vibrational excitation of the low-BE peak will overlap with the high-BE peak.

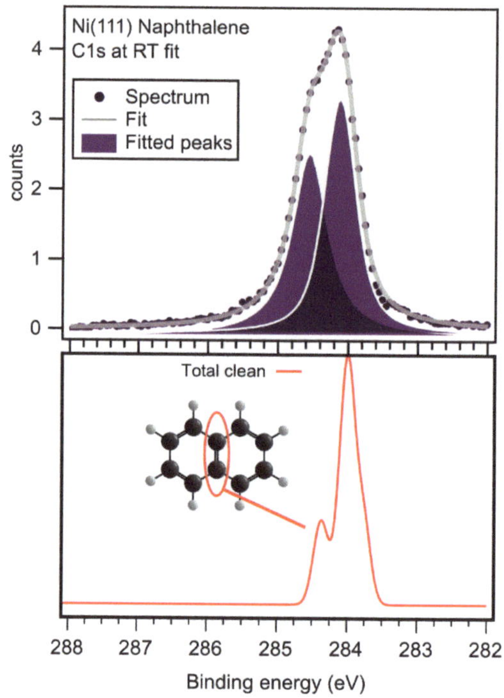

Figure 5. Experimental C1s spectrum and numerical fit, recorded from naphthalene monolayer on Ni(111) at room temperature and the corresponding DFT-calculated C1s binding energies in the lower panel.

2.1. *Adsorption of Naphthalene on Ni/O and NiO*

Figure 6 shows C1s and O1s spectra after 10 L naphthalene room temperature exposure on different surfaces. Increasing the dose of oxygen prior to naphthalene leads to decreasing C1s intensity, implying that oxygen (partially) inhibits the adsorption of naphthalene.

Adsorption of naphthalene on the low oxygen doses (from 0 to 3 L) produce C1s spectra that are rather similar. The two C1s components discussed in Figure 5 are present for all low oxygen doses; there is just a small shift of ~0.1 eV to a lower BE between 0 and 3 L of oxygen for both peaks. The relative peak area changes from 1:1.4 at 0 L to 1:2 at 3 L. Note that neither of these ratios is the expected 1:4 based on the molecular structure. As discussed above, this discrepancy is attributed to vibrational excitations. In our SFG discussion, we suggested that naphthalene on (2 × 2) leaves the CH bonds parallel to the surface. The photoelectrons were recorded in normal emission. The cross-section for vibrational excitations is smaller when the electron trajectory is perpendicular to the vibrational direction, so ideally one would expect the ideal 1:4 ratio, but the analyzer has a wide acceptance angle, allowing photoelectrons with a considerable fraction of their momentum along the surface plane and therefore able to excite CH-vibrations. The DFT calculations of the C1s spectrum from naphthalene on (2 × 2) shows a splitting of 0.4 eV between the central and the outer C atoms, which is slightly smaller than for clean Ni(111). We found only a single stable adsorption geometry where one of the central C atoms is positioned on top of a surface O with a surface Ni atom beneath the center of the rings. The molecule is at a slight angle of ~5° and only shows minor bending around its center. The nature of the adsorption is not as straight forward as for the oxygen free Ni(111) surface, but the relatively large distance from both the surface Ni and O as well as the fact that the calculated spectrum is very similar to the gas phase suggest that the naphthalene is physisorbed in this case. This is also consistent with inspection of the molecular orbitals.

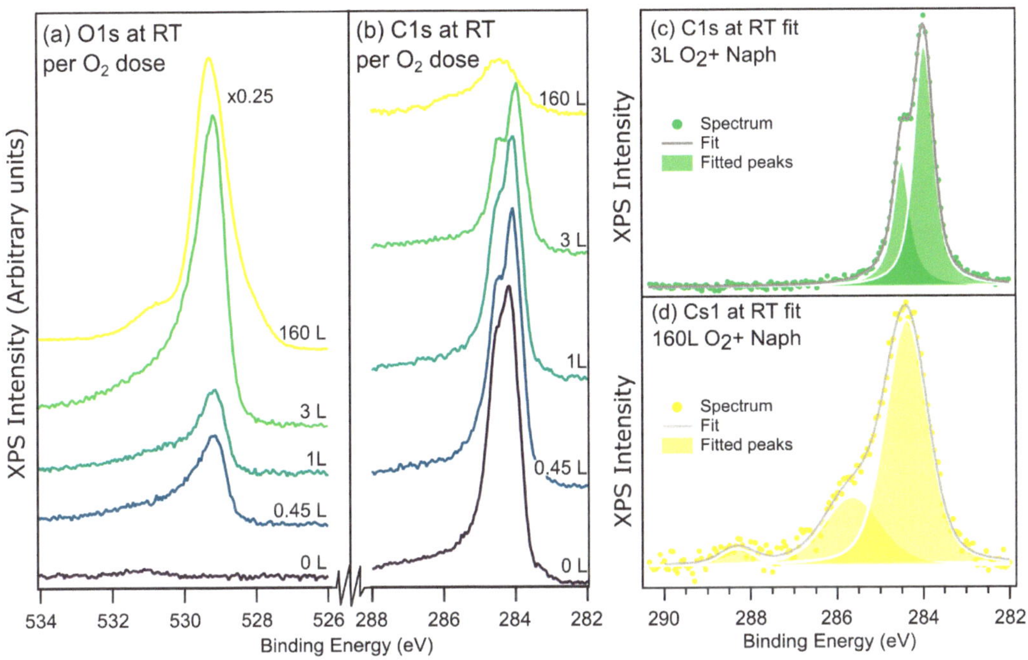

Figure 6. O1s (**a**) and C1s (**b**) spectra for a 10 L naphthalene dose on Ni(111) with increasing oxygen doses. Numerical fits of C1s after a 3 L oxygen dose (**c**) and C1s after a 160 L dose (**d**) for comparison. Peak positions in (**c**) are similar to clean nickel (Figure 5), and in (**d**), the peak positions are 288.3 eV, 285.7 eV, and 284.4 eV with widths of 1, 1.4, and 1.2 eV, respectively.

The C1s spectrum for naphthalene on NiO has a different shape as well as a significantly lower intensity (Figure 6a). Peak fitting finds peaks at 288.3 eV, 285.7 eV, and 284.4 eV with widths of 1, 1.4, and 1.2 eV, respectively. The widths are considerably larger than on the metallic Ni(111) surface. This can be explained by a lower conductivity in the oxide and a poorer screening of the core hole and is a common observation when comparing the spectra from metals and their oxides.

The peak at 284.4 eV is likely a convolution of the C-H- and C-C-bonded carbons where the overlap is such that the individual peaks can no longer be resolved. We assign the peaks at 285.7 and 288.3 eV to C-O and C=O (or O-C=O) species, respectively [20]. The presence of these species suggests the (partial) decomposition of naphthalene upon adsorption; this suggestion is strengthened by the presence of OH groups in the O1s spectra at this preparation.

2.2. Reaction of Naphthalene on Ni/O (2 × 2) and NiO

Figure 7 shows C1s and O1s spectra from naphthalene adsorbed on Ni(111) (1 × 1), on (2 × 2)-O, and on NiO as a function of temperature. In Figures 8 and 9, we show the numerical fits of selected spectra.

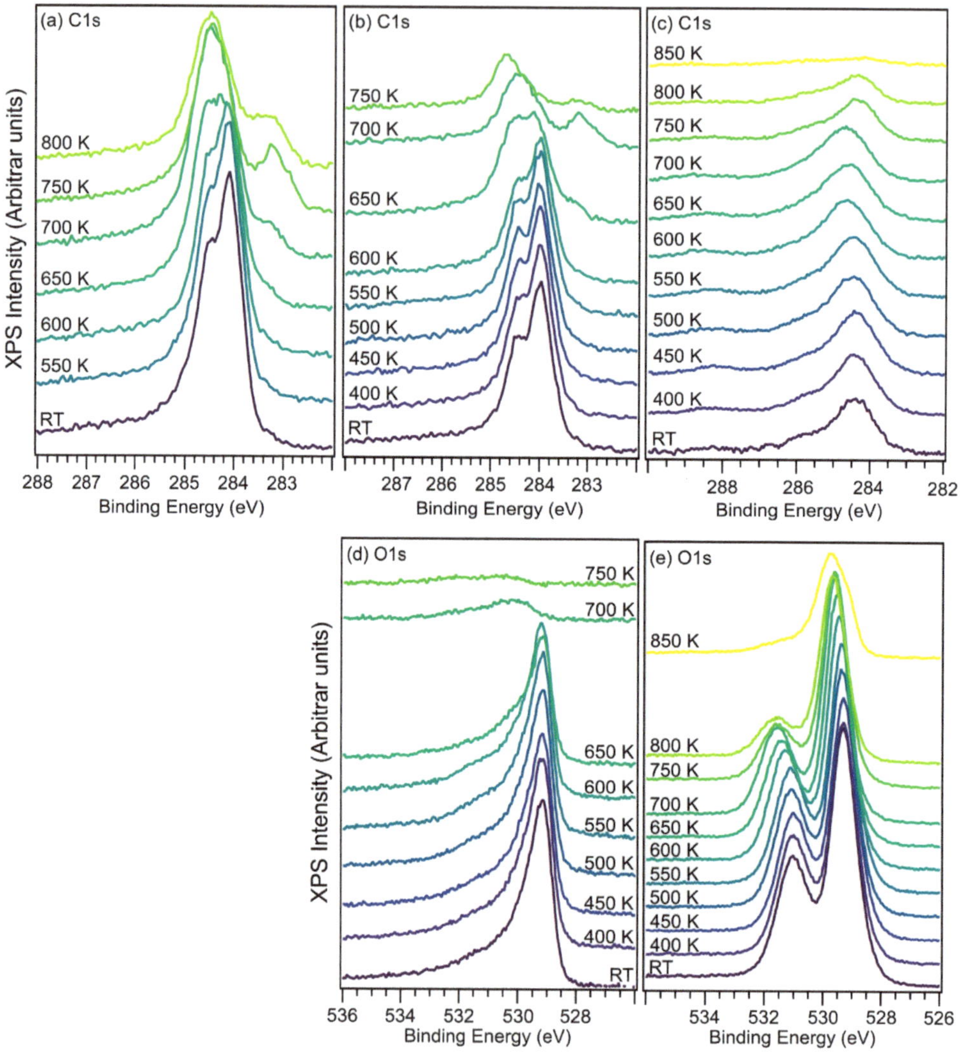

Figure 7. Temperature-dependent C1s and O1s spectra recorded after 10 L naphthalene adsorption on Ni(111) (1 × 1) (**a**), Ni(111)-O- (2 × 2) (**b**,**d**), and NiO after 160 L O_2 dose (**c**,**e**).

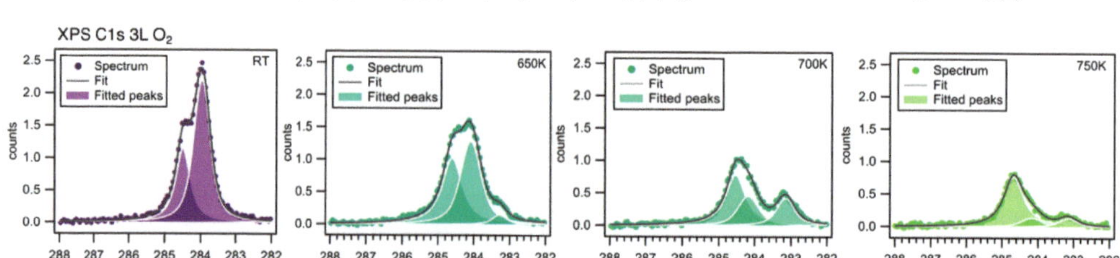

Figure 8. Fits of the C1s spectra of the 3 L preparation at four different temperatures, room temperature, 650 K, 700 K, and 750 K.

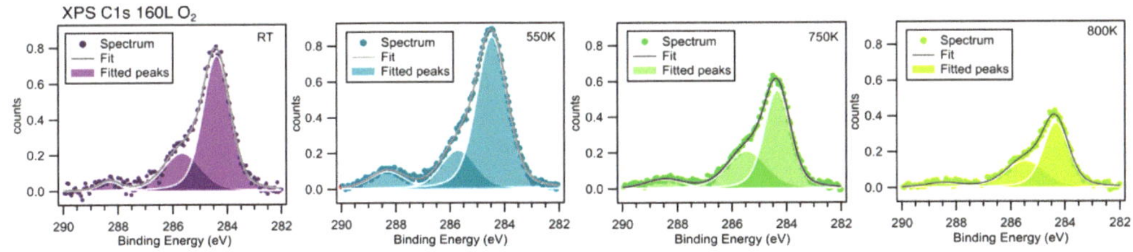

Figure 9. Fits of the C1s spectra of the 160 L preparation at four different temperatures, room temperature, 550 K, 750 K, and 850 K.

The left panel of Figure 7 presents spectra recorded from the (1 × 1) surface, going from room temperature (RT) up to 800 K. There are two important observations; around 700 K, there is a new peak appearing at lower binding energy, which was previously assigned to nickel carbide [10]. At increasing temperatures, this peak is gradually removed. Also, the main peak intensity shifts from the C-H to C-C carbon. The shifted intensity thus depicts dehydrogenation and graphene formation.

On (2 × 2), the initial C1s spectrum after RT adsorption is very similar to the spectrum from (1 × 1) at RT. The intensity is lower, indicating a lower sticking coefficient. The effect of annealing is also similar, but the change in the main peak occurs at lower temperatures. These changes happen at the same time as surface oxygen is consumed, as seen in the lower panel.

On NiO, the coverage after the 10 L dose is markedly reduced compared to (1 × 1). The C1s shape is different, but annealing does not affect the line shape until the peak disappears. O1s shows two shifted peaks: one around 529 eV and the other at 531 eV. These are assigned to NiO (529 eV) and Ni_2O_3 (531 eV) or defects [21]. Upon annealing, the Ni_2O_3 peak is reduced and shifted to higher binding energy. When the C1s signal is zero, Ni_2O_3 is consumed. Thus, Ni_2O_3 is more active than NiO.

Numerically fitted selected spectra are presented in Figures 8 and 9. Following room temperature adsorption on (2 × 2), the same two C1s peaks are present: C-C at 284.48 and C-H at 283.96 eV. They both have a width of 0.56 eV. At 650 K, these peaks have shifted by ~0.2 eV to 284.61 and 284.17 eV, and the carbidic peak appears at 283.28. Naphthalene decomposition can also be derived from the changes in intensity of the other two peaks. If we compare the fit at RT with the fit at 650 K, we can see that the C-H peak decreases to about 70% of the original intensity, while the C-C peak increases by 20%, indicating that C-H bonds are broken and C-C fragments are created. Not all intensity decreases in the C-H peak can be explained by the growth of C-C and C-Ni peaks; about 15% loss is unaccounted for in the C1s spectra, which is likely due to CO desorption, which is observed to peak at 580 K in the TPD spectra for this preparation.

Heating to 700 K increases and shifts the Ni-C feature to 283.15 eV and decreases the C-H feature even further. At 750 K, the Ni-C and C-H peaks have almost disappeared, whereas the C-C peak is remarkably stable, due to the formation of a stabilizing graphene-like structure. Chemisorbed oxygen adatoms lower the temperature where carbon is removed from the surface. They also facilitate transformation from C-H to C-C.

C1s spectra from naphthalene on NiO at 160 L O_2 are shown in Figure 9. Three peaks are required to fit the room temperature spectrum: COO^- at 288.32 eV, CO at 285.66 eV, and the largest one at 284.39 eV representing C-C and C-H with widths of 1, 1.6, and 1.2 eV, respectively. At 550 K, the shape of the spectrum is rather similar, but the intensity of the high-BE (288.3 eV) and low-BE (284.4) peaks has increased. At 750 K, the main peak has significantly decreased in area and the fit places the peaks at 288.48, 285.46, and 284.34 eV with widths of 1.6, 1.7, and 1.2 eV, respectively. At 800 K, the high-BE peak is practically gone and the low-BE peak has shifted to an even lower BE (284.12 eV). This is also the first time we see a significant intensity decrease in the middle peak from 0.35–0.45 in the previous spectra to 0.11 in this last spectrum.

3. Materials and Methods

The sample was a Ni(111) single crystal, purchased from the Surface Preparation Laboratory (SPL), Wormerveer, The Netherlands, polished and aligned to within less than 0.1° from the (111) plane. Before each experiment, the sample was cleaned by argon sputtering followed by annealing up to 1050 K in UHV.

The Ni(111) surface was pre-covered with oxygen by dosing O_2 gas using a precision leak valve (MDC Precision, Hayward, CA, USA) while keeping the sample temperature at 300 K. Oxygen was added to the surface following the preparation used by [22]: 1–3 L (Langmuir) of O_2 was dosed to produce a Ni(111)-O (2 × 2) structure (hereafter called (2 × 2)) with chemisorbed oxygen, >130 L was used to create a thin nickel oxide (NiO) film around 3 layers thick. The cleanness of the surface before dosing as well as the structure of the prepared oxygen surfaces were checked with LEED.

After pre-coverage with oxygen, the surface was exposed to a 10 L dose of naphthalene. At room temperature, this produces a monolayer (ML) [9]. For the SFG and TPD experiments, the sample was cooled to 110 K before the naphthalene dose, whereas in the XPS experiments, the sample was kept at room temperature. This difference in dosing temperature does not affect the monolayer reaction pathway under investigation in the present paper, since the dehydrogenation of naphthalene on Ni(111) starts at about 380 K. Furthermore, multilayers desorb well before room temperature [9,10].

The TPD and SFG measurements were performed in an ultrahigh vacuum (UHV) chamber (Nor-Cal Products Inc., Yreka, CA, USA) with a base pressure of 1×10^{-10} Torr. The chamber was equipped with a quadrupole mass spectrometer (QMS, Hiden Analytical, Warrington, UK), LEED instrumentation (SPECS Surface Nano Analysis GmbH, Berlin, Germany), and an ion gun (Omicron Nanotechnology, Taunusstein, Germany) for sample cleaning. The temperature of the sample during the measurements was controlled by a combination of liquid nitrogen cooling and resistive and electron bombardment heating and monitored via a thermocouple that was spot-welded onto the side of the Ni(111) sample.

The TPD measurements were performed by resistively heating the sample at a ramping rate of 50 K/min and recording the desorption rate of various masses continuously using the QMS. The masses chosen included but were not limited to naphthalene, CO, CO_2, H_2O, O_2, and H_2. The TPD data were subjected to binning and a linear background subtraction.

For the temperature-dependent SFG experiments, the resistive heating ramping rate was 3 K/min; this is less than for the TPD measurements but necessary to ensure a good signal-to-noise ratio in the SFG data. The laser light used in the SFG experiments was produced by a commercial mode-locked Ti:sapphire amplifier system (Coherent Corp., Saxonburg, PA, USA). The light had a wavelength of 800 nm and an average power of 3.7 W at a repetition rate of 1 kHz, and the pulses had ≤50 fs pulse width. The light from the laser system was split up into two parts; one part was spectrally narrowed to a bandwidth of 12 cm^{-1} using a pulse shaper, and the other part was converted to the mid-IR region using a tunable optical parametric amplifier (TOPAS) and a non-colinear difference frequency generator (n-DFG) (Light conversion, Vilnius, Lithuania). The mid-IR wavelength was set to 3300 nm to cover the C-H stretching region. The narrowed 800 nm and mid-IR beams were sent into the UHV chamber and overlapped spatially and temporally on the sample. The SFG light produced at the sample was sent into a spectrometer (Andor, Oxford Instruments Group, Abingdon, UK) with a grating centered around 640 nm and onto an iCCD camera, where the SFG spectra were recorded with a resolution of 14 cm^{-1}.

The experimental spectra were fit assuming that the second-order susceptibility ($\chi^{(2)}$) can be written as

$$\chi^{(2)} = \chi^{(2)}_{NR} + \chi^{(2)}_{R} = \chi^{(2)}_{NR} + \sum_n \frac{A_n e^{i\varphi_n}}{(\omega - \omega_n) + i\Gamma_n} \quad (1)$$

where ${\chi^{(2)}}_{NR}$ is the non-resonant contribution as measured for a clean surface, ${\chi^{(2)}}_{R}$ is the resonant contribution with A_n, ϕ_n, ω_n, and Γ_n being the amplitude, phase, frequency, and half-width of the resonance. To obtain the temperature evolution of a specific resonance, each slice of the two-dimensional temperature-dependent dataset was fit using the same fitting parameters, only allowing the amplitude to vary freely.

The XPS measurements were performed at the Photo-Emission and Atomic Resolution Laboratory (PEARL) beamline at the Swiss Light Source (SLS), Villigen PSI, Switzerland [23]. The UHV system at PEARL (Omicron Nanotechnology, Taunusstein, Germany) had a base pressure of 3×10^{-10} Torr and the preparation chamber was equipped with standard surface science equipment for Ar ion sputtering, annealing, and gas dosing. The analysis chamber at PEARL was equipped with a Scienta EW 4000 analyzer (VG Scienta, Uppsala, Sweden). Photoelectron spectra were recorded for Ni2p, O1s, and C1s using photon energies of 950 eV (Ni2p) 650 eV (O1s), and 380 eV (C1s). The settings were chosen to give a total resolution of 100 meV for C1s, 180 meV for O1s, and 250 meV for Ni2p. The temperature series was recorded by stepwise heating to increasing temperatures and measuring XP spectra after each step. The measured spectra were normalized to the number of scans per measurement as well as the background signal at the high-kinetic energy part of the spectrum. The spectra were fit using Voigt-type peaks in combination with a Shirley background.

4. Computational Details

The interaction between naphthalene and oxygen on Ni(111) surfaces was studied theoretically using spin-polarized density functional theory (DFT) calculations under periodic boundary conditions. The initial geometry optimizations and surface relaxations were carried out in the CP2K version 8.0 code [24] using the Gaussian-plane wave method [25,26] and the RPBE and Grimme's D3 van der Waals correction functional [27,28]. The electron density was expanded in a plane-wave basis set with a kinetic energy cutoff of 650 Ry, and the Kohn–Sham orbitals were described using a localized Gaussian DZVP-MOLOPT-GTH-SR 18 e- pseudopotential for Ni and TZVP-MOLOPT-GTH for O, C, and H [29] with GTH pseudopotentials [30,31]. Since the simulated surface is asymmetric, a dipole correction [32] was added to cancel the artificial electric field built up in the slab by the periodic boundary conditions. The geometry optimization was performed with 2×2 k-point sampling. The total magnetization was fixed in terms of the spin multiplicity corresponding to an average magnetization of 0.61 μ_B per Ni atom for clean and naphthalene-covered Ni(111) and (2×2) surfaces. For the NiO surface, the total magnetization was chosen assuming that the two top layers of Ni would have a net zero magnetization corresponding to the antiferromagnetic ground state of bulk NiO. The initial atomic magnetic moments of these layers were chosen similarly.

On the optimized adsorbed structures, we performed additional single-point calculations at the Gamma-point, with the addition of a Hubbard correction (DFT+U) to describe the interaction more accurately between the Ni and O for the calculation of adsorption energies and binding energies. A value of 5.7 eV was used as an effective Hubbard U parameter for the Ni top layer of the (2×2) surface and for the two uppermost layers of the NiO surface. In both cases, the lower layers were left unaffected.

The relative C1s and relative O1s binding energies were calculated using the final state approximation [33], also known as the Z + 1 approximation, in which a core electron is removed practically by replacing the excited atom with the Z + 1 element (N and F for C and O, respectively). This substitution results in the removal of a core electron to an empty valence state. Since the wave function of the final state is reoptimized, the generated core hole is screened by the valence electrons. For easy comparison to the experimental XPS spectra, the binding energies were convolved using a Gaussian function. As the Z + 1 approximation does not provide physical total energies and only relative shifts are meaningful to discuss, the calculated binding energies were shifted ad hoc to match the experimental results.

The system unit cell was modeled as a five-layer slab, each layer consisting of 4 × 4 × 4 Ni atoms with two bottom layers kept fixed during the surface relaxation. The 2 × 2 geometry was set up with O atoms in fcc sites with a total of 4 Ni atoms per O in the top layer. For NiO, the system was initialized following [34]. Starting from these two surface slabs and the clean Ni(111) slab, the naphthalene molecule was placed on top of the slab, and the system was geometry-optimized.

5. Conclusions

We have studied the effect of surface oxygen and surface oxide on Ni(111) during the dehydrogenation and decomposition of naphthalene. We combined temperature-programmed desorption, sum frequency generation spectroscopy, photoelectron spectroscopy, and density functional theory.

Chemisorbed surface oxygen lowers naphthalene adsorption and decomposition activity. The hydrogen production and desorption is shifted to higher temperatures compared to the reaction on clean Ni(111) without the presence of oxygen. CO production is observed through the reaction between naphthalene fragments and surface O. The formation of carbon residues is reduced in the presence of oxygen.

Surface oxide has a serious effect on the decomposition process, a drastically reduced adsorption and a substantial shift to higher desorption temperatures. Both hydrogen and CO start desorbing around 600 K. Ni_2O_3 is more active in naphthalene decomposition than NiO and is consumed when naphthalene decomposes and desorbs as CO.

Supplementary Materials: The following supporting information can be downloaded at https://www.mdpi.com/article/10.3390/catal14020124/s1, Figure S1: False color plot of temperature dependent SFG spectra recorded from (a) Ni(111) naphthalene, (b) Ni(111) O (2 × 2) naphthalene and (c) Ni(111) NiO naphthalene. The slices presented in figure 3 are cut from the data in this figure; Table S1: Coordinates for optimized structures; Table S2: Naphthalene on Ni(111) O (2 × 2), ground state; Table S3: Naphthalene on NiO; Table S4: Dissociated naphthalene on Ni(111)-O (2 × 2). Flat; Table S5: Dissociated naphthalene on Ni(111)-O (2 × 2). Standing.

Author Contributions: Conceptualization, H.Ö., M.O. and M.G.; Investigation, K.M., A.E., L.H., T.-E.C., M.G.Y., T.H., K.E., D.J.H., H.Ö. and M.G.; Resources, M.M., H.Ö. and M.G.; Formal analysis, K.M. and A.E.; Data curation, K.M. and T.-E.C.; Writing—original draft, K.M. and A.E.; Writing—review and editing, A.E., L.H., T.-E.C., M.M., T.H., K.E., D.J.H., H.Ö., M.O. and M.G.; Visualization, K.M. and A.E.; Project administration, K.M., H.Ö., M.O. and M.G.; Funding acquisition, K.E., D.J.H., H.Ö., M.O. and M.G. All authors have read and agreed to the published version of the manuscript.

Funding: This research was funded by Swedish research council (VR), Stiftelsen Olle Engkvist Byggmästare (SOEB), and the Swedish Foundation for Strategic Research SSF (ITM17-0236). The work was carried out within the Swedish Gasification Center (SFC) consortium. Funding from the Swedish Energy Agency and academic and industrial partners is gratefully acknowledged. The Paul Scherrer institute is kindly acknowledged for providing beamtime at the PEARL beamline. The computations were enabled by resources provided by the National Academic Infrastructure for Supercomputing in Sweden (NAISS) and the Swedish National Infrastructure for Computing (SNIC) at NSC and PDC partially funded by the Swedish Research Council through grant agreements no. 2022-06725 and no. 2018-05973.

Data Availability Statement: Data are contained within the article and Supplementary Materials.

Conflicts of Interest: The authors declare no conflict of interest.

References

1. Engvall, K.; Kusar, H.; Sjöström, K.; Pettersson, L.J. Upgrading of Raw Gas from Biomass and Waste Gasification: Challenges and Opportunities. *Top. Catal.* **2011**, *54*, 949–959. [CrossRef]
2. Okkerse, C.; van Bekkum, H. From fossil to green. *Green Chem.* **1999**, *1*, 107–114. [CrossRef]

3. Evans, R.J.; Milne, T.A. Chemistry of tar formation and maturation in the thermochemical conversion of biomass. In *Developments in Thermochemical Biomass Conversion*, 1st ed.; Bridgewater, A.V., Boocock, D.G.B., Eds.; Blackie Academic & Professional: London, UK, 1997; Volume 2, pp. 803–816.
4. Dayton, D. Review of the Literature on Catalytic Biomass Tar Destruction: Milestone Completion Report. 2002. Available online: https://www.osti.gov/biblio/15002876 (accessed on 16 January 2024).
5. Öström, H.; Triguero, L.; Nyberg, M.; Ogasawara, H.; Pettersson, L.G.M.; Nilsson, A. Bonding of Saturated Hydrocarbons to Metal Surfaces. *Phys. Rev. Lett.* **2003**, *91*, 046102. [CrossRef] [PubMed]
6. Rostrup-Nielsen, J.R.; Sehested, J.; Nørskov, J.K. Hydrogen and synthesis gas by steam- and CO_2 reforming. *Adv. Catal.* **2002**, *47*, 65–139. [CrossRef]
7. Goodman, D.R. Handling and using catalysts in the plant. In *Catalyst Handbook*, 2nd ed.; Twigg, M.V., Ed.; Wolfe: London, UK, 1989; pp. 140–188.
8. Cortazar, M.; Santamaria, L.; Lopez, G.; Alvarez, J.; Zhang, L.; Wang, R.; Bi, X.; Olazar, M. A comprehensive review of primary strategies for tar removal in biomass gasification. *Energy Convers. Manag.* **2023**, *276*, 116496. [CrossRef]
9. Yazdi, M.G.; Moud, P.H.; Marks, K.; Piskorz, W.; Öström, H.; Hansson, T.; Kotarba, A.; Engvall, K.; Göthelid, M. Naphthalene on Ni(111): Experimental and theoretical insights into adsorption, dehydrogenation, and carbon passivation. *J. Phys. Chem. C* **2017**, *121*, 22199–22207. [CrossRef]
10. Marks, K.; Yazdi, M.G.; Piskorz, W.; Simonov, K.; Stefanuik, R.; Sostina, D.; Guarnaccio, A.; Ovsyannikov, R.; Giangrisostomi, E.; Sassa, Y.; et al. Investigation of the surface species during temperature dependent dehydrogenation of naphthalene on Ni(111). *J. Chem. Phys.* **2019**, *150*, 244704. [CrossRef] [PubMed]
11. Kokalj, A. Corrosion inhibitors: Physisorbed or chemisorbed? *Corros. Sci.* **2022**, *196*, 109939. [CrossRef]
12. Rumble, J.R. *Handbook of Chemistry and Physics*, 100th ed.; Internet Version 2019; CRC Press: Boca Raton, FL, USA; Taylor & Francis: Boca Raton, FL, USA, 2019.
13. Huntley, D.R.; Jordan, S.L.; Grimm, F.A. Adsorption and thermal decomposition of benzene on nickel(110) studied by chemical, spectroscopic, and computational methods. *J. Phys. Chem.* **1992**, *96*, 1409–1417. [CrossRef]
14. Blass, P.M.; Akhter, S.; White, J.M. The adsorption and decomposition of benzene on Ni(100) and the effects of pre-adsorbed C, O, H, and Co. *Surf. Sci.* **1987**, *191*, 406–424. [CrossRef]
15. Marsh, A.L.; Gland, J.L. Mechanisms of deep benzene oxidation on the Pt(1 1 1) surface using temperature-programmed reaction methods. *Surf. Sci.* **2003**, *536*, 145–154. [CrossRef]
16. Harris, T.D.; Madix, R.J. Alkene and Arene Combustion on Pd(111). *J. Catal.* **1998**, *178*, 520–532. [CrossRef]
17. Viste, M.E.; Gibson, K.D.; Sibener, S.J. Heterogeneous Combustion of Benzene on Rh(111): Kinetics and Dynamics of CO and CO_2 Production. *J. Catal.* **2000**, *191*, 237–244. [CrossRef]
18. Arnolds, H.; Bonn, M. Ultrafast surface vibrational dynamics. *Surf. Sci. Rep.* **2010**, *65*, 45–66. [CrossRef]
19. Pearce, H.A.; Sheppard, N. Possible importance of a "metal-surface selection rule" in the interpretation of the infrared spectra of molecules adsorbed on particulate metals; infrared spectra from ethylene chemisorbed on silica-supported metal catalysts. *Surf. Sci.* **1976**, *59*, 205–217. [CrossRef]
20. Thermo Fisher Scientific: XPS Periodic Table of Elements—Carbon. Available online: https://www.thermofisher.com/se/en/home/materials-science/learning-center/periodic-table/non-metal/carbon.html (accessed on 16 January 2024).
21. Greiner, M.T.; Helander, M.G.; Wang, Z.; Tang, W.; Lu, Z. Effects of Processing Conditions on the Work Function and Energy-Level Alignment of NiO Thin Films. *J. Phys. Chem. C* **2010**, *114*, 19777–19781. [CrossRef]
22. Okazawa, T.; Nishizawa, T.; Nishimura, T.; Kido, Y. Oxidation kinetics for Ni(111) and the structure of the oxide layers. *Phys. Rev. B* **2007**, *75*, 033413. [CrossRef]
23. Muntwiler, M.; Zhang, J.; Stania, R.; Matsui, F.; Oberta, P.; Flechsig, U.; Patthey, L.; Quitmann, C.; Glatzel, T.; Widmer, R.; et al. Surface science at the PEARL beamline of the Swiss Light Source. *J. Synchrotron Rad.* **2017**, *24*, 354–366. [CrossRef] [PubMed]
24. Kühne, T.D.; Iannuzzi, M.; Del Ben, M.; Rybkin, V.V.; Seewald, P.; Stein, F.; Laino, T.; Khaliullin, R.Z.; Schütt, O.; Schiffmann, F.; et al. CP2K: An electronic structure and molecular dynamics software package—Quickstep: Efficient and accurate electronic structure calculations. *J. Chem. Phys.* **2020**, *152*, 194103. [CrossRef]
25. Hutter, J.; Iannuzzi, M.; Schiffmann, F.; VandeVondele, J. Cp2k: Atomistic simulations of condensed matter systems. *Wiley Interdiscip. Rev. Compu. Mol. Sci.* **2014**, *4*, 15–25. [CrossRef]
26. Lippert, G.; Hutter, J.; Parrinello, M. A hybrid Gaussian and plane wave density functional scheme. *Mol. Phys.* **1997**, *92*, 477–487. [CrossRef]
27. Hammer, B.; Hansen, L.B.; Nörskov, J.K. Improved adsorption energetics within density-functional theory using revised Perdew-Burke-Ernzerhof functionals. *Phys. Rev. B* **1999**, *59*, 7413–7421. [CrossRef]
28. Grimme, S.; Antony, J.; Ehrlich, S.; Krieg, H. A consistent and accurate ab initio parametrization of density functional dispersion correction (DFT-D) for the 94 elements H-Pu. *J. Chem. Phys.* **2010**, *132*, 154104. [CrossRef]
29. VandeVondele, J.; Hutter, J. Gaussian basis sets for accurate calculations on molecular systems in gas and condensed phases. *J. Chem. Phys.* **2007**, *127*, 114105. [CrossRef]
30. Hartwigsen, C.; Goedecker, S.; Hutter, J. Relativistic separable dual-space Gaussian pseudopotentials from H to Rn. *Phys. Rev. B* **1998**, *58*, 3641–3662. [CrossRef]
31. Goedecker, S.; Teter, M.; Hutter, J. Separable dual-space Gaussian pseudopotentials. *Phys. Rev. B* **1996**, *54*, 1703–1710. [CrossRef]

32. Bengtsson, L. Dipole correction for surface supercell calculations. *Phys. Rev. B* **1999**, *59*, 12301–12304. [CrossRef]
33. Köhler, L.; Kresse, G. Density functional study of CO on Rh(111). *Phys. Rev. B* **2004**, *70*, 165405. [CrossRef]
34. Flege, J.I.; Meyer, A.; Falta, J.; Krasovskii, E.E. Self-limited oxide formation in Ni(111) oxidation. *Phys. Rev. B* **2011**, *84*, 115441. [CrossRef]

catalysts MDPI

Article

3D-Printed Monoliths Based on Cu-Exchanged SSZ-13 as Catalyst for SCR of NO_x

Elisabetta M. Cepollaro [1], Stefano Cimino [1], Marco D'Agostini [2], Nicola Gargiulo [3], Giorgia Franchin [2] and Luciana Lisi [1,*]

1 Istituto di Scienze e Tecnologie per l'Energia e la Mobilità Sostenibili (STEMS), Consiglio Nazionale delle Ricerche (CNR), Via Guglielmo Marconi 4/10, 80125 Napoli, Italy; elisabettamaria.cepollaro@stems.cnr.it (E.M.C.); stefano.cimino@cnr.it (S.C.)

2 Dipartimento di Ingegneria Industriale, Università degli Studi di Padova, Via Marzolo 9, 35131 Padova, Italy; marcolorenzo.dagostini@phd.unipd.it (M.D.); giorgia.franchin@unipd.it (G.F.)

3 Centro Servizi Metrologici e Tecnologici Avanzati (CeSMA), Università di Napoli Federico II, Corso N. Protopisani, 80146 Napoli, Italy; ngargiul@unina.it

* Correspondence: luciana.lisi@cnr.it; Tel.: +39-081-7682279

Abstract: Monoliths manufactured by Direct Ink Writing containing 60% SSZ-13 (SiO_2/Al_2O_3 = 23) and SiO_2 with 10% laponite as a binder were investigated as self-standing structured catalysts for NH_3-SCR of NO_x after a short (4 h) and prolonged (24 h) ion exchange with copper and then compared with pure SSZ-13 exchanged under the same conditions. The catalysts were characterized by morphological (XRD and SEM), textural (BET and pore size distribution), chemical (ICP-MS), redox (H_2-TPR), and surface (NH_3-TPD) analyses. The silica-based binder uniformly covered the SSZ-13 particles, and copper was uniformly distributed as well. The main features of the pure Cu-exchanged SSZ-13 zeolite were preserved in the composite monoliths with a negligible contribution of the binder fraction. NH_3-SCR tests, carried out on both monolithic and powdered samples in the temperature range of 70–550 °C, showed that composite monoliths provided very good activity, and that the intrinsic activity of SSZ-13 was enhanced by the hierarchical structure of the composite material.

Keywords: structured catalysts; additive manufacturing; SSZ-13; $deNO_x$

Citation: Cepollaro, E.M.; Cimino, S.; D'Agostini, M.; Gargiulo, N.; Franchin, G.; Lisi, L. 3D-Printed Monoliths Based on Cu-Exchanged SSZ-13 as Catalyst for SCR of NO_x. *Catalysts* **2024**, *14*, 85. https://doi.org/10.3390/catal14010085

Academic Editor: Anker Degn Jensen

Received: 27 December 2023
Revised: 15 January 2024
Accepted: 17 January 2024
Published: 19 January 2024

1. Introduction

NH_3 Selective Catalytic Reduction (SCR) is the current preferred technology to remove nitrogen oxides (NO_x) from lean-burn diesel engine exhausts, using base metal-exchanged zeolite catalysts (e.g., Cu, Fe) [1]. The SCR section is usually located downstream from the diesel particulate filter (DPF) where the temperature could reach up to 650 °C. Therefore, the SCR catalyst can be easily damaged in the presence of steam generated from the combustion of diesel fuel [2,3].

The more common vanadium-based catalysts have a low hydrothermal stability and poor resistance to alkali metal, whilst zeolite-based catalysts are more tolerant to high temperatures and show a good resistance to alkali metal poisoning due to their high surface area and strong acidity, which makes the zeolite-based catalysts also suitable for $deNO_x$ in flue gas cleaning from biomass-fired applications [4].

Small-pore zeolites such as Cu-SSZ-13 have better NH_3-SCR performance due to the high concentration of active sites, higher hydrothermal stability, poisoning tolerance, and better N_2 selectivity than other zeolites. This last feature is related to the peculiar small-pore structure which limits the generation of undesired byproducts, such as N_2O, having a size larger than the 8-ring (~3.8 Å) [5]. Furthermore, the small-pore structure inhibits the dealumination of the framework and the accumulation of copper species during hydrothermal aging, and the shape selectivity of this small-pore zeolite also prevents long-chain hydrocarbons that could be present in diesel exhaust gases from entering [3]. For these reasons, this zeolite is presently the most popular NH_3-SCR catalyst in diesel exhaust purification [6,7].

Like all the zeolites with CHA topology, SSZ-13 shows a three-dimensional pore system, which is formed by stacking a hexagonal array of planar 6-membered rings (6-MRs) connected through an AABBCC scheme. Such stacking forms large chabazite cages, whose access is limited by symmetric 8-MRs (diameter: ~0.38 nm) [8]. In copper-exchanged SSZ-13, Cu ions are located in two different extra-reticular sites; indeed, the species labelled as Cu^{2+}-2Z is located in the plane of the 6-MRs, while the species labelled as $[Cu(OH)]^+$-Z is located in the 8-MR-sized window of the CHA cage [9,10]. Both species are considered as active centres for SCR reaction, although Cu^{2+}-2Z is inferred to be less active than $[Cu(OH)]^+$-Z for low-temperature SCR [9]. Cu^{2+}-2Z species prevail in zeolites with a high content of framework Al (Si/Al = 6), whereas in zeolites with low Al content (Si/Al = 35), i.e., zeolites having a low concentration of paired Al (Al atoms separated by one or two Si atoms), the main active copper species is $[Cu(OH)]^+$-Z because the presence of an extra-framework OH ligand enables the stabilization of Cu^{II} cations in the 8-MRs of CHA structure [11]. Indeed, the lack of paired Al arrangements balancing the double positive charge of Cu^{2+}-2Z results in the prevailing of $[Cu(OH)]^+$-Z species in zeolites with a high Si/Al ratio.

However, the Cu^{2+}-2Z species is more hydrothermally resistant [1,9,12] because $[Cu(OH)]^+$-Z species can react with water vapor to form CuO_x clusters. At the same time, water vapor favours the dealumination and subsequent collapse of the zeolite framework due to the formation of $CuAl_2O_4$-like species [9,13]. Moreover, CuO clusters are responsible for NH_3 overoxidation to NO or N_2O in the SCR process [14]. Copper redistribution through the HNO_3 treatment of Cu-SSZ-13 was proposed by Liu et al. [15]. They found that the acid exchange removed the unselective CuO aggregates and, at the same time, increased the fraction of isolated Cu^{2+}. The redistribution of Cu species was caused by the dealumination from outside to inside, providing the sites for the location of Cu^{2+} ions. This resulted in a larger operative SCR temperature window, providing NO_x conversion approaching or equal to 100% from 165 to 690 °C, 100% N_2 selectivity, and a very good resistance to H_2O and SO_2.

The use of purely microporous zeolites for the NH_3-SCR is limited by the slow diffusion of the reactants within a thick catalytic layer and also by the deposition of ammonium nitrates and sulphates in the pores and channels, especially at low temperatures [16]. For this reason, higher performances can be obtained with zeolite catalysts showing a hierarchical porosity. In particular, Chen et al. [17] obtained a hierarchical SSZ-13, adding a mesoporogen agent at the preparation stage in combination with the conventional Structure Directing Agent template (TMAda-OH). Therefore, the dual template method was effective to obtain a bimodal micro/meso-porosity, but also caused the increase in paired Al in the zeolite framework, which in turn promoted the presence of Cu^{2+}-2Z species. Moreover, it was found that the average distance between the framework Al sites in the hierarchical SSZ-13 shortened with increasing contents of the mesoporogen agent.

To obtain technological catalysts for environmental applications treating gas streams, the zeolite-based active phase is generally applied onto a high surface area monolithic (honeycomb) scaffold that guarantees mechanical strength and very low pressure drops [6,7]. In particular, Cu-exchanged SSZ-13 was applied as a wash coat onto cordierite honeycombs via a slurry dip-coating method, which used pseudo-boehmite as binder [18,19]. However, some common drawbacks are related to the limited zeolite loading required to prevent the loss of the active phase and/or to poor adhesion for unmatched thermal expansion behaviour with the monolithic substrate [20,21]. Moreover, conventionally extruded honeycomb monoliths with parallel channels suffer from a lack of radial mixing of reactants.

We recently reported [22,23] the preparation and testing in NH_3-SCR of 3D-printed geopolymer/Cu-ZSM-5 monoliths, possibly providing enhanced performance related to (i) the hierarchical porosity of the composite consisting of a macro-mesoporous (geopolymer) matrix and a microporous material (ZSM-5); (ii) improved radial mixing and mass transfer characteristics of 3D-printed monoliths with respect to standard honeycombs; and (iii) higher volumetric loadings of the active phase (zeolite) and good mechanical properties. In this work, following the same approach, we set out to reproduce this methodology for the better-performing zeolite with CHA topology, thus preparing self-standing structured Cu-SSZ-13 catalysts containing a high fraction of zeolite by the Direct Ink Writing method (DIW, an extrusion-based Additive Manufacturing technique). The paper describes the modification of the original methodology applied to ZSM-5 in [22,23], not trivially replicable for SSZ-13, which required the use of colloidal silica and laponite as the binder for the small-pore zeolite, therefore, still keeping a composite system with hierarchical porosity. The resulting 3D monoliths, having the target zeolite loading and good mechanical properties, were fully characterized and eventually tested for the NH_3-SCR reaction.

2. Results and Discussion

2.1. Catalysts Preparation

The first attempts to replicate the same alkaline slurry previously used for the preparation of geopolymer-ZSM-5 structured composites [22,23] were not successful and caused the almost total loss of the original crystalline structure of the SSZ-13 (Figure 1). This was due to the poor resistance of the SSZ-13 structure to the highly alkaline environment required to synthesize a geopolymer matrix. For this reason, a colloidal silica with the addition of 10% laponite was used as a binder to provide a mechanically resistant interconnecting matrix, keeping together the preformed SSZ-13 grains.

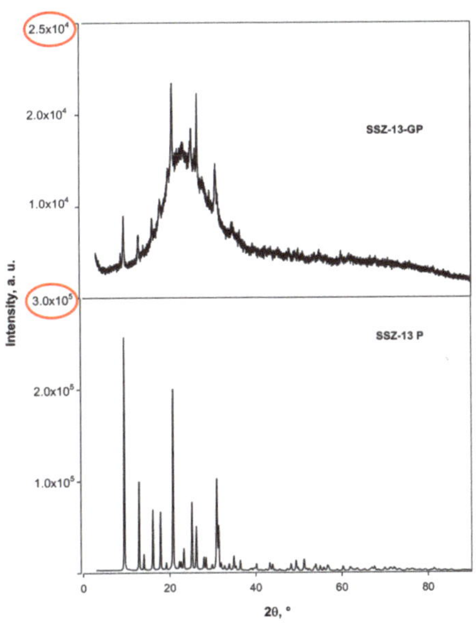

Figure 1. XRD pattern of SSZ-13 (60%) into the geopolymer matrix (**top**) compared with pattern of pure SSZ-13 (**bottom**). Red circles highlight the different order of magnitude of the two scales.

Although using colloidal silica as a binder caused a significant contraction of the structure (diameter of the filaments ranging from 680 to 720 μm) after drying when compared to the geopolymer [23], the subsequent thermal treatment in air at 550 °C produced highly resistant 3D structures with a high mesoporosity. In particular, the addition of 10% laponite clay to the binder proved successful in enhancing the mechanical resistance of the resulting 3D structures (Figure 2) during all the stages of preparation, particularly improving the stability of printed monoliths during drying.

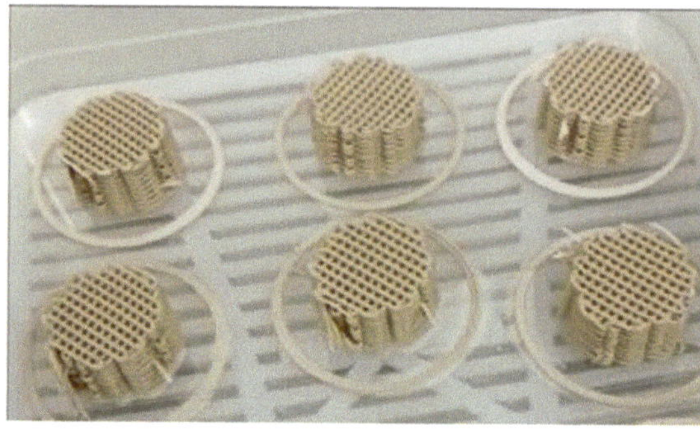

Figure 2. Image of the 3D-printed SSZ-13 monoliths at the end of the synthesis process.

2.2. Catalysts Characterization

As shown in Figure 3, XRD analysis confirmed that the original crystalline structure of the SSZ-13 was completely preserved in the 3D monoliths, using such a formulation of the binder. The small reduction of the intensity of XRD peaks for the monolith sample, with respect to the reference SSZ-13 powder, is purely related to the dilution effect, considering its zeolite loading was fixed at 60 wt%.

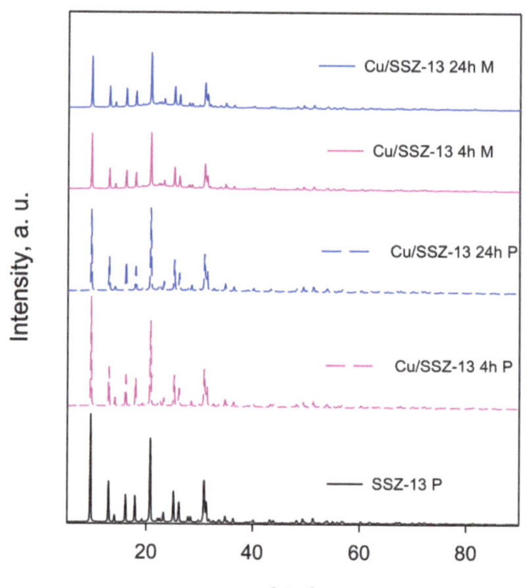

Figure 3. XRD patterns of pure SSZ-13 (black pattern) and Cu-exchanged powder catalysts as well as composite 3D monoliths with 60% SSZ-13 in the silica–laponite matrix.

Ion exchange with copper was performed directly on pre-shaped 3D monoliths after calcination and acid treatment, as described in [22,23]. The visual inspection of the cross section of structured catalysts at the end of preparation showed a uniform blue coloration, suggesting a good penetration of copper throughout the monoliths and a rather uniform distribution of the active metal, excluding any significant confinement of Cu on the external surface, commonly due to the precipitation of the corresponding insoluble hydroxides [22].

Figure 4a–d presents SEM images of 3D-printed composite monoliths at different magnifications.

Figure 4. SEM images of 3D-printed SSZ-13 composite monolith at increasing magnification (**a**) (80×), (**b**) (3000×), (**c**) (12,000×), (**d**) (24,000×)) compared with corresponding images of the parent SSZ-13 powder (**c***,**d***) at the same magnification.

Figure 4a highlights the 3D network of the monolith with the filaments with regular diameters (700 ± 20 μm), following a 0–90° sequential build-up to achieve a log-pile scaffold configuration. The filaments are composed of small particles with a cubic morphology and an average size of ≤5 μm (Figure 4b,c), which are completely embedded by a thin film overlayer (Figure 4d) made of much smaller particles. Comparison with the reference SSZ-13 powder (Figure 4c*,d*) confirms that the original morphology and size of the cubic zeolite crystals [17,24] was entirely preserved during the preparation of the 3D-printed monoliths, whereas the colloidal silica binder provided the thin yet firmly anchored interconnecting matrix among crystals.

Figure 5a presents the N_2 adsorption isotherms of both the Cu/SSZ-13 catalyst in powder form and of the 3D-printed composite monolith in comparison with the parent SSZ-13 and the binder. The N_2 physisorption isotherm of the binder calcined for 4 h at 550 °C in air confirmed the typical features of a mesoporous material with a H2 type hysteresis and a disordered interconnected porosity, whereas the isotherms of the two zeolite powders are, as expected, typical of microporous materials [25]. The composite monolith has features of both materials. The corresponding N_2 isotherm shows, in addition to the sharp increase in adsorbed gas at low p/p_0 which is typical of the zeolite, the hysteresis loop of mesoporous materials approaching $p/p_0 = 1$. The fraction and type of pores is reported in Figure 5b, which highlights the absence of micropores for the binder and the decrease in the micropores volume related to the copper exchange. The Cu/SSZ-13 4 h M has a lower fraction of micropores, as expected, due to the lower fraction of SSZ-13 in the composite material and, at the same time, larger mesopores associated with the fraction of the binder. As a matter of fact, the PSD of Cu/SSZ-13 4 h M agrees with the mathematical combination of the PSD of SSZ-13 and that of the binder.

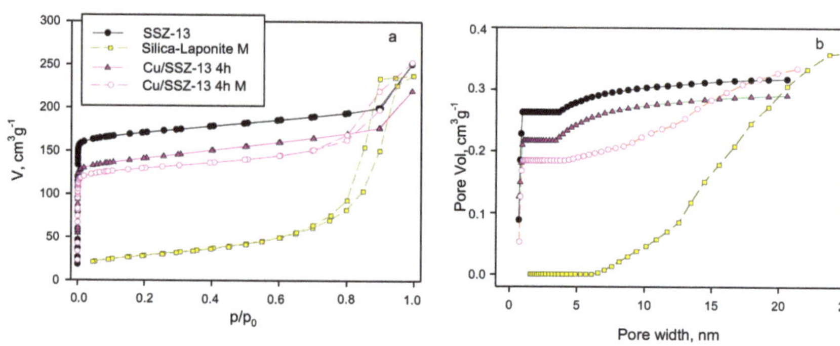

Figure 5. (**a**) N_2 physisorption isotherms and (**b**) cumulative pore size distribution of SSZ-13 powders and monoliths (4 h exchange), compared to the reference SSZ-13 powder and a 3D-printed monolith made of silica + laponite (without any zeolite filler).

On the other hand, the decrease in micropore volume related to copper exchange in the powder zeolites is higher than that expected on the basis of the metal content and should imply the occlusion of some micropores (Table 1).

Table 1. BET-specific surface area, micropore, and total pore volume of powder and monolith catalysts (evaluated by the DFT model and Gurvitch method, respectively).

Sample	Form	BET Area ($m^2\ g^{-1}$)	V Micro ($cm^3\ g^{-1}$)	V Total ($cm^3\ g^{-1}$)
SSZ-13 P	Powder	686	0.26	0.39
SSZ-13 M	3D Monolith	516	0.19	0.35
Cu-SSZ-13 P (4 h)	Powder	557	0.22	0.34
Cu-SSZ-13 M (4 h)	3D Monolith	512	0.18	0.37
Cu-SSZ-13 P (24 h)	Powder	542	0.21	0.30
Cu-SSZ-13 M (24 h)	3D Monolith	463	0.16	0.36

Table 1 reports the corresponding values of the specific surface areas which is evaluated by the BET model, as well as micropores and total volume, calculated by NLDFT model and Gurvitch methods, respectively. After copper exchange and thermal treatment, the BET area of the SSZ-13 zeolite was reduced by only 6%, which mostly corresponds to the effect predicted considering the actual metal loading in the catalyst (Table 2). Moreover, the surface area of the monolith catalyst exceeds the value calculated as the linear combination of the contributions from the two main components (zeolite + silica binder) when considering their actual content in the composite; this suggests a larger porosity of the silica matrix intercalated by the zeolite grains.

Table 2. Copper loading from ICP-MS analysis, H_2 uptake from TPR analysis, and corresponding copper content (assuming a complete Cu^{2+}-Cu^0 reduction); NH_3 amount released during TPD analysis.

Sample	Cu from ICP-MS (wt%)	H_2 Uptake (mmol g^{-1})	Cu Estimated from Total H_2 Uptake (wt%)	Desorbed NH_3 (mmol g^{-1})
SSZ-13 P	-	-	-	3.56
Cu/SSZ-13 P (4 h)	5.09	0.68	4.3	2.65
Cu/SSZ-13 M (4 h)	2.91	0.47	3.0	1.69
Cu/SSZ-13 P (24 h)	6.23	0.95	6.0	2.92
Cu/SSZ-13 M (24 h)	3.33	0.84	5.3	1.86

The comparison of micropores and total pore volume of Cu-exchanged and parent SSZ-13 showed that reduction is mostly assignable to the reduction of micropore volume, where Cu is supposedly located, whereas mesopore volume is much less affected.

The copper loading (by ICP-MS, Table 2) in the zeolite powder and in the 3D monolith after 4 h of ion exchange was equal to 5.1% and 2.9 wt%, respectively. Those values roughly scaled as the zeolite weight content in the composite, confirming that the mesoporous silica matrix does not hinder access of the metal ions to the SSZ micropores. A slight increase in the actual copper loading was achieved by extending the Cu exchange up to 24 h.

The actual copper content was significantly higher than the corresponding theoretical Cation Exchange Capacity (CEC), evaluated on the basis of the SSZ-13 unit cell and its SiO_2/Al_2O_3 ratio, which equals to ca 2.2 wt% Cu, in good agreement with Kwak et al. [26] and Clemens et al. [27]. The contribution of the laponite, having a CEC very close to that of SSZ-13 (0.79 meq/g [28]), is too small to explain the high copper content, and it only occurs for monolith samples. Therefore, it can be argued that the high metal loading is rather associated with the occurrence of an over-exchange phenomenon deriving from the formation Cu-oxo species with different oxygen/Cu ratios (e.g., $[CuO_2Cu]^{2+}$, $[CuOCu]^{2+}$, $[Cu_3O_3]^{2+}$ observed for SSZ-13 with high Si/Al ratio) [29]. However, the formation of extra-framework copper oxides cannot be excluded. In any case, this additional copper is likely also responsible for the partial occlusion of micropores, which is more evident for the pure zeolite sample than for its composite 3D—printed counterpart.

H_2-TPR experiments were used to determine the type and amount of exchanged copper. TPR results are reported in Figure 6, where profiles of Cu-exchanged SSZ-13 powders and monoliths are compared for the two exchange times. It was verified that the parent SSZ-13 powder does not give any contribution to reduction that, consequently, can be assigned only to copper in the metal-exchanged materials. Both TPR profiles of powder and monoliths exchanged for 4 h show similar features with two main signals: the first one, peaking at 252–260 °C, is sharper than the second one, reaching its peak at 501 °C and 575 °C, respectively, for the composite monolith and the powder catalyst. Increasing the Cu loading by extending the exchange time up to 24 h shifted the second reduction peak towards lower temperatures (475 °C and 520 °C, respectively, for monolith and powder). Moreover, it caused the appearance of an additional intermediate peak (at 318 °C and 340 °C), which was particularly evident for the powder sample and suggests the formation of novel copper species that cannot be associated with the presence of the binder.

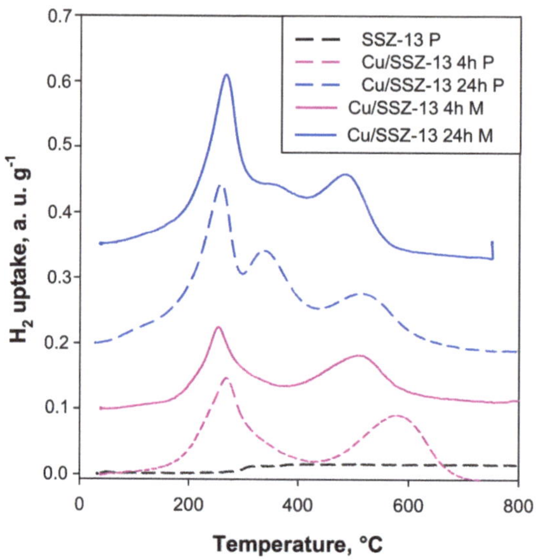

Figure 6. H_2-TPR profiles of Cu/SSZ-13 powders and composite monoliths, exchanged for 4 or 24 h, compared with the TPR profile of the parent SSZ-13.

The total H_2 consumption during TPR is reported in Table 2 together with the corresponding metal contents, estimated under the assumption of complete reduction of Cu(II) to Cu(0). Those values do not always correspond to the results from ICP-MS analysis, but they follow the same increasing trend with the exchange time.

The presence of two well-defined peaks suggests that there are two main different copper species or that a single copper species is reduced into two steps. However, there is a general agreement in the literature regarding the presence of at least two types of copper species in the SSZ-13 framework; the different areas of the two signals are in contrast with a two step reduction of a Cu^{2+} species (from Cu^{2+} to Cu^+ and from Cu^+ to Cu^0) [7]; however, they should correspond to copper located at different sites. The less easily reducible one, requiring higher temperatures, is located in the 6-MRs, whereas the more reducible copper species is located in the larger 8-MR-sized windows of the chabazite cages [7,15].

Furthermore, Chen et al. [17] found a large signal peaked at about 250 °C, which they assigned to the reduction of isolated copper ions in the form of $[Cu(OH)]^+$-Z seated at the 8-MR window site, whilst they attributed a maximum peak of about 360 °C to Cu^{2+}-2Z ions seated at the 6-MR sites.

Similarly, Liu et al. [15] assigned the peak below 300 °C to the reduction of isolated Cu^{2+} species in chabazite cages; however, the peak occurring between 300 and 400 °C was assigned to the reduction of CuO species to Cu^0, and the one cantered between 400 and 500 °C was assigned to the reduction of isolated Cu^{2+} ions located in 6-MRs. Eventually, Cu^+ ions could be reduced at even higher temperatures (>500 °C).

Kwak et al. [1] noticed that the intensity of the low temperature H_2-consumption peak increased with increasing Cu loading, while that higher temperature remained unchanged at Cu ion exchange levels $\geq$40%. They also proposed that the first Cu ions introduced into the zeolite tend to occupy the most stable and less easily reducible cationic sites (i.e., the six-membered ring-sized ones), whereas, after saturation of these sites, Cu^{2+} ions can then occupy sites inside the larger cages of the CHA structure. These results highlight that the distribution of the two different Cu^{2+} species present in SSZ-13 changes with Cu loading.

On the basis of these attributions, we reasonably assume that the first peak, common to all samples regardless of the exchange time and presence of the silica binder, is related to Cu^{2+} in the 8-MR chabazite cage. The amount of copper located in this position only slightly increases by increasing the exchange time for the powder sample, whereas a more marked enhancement was observed for the monolith, likely related to a slower penetration rate within the zeolite framework in the presence of the binder. A preliminary TPR test on a reference sample containing only Cu deposited in the silica–laponite matrix (no zeolite) showed almost no contribution.

The assignment of the signals at higher temperatures is more difficult: it can be argued that the intermediate TPR peak is most probably related to extra-framework CuO species, which are more abundant in the pure zeolite powder samples. Indeed, having already saturated the theoretical capacity of the zeolite after 4 h of exchange, any additional Cu can only be located outside of the framework. Eventually, the high temperature reduction is consistent with stable copper ions located in the 6-MRs of the SSZ-13. Accordingly, the presence of the binder surrounding zeolitic crystals appears to contrast and limit the extent of extra-framework copper in 3D composite monoliths, while also preserving easy access to the micropores.

NH_3-TPD profiles, reported in Figure 7, confirm that 3D monolith samples display a surface acidity very similar to that of powder zeolites, considering the smaller fraction of SSZ-13 in the composite material. The amount of NH_3 desorbed, evaluated from the integration of TPD trace, is reported in Table 2, and provides a quantitative check that the values calculated for monolith samples are very close to 60% of those evaluated for powder zeolites for both exchange times, indicating a negligible contribution of the silica-based binder to the ammonia adsorption.

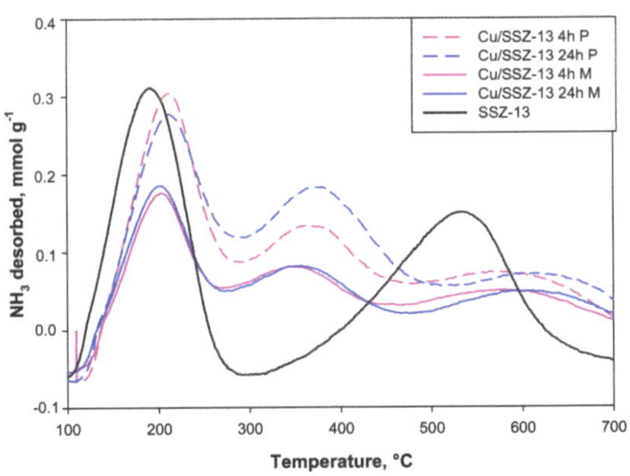

Figure 7. NH_3-TPD profiles of Cu/SSZ-13 both as powder and monoliths, exchanged for 4 or 24 h, compared with the TPR profile of the parent SSZ-13.

All NH_3-TPD profiles of exchanged materials show three desorption signals, with the first one peaking at 202–210 °C, the intermediate one peaking at ca 355–375 °C, while the high temperature peak is rather broad, extending from 500 up to 700 °C for powders and monoliths.

Fan et al. [2] found, for fresh Cu/SSZ-13, a main low-temperature desorption peak located at 190 °C and two small desorption peaks at 320 °C and 400 °C. They associated the low-temperature peak to the weakly bound NH_3 desorbed from Lewis acid sites, whilst the desorption peaks at 320 °C and 400 °C were attributed to the NH_3 adsorbed on strong Lewis acid sites, created by the isolated Cu^{2+} ions and strongly bound NH_3 on Brønsted acid sites, respectively.

Accordingly, Wang et al. [24] reported the presence of three peaks at 97, 345, and 421 °C. On the other hand, after adsorbing ammonia at 100 °C, Wang et al. [30], found only two peaks at ca 250 °C and 420 °C, associated with NH_3 desorbed from Lewis acid sites (Cu^{2+} ions) and from Brønsted acid sites (Si-O(H)-Al), respectively. The attribution of the two peaks at higher temperatures are in agreement with DRIFT results reported by

Zhao et al. [31], who detected molecular NH_3 adsorbed onto Lewis acid sites by Cu^{2+} ions at 1623 cm^{-1} and NH^{4+} formed on Brønsted acid sites at 1465/1482 cm^{-1}. In conclusion, according to these attributions, the desorption event at intermediate temperature can be unambiguously associated with copper species, as also confirmed by its absence in the TPD of the reference SSZ-13 (Figure 7). Furthermore, the more weakly bound ammonia appears to be related to the zeolite structure since the low temperature peak also appeared in the TPD of unexchanged zeolites with roughly the same intensity. Eventually, the strong Brønsted acid sites (Si-O(H)-Al), which are present in large amount in the pristine SSZ-13 zeolite, represent exchangeable sites for copper since the high temperature desorption signal became very weak for all Cu/SSZ catalysts.

Concerning the nature of the copper species associated with the intermediate signal, it is reported [3] that, compared to Cu^{2+}-2Z, $[Cu(OH)]^+$-Z is more readily reducible by NH_3 to form $Cu(NH_3)^{n+}$ and H_2O, and that it plays a pivotal role in the NH_3-SCR reaction.

2.3. NH_3-SCR Tests

Exchanged monoliths were tested under conditions described in §3.3, achieving 100% NO conversion from about 150 to 400 °C. However, due to the significantly higher amount of zeolite in the reactor when using monoliths and to some flow limitations of the lab-scale rig, it was not possible to directly compare composite 3D monoliths and their corresponding pure zeolite catalysts at the same gas hourly space velocity. Therefore, 3D monoliths were grounded, and SCR tests were repeated using the same weight and particles size as for the corresponding samples of the pure Cu-SSZ-13 catalyst. In Figure 8, the NO_x conversion as function of the reaction temperature is reported for both the pure zeolites ion-exchanged for 4 and 24 h and their corresponding composite samples derived from monoliths. A blank run over the SSZ-13 powder confirmed a negligible contribution of the pristine zeolite and the key role of the metal active sites. On the other hand, the very low NO_x conversion of the Cu-exchanged binder without zeolite confirmed the active role of copper exchanged into zeolite positions. Roughly 100% NO_x conversion was achieved over all the Cu-exchanged SSZ-13 samples (both as pure and in combination with the binder) in the 200–300 °C temperature range, which is in good agreement with the average NO_x conversion reported by others for Cu-SSZ-13 catalysts in the same range of temperature [26,30,32]. The pure zeolite sample exchanged for only 4 h provided slightly better performance when compared to its counterpart exchanged for 24 h, despite its lower copper load. The decline of NO_x conversion for temperature beyond ca 320 °C was mostly associated with the direct NO oxidation to NO_2 (shown in Figure 8b for Cu/SSZ-13 P 24 h). On the other hand, N_2O production remained below 20 ppm and progressively decreased above 200 °C. It was verified by independent NO adsorption tests under NO/O_2 mixture at 100 °C followed by heating under inert flow that the emission of NO_2 at 300–400 °C occurred by direct NO oxidation. In other words, the Cu/zeolite is able to oxidize pre-adsorbed NO at high temperatures; however, in the presence of NH_3, this unwanted side reaction is limited by the prevailing of the SCR reaction.

In Figure 9, the Arrhenius plots for the NO consumption rate on pure Cu/SSZ-13 powders and composite monoliths are compared to corresponding data for a Cu/binder sample. The domain of the kinetic regime is confirmed by the linear trend of the experimental data obtained in the low to moderate conversion range.

In order to quantitatively confirm the enhanced intrinsic activity of Cu-SSZ-13 within the mesoporous binder, Table 3 reports the values of the apparent activation energy and of the kinetic constant, evaluated at 150 °C per unit mass of catalyst and of zeolite content in the sample; kinetic constants were calculated on the basis of integral reactor conversion data, assuming an ideal isothermal plug flow behaviour and a first-order dependency on NO concentration [26,33].

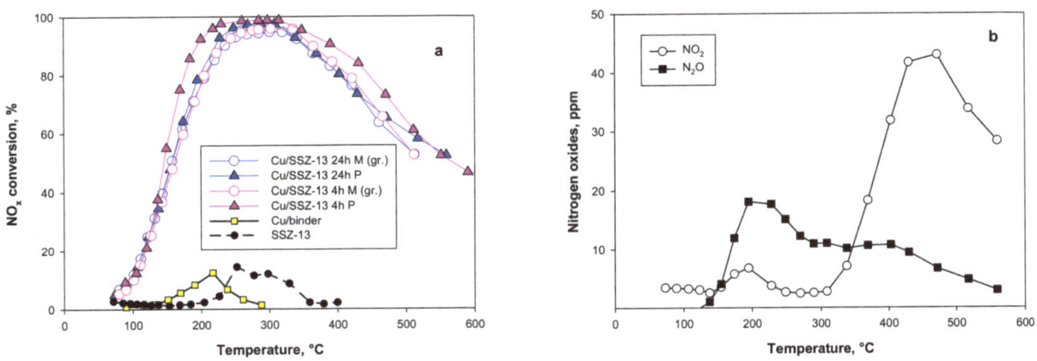

Figure 8. (**a**) NO_x conversion as a function of reaction temperature for copper exchanged (4 and 24 h) SSZ-13 powder and monolith compared with reference materials (Cu/binder and SSZ-13); (**b**) NO_2 and N_2O produced during the activity test over Cu/SSZ-13 P 24 h (Experimental conditions: Catalyst weight = 0.125 g; NO = NH_3 = 400 ppmv, 6% O_2; total flow rate = 25 Sdm^3 h^{-1}).

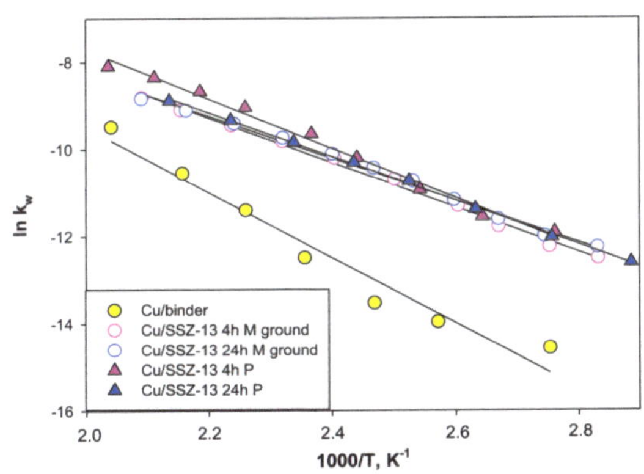

Figure 9. Arrhenius plots for the NO consumption rate per gram of catalyst during the NH_3-SCR of NO on Cu/SSZ-13 powders and monoliths and reference Cu/binder.

Table 3. Apparent activation energy (Ea) and first-order kinetic constant values for the NO removal rate by NH_3-SCR at 150 °C (k_w) per unit mass of catalyst and per unit mass of zeolite (k_{wzeol}) contained in the sample.

Catalyst	Ea (kJ mol^{-1})	$k_{wcatalyst}$ (cm^3 g^{-1} s^{-1})	k_{wzeol} (cm^3 g^{-1} s^{-1})
Cu/binder	62	4.96	-
Cu-SSZ-13 4 h P	49	58.9	58.9
Cu-SSZ-13 24 h P	42	45.2	45.2
Cu-SSZ-13 4 h M	46	44.7	74.5
Cu-SSZ-13 24 h M	44	49.4	82.3

As expected, the reference Cu/binder has a higher activation energy and a kinetic constant one order of magnitude lower than the values associated with Cu-SSZ-13 catalysts. The calculated values of the apparent reaction energy (Ea) of both monoliths and powders fall in quite a narrow range, as is in agreement with values reported for Cu/SSZ-13 by others [26,32], and lower than those evaluated for Cu/ZSM-5 (powder and monolith [23]), thus confirming the higher activity at lower temperatures for the SSZ-13-based catalyst with respect to ZSM-5 counterparts.

Notably, the intrinsic activity of the zeolite catalyst is higher when the SSZ-13 is incorporated into the silica matrix, as shown by the larger values of k_{wzeol} (last column in Table 3). In other words, both monoliths display a k_w referred to the total sample weight, which is higher than that expected summing up the two contributions proportional to the corresponding fraction of the components in the catalyst. A similar result was previously reported for a Cu/ZSM-5 in a geopolymer matrix [23], once again highlighting the positive effect of a hierarchical porosity on the SCR kinetics, rather than of a different nature of copper sites that, as shown by TPR and TPD analyses, seems to be not modified by the presence of the silica binder.

3. Materials and Methods

3.1. Catalysts Preparation

NH_4-exchanged SSZ-13 (SiO_2/Al_2O_3 = 25) was supplied by Tosoh Corporation (Tokyo, Japan) and preliminary deammoniated under He flow 3 h at 550 °C to obtain H/SSZ-13.

Zeolite-containing slurries were prepared from 59.29 wt% H-SSZ-13 powder, 35.57 wt% colloidal SiO_2 (LUDOX TM-50, Sigma-Aldrich, St. Louis, MO, USA), 3.95 wt% Laponite (Laponite-RD, BYK, Wesel, Germany), and 1.19 wt% carboxymethylcellulose (CMC Na-salt, Sigma-Aldrich, St. Louis, MO, USA) on a dry weight basis; after calcination, the composite consists of 60 wt% SSZ-13, 36 wt% SiO_2, and 4 wt% Laponite. Viscosity was adjusted by dilution with DI water, yielding an overall solid/liquid ratio of 1.53.

Slurries for control samples (without zeolite) were instead prepared from 87.38 wt% colloidal SiO_2, 9.71 wt% Laponite, 2.91 wt% CMC (dry weight basis), and DI water with an overall slurry/liquid ratio of 0.66, yielding a composite of 90 wt% SiO_2 and 10 wt% Laponite after calcination.

Slurry homogenisation was carried out on a high-energy planetary mixer (ARE-250, Thinky Corporation, Chiyoda, Japan) for five cycles (5 min, 2000 rpm) alternated by 5 min sonication cycles in an ultrasound bath. The slurries were allowed to cool down to room temperature after each mixing cycle.

The resulting homogenous pastes were 3D printed by Direct Ink Writing at room temperature on a Delta-type 3D printer (Delta 2040 Turbo, WASP, Massa Lombarda, Italy). The material was fed into the chamber of an auger-type extruder (LDM extruder 3.0, WASP, Massa Lombarda, Italy) via pneumatic pressure, from which it was extruded through an 840 μm tapered nozzle (VIEWEG GmbH, Kranzberg, Germany) with 630 μm layer height and 30 mm s^{-1} printing speed. 3D-printing toolpaths were generated through a custom algorithm implemented in the Rhinoceros/Grasshopper CAD software (Ver. 7 SR23, TLD Inc., Windsor, CT, USA) to produce cylindrical samples (Ø 19.5 mm, H 12.5 mm) with a log-pile scaffold configuration and 50% nominal design porosity.

Following Direct Ink Writing (DIW), the samples were dried for 24 h at 40 °C, and subsequently calcined at 550 °C (4 h, 5 °C min^{-1} heating ramp) in still air to remove the organic fraction and induce partial sintering of the SiO_2 nanoparticles.

An early attempt to embed the SSZ-13 into a geopolymer matrix, as already completed for ZSM-5, resulted in a very low mechanical resistance of the monolith and in the almost total loss of the crystalline structure of SSZ-13, as described in Section 2.

Before copper exchange, the 3D-printed monoliths were pre-treated in a HNO_3 aqueous solution to achieve a neutral pH, as described in [22,23], and then calcined for 4 h at 550 °C.

Then, copper was exchanged starting from a copper acetate solution (0.02 M) for 4 h or 24 h at 80 °C with a zeolite/solution volume ratio of 8 g/L for both powder and monolith SSZ-13. After the exchange, the monolith was calcined at 400 °C for 2 h.

Reference monolith samples containing 100% binder were also prepared and treated with HNO_3 then with copper solution, following the same procedure applied for monoliths exchanged for 4 h.

3.2. Catalysts Characterization

SEM analysis was carried out using an FEI Inspect scanning electron microscope (FEI Electron Optics, Eindhoven, The Netherlands).

Specific surface area measurements and micro/meso-pore analyses were performed in an Autosorb 1-C (Quantachrome Instruments, Boynton Beach, FL, USA) through N_2 adsorption at 77 K after degassing the samples for 12 h at 150 °C. The specific surface area and the pore size distribution were evaluated by the BET and NLDFT methods, respectively. The total pore volume was estimated by Gurvitch method.

The actual copper loading of both powder and monolith samples was evaluated by ICP-MS using an Agilent 7500 instrument (Santa Clara, CA, USA) after dissolving the samples in HF solution.

H_2-TPR analysis was carried out with a AutoChem 2020 (Micromeritics, Norcross, GA, USA) equipped with a TC detector. Catalyst samples (150 mg) were pre-treated in situ under air flow at 400 °C for 1 h; thereafter, 2% vol. H_2/Ar mixture (50 cm^3 min^{-1}) was passed through the sample, heating from room temperature up to 800 °C at 10 °C min^{-1}.

NH_3-TPD experiments were conducted with the same apparatus, pre-treating the catalyst at 250 °C (0.5 h) under He flow, cooling down to 100 °C, and saturating the sample at this temperature with a NH_3/He mixture for 1 h. Thereafter, ammonia was desorbed via heating up to 750 °C at 10 °C/min under He flow (50 cm^3 min^{-1}).

3.3. Catalytic Tests

SCR catalytic tests were carried out in the experimental rig described in [23] using a packed bed of powdered catalysts (0.125 g, 125–250 μm). To this aim, 3D composite monoliths were grounded and sieved to the desired size range before testing. The gas feed was obtained by mixing gas streams from high purity cylinders (1% NO in He, 0.4% NH_3 in He, O_2, He) (Air Liquide, Milano, Italy), regulated by independent, dent mass flow controllers (BROOKS MFC SLA5850S) (Brooks Instruments, Dresden, Germany) in order to achieve an inlet gas composition of NO = NH_3 = 400 ppmv (NO_2 impurity < 4 ppmv), 6% O_2. The total flow rate was fixed at 25 Sdm^3 h^{-1}, and the operating temperature ranged from 70 to 550 °C.

Catalytic tests under the same feed conditions were also repeated over 3D monoliths (d = 18 mm, l = 20 mm, average weight = 1.7 g), which were wrapped with a ceramic tape and placed into a tubular quartz reactor (d_{in} = 20 mm) [23,33].

Analysis of nitrogen oxides was performed by a continuous analyser for NO, N_2O (ND-IR), and NO_2 (UV) Emerson X-Stream XEGP (Emerson, Bussero (MI), Italy). Water and NH_3 from the product gas stream were removed using a Sycapent™ (P_2O_5) (Sigma-Aldrich, St. Louis, MO, USA) before entering the NO_x analyser.

4. Conclusions

Through an additive manufacturing approach, we successfully prepared 3D-structured monolith catalysts by embedding up to 60% weight fraction of a commercial SSZ-13 zeolite into a mesoporous SiO_2 matrix to provide both mechanical stability and hierarchical porosity. In particular, it was found that the use of silica as a binder preserved the SSZ-13 zeolite structure, as opposed to the more aggressive (strongly basic) precursors that were previously employed to produce a geopolymer matrix to embed the zeolite.

The addition of a small fraction of laponite into a silica matrix improved the mechanical stability during all the steps of a 3D-printing process by Direct Ink Writing. Ion exchange with copper was investigated to prepare active catalysts for the NH_3-SCR of NO. A prolonged copper exchange resulted in the introduction of additional inactive metal, likely not located in the main exchange positions. This effect is more evident for pure zeolite catalysts; however, it is favourably contrasted by the presence of the silica binder, which limits the undesired over-exchange phenomenon in monolith catalysts.

The main textural, acid, and redox properties of the SSZ-13 are preserved in the 3D monoliths with a negligible contribution of the binder that uniformly covers SSZ-13 particles without occluding pores, changing the nature of the copper, or neutralizing acid sites that were identified in the pristine Cu-SSZ-13. However, the binder plays a fundamental role, not only providing the required mechanical stability to the structured zeolite, but also enhancing the catalytic performance, thanks to the hierarchical porosity created by the intimate contact between the mesoporous silica and the microporous SSZ-13. While avoiding the common drawbacks of high pressure drops related to the use of a packed bed of particle catalysts, 3D-printed composite monoliths can outperform standard zeolite-wash coated honeycombs due to their higher loading of active phase, hierarchical porosity, and enhanced mass transfer coefficients.

In summary, the enhanced hydrothermal stability and sulphur tolerance of SSZ-13 relative to other SCR-active zeolites, coupled with the improved catalytic performance observed for Cu-exchanged 3D-printed silica/SSZ-13 monoliths, render these structured materials promising candidates for SCR applications in diesel engine exhaust treatment.

Author Contributions: Conceptualization, L.L., S.C. and G.F.; investigation, E.M.C., M.D. and G.F.; data curation, E.M.C. and M.D.; writing—original draft preparation, L.L.; writing—review and editing, S.C., G.F. and N.G.; visualization, S.C. and N.G.; supervision, L.L.; funding acquisition, L.L., G.F. and E.M.C. All authors have read and agreed to the published version of the manuscript.

Funding: This research was partially funded by Ministero Università e Ricerca—Italy (PRIN 2017PMR932) and European Union—NextGenerationEU under the National Recovery and Resilience Plan (NRRP), Mission 04 Component 2 Investment 3.1, Project Code: IR0000027-CUP:B33C22000710006-iENTRANCE@ENL: Infrastructure for Energy TRAnsition aNd Circular Economy @ EuroNanoLab.

Data Availability Statement: Data are contained within the article.

Conflicts of Interest: The authors declare no conflicts of interest.

References

1. Kwak, J.A.; Zhu, H.; Lee, J.H.; Peden, C.H.F.; Szanyi, J. Two different cationic positions in Cu-SSZ-13? *Chem. Commun.* **2012**, *48*, 4758–4760. [CrossRef] [PubMed]
2. Fan, C.; Chen, Z.; Pang, L.; Ming, S.; Dong, C.; Albert, K.B.; Liu, P.; Wang, J.; Zhu, D.; Chen, H.; et al. Steam and alkali resistant Cu-SSZ-13 catalyst for the selective catalytic reduction of NO_x in diesel exhaust. *Chem. Eng. J.* **2018**, *334*, 344–354. [CrossRef]
3. Shan, Y.; Du, J.; Zhang, Y.; Shan, W.; Shi, X.; Yu, Y.; Zhang, R.; Meng, X.; Xiao, F.-S.; He, H. Selective catalytic reduction of NO_x with NH_3: Opportunities and challenges of Cu-based small-pore zeolites. *Natl. Sci. Rev.* **2021**, *8*, nwab010. [CrossRef] [PubMed]
4. Putluru, S.S.R.; Riisager, A.; Fehrmann, R. Alkali resistant Cu/zeolite deNO_x catalysts for flue gas cleaning in biomass fired applications. *Appl. Catal. B Environ.* **2011**, *101*, 183–188. [CrossRef]
5. Chen, B.; Xu, R.; Zhang, R. Economical way to synthesize SSZ-13 with abundant ion-exchanged Cu^+ for an extraordinary performance in selective catalytic reduction (SCR) of NO_x by ammonia. *Environ. Sci. Technol.* **2014**, *48*, 13909–13916. [CrossRef] [PubMed]
6. Stewart, M.L.; Kamp, C.J.; Gao, F.; Wang, Y.; Engelhard, M.H. Coating Distribution in a Commercial SCR Filter. *Emiss. Control Sci. Technol.* **2018**, *4*, 260–270. [CrossRef]
7. Ma, L.; Cheng, Y.; Cavataio, G.; McCabe, R.W.; Fu, L.; Li, J. Characterization of commercial Cu-SSZ-13 and Cu-SAPO-34 catalysts with hydrothermal treatment for NH_3-SCR of NO_x in diesel exhaust. *Chem. Eng. J.* **2013**, *225*, 323–330. [CrossRef]
8. Paolucci, C.; Di Iorio, J.R.; Ribeiro, F.H.; Gounder, R.; Schneider, W.F. Chapter One—Catalysis Science of NO_x Selective Catalytic Reduction with Ammonia Over Cu-SSZ-13 and Cu-SAPO-34. In *Advances in Catalysis*; Academic Press: Cambridge, MA, USA, 2016; Volume 59, pp. 1–107.
9. Song, J.; Wang, Y.; Walter, E.D.; Washton, N.M.; Mei, D.; Kovarik, L.; Engelhard, M.H.; Prodinger, S.; Wang, Y.; Peden, C.H.F.; et al. Toward Rational Design of Cu/SSZ-13 Selective Catalytic Reduction Catalysts: Implications from Atomic-Level Understanding of Hydrothermal Stability. *ACS Catal.* **2017**, *7*, 8214–8227. [CrossRef]
10. Luo, J.; Gao, F.; Kamasamudram, K.; Currier, N.; Peden, C.H.F.; Yezerets, A. New insights into Cu/SSZ-13 SCR catalyst acidity. Part I: Nature of acidic sites probed by NH_3 titration. *J. Catal.* **2017**, *348*, 291–299. [CrossRef]
11. Daya, R.; Trandal, D.; Dadi, R.K.; Li, H.; Joshi, S.Y.; Luo, J.; Kumar, A.; Yezerets, A. Kinetics and thermodynamics of ammonia solvation on Z2Cu, ZCuOH and ZCu sites in Cu-SSZ-13—Implications for hydrothermal aging. *Appl. Catal. B Environ.* **2021**, *297*, 120444. [CrossRef]
12. Jangjou, Y.; Do, Q.; Gu, Y.; Lim, L.-G.; Sun, H.; Wang, D.; Kumar, A.; Li, J.; Grabow, L.C.; Epling, W.S. Nature of Cu Active Centers in Cu-SSZ-13 and Their Responses to SO_2 Exposure. *ACS Catal.* **2018**, *8*, 1325–1337. [CrossRef]
13. Du, J.; Shan, Y.; Sun, Y.; Gao, M.; Liu, Z.; Shi, X.; Yu, Y.; He, H. Unexpected increase in low-temperature NH_3-SCR catalytic activity over Cu-SSZ-39 after hydrothermal aging. *Appl. Catal. B Environ.* **2021**, *294*, 120237. [CrossRef]

14. Ma, Y.; Cheng, S.; Wu, X.; Ma, T.; Liu, L.; Jin, B.; Liu, M.; Liu, J.; Ran, R.; Si, Z.; et al. Improved hydrothermal durability of Cu-SSZ-13 NH_3-SCR catalyst by surface Al modification: Affinity and passivation. *J. Catal.* **2022**, *405*, 199–211. [CrossRef]
15. Liu, B.; Lv, N.; Wang, C.; Zhang, H.; Yue, Y.; Xu, J.; Bi, X.; Bao, X. Redistributing Cu species in Cu-SSZ-13 zeolite as NH_3-SCR catalyst via a simple ion-exchange. *Chin. J. Chem. Eng.* **2022**, *41*, 329–341. [CrossRef]
16. Peng, C.; Yan, R.; Peng, H.; Mi, Y.; Liang, J.; Liu, W.; Wang, X.; Song, G.; Wu, P.; Liu, F. One-pot synthesis of layered mesoporous ZSM5 plus Cu ion-exchange: Enhanced NH_3-SCR performance on Cu-ZSM5 with hierarchical pore structures. *J. Hazard. Mater.* **2020**, *385*, 121593. [CrossRef] [PubMed]
17. Chen, Z.; Ye, T.; Qu, H.; Zhu, T.; Zhong, Q. Progressive regulation of Al sites and Cu distribution to increase hydrothermal stability of hierarchical SSZ-13 for the selective catalytic reduction reaction. *Appl. Catal. B Environ.* **2022**, *303*, 120867. [CrossRef]
18. Wijayanti, K.; Xie, K.; Kumar, A.; Kamasamudram, K.; Olsson, L. Effect of gas compositions on SO_2 poisoning over Cu/SSZ-13 used for NH_3-SCR. *Appl. Catal. B Environ.* **2017**, *219*, 142–154. [CrossRef]
19. Wijayanti, K.; Leistner, K.; Chand, S.; Kumar, A.; Kamasamudram, K.; Currier, N.W.; Yezerets, A.; Olsson, L. Deactivation of Cu-SSZ-13 by SO_2 exposure under SCR conditions. *Catal. Sci. Technol.* **2016**, *6*, 2565–2579. [CrossRef]
20. Lisi, L.; Pirone, R.; Russo, G.; Stanzione, V. Cu-ZSM5 based monolith reactors for NO decomposition. *Chem. Eng. J.* **2009**, *154*, 341–347. [CrossRef]
21. Nova, I.; Bounechada, D.; Maestri, R.; Tronconi, E.; Heibel, A.K.; Collins, T.A.; Boger, T. Influence of the substrate properties on the performances of NH_3-SCR monolithic catalysts for the aftertreatment of diesel exhaust: An experimental and modeling study. *Ind. Eng. Chem. Res.* **2011**, *50*, 299–309. [CrossRef]
22. Cepollaro, E.M.; Cimino, S.; Lisi, L.; Botti, R.; Colombo, P.; Franchin, G. Cu-exchanged 3D-printed Geopolymer/ZSM-5 Monolith for Selective Catalytic Reduction of NO_x. *Chem. Eng. Trans.* **2021**, *84*, 67–72.
23. Cepollaro, E.M.; Botti, R.; Franchin, G.; Lisi, L.; Colombo, P.; Cimino, S. Cu/ZSM5-Geopolymer 3D-Printed Monoliths for the NH_3-SCR of NO_x. *Catalysts* **2021**, *11*, 1212. [CrossRef]
24. Wang, Z.; Xu, X.; Zhu, Y.; He, H.; Wang, N.; Yang, X.; Liu, L. One-pot synthesis of hierarchical MnCu-SSZ-13 catalyst with excellent NH_3-SCR activity at low temperatures. *Microporous Mesoporous Mater.* **2022**, *333*, 111720. [CrossRef]
25. Thommes, M.; Cychosz, K. Physical adsorption characterization of nanoporous materials: Progress and challenges. *Adsorption* **2014**, *20*, 233–250. [CrossRef]
26. Kwak, J.H.; Tran, D.; Szanyi, J.; Peden, C.H.F.; Lee, J.H. The Effect of Copper Loading on the Selective Catalytic Reduction of Nitric Oxide by Ammonia Over Cu-SSZ-13. *Catal. Lett.* **2012**, *142*, 295–301. [CrossRef]
27. Clemens, A.K.S.; Shishkin, A.; Carlsson, P.-A.; Skoglundh, M.; Martínez-Casado, F.J.; Matěj, Z.; Balmes, O.; Härelind, H. Reaction-driven Ion Exchange of Copper into Zeolite SSZ-13. *ACS Catal.* **2015**, *5*, 6209–6218. [CrossRef]
28. Delavernhe, L.; Pilavtepe, M.; Emmerich, K. Cation exchange capacity of natural and synthetic hectorite. *Appl. Clay Sci.* **2018**, *151*, 175–180. [CrossRef]
29. Khurana, I.; Albarracin-Caballero, J.D.; Shih, A.J. Identification and quantification of multinuclear Cu active sites derived from monomeric Cu moieties for dry NO oxidation over Cu-SSZ-13. *J. Catal.* **2022**, *413*, 111–1122. [CrossRef]
30. Wang, C.; Wang, J.; Wang, J.; Wang, Z.; Chen, Z.; Li, X.; Shen, M.; Yan, W.; Kang, X. The Role of Impregnated Sodium Ions in Cu/SSZ-13 NH_3-SCR Catalysts. *Catalysts* **2018**, *8*, 593. [CrossRef]
31. Zhao, H.; Yang, G.; Hill, A.J.; Luo, B.; Jing, G. One-step ion-exchange from Na-SSZ-13 to Cu-SSZ-13 for NH_3-SCR by adjusting the pH value of Cu-exchange solution: The effect of H^+ ions on activity and hydrothermal stability. *Microporous Mesoporous Mater.* **2021**, *324*, 111271. [CrossRef]
32. Cui, Y.; Wang, Y.; Walter, E.D.; Szanyi, J.; Wang, Y.; Gao, F. Influences of Na^+ co-cation on the structure and performance of Cu/SSZ-13 selective catalytic reduction catalysts. *Catal. Today* **2020**, *339*, 233–240. [CrossRef]
33. Gargiulo, N.; Caputo, D.; Totarella, G.; Lisi, L.; Cimino, S. Me-ZSM-5 monolith foams for the NH_3-SCR of NO. *Catal. Today* **2018**, *304*, 112–118. [CrossRef]

catalysts

MDPI

Article

Interaction of O_2 with Reduced Ceria Nanoparticles at 100–400 K: Fast Oxidation of Ce^{3+} Ions and Dissolved H_2

Kristina Chakarova , Nikola Drenchev , Mihail Mihaylov and Konstantin Hadjiivanov *

Institute of General and Inorganic Chemistry, Bulgarian Academy of Sciences, 1113 Sofia, Bulgaria; ndrenchev@svr.igic.bas.bg (N.D.); misho@svr.igic.bas.bg (M.M.)
* Correspondence: kih@svr.igic.bas.bg

Abstract: The interaction between O_2 and reduced ceria nanocubes was mainly investigated using FTIR spectroscopy. Nanorods and nanoparticles were also studied for comparison. Adsorption of O_2 at 100 K on unreduced ceria produces only O_2 molecularly adsorbed on Ce^{4+} sites. The Ce^{3+} cations on ceria reduced by H_2 at 773 K were monitored using the ${}^2F_{5/2} \rightarrow {}^2F_{7/2}$ electronic transition band at 2133–2095 cm^{-1}. This band possesses a fine structure well resolved at 100 K. The positions of the individual components depend on the Ce^{3+} environment, including the presence of nearby species such as OH groups. Even at 100 K, adsorption of O_2 on reduced ceria leads to fast oxidation of about half of the Ce^{3+} cations, including all Ce^{3+} sites bound to OH groups and carbonates, and the simultaneous formation of superoxo (O_2^-) and peroxo (O_2^{2-}) species. The remaining Ce^{3+} sites disappear upon heating up to 348 K. At higher temperatures, the peroxo species decompose directly, yielding lattice oxygen. Superoxides are converted to hydroperoxides, which then decompose into terminal OH groups. Reduced samples evacuated at T < 773 K contain sorbed H_2. Part of this hydrogen is also fast oxidized even at 100 K.

Keywords: ceria; oxidation; oxygen; nitrogen monoxide; nitrous oxide; FTIR spectroscopy

Citation: Chakarova, K.; Drenchev, N.; Mihaylov, M.; Hadjiivanov, K. Interaction of O_2 with Reduced Ceria Nanoparticles at 100–400 K: Fast Oxidation of Ce^{3+} Ions and Dissolved H_2. *Catalysts* **2024**, *14*, 45. https://doi.org/10.3390/catal14010045

Academic Editor: Soghomon Boghosian

Received: 26 November 2023
Revised: 30 December 2023
Accepted: 7 January 2024
Published: 9 January 2024

1. Introduction

Ceria has found important industrial applications for various purposes, in particular, as a catalyst, catalyst component, support, or additive [1]. It is generally recognized that one of the key advantages of ceria in catalysis is its redox chemistry: under operating conditions, Ce^{4+} cations on the surface can be easily and reversibly reduced to Ce^{3+}, thereby forming oxygen vacancies, all without changing the crystallographic structure [2–4]. Thus, ceria can store and provide oxygen in the course of redox catalytic reactions. For the design of new efficient catalysts, detailed knowledge of the mechanism of oxidation and reduction of the ceria surface is extremely important. This knowledge is also essential for the biomedical applications of ceria nanoparticles [5,6].

Stoichiometric cerium dioxide is quite stable in the absence of reducing agents. For instance, TPD experiments show no substantial desorption of O_2 from the (111) ceria plane below 1300 K [7]. Exposure of ceria to reducing agents or energetic irradiation, however, can produce Ce^{3+} sites on the surface even at much lower temperatures [4].

Many TPR studies have revealed that the reduction of ceria with hydrogen occurs in two steps [8–14]. During the first step, Ce^{4+} cations situated on the surface are reduced, and this is reflected by one or more TPR peaks in the temperature range 685–835 K. Indeed, the ceria surface oxygen is mobile and can be easily removed at low temperatures [11,15]. Several research groups have reported a correlation between CeO_2 surface area and low-temperature hydrogen consumption [8,9,11,12]. Reduction of bulk takes place at higher temperatures (TPR peaks between 1010 and 1170 K) because of the low mobility of bulk oxygen and its difficult transport to the surface [11].

Although the general picture of ceria reduction by H_2 is clear, there are many factors affecting the precise interpretation of the TPR results. Different crystal faces are exposed on the surface, and they are expected to have different reducibility. The presence of surface impurities, such as adsorbed oxygen, residual carbonates, etc. [9,12,16], may also provoke some variations in the TPR signal. Hydrogen adsorption and drop in surface area could also affect the low-temperature part of the TPR profile [12,16]. All this leads to a complex structure of the low-temperature TPR peak.

During the last two decades, many investigations have appeared on ceria nanoparticles with different well-defined shapes, e.g., nanocubes, nanorods, nanopolyhedra, etc. [5,6,17–21]. Each shape is characterized by a preferential exposure of one or more crystal facets on the surface. Although there is still disagreement in some details, these studies, in principle, allow for the estimation of the reactivity of one or another plane in a particular process.

A powerful technique for surface characterization is vibrational spectroscopy, mainly in its IR and Raman modes. It gives information on the surface hydroxyl coverage and the presence of some impurities, such as residual carbonates or nitrates. Moreover, this technique is unique for the determination of the structure of adsorbed species. Other important information, e.g., on the surface acidity, can be obtained using the so-called probe molecules. Especially for ceria, vibrational spectroscopy can also be used to follow its reduction degree. This is because the Ce^{3+} $^2F_{5/2} \rightarrow {}^2F_{7/2}$ spin-orbit electronic transition is observed in the IR spectra as a separate band around 2147–2110 cm^{-1} [20,22–29]. Although the spectral region is wide, no attempts have been made to correlate the band position with Ce^{3+} sites in different environments.

While the reduction of ceria is relatively well studied, there are fewer works considering its reoxidation. Ayastuy et al. [14] studied temperature-programmed oxidation (TPO) of ceria just after a TPR run and observed one peak at 326 K, attributed to surface oxidation. Fally et al. [13] also found that oxidation of ceria pre-reduced at 973 K occurs easily, and the process is nearly completed even at 298 K. Similar results were reported for ceria reduced at 773 K [27]. It was also noted [30] that reduced ceria surfaces are not stable at ambient conditions and exhibit very fast reoxidation kinetic.

When ceria is reduced at high temperatures, and bulk reduction has occurred, the reoxidation occurs in steps below and above 473 K [31]. Therefore, it appears that just as bulk Ce^{4+} is more difficult to reduce than surface Ce^{4+}, oxidation of Ce^{3+} in the bulk is more difficult than on the surface.

IR Spectroscopic studies have revealed that O_2 adsorption on reduced ceria leads to the formation of superoxide (O_2^-) [32–34] and peroxide (O_2^{2-}) species [33,34]. Similar results have been obtained using Raman spectroscopy [17,19,35]. Superoxide is characterized by a ν(O–O) band at 1139–1127 cm^{-1} and is stable up to 348 K [19]. Peroxide species are different kinds, and ν(O–O) was detected in the 890–830 cm^{-1} region [19,33]. They are more stable and disappear from the spectra after treatment above 473 K. A stepwise conversion between O_2, O_2^-, O_2^{2-}, O^-, and lattice O^{2-} has been proposed [19,33]. Note that the production of superoxide and peroxide species leads to immediate (fast) oxidation of Ce^{3+}. This means that the redox process occurs simultaneously with adsorption.

In light of the above considerations, many studies reporting a significant concentration of Ce^{3+} ions on ceria samples pre-exposed to air seem surprising. Analysis of these results shows that, in most cases, the conclusions were made on the basis of results from two techniques: XPS and Raman spectroscopy.

An elegant proof of the reduction of ceria during XPS analysis was recently provided [36]. The authors used sequential UPS-XPS-UPS analysis. UPS spectra of pre-oxidized ceria revealed no Ce^{3+}. However, subsequent XPS analysis indicated a rather high concentration of Ce^{3+} ions. When the UPS spectra were taken again with the same sample, Ce^{3+} ions were detected. These results unequivocally demonstrate the photoreductive effect of X-rays on ceria, and it has been pointed out that the initial Ce^{3+}/Ce^{4+} ratio can be strongly overestimated when determined using the Ce^{3+} XPS signals. In addition, it has been recently underlined that the widespread assignment of an XPS peak at 531–532 eV to oxygen vacancies is not correct [37].

Consider now the Raman results. The most important feature in the Raman spectra of stoichiometric ceria is an intense F_{2g} mode at 460 cm^{-1}. Reduction of ceria leads to the development of the so-called D1-band (D stands for defects) observed around 550 cm^{-1}. This band is attributed to Ce^{3+}–O^{2-}–Ce^{4+} modes [38] and is associated with the presence of oxygen vacancies [21,28,39–41]. It disappeared after adsorption of oxygen [17, 42], which confirms the assignment. Very often, another D-band (the so-called D2 band at ca. 600 cm^{-1}) is incorrectly also attributed to oxygen vacancies in ceria, which is consequently associated with the presence of reduced Ce^{3+} sites. Therefore, it may be concluded that in some of these studies, the Ce^{3+}/Ce^{4+} ratio is also strongly overestimated.

In this work, we investigate the reoxidation by O_2 of ceria nanoshapes reduced at 773 K. This temperature ensures that mainly Ce^{4+} cations on the surface are reduced and prevents morphological changes in the samples. We first analyzed the Ce^{3+} electronic band appearing at 2133–2095 cm^{-1} and found that it possesses a fine structure that is better resolved at 100 K. We have shown that O_2 readily oxidizes a large fraction of Ce^{3+} to Ce^{4+} even at 100 K. Some residual Ce^{3+} sites resist the fast oxidation even in the presence of gas-phase O_2, but at 400 K, the oxidation is complete. Furthermore, we found that hydrogen dissolved in ceria is also oxidized, some of it even at 100 K, indicating its high reactivity.

2. Results

2.1. Initial Characterization of the Samples

We have studied three ceria samples prepared using hydrothermal synthesis and differing in morphology: nanocubes, nanorods, and nanoparticles. Some basic characteristics, as well as the notation of the samples that will be used further on in the text, are presented in Table 1.

Table 1. Some characteristics of the samples studied.

Sample	Particle Shape	S_{BET}, m^2 g^{-1}	Mean Crystallite Size, nm [1]
CeO_2-NC	cubes	27	30
CeO_2-NR	rods	110	7
CeO_2-NP	polyhedra	140	7

[1] Average particle size determined according to Scherrer's equation.

2.1.1. TEM Studies

TEM images of the investigated samples are shown in Figure 1. The CeO_2-NC sample consists of agglomerates of nanoparticles with a cubic morphology and a typical particle size of ca. 30–35 nm, although smaller particles also exist (Figure 1a). A few particles with a polyhedral, sphere-like morphology were also noticed. Analysis showed that mainly {100} facets were exposed on the sample surface, while {111} and {110} facets were found to a small extent. The nanorods (CeO_2-NR) are between 80 and 140 nm in length and about 8 nm in diameter (Figure 1b). With this sample, the most exposed facets are {110} and {111}, while only a small fraction of {100} facets was detected. The CeO_2-NP sample is composed of polyhedral nanoparticles with a size of 2–8 nm (Figure 1c). They are packed randomly with {100}, {110} and {111} lattice fringes. In this case, the presence of traces of nanorods was also noticed.

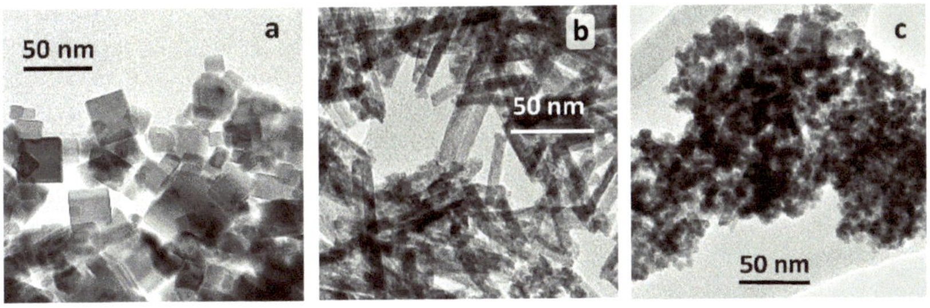

Figure 1. TEM photographs of CeO_2-NC (**a**), CeO_2-NR (**b**), and CeO_2-NP (**c**).

2.1.2. XRD Studies

Figure 2 presents the XRD patterns of our ceria samples. CeO_2-NC shows sharp peaks at 28.5°, 33.0°, 47.4°, 56.2°, 59.1°, 69.4°, 76.7°and 79°. These peaks are attributed to the (111), (200), (220), (311), (222), (400), (331), and (420) Miller indices of the cubic fluorite structure of CeO_2 (CaF_2 structural type, space group Fm3m) according to the ICSD file 53,995.

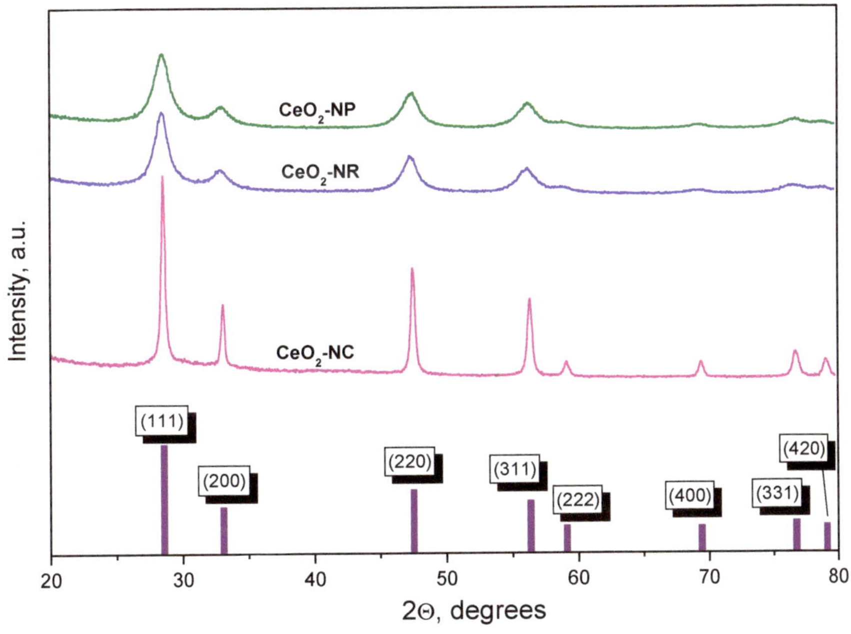

Figure 2. XRD patterns of CeO_2-NC, CeO_2-NR, and CeO_2-NP. Data from the ICSD file 53,995 are shown at the bottom for comparison.

The XRD patterns of CeO_2-NP and CeO_2-NR are very similar. Compared to the CeO_2-NC sample, the peaks are broader, which is consistent with the smaller particle size and higher specific surface area of these two samples.

2.1.3. TPR Studies

The TPR profiles obtained with the ceria samples (Figure 3) are in general agreement with previous findings [8–14]. The CeO_2-NR and CeO_2-NP samples show similar TPR profiles. The bulk reduction is associated with the peaks at 1103–1116 K, where surface reduction gives rise to composite peaks below 900 K. Computer deconvolution suggests that the low-temperature peaks consist of at least four components each. In addition, a negative peak around 870 K is also detected with both samples. This peak was attributed to the desorption of H_2 [10,43]. However, reduction of residual surface carbonates and release of CO could also contribute to the peak, as proposed by Perrichon et al. [9]. Indeed, some carbonates were detected using IR spectroscopy with all samples (see below).

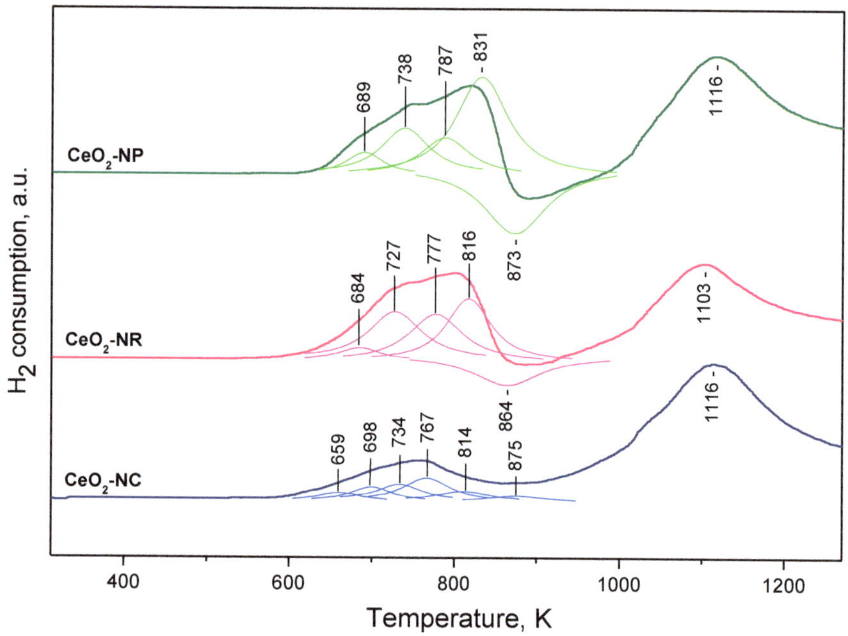

Figure 3. H_2-TPR profiles of CeO_2-NC, CeO_2-NR and CeO_2-NP.

The surface of the CeO_2-NC sample is more easily reduced. It is characterized by a less intense first peak, which is consistent with the lower specific surface of this sample. The peak appears at ca. 760 K, i.e., at a temperature lower than that for the CeO_2-NP and CeO_2-NR samples. It also has a complex character. The second peak attributed to bulk reduction is registered at 1116 K. No negative peak was discernible with this sample. However, we cannot rule out the possibility of its existence and cancelation by a positive peak around this temperature.

Data on the hydrogen consumption below and above 923 K for the different samples are provided in the Supporting Materials (Table S1).

2.2. *Background IR Spectra of Oxidized and Reduced Samples*

2.2.1. CeO_2-NC

The FTIR spectra of activated and reduced CeO_2-NC samples are compared in Figure 4. The activated sample is characterized by a typical oxidized ceria hydroxyl spectrum (see the left inset in Figure 4): a band at 3723 cm^{-1} is assigned to terminal Ce^{4+}–OH groups, while the bands at 3653 and 3635 cm^{-1} correspond to bridging hydroxyls [23,24,27,44]. A weak composite band at 3511 cm^{-1} is also visible. It is often attributed to the oxy-hydroxide phase [23,24,27,44] or triply bridging OH groups [27,44]. We support the former assignment because the frequency is too low for bridging species [45] and indicates that the hydroxyls are H–bonded. In the low-frequency region, two sharp bands at 1460 and 1384 cm^{-1} and weaker bands at 1068 and 857 cm^{-1} are detected. These bands are attributed to residual carbonate species [46].

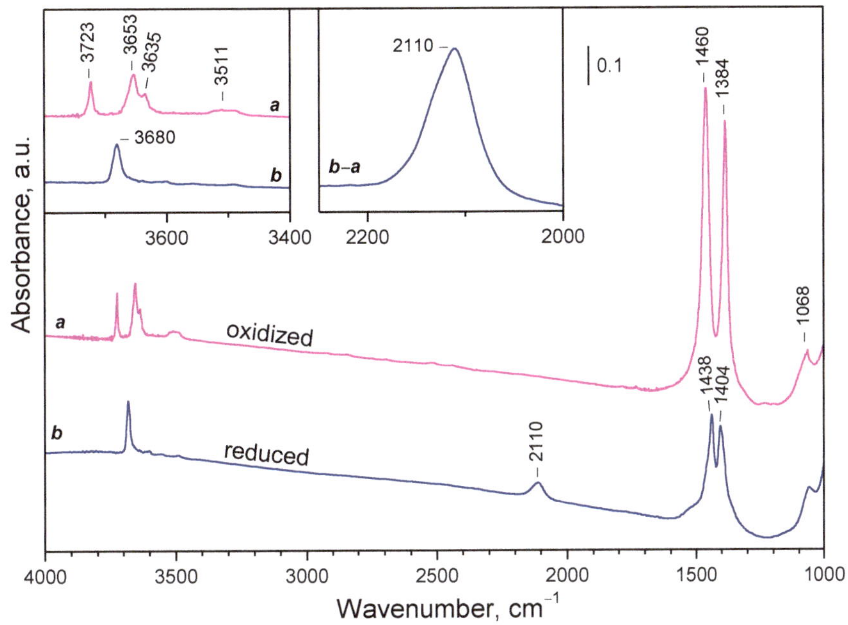

Figure 4. FTIR spectra of CeO_2-NC after activation (**a**) and reduction (**b**) at 773 K.

Reduction at 773 K, followed by evacuation at the same temperature, leads to specific changes in the IR spectrum. First, the bands due to linear and bridging OH groups disappeared, and a new hydroxyl band appeared at 3680 cm^{-1}. This band is typical of bridging hydroxyls on reduced ceria [10,24,47]. The bands of H–bonded hydroxyls (around 3500 cm^{-1}) strongly decreased in intensity. This also happened with the carbonate bands at 1460 and 1384 cm^{-1}, which also shifted to become closer to each other. This smaller band split is characteristic of higher symmetry of related carbonates [46]. It was also noticed that the carbonate bands slightly decreased in intensity after each reoxidation/reduction step. Finally, a new band at 2110 cm^{-1} emerged (see the right inset in Figure 4). This band is assigned to the $^2F_{5/2} \rightarrow {}^2F_{7/2}$ electronic transition of Ce^{3+} [20,22–27] and is used in what follows to estimate the reduction degree of our samples. The band has a complex contour, and computer deconvolution suggests it consists of several components. We have also estimated that the detection limit for Ce^{3+} in our experiments with this sample is ca. 0.03% of the total amount of cerium cations.

2.2.2. CeO_2-NR and CeO_2-NP

The FTIR spectra of the CeO_2-NR sample (Figure S1) slightly differ from those of the CeO_2-NC. With the oxidized sample, the band due to terminal OH groups is of negligible intensity, while the hydroxyl spectrum of the reduced sample is more complex. Carbonate bands were of lower intensity and detected at 1573, 1486, 1353, 1292 and 1068 cm^{-1}. For the reduced sample, only two main bands were observed at 1459 and 1371 cm^{-1}, i.e., again, they became closer in positions. The Ce^{3+} electronic band is composite, but the main component appears to be at 2128 cm^{-1}.

The spectra of CeO_2-NP (Figure S2) are rather similar to the spectra of CeO_2-NR. The main difference is the existence of both oxidized and reduced samples of a broad band centered at ca. 3400 cm^{-1}, which is attributed to occluded H–bonded hydroxyl groups.

Finally, we estimated the coverage by residual carbonates. For that purpose, we adsorbed CO_2 under 2 mbar equilibrium pressure on CeO_2-NP and then evacuated the gas phase at ambient temperature. The carbonate bands produced in the 1700–1200 cm^{-1}

region were too intense to be followed accurately. However, the deformation modes at 858 cm^{-1} appeared with optimal intensity, and it was found that the residual carbonates accounted for ca. 6% of the total amount of carbonates produced after CO_2 adsorption. Therefore, we may conclude that only a small part of the activated sample is covered by carbonates.

2.2.3. Oxidation of Reduced Ceria by Different Reagents

It has been reported that reduced ceria surface can be easily re-oxidized at ambient temperature by O_2, NO, and N_2O [25]. However, there are reports describing its oxidation by CO_2 and H_2O [48]. Although in the latter two cases, the process is considered to proceed at high temperature, we have checked the stability of Ce^{3+} in the presence of H_2O and CO_2 at room temperature. The results showed that no oxidation occurs under these conditions (Figure S3). Only a large amount of water caused erosion of the Ce^{3+} band, but it was fully recovered after subsequent evaluation at 773 K.

2.2.4. Temperature Changes of the IR Spectra

It is known that lowering the temperature could lead to better resolution of superimposed bands. Figure 5 compares the spectra in the region of Ce^{3+} electronic transition of reduced samples registered at RT, and at 100 K. First, the Ce^{3+} band increases in intensity at low temperatures. Secondly, several components become well-resolved. The low-frequency components (2105 and 2094 cm^{-1}) are emphasized for the CeO_2-NC sample, which may suggest that they are associated with the {100} facets.

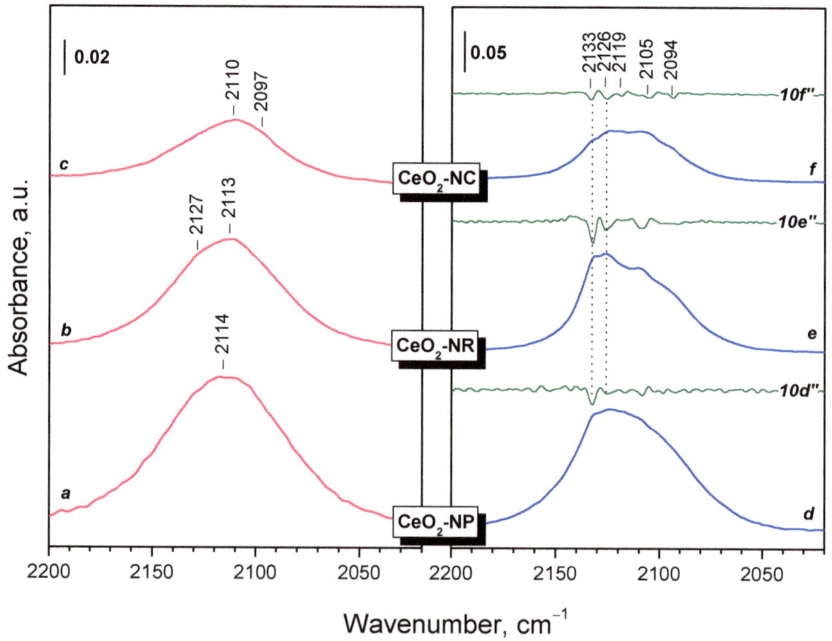

Figure 5. FTIR spectra of reduced ceria samples in the region of Ce^{3+} electronic transition: CeO_2-NP (**a**,**d**), CeO_2-NR (**b**,**e**), and CeO_2-NC (**c**,**f**). Spectra (**a–c**) are registered at ambient temperature, while spectra (**d–f**) are at 100 K. The second derivatives of the spectra (**d–f**) are multiplied by a factor of 10 and are denoted as 10d″, 10e″ and 10f″. All spectra are baseline-corrected.

The hydroxyl spectra, and to a smaller extent, the carbonate bands, are also affected by temperature (see Figure S4).

2.3. Interaction of Activated Samples with O_2 at Variable Temperatures

Although most authors are of the opinion that reactive oxygen adsorption on ceria occurs on reduced Ce^{3+} cations, there are also opinions that the adsorption sites are Ce^{4+} cations [33,34]. In order to check this, we have studied O_2 adsorption on activated (oxidized) samples. No Ce^{3+} electronic band was detected for CeO_2-NC samples activated at 573, 673, and 773 K, indicating the lack of Ce^{3+} sites. Adsorption of O_2 (2 mbar equilibrium pressure) at 100 K on these samples after each of these pretreatments led only to the appearance of a weak band at 1555–1553 cm^{-1} attributed to molecularly adsorbed O_2 (Figure S5), in agreement with previous studies. The O_2 band increases in intensity with the activation temperature of ceria due to dehydroxylation and the creation of Ce^{4+} Lewis acid sites. Warming the sample up to ambient temperature leads only to the fast disappearance of this band. The same situation was found with the other two samples. Therefore, we conclude that molecular adsorption of O_2 proceeds on Ce^{4+} sites, and for reactive O_2 adsorption, Ce^{3+} sites are necessary.

2.4. Interaction of Reduced CeO_2-NC with O_2 at Variable Temperature

In order to follow the initial stages of ceria reoxidation, we studied the interaction of reduced samples at 100 K, followed by a gradual increase in the temperature. For the unambiguous assignment of the peaks, we also performed experiments with $^{18}O_2$.

2.4.1. Interaction with $^{16}O_2$

The background spectrum of CeO_2-NC registered at 100 K is shown in Figure 6, spectrum a. As already discussed, the intensity of the Ce^{3+} electronic band is enhanced as compared to the spectrum registered at ambient temperature. The hydroxyl and carbonate bands are slightly shifted. The introduction of O_2 (2 mbar) to the sample leads to an immediate substantial drop in the intensity of the Ce^{3+} electronic band (Figure 6, spectrum b). Difference spectra (a–b) indicate that the band was substantially eroded from the higher and lower frequency sides. This indicates the lack of correlation between the band position and the reactivity of the related species. According to the integral band area, 45% of the Ce^{3+} sites were immediately oxidized. No further measurable changes occurred when the sample was allowed to stay in the O_2 atmosphere at 100 K for 10 min.

In the ν(OH) region (Figure 6, panel B), the band at 3685 cm^{-1} practically disappeared, and a new band (typical of the oxidized sample) appeared at 3653 cm^{-1}. The band at 3503 cm^{-1} remained intact. The results indicate that all Ce^{3+} sites bound to OH species (3685 cm^{-1}) are very rapidly oxidized.

A band at 1127 cm^{-1} developed upon O_2 exposure (Figure 6, panel C) and is attributed to the ν(O–O) modes of superoxo (O_2^-) species [32–34]. The respective overtone was detected at 2237 cm^{-1} (see Figure 6, pane A). This indicates that oxidation of Ce^{3+} with the formation of Ce^{4+}–O_2^- species takes place.

It should also be noted that weak bands developed in the 900–800 cm^{-1} region and could characterize peroxo species. However, the strong background changes in this region during reoxidation hinder the exact assignments. For proper assignment of these bands, we further explored the adsorption of isotopically labeled oxygen, $^{18}O_2$ (see below).

Finally, a weak band at 1553 cm^{-1}, also detected with the oxidized sample, appeared with reduced intensity. The band is of very weak stability and was already attributed to molecularly adsorbed O_2. It will not be discussed further on.

As the temperature increased, the Ce^{3+} electronic band progressively faded and disappeared at 373 K. The intensity of the superoxide band initially slightly increases (Figure 6, spectra c, d) and then starts to decrease, with the band disappearing at 400 K (Figure 6, spectra e–k). The OH band is slightly red-shifted to 3651 cm^{-1}, and weak components due to terminal hydroxyls (3722 and 3705 cm^{-1}) developed at 400 K.

Note that the intensity changes and the band shifts in the experiments performed at different temperatures should be taken with care. First, the decrease in the Ce^{3+} band in intensity is partly due to temperature increase. The shift and the slight decrease in intensity of the OH band at 3633 cm^{-1} should also be mainly related to temperature change, and this most probably applies to the position of the superoxide band.

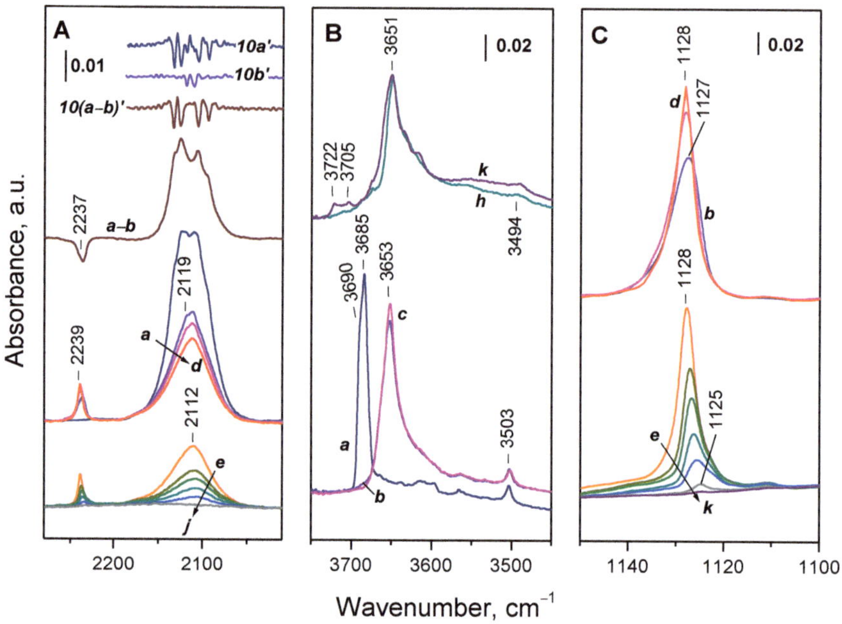

Figure 6. FTIR spectra of reduced CeO_2-NC were recorded at 100 K (**a**), after the addition of 2 mbar O_2 at 100 K (**b**), a gradual increase in temperature up to ambient one (**c**–**h**), and after 5 min. heating the sample in the oxygen atmosphere at 323 K (**i**), 373 K (**j**), and 400 K (**k**). (**A**) shows the region of Ce^{3+} electronic transition, and the spectra are baseline corrected. Some second derivatives are shown at the top part of the panel. (**B**) shows the hydroxyl stretching region. (**C**) presents the spectra in the ν(O–O) region of the superoxide anion, and the spectra are background-corrected.

2.4.2. Interaction with $^{18}O_2$

To confirm the proposed assignments of bands to superoxo and peroxo species, we performed experiments with $^{18}O_2$. In order to obtain more information on the elementary reaction steps, we first studied successive adsorption of small doses of $^{18}O_2$ (Figures 7 and 8).

Difference spectra presented in Figure 7 shows that with the increase in the amount of $^{18}O_2$ introduced into the IR cell, four components of the Ce^{3+} electronic band, at 2133, 2126, 2105, and 2095 cm^{-1}, disappear almost simultaneously.

The band at 1128 cm^{-1} observed after adsorption of $^{16}O_2$ was not detected, but a band at 1064 cm^{-1} developed instead. The isotopic shift factor of 1.06 is in excellent agreement with the expectation for the ν(O–O) modes. This confirms the assignment of the 1128 cm^{-1} band to O_2^- species and also shows that no $(^{16}O^{18}O)^-$ species were produced during the experiments. As expected, the overtone of the $^{16}O_2^-$ band was also missing. According to the isotopic shift factor, it should be observed at 2110 cm^{-1} but is superimposed with the Ce^{3+} band. Therefore, one should consider that a small contribution of the overtone as a positive band exists in the spectra presented in the left panel of Figure 7 (position marked with red segment).

Consider now the region where peroxide species are expected to absorb. The analysis is complicated because of the shift of the background band at ca. 865 cm^{-1} during sample oxidation and the sample cut-off around 870 cm^{-1}. However, a comparison of the spectra of adsorbed $^{16}O_2$ and $^{18}O_2$ (Figure 8) allows for making definite conclusions. For convenience, the results are summarized in Table 2.

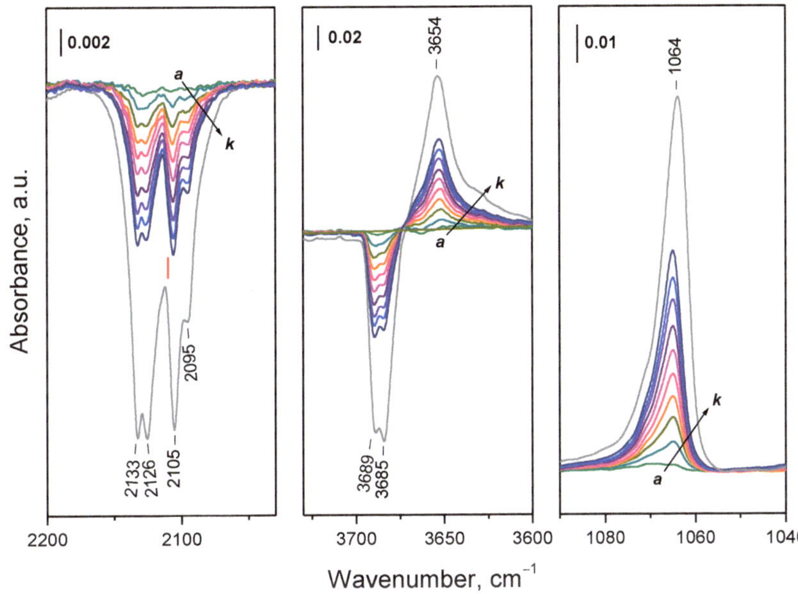

Figure 7. FTIR spectra were recorded after successive adsorption of small doses of $^{18}O_2$ at 100 K on a reduced CeO_2-NC sample (**a**–**j**) and after saturation in the presence of 2 mbar O_2 (**k**).

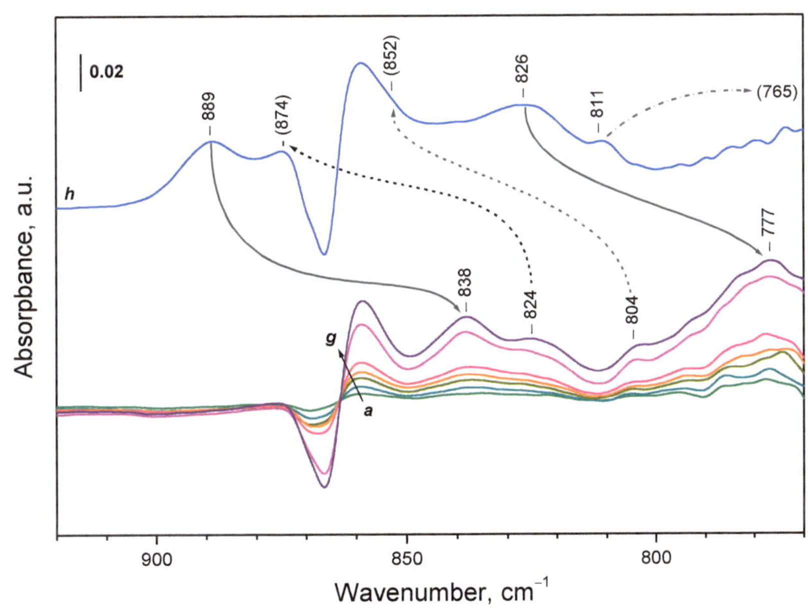

Figure 8. Selected FTIR spectra were recorded after successive adsorption of small doses of $^{18}O_2$ at 100 K on a reduced CeO_2-NC sample (**a**–**f**), in the presence of 2 mbar $^{18}O_2$ (**g**) and in the presence of 2 mbar $^{16}O_2$ (**h**).

Table 2. Positions of ^{16}O and ^{18}O superoxide and peroxide bands on CeO_2-NC and CeO_2-NP samples.

Sample	Assignment	$\nu(^{16}O–^{16}O)$, cm^{-1}	$\nu(^{18}O–^{18}O)$, cm^{-1}	i [1]
CeO_2-NC	superoxide	1128	1064	1060
CeO_2-NC	superoxide	1160	1093	1061
CeO_2-NP	superoxide	1128	1065	1059
CeO_2-NP	superoxide	1137	1072	1061
CeO_2-NC	peroxide	889	838	1061
CeO_2-NC	peroxide	874	824	1061
CeO_2-NC	peroxide	852	804	1060
CeO_2-NC	peroxide	826	777	1063
CeO_2-NC	peroxide	811	765 [2]	-
CeO_2-NP	peroxide	889	840	1058

[1] Isotopic shift factor. [2] Calculated value.

First, the two most intense bands at 889 and 826 cm^{-1}, produced after $^{16}O_2$ adsorption, are shifted to 838 and 777 cm^{-1}, respectively, when $^{18}O_2$ was utilized (isotopic shift factor 1.06). Two more bands, at 874 and 852 cm^{-1}, are not well discernible in the spectra of adsorbed $^{16}O_2$ but are well detected after adsorption of $^{18}O_2$ at 824 and 804 cm^{-1}, respectively. Another band at 811 cm^{-1} was observed only after adsorption of $^{16}O_2$, and the position of the respective $^{18}O_2$ band was below the sample cut-off. All these bands are assigned to different peroxide species, and all of them develop almost in concert.

Analysis of the hydroxyl region (Figure 7) shows that the OH bands associated with a reduced surface (3689 and 3685 cm^{-1} at 100 K) are eroded in parallel with the components of the Ce^{3+} band and a new band at 3654 cm^{-1} (typical of the oxidized surface) develops at their expense. The superoxide band at 1065 cm^{-1} and the peroxide bands also grow almost in parallel with these changes. The same is valid for the shift of the carbonate bands (Figure S6). The results show that the fast ceria oxidation at 100 K includes oxidation of all Ce^{3+} sites bound to OH groups and carbonate anions. Note, however, that this does not exclude the possibility of parallel oxidation processes, including Ce^{3+} sites that are not bound to OH groups and/or carbonates.

An increase in temperature in the presence of 2 mbar $^{18}O_2$ leads to changes similar to those described in the experiments with $^{16}O_2$, although all oxygen bands were shifted. Here, we shall only discuss the peroxide bands (Figure 9) because they were not discussed in the previous section.

The band at 889 cm^{-1} was hardly affected when the temperature was raised from 100 K to ambient one. Only a small decrease in intensity and red-shift were noticed. However, the intensity of this band is substantially reduced after heating at 373 K (Figure 9, spectrum e). At this temperature, the band at 826 cm^{-1} disappears. This means that the respective peroxide species decompose between ambient temperature and ca 373 K. In contrast, the band at 810 cm^{-1} developed, and this was more pronounced after heating at 373 K. In this case, a shoulder at 817 cm^{-1} becomes also well visible. The intensities of the other peroxide bands were too low to follow them accurately.

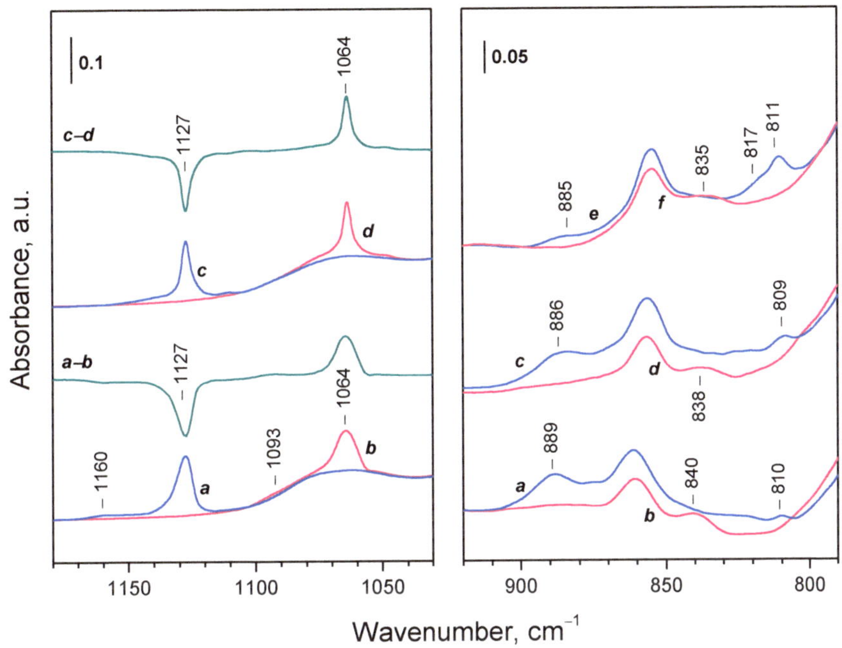

Figure 9. FTIR spectra of reduced CeO_2-NC registered after interaction with $^{16}O_2$ (**a**,**c**,**e**) or $^{18}O_2$ (**b**,**d**,**f**): addition of 2 mbar of $^{16}O_2/^{18}O_2$ to the sample at 100 K (**a**,**b**) and after increase in the temperature to 298 K (**c**,**d**) and 373 K (**e**,**f**).

2.4.3. Interaction of Hydroxylated CeO_2-NC with $^{16}O_2$

In order to obtain information on how the hydroxylation degree affects the fine structure of the Ce^{3+} electronic band and the oxidation process as a whole, we studied oxygen adsorption on a reduced sample that was evacuated only at 573 K. Preliminary experiments have shown that after this pretreatment all of the molecularly adsorbed water is eliminated (lack of δ(H_2O) band at ca. 1620 cm^{-1}). Two principal hydroxyl bands appeared in the spectrum at 3684 and 3659 cm^{-1}, together with two weaker features at 3615 and 3533 cm^{-1}. In addition, a broad band centered at ca. 3300 cm^{-1} evidences the existence of a fraction of H-bonded hydroxyls. At low temperatures, the bands shifted, and the two main bands were detected at 3690 and 3663 cm^{-1}. The carbonate bands appeared at 1475, 1464, 1440, 1407 and 1380 cm^{-1}. The second derivative indicated that the Ce^{3+} band consisted mainly of three components at 2127, 2109, and 2094 cm^{-1}.

Dosage of $^{16}O_2$ to the sample led to a gradual development of three negative bands at 2126, 2109, and 2095 $cm^{-1,}$ evidencing oxidation of Ce^{3+} sites (Figure 10). Synchronously, a superoxide band developed at 1135 cm^{-1}. The OH band at 3693 cm^{-1} fully disappeared, and the band at 3664 cm^{-1} strongly decreased in intensity. Simultaneously, a band at 3654 cm^{-1} developed together with a broad band at ca. 3200 cm^{-1}. In parallel with this, a band at 1620 cm^{-1}, assigned to δ(H_2O) modes, grew up (see the right inset in Figure 10).

Comparison with the experiments performed with the 773 K evacuated sample allows us to underline two important differences. First, the decrease in intensity of the Ce^{3+} band after dosage of the same amount of O_2 was definitely lower. Note that this cannot be due to kinetic limitations because no changes were observed with time after adding a definite amount of doses, i.e., all of the introduced oxygen was consumed. Second, the integral areas of the negative Ce^{3+} electronic bands in the two sets of experiments were practically identical after the introduction of 2 mbar of O_2 to the system, which suggests that the same amount of Ce^{3+} sites have been fast oxidized in the two experiments. We also note that the band around 1553 cm^{-1}, due to weakly bound molecular oxygen, was not discernible even in the presence of 2 mbar O_2 in the gas phase.

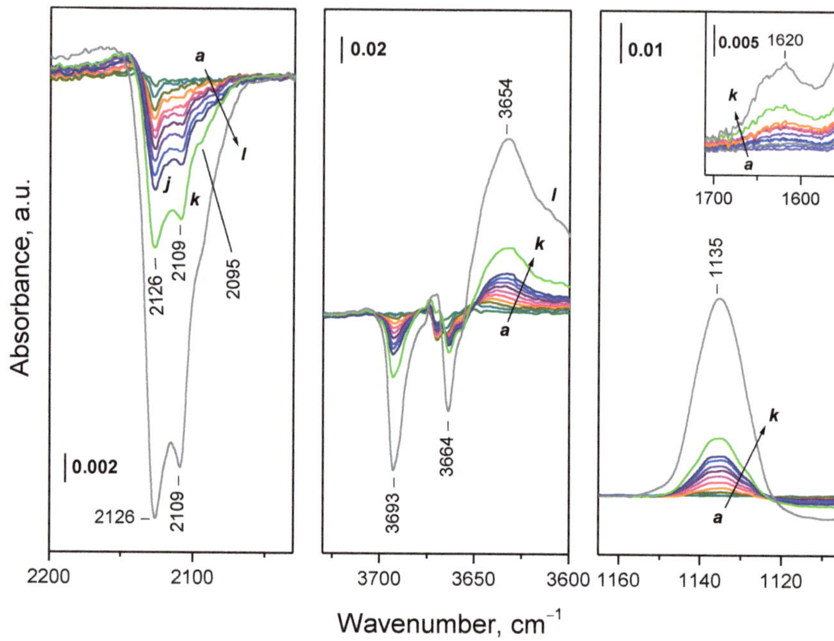

Figure 10. FTIR spectra were recorded after successive adsorption of small doses of $^{16}O_2$ at 100 K on reduced CeO_2-NC sample evacuated at 573 K (**a**–**j**), after the introduction of a bigger dose (**k**) and in the presence of 2 mbar O_2 (**l**).

The above findings can be rationalized assuming that part of the introduced oxygen is consumed for oxidation of Ce^{3+} sites and another part for oxidation of hydrogen dissolved in ceria. This explains the delayed development of the negative Ce^{3+} electronic band. When dissolved H_2 is oxidized, water is produced and is detected by the $\delta(H_2O)$ modes and the appearance of broad absorbance in the ν(OH) region due to H–bonded water. Note that the existence of dissolved hydrogen in reduced ceria has already been reported [12,16]. Thus, the results demonstrate that hydrogen dissolved in ceria can be oxidized even at 100 K in the absence of noble metals.

During temperature increase up to ambient one, the band of the superoxide (1135 cm^{-1}) initially slightly develops but disappears before reaching room temperature. Similar is the behavior of the 889 cm^{-1} peroxide band, which appeared only at intermediate temperatures and was not practically formed at 100 K. In contrast, the low-frequency peroxide bands at 843 and 830 cm^{-1} were much more intense on this hydroxylated sample than the sample evacuated at 773 K. Importantly, water was continuously produced up to ambient temperature. Based on the intensity of the $\delta(H_2O)$ band, it was roughly estimated that the amount of water formed at 100 K is ca. 15% of all water produced when the temperature reached ambient one.

2.4.4. Interaction of CeO_2-NC Reduced at Lower Temperatures with $^{16}O_2$

Finally, we studied the interaction of O_2 with CeO_2-NC reduced at 623 and 673 K and then evacuated at 773 K. Figure 11 compares the spectra in the Ce^{3+} and peroxide regions of these samples with the spectra of the sample reduced at 773 K and evacuated at different temperatures.

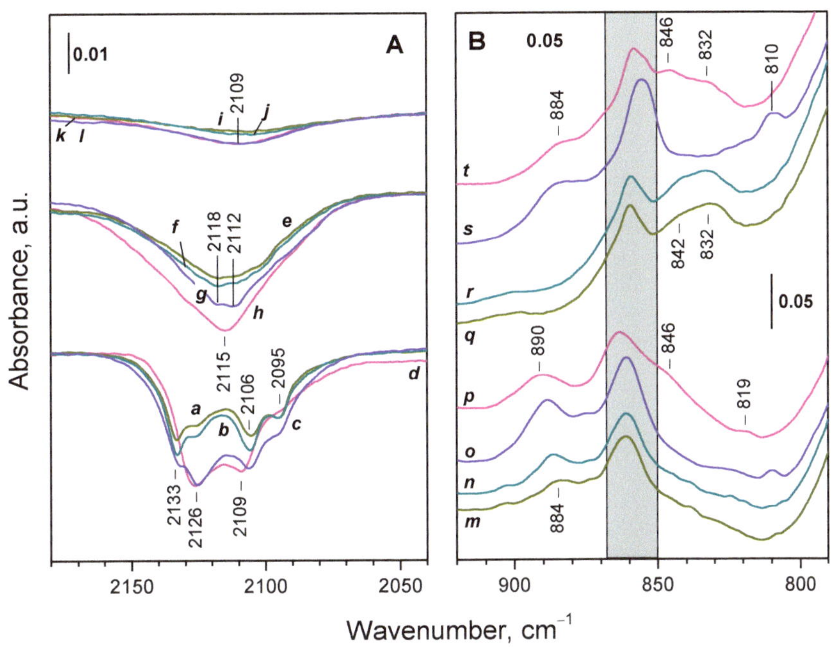

Figure 11. (**A**) Changes of the IR spectra of the CeO_2-NC sample after the introduction of 2 mbar of O_2 at 100 K (**a–d**). Sample reduced at 623 K (**a**), 673 (**b**), and 773 K (**c**,**d**); samples evacuated at 773 K (**a–c**) and at 573 K (**d**). Increase in temperature up to 293 K in the presence of O_2: spectra (**e–h**) are recorded after spectra (**a–d**), respectively. Increase in temperature up to 373 K in the presence of O_2: spectra (**i–l**) are recorded after spectra (**e–h**), respectively. The lower set of spectra corresponds to fast oxidation of Ce^{3+} at 100 K, the middle set to fast oxidation between 100 K and 293 K, and the upper set to Ce^{3+} residual to oxidation at 293 K. (**B**) IR spectra of CeO_2-NC sample recorded after introduction of 2 mbar of O_2 at 100 K (**m–p**) and after increase in temperature to 293 K (**q–t**). Samples were reduced at 623 K (**m**,**q**), 673 (**n**,**r**), and 773 K (**o**,**p**,**s**,**t**). Samples evacuated after reduction at 773 K (**m**,**n**,**o**,**q**,**r**,**s**) and at 573 K (**p**,**t**). The maxima of the peroxide species are labeled, and the grey bar shows the region of a background band.

It is seen that the reduction degree of the two samples is comparable but lower compared to the sample reduced at 773 K.

When O_2 interacted with the samples at 100 K, the main difference between the samples reduced at 623–673 K as compared to that reduced at 773 K is the low intensity of the negative component at 2125 cm^{-1} (compare spectra (a) and (b) with spectrum (c) in Figure 11A) Therefore, the respective Ce^{3+} sites are produced at higher reduction temperatures.

The spectra of the samples in the peroxide region exhibited only a weak band at 884–885 cm^{-1}, which was more intense with the sample reduced at higher temperatures (Figure 11B, spectra m, n). Only negligible features were discernible in the 850–800 cm^{-1} region. The superoxide band was slightly less intense compared to the 773 K reduced sample. A full conversion of OH groups and carbonate bands to the oxidized state was established.

When the temperature of the samples was gradually increased to 293 K, the main part of the residual Ce^{3+} sites was oxidized. Two components, at 2118 and 2112 cm^{-1}, are well discernible in the negative Ce^{3+} electronic band (Figure 11A, spectra e, f). The same components are seen with the sample reduced at 773 K (Figure 11A, spectrum g).

The superoxide band was no longer detected at 293 K. However, intense bands due to adsorbed water appeared at 1620 cm^{-1} together with a broad absorbance centered around 3400 cm^{-1} (spectra not shown). The peroxide band at 884–885 cm^{-1} also disappeared, and bands at 842 and 832 cm^{-1} grew instead (Figure 11B, spectra q, r).

In this case, again, a small fraction of Ce^{3+} resisting oxidation even at 293 K was detected (Figure 11A, spectra i, j).

2.5. Interaction of Reduced CeO_2-NP with O_2 at Variable Temperature

2.5.1. Interaction with $^{16}O_2$

As with CeO_2-NC, the introduction of O_2 (2 mbar) to the reduced CeO_2-NP sample led to an immediate decrease in the intensity of the Ce^{3+} electronic band (Figure 12, spectrum b). Difference spectra show that this happens mainly at the expense of a component at 2131 cm^{-1}, although some components at lower frequencies also disappeared. Based on the band intensities, it was estimated that ca. 23% of the Ce^{3+} sites have been oxidized at the conditions applied. This fraction is about two times lower than the CeO_2-NC sample. No noticeable changes occurred in the next 10 min, i.e., further oxidation at 100 K was restricted.

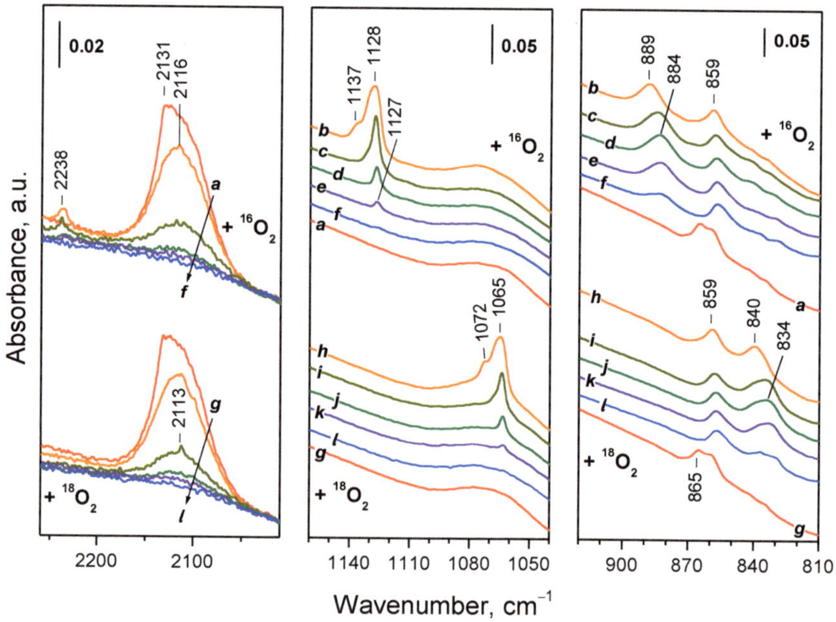

Figure 12. FTIR spectra of reduced CeO_2-NP registered at 100 K (**a**,**g**) and after interaction with $^{16}O_2$ (**b**–**f**) or $^{18}O_2$ (**h**–**l**): addition of 2 mbar of $^{16}O_2/^{18}O_2$ to the sample at 100 K (**b**,**h**) after gradual increase in temperature (**c**,**i**), at ambient temperature (**d**,**j**) and after interaction with O_2 at 323 K (**e**,**k**) and 373 K (**f**,**l**).

Simultaneously with the erosion of the Ce^{3+} electronic band, bands due to adsorbed superoxide and peroxide species emerged. The ν(O–O) band of the O_2^- species was detected at 1128 cm^{-1}, and the second derivatives clearly revealed two more components at 1131 and 1137 cm^{-1}. The peroxide species were homogeneous and characterized by a ν(O–O) band at 889 cm^{-1}. In addition, the carbonate band at 859 cm^{-1} gained intensity, and the band at 865 cm^{-1} (indicative of carbonates on a reduced surface) disappeared. A slight development of the background band at ca. 1070 cm^{-1} was also noticed.

An increase in the temperature up to ambient one led to a strong reduction and almost full disappearance of the Ce^{3+} band (Figure 12, spectra c, d), i.e., additional oxidation occurred at temperatures higher than 100 K. The band at 1128 cm^{-1} initially slightly developed with temperature increase at the expense of the component at 1137 cm^{-1} (spectra not shown), and then the higher-frequency components disappeared, and the 1227 cm^{-1} band strongly decreased in intensity. The peroxide band slightly developed and shifted to 884 cm^{-1} when the temperature increased.

Heating in oxygen at 323 and 373 K resulted in almost complete and then ultimate disappearance of the Ce^{3+} band (Figure 12, spectra e, f) and the band characterizing adsorbed O_2^- species (1127 cm^{-1}). In contrast, the peroxide species (884 cm^{-1}) were stable at 323 K, but their band was significantly eroded at 373 K.

2.5.2. Interaction with $^{18}O_2$

The changes in the spectra of the reduced CeO_2-NP sample after adsorption of $^{18}O_2$ are identical to those already described for $^{16}O_2$ adsorption, except for the red-shift of several bands (Figure 12). This demonstrates the good reproducibility of the experiments.

First, we note the shift of the weak band at 2238 cm^{-1} in the region of the Ce^{3+} electronic transition. Indeed, a band at 2113 cm^{-1} is discernible in the spectrum (i) from Figure 12.

The $^{18}O_2^-$ principal band was detected at 1065 cm^{-1}, again in line with the theoretical expectations (Figure 12, spectrum h). The same is valid for the peroxide band detected at 840–834 cm^{-1} (Figure 12, spectra h and j, respectively).

If the weak component around 1070 cm^{-1} was due to adsorbed dioxygen, a band around 1010 cm^{-1} should have appeared after the adsorption of $^{18}O_2$. However, no such band was detected, which indicates that the development of the component around 1070 cm^{-1} is due to background affection.

Let us discuss the doublet at 865 and 859 cm^{-1} registered in the background spectrum of the reduced sample (Figure 12, spectrum a). Binet et al. [23], on the basis of ^{13}C-labeling, have concluded that the band is due to residual carbonates and, as the other carbonate bands, is sensitive to reduction. However, in a recent study, a new assignment to Ce=O modes was proposed [22]. Our results do not support the latter supposition because the bands were not sensitive to exchange with ^{18}O. Moreover, the relative intensities of the two bands, at 865 and 859 cm^{-1}, could be used as an indicator for the fraction of residual carbonates attached to Ce^{3+} sites on reduced samples.

2.6. Interaction of Reduced CeO_2-NR with O_2 at Variable Temperature

The changes in the spectrum of the reduced CeO_2-NR sample (Figure 13) are generally similar to those already described with the other samples. In this case, however, two peroxide bands (889 and 831 cm^{-1}) develop at low temperatures. In the ν(OH) region, the band at 3688 cm^{-1} disappears, and two bands at 3651 and 3628 cm^{-1}, typical of oxidized samples, develop. Based on the band intensities, it was estimated that ca. 40% of the Ce^{3+} sites had been oxidized at 100 K. An increase in the temperature up to the ambient provoked the appearance of one more peroxide band at 840 cm^{-1} (not shown), while the superoxide band totally disappeared. The peroxide species at 889 and 831 cm^{-1} were still observable at 373 K.

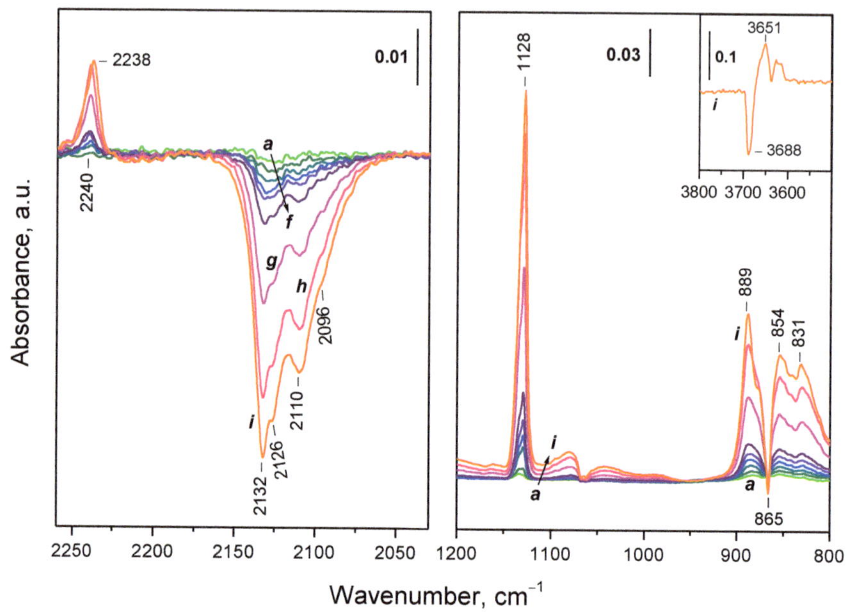

Figure 13. Changes in the FTIR spectra of reduced CeO_2-NR samples after subsequent dosing small amounts of O_2 to the sample at 100 K (**a–h**) and in the presence of 2 mbar of O_2 (**i**).

2.7. Interaction of the Samples with Oxygen at Ambient Temperature

The studies at ambient temperature allowed us to eliminate the temperature changes of the spectra, which are especially important for the Ce^{3+} electronic transition and the OH bands. However, the hydroxyl spectra of the CeO_2-NP sample were too noisy to allow a precise analysis.

2.7.1. CeO_2-NC

Small doses of O_2 were successively added to the reduced CeO_2-NC sample, and after each dose, a spectrum was recorded. Dosage of oxygen leads to erosion of the 2110 cm^{-1} band, preferentially from its higher-frequency side (Figure S7, left inset). However, even in the presence of gas-phase oxygen in the cell, the band did not totally disappear, i.e., some Ce^{3+} cations (ca. 10%) resisted fast oxidation even at ambient temperature.

The addition of oxygen also provoked the development of an O_2^- band at 1126 cm^{-1} with two satellites at 1112 and 1139 cm^{-1} (see the inset in Figure S7). The satellites are attributed to additive and subtractive combination modes of this band. The overtone is also detected at 2236 cm^{-1} (Figure S7, main panel). A careful inspection of the spectra also reveals that a very weak peroxide band at 810 cm^{-1} developed during oxygen dosage.

In the OH region, the band at 3680 cm−1 is also gradually eroded and practically disappears in the presence of O_2 in the gas phase (Figure 14, spectrum b). Bands at 3652 and 3631 cm^{-1}, typical of freshly activated samples, emerged at its expense (Figure 14, spectrum c). Therefore, we can propose that the O_2^- species are located at the same sites where the terminal OH groups were originally situated.

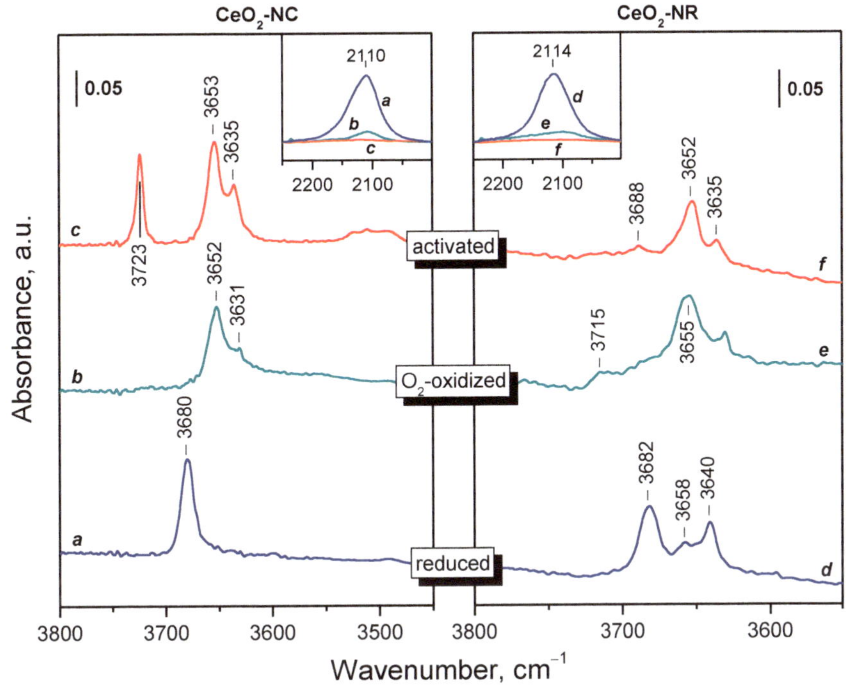

Figure 14. FTIR spectra at ambient temperature (hydroxyl stretching region) of CeO_2-NC (**a–c**) and CeO_2-NR (**d–f**). Samples were reduced with H_2 at 773 K (**a**,**d**), re-oxidized by 8 mbar O_2 (**b**,**e**), and freshly activated (oxidized) samples (**c**,**f**). The insets show the region of the Ce^{3+} electronic transition where the spectra are baseline corrected.

Oxygen dosage leads to a gradual shift of the carbonate bands to their original positions with the oxidized sample, and this conversion was full in the presence of gaseous O_2.

2.7.2. CeO_2-NR

The changes in the spectra registered after dosing oxygen to the reduced CeO_2-NR sample (See Figure S8) are similar to those already described for CeO_2-NC. Therefore, here, we will highlight only the main peculiarities.

Here, again, a small fraction of Ce^{3+} sites (ca. 13.5%) remained after the interaction of the sample with oxygen at ambient temperature. In the hydroxyl region, oxygen dosage led to gradual conversion to a spectrum typical of the oxidized sample. In this case, however, a small band due to terminal OH species appeared at 3715 cm^{-1} (Figure 14, spectrum e). We note that a similar band was detected on the oxidized sample at lower evacuation temperatures. The peroxide species were with negligible concentration. Similar results were obtained with the CeO_2-NP sample.

3. Discussion

3.1. Species Formed during Adsorption of O_2 on Oxidized and Reduced Ceria

On oxidized ceria, even after evacuation of the samples at 773 K, we have observed only molecularly adsorbed O_2 by a weak band at 1555–1553 cm^{-1}. Although some authors report that no reactive O_2 adsorption occurs on oxidized ceria [19,35], others have observed superoxides with non-reduced samples evacuated at 673 K [32]. In our experiments with oxidized samples, we treated them in O_2 immediately before evacuation at the target temperature (573–773 K). Note that ceria is very easily reduced, and even a minor amount of organic contaminants originally present in the sample or introduced during the experiments could lead to the production of a small amount of Ce^{3+} during the thermal treatment.

As a homonuclear diatomic molecule, O_2 has no IR active mode. However, after adsorption, the symmetry is reduced, and the O–O stretching modes, although with low intensity, become IR observable. The same phenomenon has been observed for similar molecules, such as N_2 and H_2 [49].

The possible active site on ceria for adsorption of molecular oxygen is coordinatively unsaturated (cus) Ce^{4+} cations and hydroxyl groups. The results presented in Figure S5 (spectra a–c) clearly show that the ν(O–O) band increases in intensity with sample dehydroxylation, which indicates that the principal adsorption sites are Ce^{4+} cations. Indeed, Ce^{4+} hydroxyls are of weak acidity, and their complexes with O_2 are expected to be very weak. The conclusion that OH groups are not active sites for adsorption of molecular O_2 is consistent with the lack of molecularly adsorbed O_2 on the hydrated sample (reduced, evacuated at 573 K, and then re-oxidized at 100 K).

The ν(O–O) band was observed at 1553 cm^{-1} with the reduced sample evacuated at 773 K and then oxidized at 100 K (Figure S5, spectrum d). Its intensity was only half of the intensity observed with the oxidized sample evacuated at the same temperature. This suggests that the fast oxidation at 100 K creates some cus Ce^{4+} sites acting as adsorption centers.

On reduced ceria, we have detected two more types of adsorbed dioxygen species: superoxide and peroxide. The $^{18}O_2$ isotopic experiments confirmed the assignment. This is in general agreement with literature data [17,19,32–35].

Superoxide ion is a monovalent radical produced as a result of one-electron reduction of dioxygen and thus has an O–O bond order of ca. 1.5. Consequently, the ν(O–O) stretching frequency is lower as compared to O_2. This is consistent with the observation of the superoxide band around 1130 cm^{-1}. Simple stoichiometric considerations show that two O_2^- radicals are necessary to formally substitute one O^{2-} anion.

There are two possible configurations of the adsorbed superoxo species: end-on (η^1) or side-on (η^2) [50], as shown in Scheme 1:

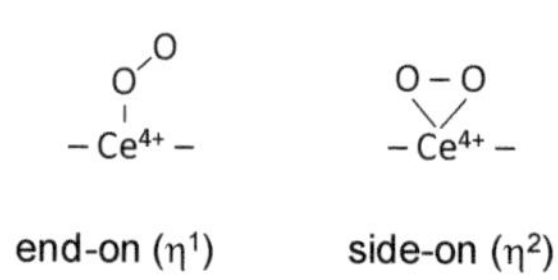

Scheme 1. Possible configurations of superoxo species on ceria.

If the end-on configuration is realized, the $Ce^{4+}(^{16}O^{18}O)^-$ species should manifest two bands, depending on the position of the labeled atom. In our experiments, we have not detected $Ce^{4+}(^{16}O^{18}O)^-$ isotopologues. However, Li et al. [34], performing co-adsorption of $^{16}O_2$ and $^{18}O_2$, succeeded in detecting $(^{16}O^{18}O)^-$ species and reported they were characterized by one band only at 1095 cm^{-1}. This indicates a side-on configuration of the superoxo species on ceria.

Consider now the peroxo species. In peroxides, the oxygen atoms are linked by a single covalent bond, and consequently, the ν(O–O) modes are observed at lower frequencies as compared to superoxo species. In contrast to superoxide and similarly to O^{2-}, it is a divalent anion. This makes peroxide anion suitable to fill oxygen vacancies on ceria.

Pushkarev et al. [35] have reported that peroxides are not formed when ceria is reduced at 573 K but appeared on 673 K reduced samples. Similar conclusions were made by Li et al. [33,34], who did not observe peroxides on ceria reduced by treatment in vacuo at 1000 K but detected them on a hydrogen-reduced sample. This is in agreement with DFT studies [51] indicating that O_2 interaction with oxygen vacancies on ceria produces peroxides. Thus, it appears that peroxides should acquire a bridging configuration.

Raman studies report two kinds of peroxo species on ceria, characterized by bands at ca. 860 and 830 cm^{-1} [19,35]. A band at 883 cm^{-1} has been reported in IR studies [21].

Interesting results have been reported by Bashir and Idriss [52]. They adsorbed H_2O_2 on ceria and detected IR bands at 890, 850, and 835 cm^{-1}. The former band was attributed to side-on peroxide, and the other two bands were attributed to bridging species. The band at 890 cm^{-1} almost disappeared after the 363 K evacuation, while the other two bands were still observable after the 433 K evacuation. The similarity of these bands to the bands appearing after O_2 adsorption on reduced ceria confirms their assignment to peroxides.

Surprisingly, the formation of Ce^{4+}–OOH^- species has not been reported. In fact, this possibility has been considered but rejected on the basis of the expectation that OOH^- species should absorb at considerably higher wavenumbers [52]. However, analysis of summarized literature data [50] indicates that OOH^- species absorb at relatively low frequencies. Thus, it has been reported that Ti–OOH species produced by H_2O_2 adsorption on Ti silicalite sieve are characterized by a ν(O–O) band at 837 cm^{-1} and a broad ν(O–O) band around 3400 cm^{-1} [53]. Indeed, H_2O_2 is a weak acid, and the protons from the OOH^- groups should tend to form H-bonds. This makes the OH mode of OOH^- species difficult to distinguish from H-bonded hydroxyls. We note that the OOH^- moiety is univalent, and, as will be shown below, the formation of Ce–OOH species will answer some open and unresolved questions in ceria surface chemistry.

To form OOH species, a source of hydrogen is necessary. One possible source is the residual OH groups. However, it seems that dissolved hydrogen is a more playable source. A look at Figure 11B shows that peroxide species absorbing in the 850–820 cm^{-1} spectral region are readily formed, with samples having a high amount of dissolved hydrogen. Therefore, we propose that these peroxide species are actually hydroperoxides. By analogy, we may suggest that the same accounts for the band at 810 cm^{-1}. In contrast, the band at 890 cm^{-1} is typical of the hydrogen-free sample, which indicates that it is due to H-free peroxide ($O_2{}^{2-}$) species.

3.2. Ce^{3+} Sites Involved in Fast Oxidation

Based on our results, we can divide the Ce^{3+} sites on reduced ceria according to their reactivity into three groups:

1. Ce^{3+} cations that are fast oxidized at 100 K;
2. Ce^{3+} cations fast oxidized between 100 and 293 K, and
3. Ce^{3+} cations resist oxidation at 293 K but oxidize at slightly higher temperatures, up to 393 K.

The fraction of the sites in the third group is relatively small, about 10–15%, depending on the pretreatment conditions. This is probably a problem of accessibility because their Ce^{3+} electronic transition band is broad, suggesting some heterogeneity. This does not exclude the possibility they are located at subsurface layers.

The fractions of the other two kinds of Ce^{3+} sites are abruptly equal. This suggests that they are surface-situated but differ in reactivity.

Consider first the sites fast oxidized at 100 K. It appears that they are of several different kinds, at least sites connected with OH groups, sites connected with residual carbonates, and sites not connected to carbonates or hydroxyls.

Looking at the spectra presented in Figure 7, one could conclude that the Ce^{3+} sites attached to OH groups are fast oxidized with simultaneous production of $O_2{}^-$ species according to the following hypothetic reaction:

$$Ce^{3+}\text{-(OH)-}Ce^{3+} + O_2 \rightarrow Ce^{3+}\text{-(OH)-}Ce^{4+}\text{-}O_2{}^- \quad (1)$$

However, a careful analysis of the spectra indicates that the situation is much more complicated. Thus, if Reaction (1) occurred, the OH band should undergo an additional shift at full oxidation (including the two Ce^{3+} sites), which was not observed in the experiments.

Note that during the fast oxidation of Ce^{3+} sites, the formation of superoxides and peroxides, as well as the concomitant changes (e.g., in the hydroxyl or carbonate regions), proceed in parallel. We attribute this to the so-called "wall effect". Reactive adsorption of oxygen is irreversible, and O_2 molecules interact with the first appropriate site they meet along the way. Thus, the ratio between the different Ce^{3+} sites oxidized at 100 K will not be determined by the differences in their affinity for oxygen. From this, we may conclude that all Ce^{3+} sites participating in the fast oxidation are accessible and situated on the surface.

Consequently, an important conclusion on the location of surface carbonates can be made. There are some arguments that they are situated not on the subsurface [25,27]. However, it appears that on reduced samples, at least part of them are connected to surface Ce^{3+} sites. This could happen if the carbonates are located on the surface or in the subsurface layer.

Bridging hydroxyls on ceria are of II or III type, i.e., connected with two or three cerium cations. However, it is important to note that for samples evacuated at 773 K, the regular crystal planes are dehydroxylated, and the residual hydroxyls are located at edges, corners, or other defect sites. Therefore, the Ce^{3+} ions connected to OH groups should be highly accessible.

A widely accepted reaction of ceria reduction is that leading to the formation of oxygen vacancies:

$$Ce^{4+}\text{-}O^{2-}\text{-}Ce^{4+} + H_2 \rightarrow Ce^{3+} \square Ce^{3+} + H_2O \quad (2)$$

where □ denotes an oxygen vacancy.

Reoxidation needs the formation of a divalent anion, and this can be easily achieved according to the reaction:

$$Ce^{3+} \square Ce^{3+} + O_2 \rightarrow Ce^{4+}\text{-}O_2^{2-}\text{-}Ce^{4+} \quad (3)$$

At temperatures above ca. 473 K, peroxides completely decompose, and the original state typical of oxidized ceria surface is restored:

$$Ce^{4+}\text{-}O_2^{2-}\text{-}Ce^{4+} \rightarrow Ce^{4+}\text{-}O^{2-}\text{-}Ce^{4+} + ½\, O_2 \quad (4)$$

The most probably analogous reaction occurs when Ce^{3+} sites from bridging hydroxyls are oxidized because it was already noted that no step oxidation of the two Ce^{3+} sites was detected.

It was noted that the formation of superoxide species cannot be explained by the model of surface oxygen vacancy and is rather indicative of the presence of low-coordinated Ce^{3+} sites [51]. We propose that such Ce^{3+} sites can be produced by reduction of Ce^{4+} sites connected to terminal OH groups:

$$Ce^{4+}\text{-}OH + ½\, H_2 \rightarrow Ce^{3+} + H_2O \quad (5)$$

Indeed, the results presented in Figure 6, Figure 7, and Figure 13 show that after the fast oxidation, the terminal OH groups on the CeO_2-NC sample are not regenerated. It is logical to expect that the charge balance is achieved by the formation of a Ce^{4+}–O_2^- complex, e.g., according to the following reaction:

$$Ce^{3+} + O_2 \rightarrow Ce^{4+}\text{-}O_2^- \quad (6)$$

Thus, one of the possible locations of the Ce^{4+}–O_2^- species is in the place of the original terminal Ce^{4+}–OH groups.

It was established that oxidation at 100 K leads to obtaining a highly stable semi-oxidized state of ceria. A large part of Ce^{3+} sites is additionally oxidized between 100 and 293 K. First, some extra superoxide species were produced in this way. This indicates the proceeding of reaction (6) and evidences the existence of a fraction of less reactive isolated Ce^{3+} sites. However, the superoxide band started to decline with the rise in temperature, and particular peroxide species were produced at its expense.

Based on similar observations, Li et al. [33] proposed the following evolution of the oxygen adspecies on ceria:

$$O_{2\,ads} \rightarrow O_2{}^{-}{}_{ads} \rightarrow O_2{}^{2-}{}_{ads} \rightarrow 2\,O^{-}{}_{ads} \rightarrow 2\,O^{2-}{}_{lattice} \quad (7)$$

According to our results, these consecutive conversions could be divided into two independent pathways: evolution of peroxo and evolution of superoxo species.

It was already noted that peroxides could be directly decomposed, thus producing lattice oxygen in the place of the oxygen vacancy.

Consider now the superoxide species. The fact that they form on isolated ex-Ce^{3+} sites indicates that they must evolve by forming monovalent anions. A hypothetic candidate for this is the O^- anion. However, we have no indication of its formation. In contrast, we already proposed that the bands below 850 cm^{-1} characterize hydroperoxo species. Thus, a possible reduction of the Ce^{4+}–$O_2{}^-$ species is according to the following reaction:

$$Ce^{4+}\text{-}O_2{}^{-} + ½\,H_2 \rightarrow Ce^{4+}\text{-OOH} \quad (8)$$

In this case, the interaction proceeds with hydrogen dissolved in ceria. Indeed, it was established that it promotes the formation of hydroperoxides. However, we cannot exclude the possibility of some hydroperoxides being formed with the participation of hydrogen from surface OH groups.

After the decomposition of the Ce^{4+}–OOH groups, the original terminal hydroxyls typical of oxidized surface are restored:

$$Ce^{4+}\text{-OOH} \rightarrow Ce^{4+}\text{-OH} + ½\,O_2 \quad (9)$$

The initial stage of this reaction was documented in our IR results (see Figure 6B, spectrum k).

The O^- radicals are invisible in IR but have been observed on some oxides by ESR [54]. Although suggested [33], these species have not been reported for pure ceria. In any case, we cannot totally rule out the possibility that, under hydrogen-deficient conditions, direct decomposition of $O_2{}^-$ to O^- occurs, the latter ensuring the charge balance instead of OH^- or OOH^-.

In conclusion, the evolution of the adsorbed oxygen species on isolated Ce^{3+} sites and on oxygen vacancies follows two different and independent routes.

3.3. Fine Structure of Ce^{3+} Band

In this study, we report for the first time a fine structure of the Ce^{3+} electronic band. This can explain the different values of the band position reported in the literature. This work is the first attempt to assign the different components of the band. We realize that additional work has to be performed for the precise attribution, but our results allow us to make some definite conclusions.

At first, it was unambiguously established that the position of the sub-bands strongly depends on the presence of OH group(s) in the first coordination sphere of Ce^{3+} cations. Comparison between the spectra of reduced sample evacuated at 773 K and 573 K (Figure 11A, spectra c, and d, respectively) clearly shows that the components around 2133 and 2096–2095 cm^{-1} consumed during the fast oxidation at 100 K are much more intense with the highly dehydroxylated sample and therefore should be associated with Ce^{3+} sites having no OH groups in the vicinity. The same seems to be also valid for the component at 2105 cm^{-1}. In contrast, the component at 2109 cm^{-1} appears with enhanced intensity with the hydroxylated sample and thus can be associated with Ce^{3+} cations connected with the extra OH groups on this sample (3664 cm^{-1}).

On the basis of the above considerations, one could expect that Ce^{3+} cations bound on residual carbonates should also possess a specific spectral behavior. Unfortunately, on the basis of our experiments, we are not able to draw any definite conclusions at this stage.

Comparison between the samples reduced at different temperatures (Figure 11A, spectra a–c) clearly shows that the component at 2126 cm^{-1} corresponds to Ce^{3+} sites that are produced more difficult during reduction.

The sites that resist fast oxidation at 100 K but are oxidized at higher temperatures up to 293 K give rise to a broad Ce^{3+} band with two resolved components at 2118 and 2112 cm^{-1}.

Finally, the small fraction of sites residual to oxidation at 293 K produces a broad band at 2109 cm^{-1}. Since the band does not differ significantly from the above-discussed band, we could suggest that the inertness of these sites could be due to low accessibility. It is not excluded as they are in the subsurface layer.

At the beginning of this study, we aimed to establish a clear relationship between the band components and the exposed crystal facets. However, it appears that the positions of the components depend on too many factors, and at this stage, we desist from making definite conclusions. Further specially designed experiments are necessary to answer this question and to establish eventual dependence between the component position and location of the Ce^{3+} on different crystal planes and edges.

3.4. Oxidation of Sorbed H_2

It was reported that hydrogen is stored on ceria below 773 K and released above this temperature [16,55]. A surprising observation in this study is the ability of O_2 to oxidize dissolved hydrogen at 100 K in the absence of any noble metal. Although the fraction of this hydrogen is relatively small, the phenomenon could be of great importance for catalysis because it demonstrates the high reactivity of hydrogen in ceria. We also found that this hydrogen strongly affects the conversion pathway of superoxide species. Since it is invisible in IR, we suggest that some of the results reported in the literature have been affected by dissolved hydrogen. Our results indicate that hydrogen is more stable on partly reduced samples, and thus, it can be present in some amount even after evacuation at $T > 773$ K. Therefore, dissolved hydrogen could exist during the proceeding of some catalytic redox reactions even at temperatures higher than 773 K.

The nature of hydrogen dissolved in ceria is still not well known. It has recently been pointed out that the question of whether H atoms enter the bulk or are chemisorbed on the surfaces of CeO_2 is a subject of debate [56]. Our results rather support the view that at least part of this hydrogen is in the bulk and needs to migrate to the surface in order to be oxidized. Indeed, the major part of the dissolved hydrogen is oxidized at temperatures up to ambient one.

4. Materials and Methods

4.1. Synthesis of the Samples

A hydrothermal method was used for the synthesis of nanosized samples of CeO_2 with different morphologies (cubes, rods, and polyhedra) [57,58]. 85 mL aqueous solution of 5 g $Ce(NO_3)_3{\cdot}6H_2O$ (Fluka, Buchs, Switzerland, 99% purity) was added with vigorous stirring to 150 mL alkaline solution of 54 g NaOH (Merck, Darmstadt, Germany, 99% purity) in water. After obtaining a suspension, stirring was continued for another 30 min, after which the reaction mixture was transferred to an autoclave. Nanocubes (sample CeO_2-NC) and nanorods (sample CeO_2-NR) were obtained after "aging" in an autoclave for 24 h at 453 and 373 K, respectively. The conditions for the synthesis of the nanopolyhedra (sample CeO_2-NP) are identical to those of the nanorods, but the alkaline solution used was more diluted (9 g NaOH in 150 mL water). Finally, the suspensions were centrifuged, and the precipitates were washed with deionized water, dried at 393 K, and calcined at 673 K for 2 h.

4.2. Gases

All non-labeled gases used for the IR and TPR experiments were provided by Messer (Bad Soden, Germany): hydrogen (>99.999%); helium (99.999%); oxygen (99.999%), 5% O_2/He (O_2 99.95%) and 10% H_2/Ar (H_2 99.95%). Labeled oxygen, $^{18}O_2$, was provided by ISOTEC (Miamisburg, OH, USA) and had an isotopic purity of 99 atom % ^{18}O.

4.3. Methods

4.3.1. FTIR Spectroscopy

FTIR spectra were registered with a Thermo Scientific Nicolet 6700 FTIR Spectrometer equipped (Madison, WI, USA) with an MCT/A detector. The spectral resolution was 2 cm^{-1}, and 64 scans were accumulated for each spectrum. Self-supporting wafers (ca. 15 mg cm^{-2}) were prepared from the sample powders and treated directly in a purpose-made IR cell connected to a vacuum-adsorption apparatus with a residual pressure below 10^{-4} Pa. Before the experiments, the samples were activated by heating them in oxygen (100 mbar) for 30 min at 773 K, followed by 30 min of evacuation at the same temperature.

To obtain an oxidized sample, the pellet was treated at the desired activation temperature for 10 min in the presence of 2 mbar of O_2 and then evacuated for 30 min at the same temperature.

To obtain reduced samples, the pellets were treated in 100 mbar of H_2 for 1 h at 773 K, followed by 1 h evacuation at the desired temperature, typically at 773 K. In this way, all eventually stored hydrogen (characterized by a TPD peak at 580 K [43]) was removed. When lower evacuation temperatures were applied, this was specially noted.

For all low-temperature measurements, the IR cell was initially cooled using liquid nitrogen, and then 2 mbar of He was added to the system in order to improve heat transfer.

Low-temperature adsorption of O_2 on reduced samples was performed either by successive addition of small doses of O_2 or by direct introduction of oxygen at 2 mbar pressure. Then, the sample temperature was gradually raised. Interaction with O_2 at 313–400 K was studied by heating the sample at the target temperature, followed by cooling to ambient temperature.

For the experiments performed at ambient temperature, small doses of O_2 were successively added to the system.

4.3.2. Temperature-Programmed Reduction

Temperature-programmed reduction with hydrogen (H_2-TPR) was carried out using a ChemBET TPR/TPD apparatus (Quantachrome Instruments, Boynton Beach, FL, USA) equipped with a thermal conductivity detector. A nitrogen/ethanol cold trap (placed before the detector) was used to eliminate water vapor from the gas flow. Before the TPR, the samples were heated to 773 K in a flowing stream of 5% O_2/He. TPR curves were recorded while reducing the samples (~40 mg) in a 10% H_2/Ar mixture at a flow rate of 20 mL min^{-1} and increasing the temperature from room temperature to 1273 K at a ramp rate of 10 K min^{-1}.

4.3.3. Transmission Electron Microscopy

The Transmission Electron Microscopy (TEM) study was performed with a JEOL 2100 transmission electron microscope (TEM, Tokyo, Japan) at an accelerating voltage of 200 kV equipped with X-ray energy-dispersive spectrometer (XEDS) Oxford Instruments (Abingdon, UK), X-MAX N 80 T; CCD Camera ORIUS 1000, 11 Mp, GATAN (Pleasanton, CA, USA). Before examination, each sample was suspended in ethanol solution and dripped onto a standard holey carbon/Cu grid. The data analysis was carried out using the Digital Micrograph™ software (version 3.7.4) from Gatan.

4.3.4. X-ray Diffraction

Powder X-ray diffraction (XRD) analysis was performed with a Bruker D8 Advance diffractometer with CuKa radiation and a LynxEye solid position-sensitive detector (Bruker AXS, Karlsruhe, Germany). XRD patterns were recorded in the range of 5.3 to 80° 2θ with a constant step of 0.02° 2θ and a counting time of 17.5 s per step. The mean crystallite size was calculated according to Scherrer's equation with Topas v.4.2 software (Bruker AXS, Karlsruhe, Germany).

4.3.5. BET Surface Area

Brunauer-Emmett-Teller (BET) surface areas were determined using low-temperature (77.4 K) nitrogen adsorption in a NOVA 1200e apparatus (Quantachrome Instruments, Boynton Beach, FL, USA). The samples were degassed overnight at 473 K under vacuum to ensure a clean, dry surface.

5. Conclusions

- No autoreduction in Ce^{4+} to Ce^{3+} occurs during the evacuation of pure ceria nanoparticles at 573–773 K.
- Reduction in ceria with H_2 at 773 K leads to the formation of Ce^{3+} cations on the surface, which are monitored in the IR spectra by a band at 2133–2094 cm^{-1}. This band possesses a fine structure, well resolved at 100 K. The positions of the individual components depend on the environment of Ce^{3+}, including the presence of nearby OH groups and likely residual carbonates.
- Even at 100 K, part of the Ce^{3+} sites on reduced ceria are quickly oxidized by O_2. These sites are situated on the surface and include all Ce^{3+} cations bound to OH groups and carbonates.
- Depending on the location of the Ce^{3+} sites, O_2^- or O_2^{2-} are produced during the fast oxidation of reduced ceria at 100 K.
- Some Ce^{3+} sites resist oxidation at 100 K but are oxidized at higher temperatures, between 100–400 K. These sites are also assumed to be surface situated, but a location in subsurface layers is not excluded.
- Peroxide (O_2^{2-}) species decompose to give lattice oxygen, while superoxides first convert to hydroperoxides (OOH^-) and then to terminal OH groups.
- H_2 dissolves in reduced ceria and is not completely removed upon evacuation at temperatures <773 K. Part of this hydrogen is also fast oxidized at 100 K, which demonstrates its high reactivity.

Supplementary Materials: The following supporting information can be downloaded at: https://www.mdpi.com/article/10.3390/catal14010045/s1: Figure S1: Background spectra of the CeO_2-NR sample; Figure S2: Background spectra of the CeO_2-NP sample; Figure S3: Spectra of the Ce^{3+} electronic band in presence of CO_2 and H_2O; Figure S4: Temperature effect on the spectra in the hydroxyl and carbonate regions; Figure S5: IR spectra of molecularly adsorbed O_2; Figure S6: Effect of oxidation of CeO_2-NC on the position of the carbonate bands; Figure S7: IR spectra of oxygen adsorbed on CeO_2-NC at ambient temperature; Figure S8: IR spectra of oxygen adsorbed on CeO_2-NR at ambient temperature; Table S1: Consumption of H_2 during TPR of the samples.

Author Contributions: Conceptualization, M.M. and K.H.; methodology, K.H.; investigation, K.C. and N.D.; writing—original draft preparation, K.H.; writing—review and editing, K.C., N.D., M.M. and K.H; visualization, K.C. and K.H.; funding acquisition, M.M. and K.H. All authors have read and agreed to the published version of the manuscript.

Funding: This research was funded by the Bulgarian Science Fund, grant number KП-06-H-59/5/2021.

Data Availability Statement: The data presented in this study are available in the article and the Supplementary Material.

Acknowledgments: Research equipment of the National Centre for Mechatronics and Clean Technologies was used for the experiments (project BG05M2OP001-1.001-0008, funded by the European Regional Development Fund under the Operational Program "Science and Education for Smart Growth 2014–2020"). Thanks are also due to: D. Kovacheva for XRD analysis, L. Mihaylov for TEM analysis, and L. Dimitrov for sample synthesis.

Conflicts of Interest: The authors declare no conflicts of interest.

References

1. Trovarelli, A.; de Leitenburg, C.; Boaro, M.; Dolcetti, G. The utilization of ceria in industrial catalysis. *Catal. Today* **1999**, *50*, 353–367. [CrossRef]

2. Chung, C.-H.; Tu, F.-Y.; Chiu, T.-A.; Wu, T.-T.; Yu, W.-Y. Critical roles of surface oxygen vacancy in heterogeneous catalysis over ceria-based materials: A selected review. *Chem. Lett.* **2021**, *50*, 856–865. [CrossRef]
3. Chuen, W.C.; Haile, S.M. A thermochemical study of ceria: Exploiting an old material for new modes of energy conversion and CO_2 mitigation. *Phil. Trans. R. Soc. A* **2010**, *368*, 3269–3294.
4. Ganduglia-Pirovano, M.V.; Hofmann, A.; Sauer, J. Oxygen vacancies in transition metal and rare earth oxides: Current state of understanding and remaining challenges. *Surf. Sci. Rep.* **2007**, *62*, 219–270. [CrossRef]
5. Rozhin, P.; Melchionna, M.; Fornasiero, P.; Marchesan, S. Nanostructured ceria: Biomolecular templates and (bio)applications. *Nanomaterials* **2021**, *11*, 2259. [CrossRef]
6. Song, G.; Cheng, N.; Zhang, J.; Huang, H.; Yuan, Y.; He, X.; Luo, Y.; Huang, K. Nanoscale cerium oxide: Synthesis, biocatalytic mechanism, and applications. *Catalysts* **2021**, *11*, 1123. [CrossRef]
7. Putna, E.S.; Vohs, J.M.; Gorte, R.J. Evidence for weakly bound oxygen on ceria films. *J. Phys. Chem.* **1996**, *100*, 17862–17865. [CrossRef]
8. Yao, H.C.; Yao, Y.F.Y. Ceria in automotive exhaust catalysts. I. Oxygen storage. *J. Catal.* **1984**, *86*, 254–265. [CrossRef]
9. Laachir, A.; Perrichon, V.; Badri, A.; Lamotte, J.; Catherine, E.; Lavalley, J.C.; El Fallah, J.; Hilaire, L.; Le Normand, F.; Quéméré, E.; et al. Reduction of CeO_2 by hydrogen. Magnetic susceptibility and Fourier-transform infrared, ultraviolet and X-ray photoelectron spectroscopy measurements. *J. Chem. Soc. Faraday Trans.* **1991**, *87*, 1601–1609. [CrossRef]
10. Holgado, J.P.; Munuera, G. XPS/TPR study of the reducibility of M/CeO_2 catalysts (M = Pt, Rh): Does junction effect theory apply? In *Studies in Surface Science and Catalysis*; Frennet, A., Bastin, J.M., Eds.; Elsevier: Amsterdam, The Netherlands, 1995; Volume 96, pp. 109–122.
11. Rao, G.R. Influence of metal particles on the reduction properties of ceria-based materials studied by TPR. *Bull. Mater. Sci.* **1999**, *22*, 89–94. [CrossRef]
12. Giordano, F.; Trovarelli, A.; Leitenburg, C.; Giona, M. A model for the temperature-programmed reduction of low and high surface area ceria. *J. Catal.* **2000**, *193*, 273–282. [CrossRef]
13. Fally, F.; Perrichon, V.; Vidal, H.; Kaspar, J.; Blanco, G.; Pintado, J.M.; Bernal, S.; Colon, G.; Daturi, M.; Lavalley, J.C. Modification of the oxygen storage capacity of CeO_2–ZrO_2 mixed oxides after redox cycling aging. *Catal. Today* **2000**, *59*, 373–386. [CrossRef]
14. Ayastuy, J.L.; Gil-Rodríguez, A.; González-Marcos, M.P.; Gutiérrez-Ortiz, M.A. Effect of process variables on Pt/CeO_2 catalyst behaviour for the PROX reaction. *Int. J. Hydrogen Energy* **2006**, *31*, 2231–2242. [CrossRef]
15. Lee, J.; Ryou, Y.S.; Chan, X.; Kim, T.J.; Kim, D.H. How Pt interacts with CeO_2 under the reducing and oxidizing environments at elevated temperature? The origin of improved thermal stability of Pt/CeO_2 compared to CeO_2. *J. Phys. Chem C* **2016**, *120*, 25870–25879. [CrossRef]
16. Agarwal, S.; Mojet, L.L.; Lefferts, L.; Datye, A.K. Ceria nanoshapes-structural and catalytic properties. In *Catalysis by Materials with Well-Defined Structures*; Elsevier: Amsterdam, The Netherlands, 2015; pp. 31–70.
17. Qiao, Z.-A.; Wu, Z.; Dai, S. Shape-controlled ceria-based nanostructures for catalysis applications. *ChemSusChem* **2013**, *6*, 1821–1833. [CrossRef]
18. Sun, C.; Li, H.; Chen, L. Nanostructured ceria-based materials: Synthesis, properties, and applications. *Energy Environ. Sci.* **2012**, *5*, 8475–8505. [CrossRef]
19. Wu, Z.; Li, M.; Howe, J.; Meyer, H.M.; Overbury, S.H. Probing defect sites on CeO_2 nanocrystals with well-defined surface planes by Raman spectroscopy and O_2 adsorption. *Langmuir* **2010**, *26*, 16595–16606. [CrossRef]
20. Wu, W.; Savereide, L.M.; Notestein, J.; Weitz, E. In-situ IR spectroscopy as a probe of oxidation/reduction of Ce in nanostructured CeO_2. *Appl. Surf. Sci.* **2018**, *445*, 548–554. [CrossRef]
21. Zotin, F.M.Z.; Tournayan, L.; Varloud, J.; Perrichon, V.; Frety, R. Temperature-programmed reduction: Limitation of the technique for determining the extent of reduction of either pure ceria or ceria modified by additives. *Appl. Catal. A* **1993**, *98*, 99–114. [CrossRef]
22. Afrin, S.; Bollini, P. On the utility of Ce^{3+} spin-orbit transitions in the interpretation of rate data in ceria catalysis: Theory, validation, and application. *J. Phys. Chem. C* **2023**, *127*, 234–247. [CrossRef]
23. Binet, C.; Badri, A.; Lavalley, J.-C. A spectroscopic characterization of the reduction of ceria from electronic transitions of intrinsic point defects. *J. Phys. Chem.* **1994**, *98*, 6392–6398. [CrossRef]
24. Binet, C.; Daturi, M.; Lavalley, J.-C. IR study of polycrystalline ceria properties in oxidised and reduced states. *Catal. Today* **1999**, *50*, 207–225. [CrossRef]
25. Mihaylov, M.Y.; Ivanova, E.Z.; Aleksandrov, H.A.; Petkov, P.S.; Vayssilov, G.N.; Hadjiivanov, K.I. FTIR and density functional study of NO interaction with reduced ceria: Identification of N_3^- and NO^{2-} as new intermediates in NO conversion. *Appl. Catal. B* **2015**, *176–177*, 107–119. [CrossRef]
26. Mihaylov, M.Y.; Ivanova, E.Z.; Aleksandrov, H.A.; Petkov, P.S.; Vayssilov, G.N.; Hadjiivanov, K.I. Formation of N_3^- during interaction of NO with reduced ceria. *Chem. Commun.* **2015**, *51*, 5668–5671. [CrossRef]
27. Mihaylov, M.Y.; Ivanova, E.Z.; Vayssilov, G.N.; Hadjiivanov, K.I. Revisiting ceria-NO_x interaction: FTIR studies. *Catal. Today* **2020**, *357*, 613–620. [CrossRef]
28. Li, H.; Zhang, P.; Li, G.; Lu, J.; Wu, Q.; Gu, Y. Stress measurement for nonstoichiometric ceria films based on Raman spectroscopy. *J. Alloys Compd.* **2016**, *682*, 132–137. [CrossRef]

29. Bozon-Verduraz, F.; Bensalem, A. IR studies of cerium dioxide: Influence of impurities and defects. *J. Chem. Soc. Faraday Trans.* **1994**, *90*, 653–657. [CrossRef]
30. Chen, L.; Fleming, P.; Morris, V.; Holmes, J.D.; Morris, M.A. Size-related lattice parameter changes and surface defects in ceria nanocrystals. *J. Phys. Chem. C* **2010**, *114*, 12909–12919. [CrossRef]
31. Zhang, B.; Zhang, S.; Liu, B. Effect of oxygen vacancies on ceria catalyst for selective catalytic reduction of NO with NH_3. *Appl. Surf. Sci.* **2020**, *529*, 147068. [CrossRef]
32. Descorme, C.; Madier, Y.; Duprez, D. Infrared study of oxygen adsorption and activation on cerium–zirconium mixed oxides. *J. Catal.* **2000**, *196*, 167–173. [CrossRef]
33. Li, C.; Domen, K.; Maruya, K.; Onishi, T. Dioxygen adsorption on well-outgassed and partially reduced cerium oxide studied by FT-IR. *J. Am. Chem. Soc.* **1989**, *111*, 7683–7687. [CrossRef]
34. Li, C.; Domen, K.; Maruya, K.-I.; Onishi, T. Oxygen exchange reactions over cerium oxide: An FT-IR study. *J. Catal.* **1990**, *123*, 436–442. [CrossRef]
35. Pushkarev, V.V.; Kovalchuk, V.I.; d'Itri, J.L. Probing defect sites on the CeO_2 surface with dioxygen. *J. Phys. Chem. B* **2004**, *108*, 5341–5348. [CrossRef]
36. Cardenas, L.; Molinet-Chinaglia, C.; Loridant, S. Unraveling Ce^{3+} detection at the surface of ceria nanopowders by UPS analysis. *Phys. Chem. Chem. Phys.* **2022**, *24*, 22815–22822. [CrossRef] [PubMed]
37. Morgan, D.J. Photoelectron spectroscopy of ceria: Reduction, quantification and the myth of the vacancy peak in XPS analysis. *Surf. Interface Anal.* **2023**, *55*, 845–850. [CrossRef]
38. Xu, Y.; Wang, F.; Liu, X.; Liu, Y.; Luo, M.; Teng, B.; Fan, M.; Liu, X. Resolving a decade-long question of oxygen defects in Raman spectra of ceria-based catalysts at atomic level. *J. Phys. Chem. C* **2019**, *123*, 18889–18894. [CrossRef]
39. Filtschew, A.; Hofmann, K.; Hess, C. Ceria and its defect structure: New insights from a combined spectroscopic approach. *J. Phys. Chem. C* **2016**, *120*, 6694–6703. [CrossRef]
40. Schmitt, R.; Nenning, A.; Kraynis, O.; Korobko, R.; Frenkel, A.I.; Lubomirsky, I.; Haile, S.M.; Rupp, J.L.M. A review of defect structure and chemistry in ceria and its solid solutions. *Chem. Soc. Rev.* **2020**, *49*, 554–592. [CrossRef]
41. Wang, X.; Li, M.; Wu, Z. In situ spectroscopic insights into the redox and acid-base properties of ceria catalysts. *Chin. J. Catal.* **2021**, *42*, 2122–2140. [CrossRef]
42. Loridant, S. Raman spectroscopy as a powerful tool to characterize ceria-based catalysts. *Catal. Today* **2021**, *373*, 98–111. [CrossRef]
43. Bruce, L.A.; Hoang, M.; Hughes, A.E.; Turney, T.W. Surface area control during the synthesis and reduction of high area ceria catalyst supports. *Appl. Catal. A* **1996**, *134*, 351–362. [CrossRef]
44. Badri, A.; Binet, C.; Lavalley, J.-C. An FTIR study of surface ceria hydroxy groups during a redox process with H_2. *J. Chem. Soc. Faraday Trans.* **1996**, *92*, 4669–4673. [CrossRef]
45. Tsyganenko, A.A.; Filimonov, V.N. Infrared spectra of surface hydroxyl groups and crystalline structure of oxides. *J. Mol. Struct.* **1973**, *19*, 579–589. [CrossRef]
46. Vayssilov, G.N.; Mihaylov, M.; Petkov, P.S.; Hadjiivanov, K.I.; Neyman, K.M. Reassignment of the vibrational spectra of carbonates, formates, and related surface species on ceria: A combined density functional and infrared spectroscopy investigation. *J. Phys. Chem. C* **2011**, *115*, 23435–23454. [CrossRef]
47. Hadjiivanov, K. Identification and characterization of surface hydroxyl groups by infrared spectroscopy. *Adv. Catal.* **2014**, *47*, 99–318.
48. Otsuka, K.; Hatano, M.; Morikawa, A. Hydrogen from water by reduced cerium oxide. *J. Catal.* **1983**, *79*, 493–496. [CrossRef]
49. Zecchina, A.; Otero Areán, C. Diatomic molecular probes for mid-IR studies of zeolites. *Chem. Soc. Rev.* **1996**, *25*, 187–197. [CrossRef]
50. Nakamoto, K. *Infrared and Raman Spectra of Inorganic and Coordination Compounds, Part B*, 6th ed.; John Wiley & Sons: Hoboken, NJ, USA, 2009.
51. Preda, G.; Migani, A.; Neyman, K.M.; Bromley, S.T.; Illas, F.; Pacchioni, G. Formation of superoxide anions on ceria nanoparticles by interaction of molecular oxygen with Ce^{3+} sites. *J. Phys. Chem. C* **2011**, *115*, 5817–5822. [CrossRef]
52. Bashir, S.M.; Idriss, H. The reaction of propylene to propylene-oxide on CeO_2: An FTIR spectroscopy and temperature programmed desorption study. *J. Chem. Phys.* **2020**, *152*, 044712. [CrossRef]
53. Lin, W.Y.; Frei, H. Photochemical and FT-IR probing of the active site of hydrogen peroxide in Ti silicalite sieve. *J. Am. Chem. Soc.* **2002**, *124*, 9292–9298. [CrossRef]
54. Anpo, M.; Costentin, G.; Giamello, E.; Lauron-Pernot, H.; Sojka, J. Characterisation and reactivity of oxygen species at the surface of metal oxides. *J. Catal.* **2021**, *393*, 259–280. [CrossRef]
55. Bernal, S.; Calvino, J.J.; Cifredo, G.A.; Gatica, J.M.; Perez Omil, H.; Pintado, J.M. Hydrogen chemisorption on ceria: Influence of the oxide surface area and degree of reduction. *J. Chem. Soc. Faraday Trans.* **1993**, *89*, 3499–3505. [CrossRef]
56. Kammert, J.; Moon, J.; Wu, Z. A review of the interactions between ceria and H_2 and the applications to selective hydrogenation of alkynes. *Chin. J. Catal.* **2020**, *41*, 901–914. [CrossRef]

57. Zabilskiy, M.; Djinović, P.; Tchernychova, E.; Tkachenko, O.P.; Kustov, L.M.; Pintar, A. Nanoshaped CuO/CeO_2 materials: Effect of the exposed ceria surfaces on catalytic activity in N_2O decomposition reaction. *ACS Catal.* **2015**, *5*, 5357–5365. [CrossRef]
58. Mai, H.-X.; Sun, L.-D.; Zhang, Y.-W.; Si, R.; Feng, W.; Zhang, H.-P.; Liu, H.-C.; Yan, C.-H. Shape-selective synthesis and oxygen storage behavior of ceria nanopolyhedra, nanorods, and nanocubes. *J. Phys. Chem. B* **2005**, *109*, 24380–24385. [CrossRef]

catalysts

MDPI

Article

Plasma-Deposited CoO–(Carbon Matrix) Thin-Film Nanocatalysts: The Impact of Nanoscale *p-n* Heterojunctions on Activity in CO_2 Methanation

Niloofar Mohammadpour [1], Hanna Kierzkowska-Pawlak [1,*], Jacek Balcerzak [1], Paweł Uznański [2] and Jacek Tyczkowski [1]

1 Department of Molecular Engineering, Faculty of Process and Environmental Engineering, Lodz University of Technology, Wolczanska 213, 93-005 Lodz, Poland; niloofar.mohammadpour@dokt.p.lodz.pl (N.M.); jacek.balcerzak@p.lodz.pl (J.B.); jacek.tyczkowski@p.lodz.pl (J.T.)

2 Centre of Molecular and Macromolecular Studies Polish Academy of Sciences, Sienkiewicza 112, 90-363 Lodz, Poland; pawel.uznanski@cbmm.lodz.pl

* Correspondence: hanna.kierzkowska-pawlak@p.lodz.pl

Abstract: Addressing the challenges associated with the highly exothermic nature of CO_2 methanation, there is considerable interest in innovative catalyst designs on structural metallic supports. One promising solution in this regard involves thin films containing cobalt oxide within a carbon matrix, fabricated using the cold plasma deposition method (PECVD). The objective of this study was to search for a relationship between the molecular structure, nanostructure, and electronic structure of such films and their catalytic activity. The investigations employed various techniques, including X-ray photoelectron spectroscopy (XPS), ultraviolet photoelectron spectroscopy (UPS), X-ray diffraction (XRD), UV-VIS absorption, and catalytic tests in the CO_2 methanation process. Three types of films were tested: untreated as-deposited (ad-CoO), thermally post-treated (TT-CoO), and argon plasma post-treated (PT-CoO) films. Among these, TT-CoO exhibited the most favorable catalytic properties, demonstrating a CO_2 conversion rate of 83%, CH_4 selectivity of 98% at 400 °C, and stability during the catalytic process. This superior performance was attributed to the formation of nanoscale heterojunctions in the TT-CoO film, where *p*-type CoO nanocrystallites interacted with the *n*-type carbon matrix. This work provides compelling evidence highlighting the key role of nanoscale heterojunctions in shaping the properties of nanocatalysts in thermal catalysis. These findings suggest promising prospects for designing new catalytic systems by manipulating interactions at the nanoscale.

Keywords: thin-film nanocatalysts; cold plasma deposition; carbon matrix; cobalt oxide; CO_2 methanation; nanoscale heterojunctions

Citation: Mohammadpour, N.; Kierzkowska-Pawlak, H.; Balcerzak, J.; Uznański, P.; Tyczkowski, J. Plasma-Deposited CoO–(Carbon Matrix) Thin-Film Nanocatalysts: The Impact of Nanoscale *p-n* Heterojunctions on Activity in CO_2 Methanation. *Catalysts* **2024**, *14*, 38. https://doi.org/10.3390/catal14010038

Academic Editors: Leonarda Liotta, Narendra Kumar and Konstantin Ivanov Hadjiivanov

Received: 21 November 2023
Revised: 1 January 2024
Accepted: 3 January 2024
Published: 4 January 2024

1. Introduction

The demand for highly efficient and stable supported catalysts has been increasing due to the growing need to address various energy and environmental concerns. Among different forms of catalysts, those applied as coatings of structured packings such as meshes, open-cell foams, and honeycomb monoliths deserve special attention due to their versatility and potential for improving process efficiency when applied in heterogeneous reactors [1,2]. Catalytic structured packings can be successfully employed in various processes, such as automotive gas treatment, the combustion of volatile organic compounds, CO_2 methanation, methane/methanol steam reforming, and many others [3]. Compared to the conventional packed beds of particle catalysts, structured catalysts offer lower pressure drop, better mass and heat transfer, and vast possibilities for tailoring their design to specific applications, providing great potential for process intensification in demanding applications, such as CO_2 valorization [4].

Recently, we developed an effective method for fabricating a novel and sophisticated CoO-based nanocatalyst. This catalyst is in the form of a thin film deposited on a wire-gauze support, with the aim of CO_2 hydrogenation to methane [5]. Using the plasma-enhanced chemical vapor deposition (PECVD) technique and volatile organometallic complex precursors enabled the creation of thin-film catalytic coatings that did not alter the geometry of the fine support, which is a critical factor for future applications. Thin-film catalysts prepared in this manner differ significantly from analogous conventional catalytic systems, primarily in that they cover the entire surface of the support, thus making it inert to the ongoing catalytic process. Moreover, the material deposited from the organometallic precursor in the plasma process always contains carbon, which, for example, in the case of the above-mentioned CoO-based nanocatalysts, allows one to obtain a unique nanostructure of cobalt oxide embedded in a carbon matrix. This specific structure has demonstrated very high activity and selectivity towards CH_4 in CO_2 hydrogenation [6]. Yet, understanding the nature of this structure and its impact on catalytic properties remains an open problem.

The active form of cobalt in conventional catalysts for CO_2 hydrogenation has been an ongoing subject of debate within the scientific community [7]. Some researchers claim that it exists in a metallic form [8], while others argue that it is present in an oxidized state [9], which was also confirmed in the case of our plasma-prepared CoO-based catalyst [6]. Moreover, there are assertions that both metallic and oxidized forms of Co coexist and are necessary for efficient catalytic performance [10,11]. Importantly, the interaction of Co metal or its oxide forms with other materials can significantly affect the catalytically active states. In addition, due to the interconversion of various forms under reaction conditions, identifying the exact nature of the catalytically active species is still a challenge [12].

While cobalt-based conventional catalysts used for CO_2 methanation are mainly oxide-supported, recent studies have shown that carbon-based materials are gaining attention due to their high thermal and chemical stability, as well as the ability to enhance catalytic performance through metal–support interactions [13]. It was reported that carbon, in the amorphous and nanostructured forms, participated in various ways to improve the activity and selectivity of the catalyst in CO_2 hydrogenation compared to the corresponding catalysts supported on traditional oxidic supports [14]. Particularly, it could provide a surface that promotes the activation of H_2 molecules and the subsequent formation of hydrocarbons [15,16]. The presence of carbon also improved the dispersion and stability of Co metal nanoparticles or Co oxides, preventing the sintering and deactivation of the catalyst [17]. It is therefore obvious that we should not forget about the possible role of the carbon matrix in our plasma-prepared catalysts.

In this work, we aimed to research the essential aspects of the structure of plasma-produced Co-based thin films, placing particular emphasis on the carbon matrix present in these films, and establishing connections with their catalytic activity in CO_2 methanation. It is assumed that establishing such a relationship may have far-reaching consequences in the design of thin-film nanocatalysts for their desired properties.

2. Results

2.1. Catalytic Performance

The performance of different thin-film CoO-based catalysts, namely ad-CoO, TT- CoO, and PT-CoO, which are referred to as as-deposited (ad-); calcined at 400 °C in an argon atmosphere for 30 min (TT-); and argon plasma post-treated for 2 h (PT-), were evaluated in the CO_2 methanation reaction. Additionally, solely carbon matrix thin films, as-deposited (ad-CM) and thermally treated (TT-CM) in the same way as the TT-CoO, were prepared and also tested. The comparative assessment was conducted in terms of CO_2 conversion and CH_4 selectivity over a temperature range of 250–400 °C.

As can be seen in Figure 1a, the TT-CoO film is active at 250 °C with 84% selectivity toward methane, while both the ad-CoO and PT-CoO films were inactive at this temperature. As the temperature increases, the activity of all the tested catalysts appears, among which the ad-CoO and TT-CoO films achieve CH_4 selectivity of approximately 98% after reaching a temperature of 400 °C. On the other hand, the PT-CoO film apparently exhibits poorer performance. At 400 °C, both CO_2 conversion and CH_4 selectivity are lower by about 17% and 8%, respectively, compared to the best TT-CoO catalytic film.

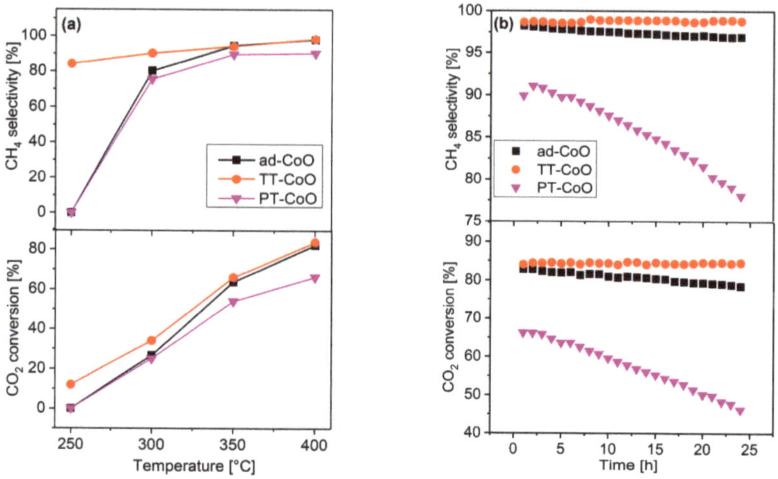

Figure 1. Performance of CoO-based thin films in CO_2 methanation reaction as a function of (**a**) temperature; (**b**) time-on-stream at 400 °C.

The differences in the performance of the CoO-based films become evident when analyzing the time-on-stream catalytic test at 400 °C. According to the experimental findings depicted in Figure 1b, the only catalyst that is stable over 24 h in the CO_2 methanation reaction is the thermally treated CoO-based thin film (TT-CoO). Both the CO_2 conversion and selectivity to CH_4 do not change over time. On the contrary, the as-deposited catalyst without any treatment (ad-CoO) demonstrates a gradual loss of activity over time. The PT-CoO film not only shows the lowest activity at 400 °C but also exhibits very poor stability. Over the course of 24 h, the CO_2 conversion achieved over this catalyst decreased from 66% to 46%, while the CH_4 selectivity decreased from 90% to 78%. The performance of the carbon matrix thin films (ad-CM, TT-CM) when subjected to the reaction mixture of CO_2 and H_2 is not shown in Figure 1. This is because no products were detected by the GC analysis up to 350 °C, indicating no catalytic activity. However, at 400 °C, a small peak of CO appeared for ad-CM and TT-CM, suggesting some chemical processes occurring over both carbon matrix thin films at this temperature. The estimated conversion of CO_2 was approximately 1.5% for ad-CM and 2% for TT-CM, with a relative uncertainty of about 25% due to the low precision associated with the low concentration levels. During these catalytic tests, no water production was detected, which proves that this is not a CO_2 hydrogenation process.

2.2. *Chemical Structure of CoO-Based Thin Films*

The surface composition and molecular structure of the studied CoO-based catalysts, namely ad-CoO, TT-CoO, and PT-CoO, which exhibited different activities and stabilities in CO_2 methanation, were investigated by X-ray photoelectron spectroscopy (XPS). Additionally, the spent samples of these catalysts obtained after subjecting them to a catalytic test at a temperature of 400 °C were examined to determine possible changes in the molecular structure of their surface during the CO_2 methanation reaction.

As already mentioned in the Introduction, thin films prepared using the PECVD technique from organometallic precursors always contain, in addition to the metal-based phase, a carbon phase, which we call the carbon matrix. The XPS analysis of the elemental composition of the tested catalytic films revealed the presence of cobalt (Co), oxygen (O), and carbon (C) on their surface. The atomic ratios of carbon to cobalt (C/Co) and oxygen to cobalt (O/Co) are shown in Table 1.

Table 1. Atomic ratio of elements and their various forms in fresh and spent CoO-based films.

Catalyst	ad-CoO Fresh	ad-CoO Spent	TT-CoO Fresh	TT-CoO Spent	PT-CoO Fresh	PT-CoO Spent
			Atomic Ratio			
C/Co	6.6 ± 0.2	4.8 ± 0.1	6.4 ± 0.3	6.4 ± 0.3	3.3 ± 0.4	8.3 ± 0.9
O/Co	2.4 ± 0.1	2.0 ± 0.2	2.3 ± 0.2	2.0 ± 0.2	1.4 ± 0.1	1.3 ± 0.1
Co^{2+}/Co^0	5.1 ± 0.4	26 ± 3	10.1 ± 0.9	10.9 ± 0.9	2.4 ± 0.3	2.8 ± 0.3
$C(sp^2)/C(sp^3)$	2.6 ± 0.2	1.8 ± 0.2	7.4 ± 0.3	7.6 ± 0.4	0.5 ± 0.1	10 ± 1

As can be seen, the fresh calcined films (TT-CoO) show very similar contents of both carbon and oxygen (relative to cobalt) compared to the fresh as-deposited films (ad-CoO). Meanwhile, in the case of the fresh PT-CoO films, a distinct decrease in the C/Co and O/Co values is revealed compared to the ad-CoO films. This can be considered to be the result of the argon plasma treatment inducing the preferential etching of carbon and oxygen from the film surface. After the catalytic process (spent catalysts), the carbon content (C/Co) in the ad-CoO film decreases clearly; in the TT-CoO film it remains at the same level; while in the PT-CoO film, this content increases significantly. These effects will be discussed further in Section 3. On the other hand, no significant differences were observed in the oxygen content (O/Co) between the fresh and spent catalytic films.

To take a deeper look at the molecular structure of the studied catalytic films, the regions corresponding to Co 2p and C 1s in the XPS spectra were analyzed in detail. Figure 2 shows the Co 2p spectra for the as-CoO, TT-CoO, and PT-CoO catalyst films before the catalytic process (fresh) and after this process (spent). These spectra consist of a 2p3/2 and 2p1/2 spin-orbital doublet, from which the more intense 2p3/2 spectrum is usually analyzed. Its numerical deconvolution allowed for the identification of three bands, which were assigned to the appropriate chemical states of cobalt [6,18,19]: a narrow band (Co^0) in the range of 778.2–778.7 eV belonging to the Co–Co and Co–C bonds, a broad band in the range of 780.2–781.0 eV assigned to CoO (Co^{2+}), and a band in the range of 785.6–786.2 eV representing the Co^{2+} shake-up satellite (sat.).

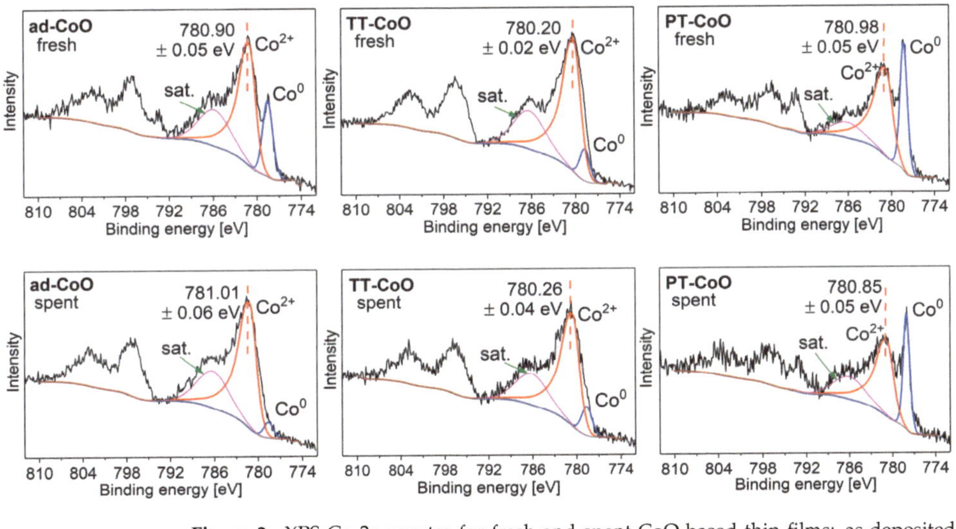

Figure 2. XPS Co 2p spectra for fresh and spent CoO-based thin films: as-deposited (ad-CoO); thermally treated in Ar (TT-CoO); Ar plasma-treated for 2 h (PT-CoO).

In turn, Figure 3 shows the C 1s spectra for the above cases. As a result of the deconvolution of these spectra, we distinguished six bands which, based on the literature [6,19–21], were assigned in the order of increasing binding energy to cobalt carbide (Co–C) at 283.5 eV, sp^2 carbon (C=C) at 284.6 eV (used to calibrate the spectra), sp^3 carbon (C–C, C–H) at 285.2–285.4 eV, carbon single-bonded to oxygen (C–O) at 286.3–287.2 eV, carbon doubly bonded to oxygen (C=O) at 287.9–288.8 eV, and strongly oxidated carbon (C–OOR) at approx. 290 eV. The highest part of the carbon species in all the studied films is sp^2 and sp^3 carbon.

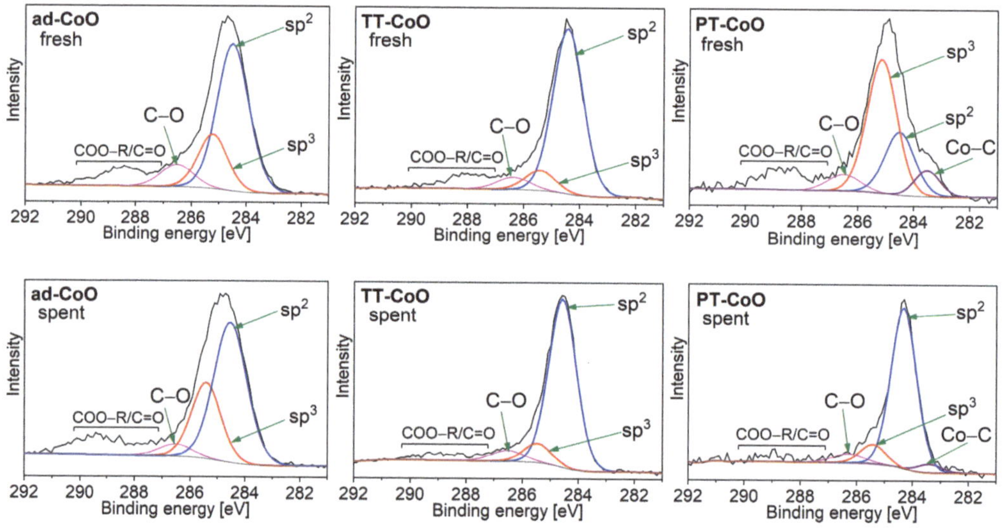

Figure 3. XPS C 1s spectra for fresh and spent CoO-based thin films: as-deposited (ad-CoO); thermally treated in Ar (TT-CoO); Ar plasma-treated for 2 h (PT-CoO).

According to Figure 2, in all the tested catalytic films, both before and after the CO_2 hydrogenation process, cobalt atoms occur in the form of CoO (Co^{2+}) and Co^0, which can be attributed to the metallic form (Co–Co bonds) or carbide form (Co–C bonds). The latter form occurs solely in the PT-CoO film, because only for this film the C 1s spectrum reveals the Co–C band (Figure 3). Thus, in this case, a mixture of Co–Co and Co–C bonds may be present. However, a rough calculation of the content of the Co–C based on the C 1s and Co 2p spectra for PT-CoO (fresh) indicates that only cobalt carbide in the CoC form is present here. This unusual cobalt monocarbide form is already known, among others, as a product of the reaction of cobalt vapor and methane in a glow discharge [22], which can be considered a situation similar to the plasma treatment we perform. A similar situation concerns the PT-CoO (spent) film, for which the Co^{2+}/Co^0 ratio is maintained within the error limits compared to PT-CoO (fresh) (Table 1). However, the quantitative interpretation of the C 1s spectrum is limited here due to the sp^2 carbon deposit covering the surface of the film, which is produced during the catalytic process (Figure 3). In all other cases, the Co^0 band should be associated with the metallic Co–Co form, the relation of which with the Co^{2+} form (Co^{2+}/Co^0) is shown in Table 1.

Analyzing the Co^{2+}/Co^0 ratio for the ad-CoO and TT-CoO films, it is easy to notice that the calcination of the ad-CoO films causes an approximately two-fold increase in this ratio, which should be associated with the rearrangement involving the oxidation of Co^0 to CoO with the participation of oxygen initially bonded with carbon. The catalytic process causes the much more advanced oxidation of Co^0 in the ad-CoO films, this time with the participation of the reaction mixture (CO_2, H_2O). Interestingly, however, for the TT-CoO films, no change in the Co^{2+}/Co^0 ratio is observed within the error limits, which proves the high stability of the chemical structure of these films.

The stability of the chemical structure of the TT-CoO films in the catalytic CO_2 hydrogenation process is also confirmed by the analysis of the ratio of sp^2 to sp^3 carbon content, which does not change between the fresh and spent films within the limits of error (Table 1). On the contrary, for the ad-CoO film, there is a visible decrease in the $C(sp^2)/C(sp^3)$ ratio and, as already mentioned, a simultaneous decrease in the carbon content on the film surface between the fresh and spent samples. This clearly confirms the instability of the chemical structure of such films during the catalytic process.

It should also be noted that the calcination process causes a significant increase in the sp^2 carbon content in relation to sp^3 carbon (ad-CoO fresh and TT-CoO fresh), which indicates the transformation of the carbon matrix into a more graphite-like structure. This was already noticed in our previous studies [6]. In turn, plasma treatment of the ad-CoO film (PT-CoO fresh) causes a sharp increase in the sp^3 carbon content, which, apart from the previously mentioned preferential etching, is the result of the action of argon plasma on the carbon matrix [23]. After the catalytic process, these films reveal a huge increase in the carbon content (C/Co), including sp^2 carbon (sp^2/sp^3 ratio increases), which, however, should not be associated with changes in the chemical structure of the film surface, but with the formation of the carbon deposit.

One more important result requires emphasizing, namely the comparison of the positions of the Co^{2+} band maximum for different films, when all the XPS spectra were normalized to the same value (284.6 eV) for the C 1s sp^2 band. It is evident (Figure 2) that for the TT-CoO, both the fresh and spent films, this maximum is shifted towards a lower binding energy (780.2 eV) compared to the ad-CoO and PT-CoO (fresh and spent) films, for which the Co^{2+} band is approximately in the same position at 780.8–781.0 eV. Thus, the observed shift is about −0.7 eV.

2.3. Nanostructure of CoO-Based Thin Films

To better understand the reasons for the different catalytic performances exhibited by the plasma-prepared CoO-based thin films, their nanostructure was examined using X-ray diffraction (XRD). The raw XRD patterns for the ad-CoO and TT-CoO films, both fresh and spent, are shown in Figure 4a. As one can see, the fresh ad-CoO film is completely amorphous, unlike the fresh TT-CoO film, where a clear nanocrystalline fraction formed during the calcination process is visible. This behavior of CoO-based films is known and has been analyzed in detail (in the analogous case of the calcination of these films in air and the formation of Co_3O_4 spinel nanocrystallites in them) in Ref. [24]. After the catalytic process, the nanocrystalline structure in the TT-CoO film remains virtually unchanged, while in the spent ad-CoO film, a trace appearance of the nanocrystalline fraction in the amorphous matrix can be observed. A detailed analysis of the XRD pattern for the spent ad-CoO reveals its considerable similarity to the XRD patterns for the fresh and spent TT-CoO. As an example, the most intense band at 42.4 deg appropriately magnified for the spent ad-CoO is shown in Figure 4b.

The numerical analysis of the XRD pattern for the TT-CoO films (Figure 4c) clearly indicates (according to the ICDD-PDF card no. 48-1719) the most well-known cubic rock-salt structure of CoO, revealing reflections at 36.5, 42.4, 61.5, 73.8, and 77.4 deg. In this case, the size of the CoO nanocrystallites determined using the Scherrer equation is approximately 11 nm. The CoO nanocrystallite size estimated in the same way for the spent ad-CoO film based on Figure 4b is approximately 3 nm.

It should be added that since Ar plasma treatment (PT-CoO) causes changes only in a very thin surface layer, much thinner (several nanometers) than the thickness of the film (several hundred nanometers), XRD tests (analyzing the entire mass of the film) are unjustified in this case.

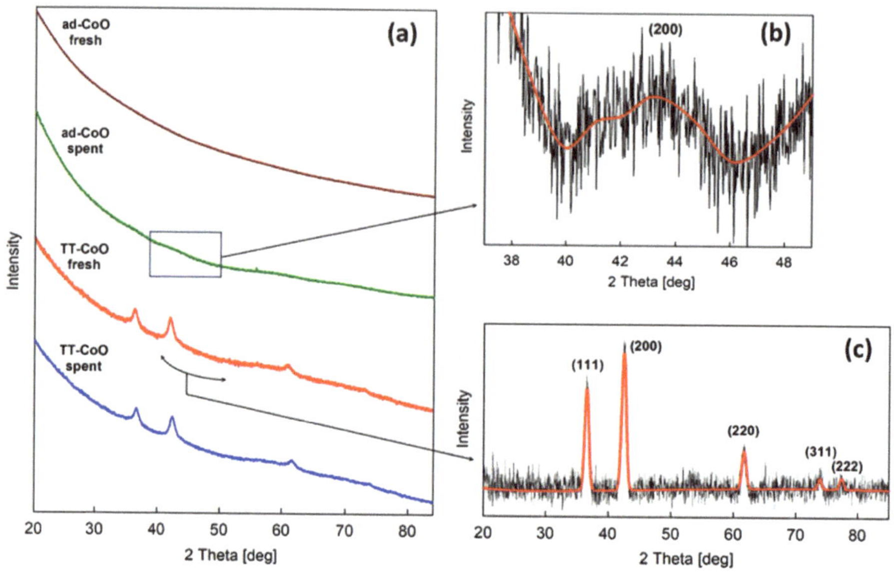

Figure 4. XRD patterns for (**a**) ad-CoO and TT-CoO films—raw data; (**b**) ad-CoO spent—analysis; (**c**) TT-CoO—analysis.

2.4. Molecular and Electronic Structure of Carbon Matrix

As already presented above, the nature of plasma deposition from the cobalt organometallic precursor leads to the formation of films containing a carbon matrix with a CoO fraction. The molecular structure and nanostructure of such a system turned out to have a significant impact on the catalytic properties of these films (Sections 2.1–2.3). To determine the role of the carbon matrix more precisely in the catalytic effect, films containing only this matrix (without Co) were investigated. These films were prepared under the same plasma process conditions as the ad-CoO films, using a mixture of cyclopentadiene and carbon monoxide in a 2:1 ratio as a precursor (the same ratio in which analogous ligands are present in the organometallic cobalt complex $CpCo(CO)_2$ used to deposit the ad-CoO films).

Figure 5 shows the XPS C 1s spectra for an as-deposited carbon film (ad-CM) and such a film thermally treated (TT-CM). The carbon species composition for the fresh films shows similarity with the corresponding chemical structure of the carbon matrix in CoO-based thin films (Figure 3). A comparison of the C 1s spectra between ad-CM and TT-CM reveals that after the thermal treatment, the $C(sp^2)/C(sp^3)$ ratio increases (Table 2), analogously to what is observed in the case of changes between the TT-CoO and ad-CoO fresh films (Table 1). Similarly to the ad-CoO spent film (Table 1), after treatment with the reaction mixture of CO_2 and H_2 at 400 °C, an increase in the $C(sp^3)$ content compared to $C(sp^2)$ was observed in the ad-CM film, but also in the TT-CM film (Figure 5 and Table 2), which is not the case in the spent TT-CoO film (Table 1).

Table 2. $C(sp^2)/C(sp^3)$ ratio for the as-deposited and thermal-treated (fresh and spent) carbon matrix films.

Carbon Matrix	ad-CM Fresh	ad-CM Spent	TT-CM Fresh	TT-CM Spent
		Atomic Ratio		
$C(sp^2)/C(sp^3)$	4.1 ± 0.4	2.9 ± 0.4	6.0 ± 0.5	2.7 ± 0.5

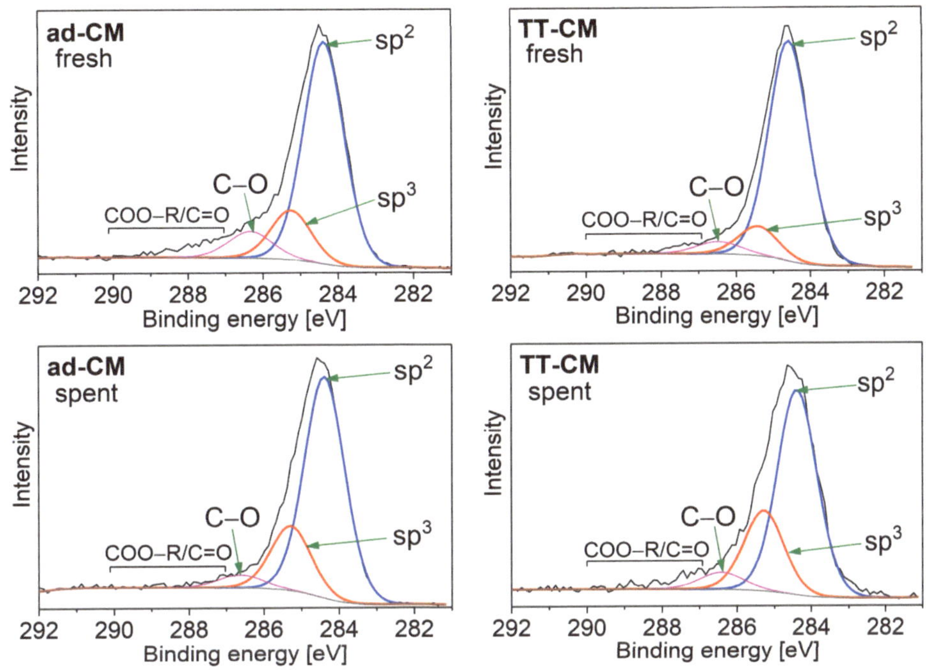

Figure 5. XPS C 1s spectra of the carbon matrix films as-deposited (ad-CM) and thermally treated in Ar (TT-CM).

Unfortunately, the ad-CM films treated in Ar plasma under the same conditions as the ad-CoO films, unlike them, were quickly etched and completely removed after the same treatment time of 2 h, making them impossible to study further. This result clearly indicates the role of the CoO fraction in stabilizing the carbon matrix in the plasma-treated catalytic films (PT-CoO).

To determine the electronic structure of the ad-CM and TT-CM carbon films, which constitute carbon matrices in the ad-CoO and TT-CoO catalytic films, respectively, ultraviolet photoelectron spectroscopy (UPS) and UV-VIS absorption spectroscopy studies were performed. Figure 6a shows the measured UPS spectra for the ad-CM and TT-CM films. The interpretation of these spectra was carried out according to Ref. [25]. On this basis, two important parameters from the point of view of the electronic structure of the films were determined. The first one is the valence band edge value (E_{BE}), which characterizes the energy distance of the upper edge of the valence band from the Fermi level. As can be seen from the inset in Figure 6a, this value is virtually the same for both films and is 0.8 eV. The second parameter is the secondary electron cutoff (*SECO*), the value of which was set at 16.7 eV and 16.3 eV for the TT-CM and ad-CM films, respectively. Using the determined parameters, it is possible to estimate the elements of the electronic structure, such as work function (W) and ionization potential (J), according to the following equations:

$$W = E_{h\nu} - SECO, \quad (1)$$

where $E_{h\nu}$ is the impinging photon energy (in our case, it is the He I emission line with an energy of 21.2 eV), and

$$J = E_{BE} + W. \quad (2)$$

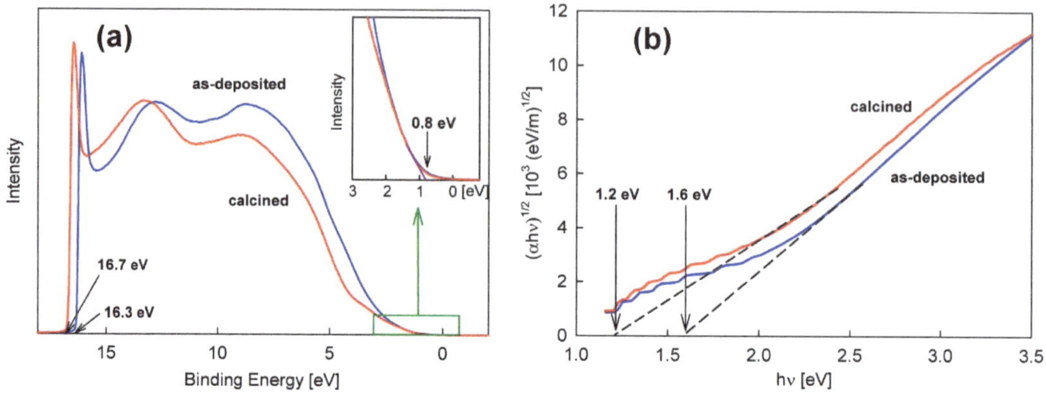

Figure 6. UPS (**a**) and optical absorption spectra (**b**) for as-deposited (ad-CM) and calcined (TT-CM) carbon matrix films.

To complete the band model of the electronic structure of the films, optical gap (E_{OPT}) values were determined from UV-VIS absorption measurements. For amorphous materials, which include the tested carbon films, E_{OPT} is most frequently calculated according to the Tauc relation [26]:

$$\alpha\, h\nu = B\,(h\nu - E_{OPT})^n, \tag{3}$$

where α is the absorption coefficient, $h\nu$ is the energy of absorption light, n is a parameter connected with the density-of-states (DOS) distribution in the band gap, and B is a proportionality factor that also provides information about the DOS distribution.

The optical gap for the plasma-deposited semiconductor thin films is usually determined by assuming a parabolic function for the DOS distribution in the band tails (n = 2 in Equation (3)), and in this case, it is assumed that no large error is made by taking the E_{OPT} approximately equal to the band gap (E_G) [27]. Figure 6b, based on Equation (3), shows the measured UV-VIS optical absorption for the ad-CM and TT-CM films as a dependence of $(\alpha\, h\nu)^{1/2}$ on $h\nu$. The E_{OPT} determined in this way is 1.6 eV and 1.2 eV, respectively.

Knowing the values of the band gap (E_G) and the ionization potential (J), it is possible to calculate another important element of the electronic structure, which is the electron affinity (χ):

$$\chi = J - E_G. \tag{4}$$

In turn, by subtracting the measured E_{BE} value from E_{OPT}, we obtain the energy distance of the Fermi level from the lower edge of the conduction band.

Using the determined elements of the electronic structure, band models for the ad-CM and TT-CM films were proposed and presented in Figure 7. While the electronic structure parameters of both films are similar and generally align with the values reported in the literature—such as the work function (from 4.55 to 4.85 eV, [28]) and the optical gap (from 1.09 to 2.09 eV, [29])—there exists a notable distinction between them. This distinction is characterized by the transition from an intrinsic semiconductor for the ad-CM film to the n-type semiconductor for the TT-CM film, which is defined by the position of the Fermi level.

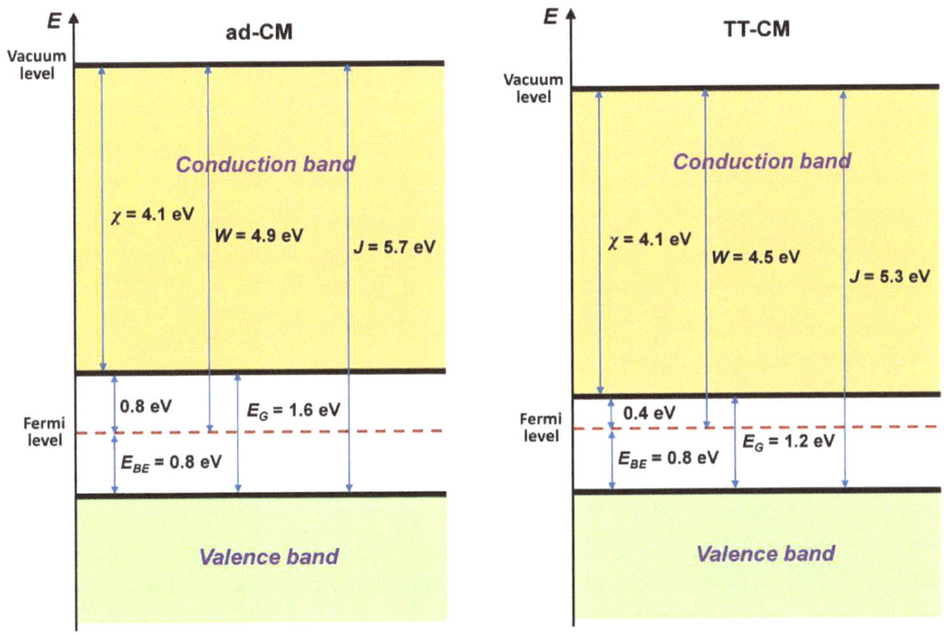

Figure 7. Band model for the ad-CM and TT-CM films.

3. Discussion

Three types of plasma-deposited CoO-based films (ad-CoO, TT-CoO, and PT-CoO) were investigated for their catalytic performance in the CO_2 methanation process and the relationship between this performance and the composition and structure of the films. Among them, the TT-CoO films turned out to be the best in terms of CO_2 conversion, selectivity to CH_4, and stability during the process (Figure 1). These films also demonstrated the complete stability in their molecular structure in the CO_2 methanation process, unlike the other two types of films (Table 1).

In the ad-CoO films, following the catalytic process, the carbon content decreases, and at the same time the content of its sp^3 form increases compared to sp^2. Interestingly, maintaining these films at a temperature of 400 °C during the process causes only a slight change in the amorphous nanostructure composed of Co, C, and O atoms, consisting of the formation of a very small amount of CoO nanocrystallites with a size of about 3 nm (Section 2.3). Meanwhile, by preparing TT-CoO films by the thermal treatment of ad-CoO films at the same temperature, however, not in the reaction mixture, but in an argon atmosphere, we create a stable structure containing a significant amount of CoO nanocrystallites with a size of approximately 11 nm in an amorphous carbon matrix. This means that in the case of ad-CoO films, the reaction mixture diffuses into their bulk and influences the structure transformation process. The phenomenon of the permeability of various gases, such as CO_2, through films of amorphous carbon produced by PECVD, is widely known [30,31], which confirms the correctness of the above assumption.

As a probable reaction, referring to the reverse Boudouard reaction [32], an electrophilic attack of a carbon atom with a partial positive charge (δ+) in a CO_2 molecule on a double bond in the carbon matrix is proposed:

$$\underset{\text{carbon matrix}}{>\overset{sp^2}{C}=\overset{sp^2}{C}<} + \underset{\text{gas}}{\overset{\delta+}{CO_2}} \rightarrow \underset{\text{carbon matrix}}{>\overset{sp^3}{C}<} + \underset{\text{gas}}{2CO} \qquad (5)$$

As a result of this reaction, which occurs simultaneously with the CO_2 methanation process, the carbon content in the matrix decreases by removing it as CO, and the concentration of carbon in the sp^3 form increases (Table 1; compare fresh and spent ad-CoO films). The alternation of the carbon matrix in the ad-CoO film appears to be a plausible explanation for the substantial decrease in the formation of CoO nanocrystallites, which are readily produced at the same temperature (400 °C) in an argon atmosphere. It is noteworthy that the TT-CoO film, produced in this manner, maintains its structure unaffected by the reaction mixture (Table 1; compare fresh and spent TT-CoO films). Contrastingly, the pure carbon matrix thermally treated at 400 °C (TT-CM) exhibits the evolution of CO under these conditions (Section 2.1) and a rise in the Csp^3 content in comparison to Csp^2 (Table 2), providing evidence of the reaction (5) progression. This clearly highlights the key role of CoO nanocrystallites in stabilizing the carbon matrix within TT-CoO.

Thus, a fundamental question arises: what role do CoO nanocrystallites play in stabilizing the structure of the TT-CoO film during the CO_2 methanation process, particularly as this structure demonstrates superior catalytic performance among all the tested films? The concept of nanoscale heterojunctions turns out to be crucial in addressing this query, having been previously invoked several times to elucidate catalytic phenomena in thermal catalysis processes [33–36]. In the case of the TT-CoO films under discussion, we hypothesize that such heterojunctions form between CoO nanocrystallites and the carbon matrix.

It has been extensively documented in the literature that undoped CoO nanocrystallites function as a p-type semiconductor [37,38], exhibiting a band gap within the range of 2.2–2.8 eV [39,40]. Based on the recent research by Zhai et al. [41], who delved into the electronic structure of cubic CoO nanocrystallites using UPS and UV-VIS spectroscopy, very similar to our investigation of the electronic structure of the carbon matrix, we adopted their E_G = 2.43 eV and E_{BE} = 0.74 eV for further analysis. Taking these parameters of the electronic structure of CoO and the band model of the TT-CM carbon matrix (Figure 7), a band model of the CoO–TM-CM heterojunction was proposed, which is shown in Figure 8a. This heterojunction is of the straddling gap type (type I-2) [42].

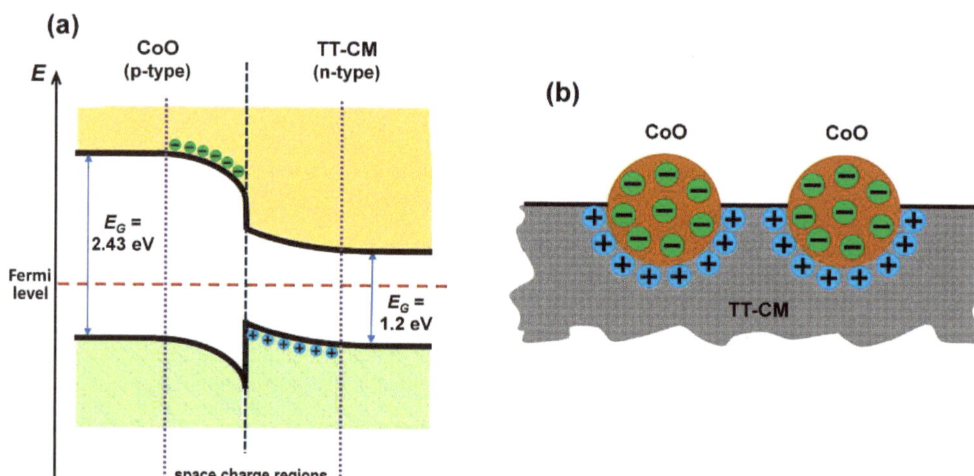

Figure 8. Heterojunction between CoO nanocrystallite and carbon matrix: (**a**) band model of the heterojunction (straddling gap (type I-2)); (**b**) a model of the TT-CoO film surface.

As a result of achieving thermodynamic equilibrium within the created heterojunction, marked by the alignment of the Fermi level, regions of negative and positive space charge are established on either side of the heterojunction, specifically in the p-type (CoO) and n-type (TT-CM) semiconductors, respectively. The dimensions of these space charge regions often span several dozen nanometers [43]. Consequently, when nano-sized entities interact within the heterojunction, these space charge regions might entirely fill them (Figure 8b), thereby altering their electronic characteristics. This transformation subsequently modifies the nature of the active catalytic centers on their surfaces, thereby influencing the catalytic properties.

The presence of space charge regions in the TT-CoO film has been confirmed by the X-ray photoelectron spectroscopy (XPS) studies, indicating a noticeable shift in the maximum position of the Co^{2+} (2p3/2) band (Section 2.2). The maximum position for both the fresh and spent ad-CoO and PT-CoO films remains virtually constant, at approximately 780.8–781.0 eV (Figure 2). Moreover, according to the literature, the maximum of the Co^{2+} band for the pure CoO nanocrystallites also has a very similar position, typically at 780.9 eV [44] or 780.6 [45]. However, within the TT-CoO film, where CoO nanocrystallites exist and presumed heterojunctions form between them and the carbon matrix, the Co^{2+} band experiences a shift towards a lower binding energy (BE), settling at approximately 780.2 eV. The determined shift of about 0.7 eV is a combined effect of the Co^{2+} band's shift towards a lower binding energy and the expected shift of the sp^2 band towards a higher binding energy. However, the position of the sp^2 band remains fixed at 284.6 eV. These shifts result from the presence of a space charge, which, in the case of a negative charge, diminishes the binding energy, while in the case of a positive charge, elevates the binding energy of the photoelectron emitted from the atom during XPS measurements [46].

The presence of a negative charge in the CoO nanocrystallites enhances their basic nature (catalytic basic sites), thus promoting CO_2 hydrogenation towards CH_4 [47]. In turn, the positive charge on TT-CM inhibits the electrophilic attack of CO_2, effectively impeding the reaction (5), and thereby stabilizing the carbon matrix. This mechanism, reliant on the formation of nanoscale heterojunctions, adequately explains the differences in the catalytic performance and structural changes between the ad-CoO and TT-CoO films within the reaction mixture.

A distinct topic concerns plasma-treated film (PT-CoO). Contrary to the initial anticipation that subjecting ad-CoO films to argon plasma treatment would serve as an energy-efficient alternative to the more intensive calcination process, the PT-CoO films turned out to be unstable in the CO_2 hydrogenation process, revealing a clear decline in CO_2 conversion over time and a shift in selectivity towards CO production rather than CH_4 (Figure 1). The XPS investigations revealed, similar to the ad-CoO films, the absence of heterojunctions in the PT-CoO film. However, the plasma treatment caused significant changes in the molecular structure of its surface, resulting in the formation of the monocarbide (CoC) form and a marked increase in the sp^3 carbon content compared to sp^2. This transformation likely contributes to the altered selectivity, favoring CO production. The resultant CO appears to be the source of carbon deposit formation [36], as observed in this case (Figure 3, spent PT-CoO), which impedes the conversion of CO_2. Nonetheless, due to the unattractive catalytic properties exhibited by the PT-CoO films, further comprehensive studies on them were not pursued.

In a summary of the discussion, the main findings from the conducted research are condensed in Table 3.

Table 3. Summary of findings—the impact of CoO-based thin films structure on activity in CO_2 methanation.

Catalyst	ad-CoO		TT-CoO		PT-CoO	
Electronic structure	No formation of heterojunctions		Formation of nanoscale *p-n* heterojunctions between CoO (*p*-type) and CM (*n*-type)		No formation of heterojunctions	
Performance at 400 °C (initial values)	X_{CO2} = 82% S_{CH4} = 97%	↘	X_{CO2} = 83% S_{CH4} = 98%	→	X_{CO2} = 66%S_{CH4} = 90%	↘

Table 3. *Cont.*

Catalyst	ad-CoO	TT-CoO	PT-CoO
Performance stability over time	Moderate	Excellent	Very poor
Reasons of deactivation	Unstable CM structure: carbon consumption and transition from sp^2 to sp^3 in reaction with CO_2	n/a	Poisonous carbon deposit

4. Materials and Methods

4.1. Preparation of Thin Films

The plasma-enhanced chemical vapor deposition (PECVD) technique was employed to fabricate CoO-based and carbon matrix thin films. The deposition process took place in a parallel-plate radio-frequency (RF 13.56 MHz) plasma reactor, which is a home-made device. The reactor construction and film fabrication procedure are detailed in our previous papers [6,48]. For the preparation of CoO-based catalytic films, cyclopentadienyl (dicarbonyl) cobalt(I) ($CpCo(CO)_2$, Stream Chemicals, Newburyport, MA, USA) served as the deposition process precursor, with a vapor flow rate of 0.083 sccm. Argon (99.999%, Linde Gas, Cracow, Poland) was used as a carrier gas with a flow rate of 0.71 sccm. The reactor chamber maintained a total pressure of 4–5 Pa. The deposition was carried out at a discharge power of 60 W for 10–30 min, resulting in films with a thickness of approximately 280–850 nm.

Thin films with a pure carbon matrix were prepared using the same plasma process parameters as those for the CoO-based films: a discharge power of 60 W, an argon flow rate of 0.71 sccm, and a total pressure in the reactor chamber of 4–5 Pa. The precursor for the deposition process was a mixture of CO (99.99%, Linde Gas, Cracow, Poland) and cyclopentadiene (C_5H_6, Warchem, Poland) in a molar ratio of 2:1, corresponding to the ligand content in the $CpCo(CO)_2$ complex. The flow rates of CO and C_5H_6 were set to 0.055 and 0.028 sccm, respectively. The films were deposited for 6–10 min, resulting in thicknesses of 190–320 nm.

Kanthal steel meshes, in the form of knitted wire gauze (TermTech, Warsaw, Poland), sized at 2.3 cm diameter with a 1 cm hole in the center, which corresponded to the size of the catalytic reactor, were utilized as supports for catalytic tests. Plates made of the same Kanthal steel, measuring 1.5 × 1.5 cm, were employed for studying the chemical structure of the films through XPS spectroscopy. Before plasma deposition, the Kanthal supports underwent calcination in air at 900 °C for 48 h to ensure high surface development [6,48]. For UV-VIS absorption measurements, the films were deposited on quartz wafers, while for XRD diffraction studies, zero background Si wafers (Si 510, Institute of Electronic Materials Technology—Łukasiewicz Research Network, Warsaw, Poland) were used as substrates.

Three types of CoO-based films were prepared for investigations: as-deposited films without any post-treatment (ad-CoO); as-deposited films thermally treated (calcined) in argon with a flow rate of 2 L/min and a heating rate of 18 °C/min, maintained at 400 °C for 30 min (TT-CoO); and as-deposited films treated in plasma at a constant argon flow rate of 0.71 sccm, with a discharge power of 80 W for 2 h. For the carbon matrix films, two forms were prepared for further studies: as-deposited films without any post-treatment (ad-CM) and calcined films (TT-CM), following the same procedure as the TT-CoO films.

4.2. Catalytic Tests

Catalytic tests were conducted at atmospheric pressure using a tube-in-tube fixed-bed quartz reactor, as described in detail in a previous paper [35]. For each experiment, six circular pieces of catalyst (disc of mesh with OD = 2.3 cm and a hole ID = 1 cm), providing a geometric surface area of 38 cm^2, were loaded into the reactor. To maintain data comparability, the position of the meshes constituting a catalytic bed was kept the same in all experiments. After purging the lines and the reactor with helium (99.999%), a mixture of H_2 (99.999%) and CO_2 (99.99%) gases (all gases supplied by Linde Gas, Cracow, Poland) with a total flow rate of 25 sccm and the ratio of $H_2/CO_2 = 4$ was introduced into the reactor. The temperature was raised at a rate of 5 °C/min in stages of 50 °C from 250 °C to 400 °C. On the way of the outlet reaction mixture to a gas chromatography analyzer (GC Master, Dani Instruments, Cologno Monzese, Italy), a cold trap was inserted to collect the condensed water. The composition of outlet dry gas was analyzed using a thermal conductivity detector (TCD) and a ShinCarbon ST packed column (Restek, Bellefonte, PA, USA), with measurements taken at 10 and 20 min after reaching the set temperatures. At each temperature, the catalyst was kept for 30 min. Utilizing the calibration curves for CO_2, CO, and CH_4, the CO_2 conversion and selectivity to CO and CH_4 were determined. The stability of the catalyst was investigated over a period of 24 h at 400 °C.

4.3. Chemical Structure, Electronic Structure, and Nanostructure of the Films

The chemical structure of both the CoO-based and carbon matrix films was investigated by X-ray photoelectron spectroscopy (XPS) using a Kratos AXIS Ultra spectrometer (Kratos Analytical Ltd., Manchester, UK) with a monochromatic Al-Kα X-ray source (1486.6 eV). Spectra were obtained from an analysis area of 300 × 700 μm. The anode power was set at 180 W, and the hemispherical electron energy analyzer was operated at a pass energy of 20 eV for all high-resolution measurements. XPS spectra were analyzed using Kratos Vision 2.2.10 software, calibrated by setting the C 1s carbon peak assigned to the sp^2 carbon at 284.6 eV. The Shirley algorithm was used for background subtraction.

The same Kratos AXIS Ultra spectrometer, equipped with a helium lamp with a He I line (21.21 eV), was used for ultraviolet photoelectron spectroscopy (UPS). Valence band spectra, providing information about the electronic structure of the films, were collected from 55 μm diameter spots on the sample surface at 50 meV resolution. Data analysis was performed using Kratos Vision 2 software (ver. 2.2.10, Kratos Analytical Ltd., Manchester, UK, 2012). To complement this information about the band gap value, optical absorption measurements were conducted using a Shimadzu UV-1800 spectrophotometer (Shimadzu Corporation, Kyoto, Japan). Measurements were made in the transmission mode in the wavelength range of 200–800 nm with a 1 nm sampling interval.

To determine the nanocrystalline structure of the films, the X-ray diffraction technique was employed. The measurement used an Empyrean diffractometer (Malvern Panalytical Ltd., Malvern, UK) operating with Cu Kα radiation (λ = 1.541874 Å) at 45 kV and 40 mA, utilizing a reflection Bragg–Brentano mode. The diffractometer was equipped with a PIXcel3D detector with 255 active channels. Diffractograms were obtained in the 2Theta range of 20–85 deg using a continuous scan mode with a step size of 0.0263 deg and a counting time of 176 s. Samples were rotated with a rotation time of 8 s.

5. Conclusions

Considering the vast potential of plasma-deposited thin-film CoO-based nanocatalysts for CO_2 methanation, this work aimed to investigate the correlation between the molecular structure, nanostructure, and their catalytic activity. Three types of catalysts were analyzed: as-deposited (ad-CoO) films, films subjected to calcination in argon after plasma deposition (TT-CoO), and films post-treated with argon plasma (PT-CoO). The TT-CoO films exhibited the most favorable catalytic properties, showcasing the highest CO_2 conversion rate, selectivity for CH_4, and structural stability during the catalytic process.

The results indicate that the remarkable catalytic properties of the TT-CoO films arise from the formation of nanoscale heterojunctions between the CoO nanocrystallites and the carbon matrix. These heterojunctions create spatial charge regions—negative on the CoO nanocrystallites and positive on the carbon matrix—enhancing and stabilizing the catalytic activity towards CH_4 production. These findings are in line with the traditional concept of acidic and basic catalytically active sites. Positively charged acidic sites are known to impede the conversion to CH_4, while increased basicity promotes CH_4 production. In our study, we have shown that a negative charge on CoO nanocrystallites, consistent with the concept of basic active centers, enhances the CO_2 hydrogenation toward methane.

This work provides strong evidence for the key role of nanoscale heterojunctions in determining the properties of nanocatalysts. It emphasizes how this phenomenon can be used for a more rational design of novel catalytic systems for targeted products.

Author Contributions: Conceptualization, H.K.-P. and J.T.; methodology, H.K.-P., J.B. and J.T.; formal analysis, N.M., H.K.-P., J.B., P.U. and J.T.; investigation, N.M., J.B. and P.U.; resources, H.K.-P. and J.T.; data curation, H.K.-P., J.B., P.U. and J.T.; writing, H.K.-P. and J.T.; visualization, N.M., H.K.-P., J.B. and J.T.; supervision, H.K.-P. and J.T.; project administration, H.K.-P. and J.T. All authors have read and agreed to the published version of the manuscript.

Funding: This research received no external funding.

Data Availability Statement: The data presented in this study are available on request from the corresponding author.

Conflicts of Interest: The authors declare no conflicts of interest.

References

1. Kapteijn, F.; Moulijn, J.A. Structured Catalysts and Reactors–Perspectives for Demanding Applications. *Catal. Today* **2022**, *383*, 5–14. [CrossRef]
2. Landi, G. Novel Structured Catalytic Reactors. *Catalysts* **2021**, *11*, 1472. [CrossRef]
3. Balzarotti, R.; Ambrosetti, M.; Beretta, A.; Groppi, G.; Tronconi, E. Recent Advances in the Development of Highly Conductive Structured Supports for the Intensification of Non-Adiabatic Gas-Solid Catalytic Processes: The Methane Steam Reforming Case Study. *Front. Chem. Eng.* **2022**, *3*, 811439. [CrossRef]
4. Navarro, J.C.; Centeno, M.A.; Laguna, O.H.; Odriozola, J.A. Policies and Motivations for the CO_2 Valorization through the Sabatier Reaction Using Structured Catalysts. A Review of the Most Recent Advances. *Catalysts* **2018**, *8*, 578. [CrossRef]
5. Tyczkowski, J.; Kierzkowska-Pawlak, H.; Kapica, R. Method of Producing a Thin Layer of Catalyst on Structured Packing of Reactors for CO_2 Methanation. Patent PL241342B1, 12 September 2022. Available online: https://worldwide.espacenet.com/patent/search/family/072826561/publication/PL241342B1?q=PL241342B1 (accessed on 12 September 2022).
6. Kierzkowska-Pawlak, H.; Tyczkowski, J.; Balcerzak, J.; Tracz, P. Advances in Plasma Produced CoO_x-Based Nanocatalysts for CO_2 Methanation. *Catal. Today* **2019**, *337*, 162–170. [CrossRef]
7. Puga, A.V. On the Nature of Active Phases and Sites in CO and CO_2 Hydrogenation Catalysts. *Catal. Sci. Technol.* **2018**, *8*, 5681–5707. [CrossRef]
8. Liu, Z.; Gao, X.; Liu, B.; Song, W.; Ma, Q.; Zhao, T.; Wang, X.; Bae, J.W.; Zhang, X.; Zhang, J. Highly Stable and Selective Layered Co–Al–O Catalysts for Low-Temperature CO_2 Methanation. *Appl. Catal. B Environ.* **2022**, *310*, 121303. [CrossRef]
9. Melaet, G.; Ralston, W.T.; Li, C.-S.; Alayoglu, S.; An, K.; Musselwhite, N.; Kalkan, B.; Somorjai, G.A. Evidence of Highly Active Cobalt Oxide Catalyst for the Fischer–Tropsch Synthesis and CO_2 Hydrogenation. *J. Am. Chem. Soc.* **2014**, *136*, 2260–2263. [CrossRef]
10. Have, I.C.t.; Kromwijk, J.J.G.; Monai, M.; Ferri, D.; Sterk, E.B.; Meirer, F.; Weckhuysen, B.M. Uncovering the Reaction Mechanism behind CoO as Active Phase for CO_2 Hydrogenation. *Nat. Commun.* **2022**, *13*, 324. [CrossRef]
11. Zhao, K.; Calizzi, M.; Moioli, E.; Li, M.; Borsay, A.; Lombardo, L.; Mutschler, R.; Luo, W.; Züttel, A. Unraveling and Optimizing the Metal-Metal Oxide Synergistic Effect in a Highly Active $Co_x(CoO)_{1-x}$ Catalyst for CO_2 Hydrogenation. *J. Energy Chem.* **2021**, *53*, 241–250. [CrossRef]
12. Liu, S.; He, Y.; Fu, W.; Chen, J.; Ren, J.; Liao, L.; Sun, R.; Tang, Z.; Mebrahtu, C.; Zeng, F. Hetero-Site Cobalt Catalysts for Higher Alcohols Synthesis by CO_2 Hydrogenation: A Review. *J. CO2 Util.* **2023**, *67*, 102322. [CrossRef]
13. Hu, M.; Yao, Z.; Wang, X. Department of Chemical, Biological and Pharmaceutical Engineering, New Jersey Institute of Technology, Newark, NJ 07102, USA Characterization Techniques for Graphene-Based Materials in Catalysis. *AIMS Mater. Sci.* **2017**, *4*, 755–788. [CrossRef]
14. Furimsky, E. CO_2 Hydrogenation to Methanol and Methane over Carbon-Supported Catalysts. *Ind. Eng. Chem. Res.* **2020**, *59*, 15393–15423. [CrossRef]
15. Meijer, R.; van Doorn, R.; Kapteijn, F.; Moulijn, J.A. Methane Formation in H_2, CO Mixtures over Carbon-Supported Potassium Carbonate. *J. Catal.* **1992**, *134*, 525–535. [CrossRef]
16. Sazama, P.; Pastvova, J.; Rizescu, C.; Tirsoaga, A.; Parvulescu, V.I.; Garcia, H.; Kobera, L.; Seidel, J.; Rathousky, J.; Klein, P.; et al. Catalytic Properties of 3D Graphene-Like Microporous Carbons Synthesized in a Zeolite Template. *ACS Catal.* **2018**, *8*, 1779–1789. [CrossRef]
17. Zhao, T.; Hui, Y.; Niamatullah; Li, Z. Controllable preparation of ZIF-67 derived catalyst for CO_2 methanation. *Mol. Catal.* **2019**, *474*, 110421. [CrossRef]

18. Biesinger, M.C.; Payne, B.P.; Grosvenor, A.P.; Lau, L.W.M.; Gerson, A.R.; Smart, R.S.C. Resolving surface chemical states in XPS analysis of first row transition metals, oxides and hydroxides: Cr, Mn, Fe, Co and Ni. *Appl. Surf. Sci.* **2011**, *257*, 2717–2730. [CrossRef]
19. Fan, Q.; Guo, Z.; Li, Z.; Wang, Z.; Yang, L.; Chen, Q.; Liu, Z.; Wang, X. Atomic layer deposition of cobalt carbide thin films from cobalt amidinate and hydrogen plasma. *ACS Appl. Electron. Mater.* **2019**, *1*, 444–453. [CrossRef]
20. Speranza, G.; Laidani, N. Measurement of the relative abundance of sp2 and sp3 hybridised atoms in carbon-based materials by XPS: A critical approach. Part II. *Diam. Relat. Mater.* **2004**, *13*, 451–458. [CrossRef]
21. Lesiak, B.; Kövér, L.; Tóth, J.; Zemek, J.; Jiricek, P.; Kromka, A.; Rangam, N. C sp2/sp3 hybridisations in carbon nanomaterials-XPS and (X)AES study. *Appl. Surf. Sci.* **2018**, *452*, 223–231. [CrossRef]
22. Brewster, M.A.; Ziurys, L.M. The millimeter-wave spectrum of NiC ($X^1\Sigma^+$) and CoC ($X^2\Sigma^+$). *ApJ* **2001**, *559*, L163–L166. [CrossRef]
23. Mendez-Linan, J.I.; Ortiz-Ortega, E.; Jimenez-Moreno, M.F.; Mendivil-Palma, M.I.; Martínez-Guerra, E.; Aguirre-Tostado, F.S.; Martinez-Chapa, S.O.; Hosseini, S.; Madou, M.J. Aging Effect of Plasma-Treated Carbon Surfaces: An Overlooked Phenomenon. *Carbon* **2020**, *169*, 32–44. [CrossRef]
24. Tyczkowski, J.; Kapica, R.; Kozanecki, M.; Kierzkowska-Pawlak, H.; Sielski, J.; Aoki, T.; Mimura, H. Tailoring the nanostructure of plasma-deposited CoO_x-based thin films for catalytic applications–A step forward in designing nanocatalysts. *Mater. Des.* **2022**, *222*, 111095. [CrossRef]
25. Whitten, J.E. Ultraviolet photoelectron spectroscopy: Practical aspects and best practices. *Appl. Surf. Sci. Appl.* **2023**, *13*, 100294. [CrossRef]
26. Tauc, J. (Ed.) Optical Properties of Amorphous Semiconductors. In *Amorphous and Liquid Semiconductors*; Springer: Boston, MA, USA, 1974; pp. 159–220. [CrossRef]
27. Tyczkowski, J. Electrical and optical properties of plasma polymers. In *Plasma Polymer Films*; Biederman, H., Ed.; Imperial College Press: London, UK, 2004; pp. 143–216. [CrossRef]
28. Theodosiou, A.; Spencer, B.F.; Counsell, J.; Jones, A.N. An XPS/UPS study of the surface/near-surface bonding in nuclear grade graphites: A comparison of monatomic and cluster depth-profiling techniques. *Appl. Surf. Sci.* **2020**, *508*, 144764. [CrossRef]
29. Konshina, E.A. Absorption and the optical gap of a-C:H films produced from acetylene plasmas. *Semiconductors* **1999**, *33*, 451–456. [CrossRef]
30. Nagasawa, H.; Kanezashi, M.; Yoshioka, T.; Tsuru, T. Plasma-enhanced chemical vapor deposition of amorphous carbon molecular sieve membranes for gas separation. *RCS Adv.* **2016**, *6*, 59045. [CrossRef]
31. Kyaw, M.; Dugos, N.P.; Mori, S.; Roces, S.A.; Beltran, A.B.; Suzuki, S. Gas permeation properties and preparation of carbon membrane by PECVD method using indene as precursor. *J. Phys. Conf. Ser.* **2019**, *1295*, 012057. [CrossRef]
32. Blackwood, J.D.; Ingeme, A.J. The reaction of carbon with carbon dioxide at high pressure. *Aust. J. Chem.* **1960**, *13*, 104–209. [CrossRef]
33. Jia, J.; Qian, C.; Dong, Y.; Li, Y.F.; Wang, H.; Ghoussoub, M.; Butler, K.T.; Walsh, A.; Ozin, G.A. Heterogeneous catalytic hydrogenation of CO_2 by metal oxides: Defect engineering–perfecting imperfection. *Chem. Soc. Rev.* **2017**, *46*, 4631–4644. [CrossRef]
34. Wang, H.H.; Zhang, S.N.; Zhao, T.J.; Liu, Y.X.; Liu, X.; Su, J.; Li, X.H.; Chen, J.S. Mild and selective hydrogenation of CO_2 into formic acid over electron-rich MoC nanocatalysts. *Sci. Bull.* **2020**, *65*, 651–657.
35. Kierzkowska-Pawlak, H.; Ryba, M.; Fronczak, M.; Kapica, R.; Sielski, J.; Sitarz, M.; Zając, P.; Łyszczarz, K.; Tyczkowski, J. Enhancing CO_2 Conversion to CO over Plasma-Deposited Composites Based on Mixed Co and Fe Oxides. *Catalysts* **2021**, *11*, 883. [CrossRef]
36. Panek, B.; Kierzkowska-Pawlak, H.; Uznański, P.; Nagy, S.; Nagy-Trembošová, V.; Tyczkowski, J. The Role of Carbon Nanotube Deposit in Catalytic Activity of FeO_x-Based PECVD Thin Films Tested in RWGS Reaction. *Catalysts* **2023**, *13*, 1302.
37. Wang, Y.; Ge, H.X.; Chen, Y.P.; Meng, X.Y.; Ghanbaja, J.; Horwat, D.; Pierson, J.F. Wurtzite CoO: A direct band gap oxide suitable for a photovoltaic absorber. *Chem. Commun.* **2018**, *54*, 13949–13952. [CrossRef]
38. Li, B.; Rui, Y.; Xu, J.; Wang, Y.; Yang, J.; Zhang, Q.; Müller-Buschbaum, P. Solution-processed p-type nanocrystalline CoO films for inverted mixed perovskite solar cells. *J. Colloid Interface Sci.* **2020**, *573*, 78–86. [CrossRef]
39. Xu, Y.; Schoonen, M.A.A. The absolute energy positions of conduction and valence bands of selected semiconducting minerals. *Am. Min.* **2000**, *85*, 543–556. [CrossRef]
40. Alidoust, N.; Lessio, M.; Carter, E.A. Cobalt (II) oxide and nickel (II) oxide alloys as potential intermediate-band semiconductors: A theoretical study. *J. Appl. Phys.* **2016**, *119*, 025102.
41. Zhai, L.; She, X.; Zhuang, L.; Li, Y.; Ding, R.; Guo, X.; Zhang, Y.; Zhu, Y.; Xu, K.; Fan, H.J.; et al. Modulating built-in electric field via variable oxygen affinity for robust hydrogen evolution reaction in neutral media. *Angew. Chem.* **2022**, *61*, e202116057. [CrossRef]
42. Zhang, L.; Jaroniec, M. Toward designing semiconductor-semiconductor heterojunctions for photocatalytic applications. *Appl. Surf. Sci.* **2018**, *430*, 2–17. [CrossRef]
43. Sze, S.M.; Ng, K.K. *Physics of Semiconductor Devices*, 3rd ed.; John Wiley & Sons, Inc.: Hoboken, NJ, USA, 2007; pp. 80–90.
44. Tian, Y.; Liu, X.; Xu, L.; Yuan, D.; Qiu, J.; Li, H.; Ma, J.; Wang, Y.; Su, D.; Zhang, S. Engineering crystallinity and oxygen vacancies of Co(II) oxide nanosheets for high performance and robust rechargeable Zn–air batteries. *Adv. Funct. Mater.* **2021**, *31*, 210239. [CrossRef]

45. Kuang, X.; Kang, B.; Wang, Z.; Gao, L.; Guo, C.; Lee, J.Y.; Sun, X.; Wei, Q. Sulfur-doped CoO nanoflakes with loosely packed structure realizing enhanced oxygen evolution reaction. *Chem. Eur. J.* **2018**, *24*, 17288–17292. [CrossRef] [PubMed]
46. Wang, L.; Chen, B.; Zhang, L.; Yu, J. In situ irradiated XPS investigation on S-scheme TiO_2@$ZnIn_2S_4$ photocatalyst for efficient photocatalytic CO_2 reduction. *Small* **2021**, *17*, 2103447. [CrossRef] [PubMed]
47. Gao, X.; Cai, P.; Wang, Z.; Lv, X.; Kawi, S. Surface acidity/basicity and oxygen defects of metal oxide: Impacts on catalytic performances of CO_2 reforming and hydrogenation reactions. *Top. Catal.* **2023**, *66*, 299–325. [CrossRef]
48. Tyczkowski, J.; Kierzkowska-Pawlak, H.; Kapica, R.; Balcerzak, J.; Sielski, J. Cold Plasma—A Promising Tool for the Production of Thin-Film Nanocatalysts. *Catal. Today* **2019**, *337*, 44–54. [CrossRef]

catalysts

MDPI

Article

Construction of Spinel/Perovskite Heterojunction for Boosting Photocatalytic Performance for Polyacrylamide

Qinghan Zhu [1], Yuxue Luo [1], Ke Yang [1], Guangbo Che [2], Haiwang Wang [1,*] and Jian Qi [3,4,*]

1 Key Laboratory of Dielectric and Electrolyte Functional Material Hebei Province, Northeastern University at Qinhuangdao, Qinhuangdao 066004, China; zhuqinghan0110@163.com (Q.Z.); luoyuxue2001@163.com (Y.L.); y1210620501@163.com (K.Y.)
2 College of Chemistry, Baicheng Normal University, Baicheng 137000, China; guangboche@bcnu.edu.cn
3 State Key Laboratory of Biochemical Engineering, Institute of Process Engineering, Chinese Academy of Sciences, Beijing 100190, China
4 School of Chemical Engineering, University of Chinese Academy of Sciences, Beijing 100049, China
* Correspondence: whwdbdx@neuq.edu.cn (H.W.); jqi@ipe.ac.cn (J.Q.)

Abstract: The use of photocatalytic technology to degrade polyacrylamide in crude oil extraction wastewater is a promising approach, but there have been few reports so far. In this study, $ZnFe_2O_4/Ba_{0.7}Sr_{0.3}TiO_3$ heterogeneous composite materials of a spinel/perovskite type with different proportions were synthesized. The composite materials with 31% $ZnFe_2O_4$ content exhibited a maximum polyacrylamide degradation efficiency of 46.54%, which demonstrated the unique role of the spinel/perovskite heterogeneous structure. When Ag nanoparticles were grown in situ on the surface of $ZnFe_2O_4/Ba_{0.7}Sr_{0.3}TiO_3$, the photocatalytic degradation efficiency reached 81.28%. The main reason was that the introduction of Ag nanoparticles not only increased the active sites and enhanced light absorption capacity but also accelerated the separation of photo-generated charges. This work provides new ideas for the construction of spinel/perovskite heterogeneous composite materials and has reference significance for the application of photocatalytic degradation in the treatment of wastewater-containing polymers.

Keywords: zinc ferrite; strontium barium titanate; spinel/perovskite heterogeneous composite; in situ growth of Ag nanoparticles; photocatalytic degradation of polyacrylamide

Citation: Zhu, Q.; Luo, Y.; Yang, K.; Che, G.; Wang, H.; Qi, J. Construction of Spinel/Perovskite Heterojunction for Boosting Photocatalytic Performance for Polyacrylamide. *Catalysts* **2023**, *13*, 1424. https://doi.org/10.3390/catal13111424

Academic Editors: Leonarda Liotta, Narendra Kumar and Konstantin Ivanov Hadjiivanov

Received: 11 September 2023
Revised: 30 October 2023
Accepted: 3 November 2023
Published: 8 November 2023

1. Introduction

In oil recovery techniques, polyacrylamide (PAM) can be dissolved in water in any proportion. Due to its good thickening, flocculation, and rheological adjustment properties, its aqueous solution can significantly improve the oil recovery rate and is widely used in petroleum extraction. However, while PAM enhances the oil recovery rate, it also generates high-viscosity wastewater-containing polymers, which can be harmful to the environment when discharged. Through migration, adsorption in formations, and other processes, PAM can pose long-term hazards to soil and water bodies. In addition, PAM can spontaneously degrade into acrylamide, and the monomer acrylamide is toxic and harmful to the human nervous system. Direct contact can cause symptoms such as eye inflammation, dizziness, and headaches, and it can also affect lung function [1]. Therefore, the effective degradation of PAM has important practical significance.

Currently, there are mainly physical degradation methods for the degradation of PAM [2] (mechanical degradation [3], ultrasonic degradation [4–6], thermal degradation [7,8], photodegradation [9,10], laser pulse degradation [11], low-temperature plasma degradation [12,13]) and chemical degradation methods (oxidative degradation [14], photocatalytic degradation [15]). Physical degradation methods often require the application of energy such as light and heat, which increases energy consumption and does not fully degrade PAM. Chemical degradation methods include oxidative degradation and photocatalytic degradation. Oxidative degradation requires the addition of oxidants, which can achieve complete degradation but may result in secondary pollution. Photocatalytic degradation, on the other hand, is an environmentally friendly water treatment technology that does not cause secondary pollution and can completely degrade pollutants such as PAM [15]. However, there are fewer reports on this topic. For example, Zhu et al. [16] synthesized Cu_2O@ZnO using the immersion-reduction air oxidation method and prepared rGO@Fe_3O_4/Cu_2O@ZnO magnetic recyclable composite material via a hydrothermal method with GO@Fe_3O_4 prepared using a chemical co-precipitation method. The composite of Cu_2O and ZnO can improve the range of light response and solar energy utilization and form a heterogeneous (p-n type) structure with ZnO, further promoting the separation of light-induced charge carriers, thereby improving the catalytic performance. The high conductivity of rGO further enhances the catalytic activity of Cu_2O/ZnO. Rong et al. [17] prepared an Ag-AgBr/TiO_2-graphene (AATG) composite photocatalyst by modifying TiO_2 with an Ag/AgBr semiconductor and graphene using the deposition–precipitation method. The Ag-AgBr and graphene reduce the bandgap of AATG, thereby enhancing the visible light utilization and widening the light-response range of the composite material. Simultaneously, the addition of graphene enhances the surface adsorption effect [18], as well as the effective separation of photoelectrons and holes, leading to a significant enhancement of the photocatalytic activity of TiO_2.

As is well-known, the construction of heterogeneous junctions can effectively promote the transfer of photo-generated charges and suppress the recombination of photo-generated electrons and holes in photocatalytic reactions [19–21]. Currently, there are many reports on the construction of noble metal/spinel heterojunctions [22] and noble metal/perovskite heterojunctions. For example, Wang et al. [23] constructed noble metal/spinel heterojunctions with Au and Co_3O_4, which have multiple bandgaps, expanding the light-response range and exhibiting a clear surface plasmon resonance (SPR) peak. SPR generates additional photo-generated electrons, thereby hindering the recombination rate of electron–hole pairs and improving photocatalytic performance. Cho et al. [24] synthesized the noble metal/perovskite heterojunction of Rh and $SrTiO_3$, reducing the consumption of CB electrons by protons and other intermediate products, and improving the overall dual photocatalytic activity. However, there are few reports on spinel/perovskite heterojunctions.

In this study, we used zinc ferrite with a spinel structure and strontium barium titanate with a perovskite structure to form heterojunctions. By regulating the ratio of the two to form a good, solid solution structure, the efficiency of electron–hole separation was enhanced, and the photocatalytic activity of PAM degradation was increased. Furthermore, Ag nanoparticles with a better ability to transfer electrons were introduced on the surface of the composite material. The loading of Ag nanoparticles improves the efficiency of electron–hole separation, reduces the bandgap width, and the energy required for electron transition reactions, and further enhances photocatalytic activity.

2. Results

2.1. Phase Structure Analysis of $ZnFe_2O_4$/$Ba_{0.7}Sr_{0.3}TiO_3$

Figure 1 shows the XRD diffraction patterns of $ZnFe_2O_4$, $Ba_{0.7}Sr_{0.3}TiO_3$, and $ZnFe_2O_4$/$Ba_{0.7}Sr_{0.3}TiO_3$ with different component contents (ZFO and BST are abbreviations for $ZnFe_2O_4$ and $Ba_{0.7}Sr_{0.3}TiO_3$, respectively). Characteristic diffraction peaks of $Ba_{0.7}Sr_{0.3}TiO_3$ appear at 22.4°, 31.9°, 39.3°, 45.6°, 51.4°, 56.7°, 66.4°, and 75.4°, corresponding to crystal planes (100), (110), (111), (200), (210), (211), (220), and (310), respectively. Characteristic diffraction peaks of $ZnFe_2O_4$ were observed at 30°, 35.2°, 42.8°, 56.7°, and 62.2°, corresponding to crystal planes of (220), (311), (400), (511), and (440), respectively. This indicates the successful combination of $ZnFe_2O_4$ and $Ba_{0.7}Sr_{0.3}TiO_3$. The characteristic peaks of $ZnFe_2O_4$/$Ba_{0.7}Sr_{0.3}TiO_3$ are obvious and no impurity peaks appear, and their diffraction peaks are completely consistent with the standard XRD pattern in the standard card (card numbers ZFO PDF # 22-1012, BST PDF # 44-0093), indicating that the synthesized $ZnFe_2O_4/Ba_{0.7}Sr_{0.3}TiO_3$ composite material has high purity. As the proportion of the $ZnFe_2O_4$ component increases, the positions and configurations of each diffraction peak on the XRD spectrum are consistent, indicating that changes in the composition of the composite material will not affect its crystal structure. Meanwhile, comparing the XRD diffraction patterns of $ZnFe_2O_4$, $Ba_{0.7}Sr_{0.3}TiO_3$, and $ZnFe_2O_4/Ba_{0.7}Sr_{0.3}TiO_3$ samples, it was found that the XRD diffraction peak widths of the samples after the composition with $Ba_{0.7}Sr_{0.3}TiO_3$ narrowed, indicating that the grain size of the samples after the composition was larger than before the composition. In addition, Figures S1 and S2 show the physical and chemical properties of pure $ZnFe_2O_4$ and $Ba_{0.7}Sr_{0.3}TiO_3$, respectively. From Figure S1a, it can be observed that the morphology of the $ZnFe_2O_4$ sample exhibits an irregular layered shape, with an unsmooth surface and particle sizes ranging from 10 to 30 μm, and Figure S1b further demonstrates the obvious pore structure inside the material, which is conducive to the increase of active sites. The average grain size of pure-phase $ZnFe_2O_4$ particles calculated by combining Figure S1c and Scheler's formula is 19.9 nm, with a lattice constant of 8.440 Å, which is slightly smaller than the standard 8.443 Å [25]; this is due to the redistribution of Zn^{2+} and Fe^{3+} at tetrahedral and octahedral positions when the particle size decreases to the nanoscale. The infrared spectrum (Figure S1d) shows that the absorption peaks at 3452.34 cm^{-1} and 1635.52 cm^{-1} belong to the O–H stretching and bending vibration peaks in the surface water of $ZnFe_2O_4$, while 541.96 cm^{-1} and 420.45 cm^{-1} represent the stretching vibration of Zn–O bonds in tetrahedral positions and Fe–O bonds in octahedral positions in $ZnFe_2O_4$ [25]. In addition, the photoluminescence spectrum (PL spectrum) of $ZnFe_2O_4$ was tested using an excitation wavelength of 259 nm (Figure S1e), and it was found that the emission center of the $ZnFe_2O_4$ sample was concentrated around 468 nm, indicating a high PL intensity of $ZnFe_2O_4$, and the electron–hole pair recombination efficiency of the $ZnFe_2O_4$ sample is high, which is not conducive to the separation of electron–hole pairs. Figure S1f shows that the nitrogen adsorption–desorption isotherm type of $ZnFe_2O_4$ is type IV, with an H3-type hysteresis loop, originating from the mesoporous structure of material stacking [26]. Its pore size is concentrated at 13.52 nm (Figure S1g), and its specific surface area is 17.97 $m^2{\cdot}g^{-1}$. From Figure S2a,b, it can be observed that the morphology of $Ba_{0.7}Sr_{0.3}TiO_3$ consists of spherical nanoparticles, with an obvious agglomeration phenomenon. Small grains easily adhere to each other to form larger particles, which may affect photocatalytic activity. The average grain size of pure-phase $Ba_{0.7}Sr_{0.3}TiO_3$ particles was calculated to be 6.3 nm, based on Figure S2c and Scheler's formula. The infrared spectrum (Figure S2d) shows that 547.74 cm^{-1} and 435.88 cm^{-1} are the tensile vibrations of the Ti–O and Ba–O bonds [27], respectively. The photoluminescence spectrum (Figure S2e) shows that the emission center of the $Ba_{0.7}Sr_{0.3}TiO_3$ sample is concentrated at around 470 nm. The $Ba_{0.7}Sr_{0.3}TiO_3$ sample exhibited a lower PL intensity compared with $ZnFe_2O_4$, indicating a lower recombination efficiency of electron–hole pairs. This suggests that the separation of electron–hole pairs will be facilitated in the $Ba_{0.7}Sr_{0.3}TiO_3$ sample.

The $ZnFe_2O_4/Ba_{0.7}Sr_{0.3}TiO_3$ composite material prepared by the sol–gel method was analyzed for functional group changes before and after calcination at 1100 °C using infrared absorption spectroscopy, as shown in Figure 2. After the calcination treatment, the C-H, C=O, and C–O–Ti bonds in the dried gel disappeared, indicating the decomposition of carbonates after the heat treatment. Figure S3 shows the infrared spectra of $ZnFe_2O_4/Ba_{0.7}Sr_{0.3}TiO_3$ with different compositions of $ZnFe_2O_4$. By comparison, there is no significant difference in the infrared spectra of $ZnFe_2O_4/Ba_{0.7}Sr_{0.3}TiO_3$ with different compositions. Figure S4 displays the TEM image of $ZnFe_2O_4/Ba_{0.7}Sr_{0.3}TiO_3$ (ZFO wt% = 31), where the interplanar spacing corresponding to the (220) crystal plane of $ZnFe_2O_4$ is 0.168 nm, and the interplanar spacing corresponding to the (210) crystal plane of $Ba_{0.7}Sr_{0.3}TiO_3$ is 0.285 nm.

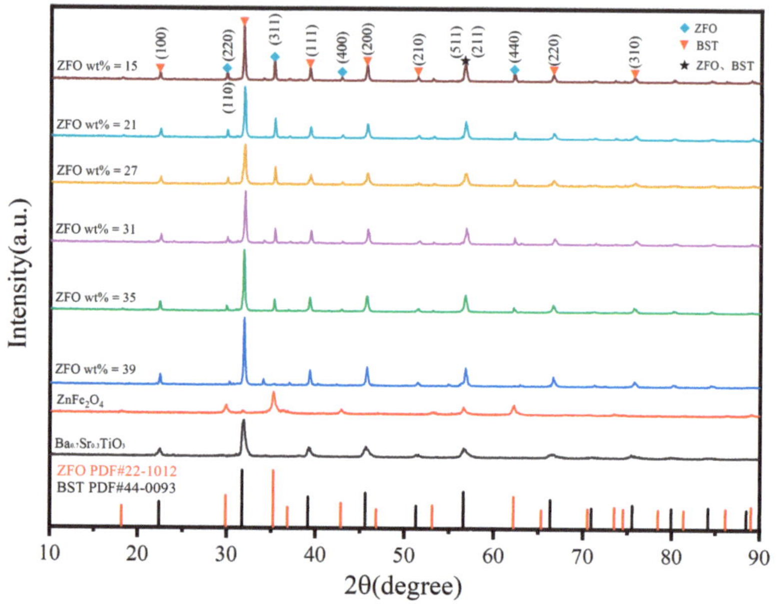

Figure 1. XRD patterns of $ZnFe_2O_4$, $Ba_{0.7}Sr_{0.3}TiO_3$, and $ZnFe_2O_4/Ba_{0.7}Sr_{0.3}TiO_3$ with different components.

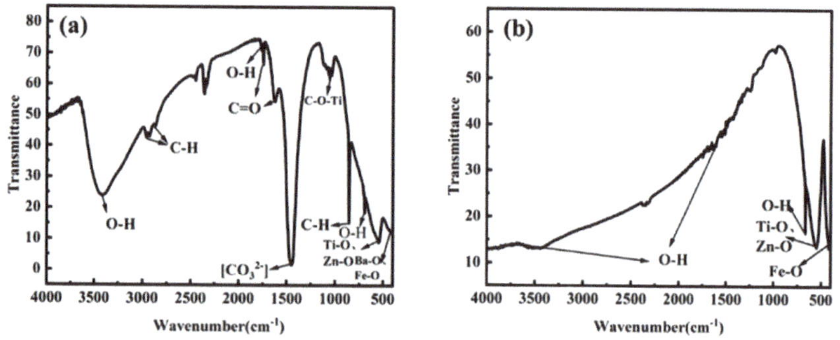

Figure 2. FT-IR spectra of $ZnFe_2O_4/Ba_{0.7}Sr_{0.3}TiO_3$ (ZFO wt% = 31) sample, (**a**) before calcination and (**b**) after calcinations at 1100 °C.

Figure 3 shows the EDS spectra of $ZnFe_2O_4/Ba_{0.7}Sr_{0.3}TiO_3$ (ZFO wt% = 31), indicating the presence of Ti, Fe, Zn, Sr, Ba, and O elements in the sample. Ba has the highest elemental concentration, followed by Fe and Ti elements, and the distribution of elements in the grains is relatively uniform [28].

The heat treatment process of $ZnFe_2O_4/Ba_{0.7}Sr_{0.3}TiO_3$ (ZFO wt% = 31) dry gel involves the volatilization of crystal water, the combustion of organic matter, the decomposition of chelates, the crystallization of perovskite-phase BST, and the growth of grains, which are accompanied by endothermic and exothermic phenomena. In order to further investigate the reaction mechanism of the $ZnFe_2O_4/Ba_{0.7}Sr_{0.3}TiO_3$ composite material during the heating process, a TG-DSC analysis was conducted [29], as shown in Figure S5. The weight loss of the dried gel can be divided into three stages in the temperature range from 90 °C to 1100 °C. The first stage occurs before 200 °C and is mainly attributed to the volatilization of absorbed water and ethanol solvents. An endothermic peak is observed around 192 °C on the DSC curve, with a weight loss of approximately 3.5%. The second stage takes place between 200 °C and 500 °C, during which the organic functional groups, such as the butoxy groups in tetra-*n*-butyl titanate, undergo combustion, decomposition, and hydrogen bond cleavage, resulting in the evaporation of H_2O and CO_2 gases from the system. A significant exothermic peak occurs at around 270 °C on the DSC curve, with a weight loss of approximately 6%. The third stage occurs from 500 °C to 900 °C. In this stage, the weight of the material rapidly decreases due to the burning of organic residues in the powder of the dried gel, as well as the reaction between $SrCO_3$ and $BaCO_3$ with TiO_2, leading to the formation of $SrTiO_3$ and $BaTiO_3$ and the release of CO_2. The characteristics of this stage are the endothermic peaks at around 771 °C and 842 °C on the DSC curve, as well as the exothermic peak at around 860 °C, with a weight loss of approximately 7.5%.

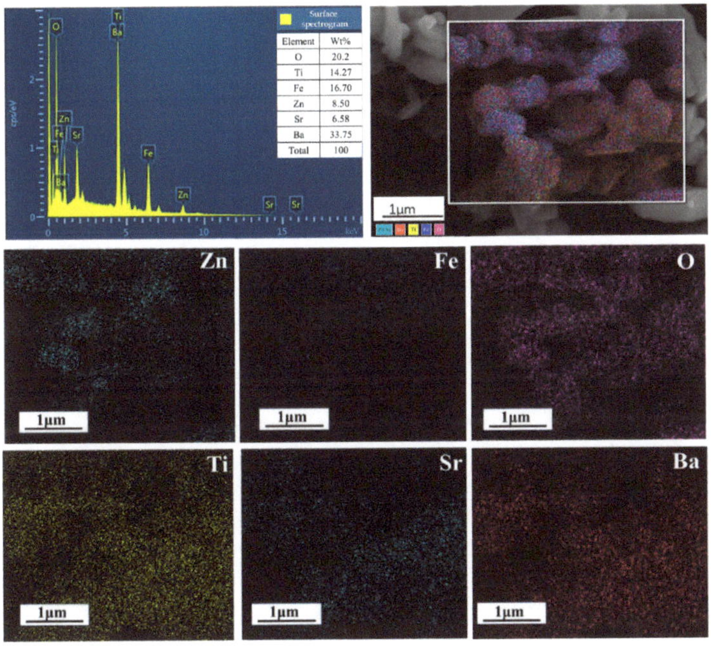

Figure 3. EDS spectra of $ZnFe_2O_4/Ba_{0.7}Sr_{0.3}TiO_3$ (ZFO wt% = 31).

2.2. Composition Structure and Surface State of $ZnFe_2O_4/Ba_{0.7}Sr_{0.3}TiO_3$

Qualitative and semi-quantitative analysis of the $ZnFe_2O_4/Ba_{0.7}Sr_{0.3}TiO_3$ with different composition ratios was performed using X-ray fluorescence spectroscopy (XRF), and the results are shown in Table S1. It can be observed that the composition ratios of $ZnFe_2O_4/Ba_{0.7}Sr_{0.3}TiO_3$ are close to the theoretical values, which is further confirmed by X-ray diffraction (XRD) characterization, indicating the successful preparation of $ZnFe_2O_4/Ba_{0.7}Sr_{0.3}TiO_3$ with different composition ratios.

Elemental and valence state information of $ZnFe_2O_4/Ba_{0.7}Sr_{0.3}TiO_3$ was analyzed using X-ray photoelectron spectroscopy (XPS) [25,30]. As shown in Figure 4a, the high-resolution spectrum of Ba 3d exhibits two peaks at 794.21 eV and 779.21 eV, corresponding to the Ba $3d_{3/2}$ and Ba $3d_{5/2}$ electrons, respectively. In Figure 4b, the high-resolution spectrum of Fe 2p shows peaks at binding energies of 725.21 eV and 711.21 eV, corresponding to the Fe $2p_{1/2}$ and Fe $2p_{3/2}$ electrons, respectively. Additionally, a satellite peak of Fe $2p_{3/2}$ is observed at 720.21 eV, indicating the presence of Fe^{3+}. In Figure 4c, the O 1s spectrum of $ZnFe_2O_4/Ba_{0.7}Sr_{0.3}TiO_3$ consists of three oxygen components, with a peak at approximately 530.11 eV attributed to lattice oxygen, a peak at approximately 530.21 eV corresponding to oxygen vacancies, and a peak at approximately 532.45 eV corresponding to adsorbed H_2O. In Figure 4d, the high-resolution spectrum of Sr 3d exhibits two peaks at 134.01 eV and

132.11 eV, corresponding to Sr $3d_{3/2}$ and Sr $3d_{5/2}$ electrons, respectively. The Ti 2p spectrum in Figure 4e consists of peaks from the Ti $2p_{1/2}$ and Ti $2p_{3/2}$ electrons, with binding energies of 463.21 eV and 457.21 eV, respectively, which are close to the reported binding energies of α-Ti in the BST perovskite structure, indicating the presence of a Ti^{4+} oxidation state. In Figure 4f, the peaks of Zn 2p appear at 1044.21 eV and 1021.21 eV, corresponding to the Zn $2p_{1/2}$ and Zn $2p_{3/2}$ electrons, respectively. From Figure S6, it can be observed that the Zn, Ba, Fe, Ti, Sr, and O elements are present, while the presence of carbon may be attributed to equipment contamination during the measurement.

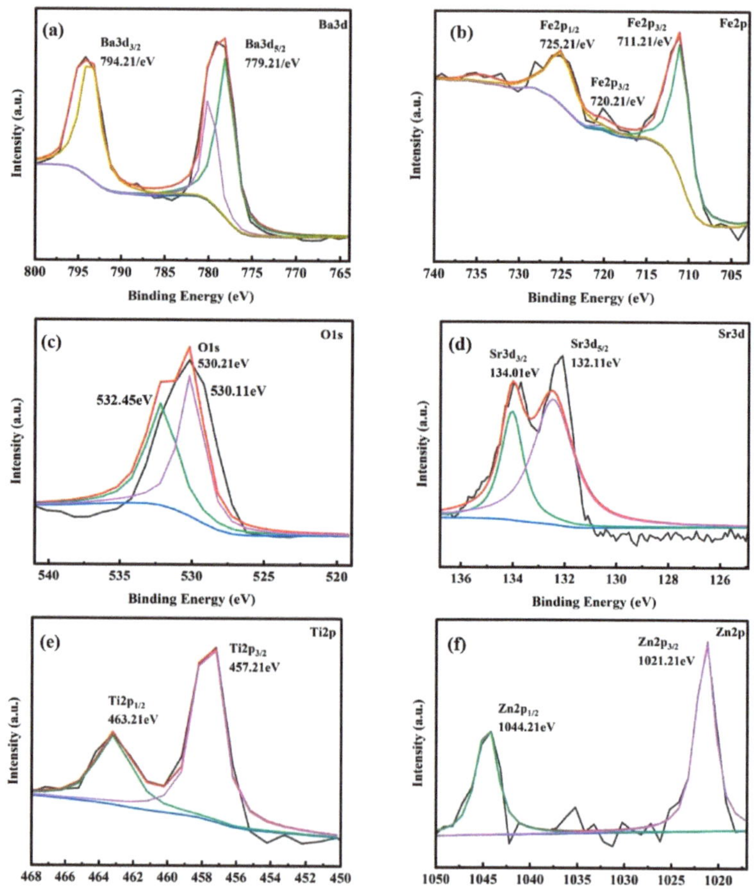

Figure 4. XPS spectra of $ZnFe_2O_4/Ba_{0.7}Sr_{0.3}TiO_3$ (ZFO wt% = 31), (**a**) Ba 3d, (**b**) Fe 2p, (**c**) O 1s, (**d**) Sr 3d, (**e**) Ti 2p and (**f**) Zn 2p.

SEM images of $ZnFe_2O_4/Ba_{0.7}Sr_{0.3}TiO_3$ with ZFO wt% of 31 and 35 are shown in Figure 5, with the magnification indicated at the bottom of each image. Figure 5a–d represent the SEM images of $ZnFe_2O_4/Ba_{0.7}Sr_{0.3}TiO_3$ with ZFO wt% of 31, and with ZFO wt% of 35, respectively. It can be observed that the $ZnFe_2O_4/Ba_{0.7}Sr_{0.3}TiO_3$ samples present a flat and irregular particle shape, with a particle size ranging from 20 to 200 nm. The particle size is uneven, and some particles show an agglomeration phenomenon. The change in the proportion of $ZnFe_2O_4$ and $Ba_{0.7}Sr_{0.3}TiO_3$ components has little effect on the morphology of the $ZnFe_2O_4/Ba_{0.7}Sr_{0.3}TiO_3$ samples.

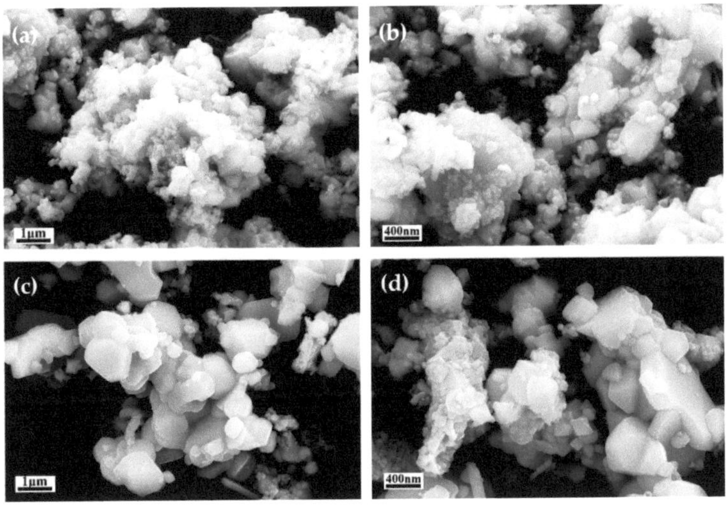

Figure 5. SEM images of $ZnFe_2O_4/Ba_{0.7}Sr_{0.3}TiO_3$ with different compositions, (**a**,**b**) ZFO wt% = 31 and (**c**,**d**) ZFO wt% = 35.

2.3. Analysis of Specific Surface Area of $ZnFe_2O_4/Ba_{0.7}Sr_{0.3}TiO_3$

According to Table S2, the specific surface area of $ZnFe_2O_4/Ba_{0.7}Sr_{0.3}TiO_3$ (ZFO wt% = 31) is 1.85 $m^2 \cdot g^{-1}$, with an average pore size of 14.55 nm. In Figure S7a, based on the IUPAC definition, the nitrogen adsorption–desorption isotherm of $ZnFe_2O_4/Ba_{0.7}Sr_{0.3}TiO_3$ is of type IV, exhibiting a hysteresis loop of type H3, indicating that the material contains a mesoporous structure. Figure S7b represents the pore-size distribution of the sample, showing that the pore sizes mainly range from 9.45 nm to 26.72 nm, with the peak at 12.27 nm. The pore-size distribution of the sample is moderately uniform. Compared to pure $ZnFe_2O_4$, the specific surface area of the $ZnFe_2O_4/Ba_{0.7}Sr_{0.3}TiO_3$ composite sample is reduced, primarily due to the higher sintering temperature, resulting in decreased surface roughness, which has a certain influence on the transfer efficiency of the photo-generated electron to the surface.

2.4. Analysis of Electron–Hole Pair Separation Efficiency in $ZnFe_2O_4/Ba_{0.7}Sr_{0.3}TiO_3$

The photoluminescence (PL) spectra of $ZnFe_2O_4/Ba_{0.7}Sr_{0.3}TiO_3$ samples can be used to qualitatively analyze the photo-generated electron and hole recombination probabilities in samples of different components [17], as shown in Figure 6. It is observed that the emission centers of different component samples are concentrated around 390 nm. With an increase in the ZFO component content, the peak of the PL spectra initially increases, then decreases, and finally increases again. Therefore, the result within this range of ZFO content is that, as the ZFO content increases, the recombination probability of electrons and holes first increases, then decreases, and finally increases again. When the ZFO wt% is 31, the recombination probability of electrons and holes is the lowest, and the separation efficiency is the highest, which is beneficial for improving photocatalytic performance. In addition, the UV-visible absorption spectrum of $ZnFe_2O_4/Ba_{0.7}Sr_{0.3}TiO_3$ (ZFO wt% = 31) exhibits strong absorption in the visible light range [31], with the maximum absorbance occurring at a wavelength of approximately 420 nm (Figure S8a). The bandgap width of the sample is estimated to be 1.62 eV based on the Kubelka–Munk equation plot $(\alpha h\nu)^2$ -hv curve shown in Figure S8b [32].

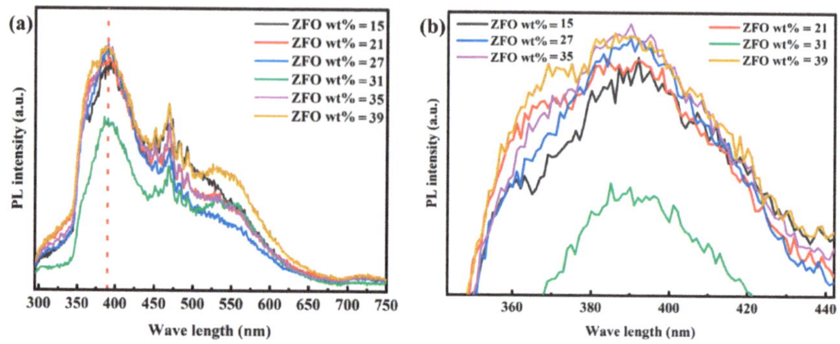

Figure 6. PL spectra of $ZnFe_2O_4/Ba_{0.7}Sr_{0.3}TiO_3$ with different ZFO/BST ratios, (**a**) photoluminescence spectra; (**b**) partial enlarged view of (**a**).

2.5. Degradation Mechanism and Products of PAM

The degradation mechanism of PAM can be described as follows: •OH is the main active species for oxidative degradation of PAM. Under light conditions, the photo-generated electrons generated by ZFO-BST migrate to the surface, while Ag nanoparticles can accelerate their migration and leave photo-generated holes. This process is accompanied by the formation of $\bullet O_2^-$ and further forms •OH, which is then quickly transferred to the PAM adsorbed on the catalyst surface and destroys its chain structure and decomposes it into acrylic acid and acrylamide, which in turn degrade into CO_2 and H_2O [17]. The degradation product process is shown in Figure S9, and the degradation mechanism is similar to that reported in the literature [33].

3. Discussion

3.1. Study on the Photocatalytic Performance of $ZnFe_2O_4/Ba_{0.7}Sr_{0.3}TiO_3$

ZFO exhibits a narrow bandgap (about 1.9 eV), more pores, and a larger specific surface area, making it easily excited by visible light, and a large number of photo-generated electrons and holes can be generated, while it also possesses good optical stability. However, its valence band potential is low, resulting in a poor photoelectric conversion performance [34]. The built-in electric field generated by the spontaneous polarization of BST in composite materials helps to promote the separation of photo-generated electron–hole pairs at the interface and improve visible light catalytic performance, but particles are prone to aggregation [35]. Therefore, when these two materials are used as photocatalysts alone, the improvement in their photocatalytic performance is limited.

This is the equation for degradation efficiency:

$$\text{Degradation efficiency} = (C_0 - C_t)/C_0 \times 100\%, \quad (1)$$

C_0 and C_t are the initial concentration of PAM solution and the concentration after irradiation time, respectively.

Figure S10 shows the degradation efficiencies of pure ZFO and pure BST in the photocatalytic degradation of PAM, which is 43.67% and 33.49%, respectively. Subsequently, the degradation efficiency of $ZnFe_2O_4/Ba_{0.7}Sr_{0.3}TiO_3$ was studied, as shown in Figure 7a. When the proportion of ZFO in the $ZnFe_2O_4/Ba_{0.7}Sr_{0.3}TiO_3$ composite catalyst is different, the photocatalytic degradation performance of the PAM solution shows an initial increase, followed by a decrease. The photocatalytic performance of $ZnFe_2O_4/Ba_{0.7}Sr_{0.3}TiO_3$ is best when the ZFO wt% is 31, with a degradation efficiency of 46.54% after 7 h of irradiation (Figure 7b). After being reused five times, the PAM degradation efficiency of the $ZnFe_2O_4/Ba_{0.7}Sr_{0.3}TiO_3$ (ZFO wt% = 31) composite material can still be maintained at around 46%, which means that the composite material has a good photocatalytic cycling stability and can be reused multiple times, as shown in Figure S11. When the proportion of ZFO in the composite catalyst is low, the sample shows a slight inhibition in the photocatalytic degradation of the PAM solution. This may be due to the low ZFO content in the sample, where BST plays the dominant role. However, BST is prone to aggregation, which reduces the contact area and blocks the light source, leading to the inhibition of PAM photocatalytic degradation. When the proportion of ZFO is too high, the electron–hole recombination probability of the composite catalyst will increase, resulting in a decrease in the separation efficiency and affecting the photocatalytic activity.

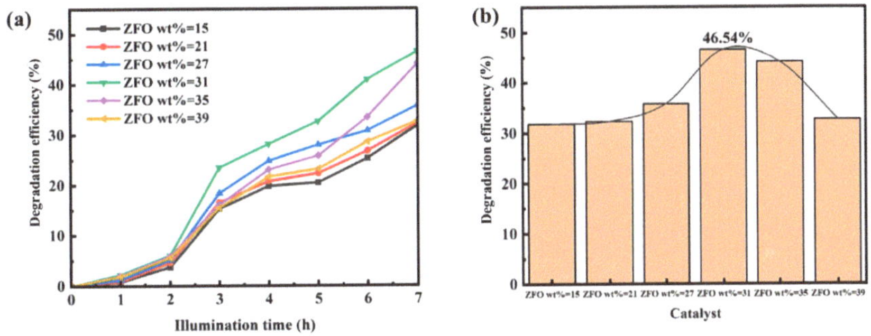

Figure 7. (**a**) Photodegradation of PAM aqueous solution with different components of $ZnFe_2O_4/Ba_{0.7}Sr_{0.3}TiO_3$; (**b**) total degradation efficiency of different components.

3.2. Study on the Photocatalytic Performance of $ZnFe_2O_4/Ba_{0.7}Sr_{0.3}TiO_3/Ag$

It is well-known that Ag nanoparticles have excellent electron transfer abilities, facilitating the effective separation of electrons and holes and improving the absorption and utilization of visible light by the catalyst [17,36]. Here, Ag nanoparticles are loaded onto the surface of $ZnFe_2O_4/Ba_{0.7}Sr_{0.3}TiO_3$ (ZFO wt% = 31) by a chemical reduction method. And its crystal phase, microstructure, surface state, specific surface area, element distribution, electron hole recombination probability, and light absorption performance were characterized (Figures S12–S17). The photocatalytic performance of $ZnFe_2O_4/Ba_{0.7}Sr_{0.3}TiO_3/Ag$ in the treatment of PAM wastewater was studied, and the influencing factors of photocatalytic degradation of PAM were explored. Figure S12 shows the XRD patterns of $ZnFe_2O_4/Ba_{0.7}Sr_{0.3}TiO_3/Ag$ with different Ag content. The characteristic peaks of Ag appear in all the samples, and the corresponding crystal plane is (008). Figure S13 shows the TEM images of $ZnFe_2O_4/Ba_{0.7}Sr_{0.3}TiO_3/Ag$ (Ag wt% = 4). The nanoparticles inside the dashed line are Ag nanoparticles attached to the crystal surface. The interplanar spacing corresponding to the (100) plane of $Ba_{0.7}Sr_{0.3}TiO_3$ is 0.110 nm, and that of the (400) plane of $ZnFe_2O_4$ is 0.222 nm. Figure S14 shows the XPS spectrum of $ZnFe_2O_4/Ba_{0.7}Sr_{0.3}TiO_3/Ag$, where Figure S14a–g are similar to the XPS spectrum of $ZnFe_2O_4/Ba_{0.7}Sr_{0.3}TiO_3$, and Figure S14h is the high-resolution spectrum of Ag 3d, showing two peaks at 373.12 eV and 367.12 eV, respectively corresponding to Ag $3d_{3/2}$ and Ag $3d_{5/2}$ electrons. Table S3 shows that the specific surface area of $ZnFe_2O_4/Ba_{0.7}Sr_{0.3}TiO_3/Ag$ is 2.59 $m^2 \cdot g^{-1}$, and the average pore size is 14.65 nm, which is higher than that before loading Ag, resulting in an increase in the active sites of the composite material, which will help enhance photocatalytic activity. Figure S15 shows the EDS spectrum of $ZnFe_2O_4/Ba_{0.7}Sr_{0.3}TiO_3/Ag$ (Ag wt% = 4), confirming the presence of Ag in the sample. As shown in Figure S16, the emission centers of different component samples are concentrated around 470 nm. When the Ag content changes, the efficiency of electron and hole separation will be inhibited if the Ag content is too low or too high. However, when the Ag content is between 2% and 6%, the variation in Ag content has little effect on the efficiency of electron and hole separation. Figure S17 shows that the bandgap width of $ZnFe_2O_4/Ba_{0.7}Sr_{0.3}TiO_3/Ag$ (Ag wt% = 4) is 1.60 eV, which is narrower than that of $ZnFe_2O_4/Ba_{0.7}Sr_{0.3}TiO_3$ (1.62 eV). This means that less energy is required for electron transition, which will be beneficial for improving catalytic activity.

By comparing Figures S17 and S18, it was found that the two catalysts mainly absorb light with wavelengths above 350 nm, thereby achieving photocatalytic degradation.

Next, the photocatalytic performance of $ZnFe_2O_4/Ba_{0.7}Sr_{0.3}TiO_3/Ag$ and influencing factors were systematically studied. As shown in Figure 8a, the photocatalytic performance of the catalysts first increases and then decreases with the increase of Ag content. When the content of Ag is 4%, the highest degradation efficiency of photocatalytic degradation of PAM can reach 76.67% (Figure 8b). This is because, when the Ag content is too high, the excess Ag in the sample covers the active sites on the surface of $ZnFe_2O_4/Ba_{0.7}Sr_{0.3}TiO_3$, resulting in reducing photocatalytic activity. However, when the Ag content is too low, the formation of heterojunctions between Ag and $ZnFe_2O_4/Ba_{0.7}Sr_{0.3}TiO_3$ is limited, which is not conducive to the transfer of electrons and the separation of electron–hole pairs, resulting in low photocatalytic activity. In Figure 8c,d, the best photocatalytic degradation performance is achieved when the catalyst dosage is 50 mg. This is because the increase of the catalyst dosage accelerates the photocatalytic degradation efficiency, thereby improving the photocatalytic efficiency. However, when the catalyst dosage is too high, the transmittance of the solution decreases, leading to a decrease in photocatalytic efficiency. From Figure 8e,f, it can be observed that the initial concentration of the reaction system has a certain influence on the photocatalytic degradation performance. The best degradation efficiency (80.97%) for the photocatalytic degradation of PAM is observed when the initial mass concentration of the PAM solution is 75 mg/L. This may be due to the lower viscosity of the solution at lower concentrations, which can cause some catalysts to precipitate and weaken the degradation performance. On the other hand, when the initial concentration is higher, the overall catalytic degradation capacity is limited, resulting in a less pronounced photocatalytic degradation effect.

Figure 8g,h shows the degradation performance under different pH values. As the pH value of the PAM solution increases from acidic to alkaline, the performance of the catalyst in the photocatalytic degradation of PAM initially increases and then decreases. The catalyst exhibits a better degradation performance under acidic and neutral conditions, with optimal performance at pH 6. This is because when the pH value is low, protonation occurs on the surface of the sample and positive charges are generated on its surface, resulting in enhancing the transfer of photo-generated electrons to the sample surface. When the solution is neutral, the water molecules react with photo-generated holes to form •OH and protons. When the pH is high, OH^- adsorbs on the surface of the sample and forms a negative charge, thereby increasing the efficiency of hole transfer from the inside of the particle to the surface of the particle [37]. Zete potential testing was conducted on the catalysts, and the results showed that both $ZnFe_2O_4/Ba_{0.7}Sr_{0.3}TiO_3$ and $ZnFe_2O_4/Ba_{0.7}Sr_{0.3}TiO_3/Ag$ displayed negative surface charges, and their values were −7.22 mV and −18.29 mV, respectively. In the reaction system with the solution pH value of 6, PAM has a positive charge on its surface. Due to electrostatic attraction, PAM is more conducive to being adsorbed on the catalyst surface and accelerating the catalytic reaction process. Therefore, the optimal experimental conditions for the reaction are as follows: selecting a catalyst sample with 4 wt% Ag content, taking 50 mg of the catalyst, preparing a PAM solution with a mass concentration of 75 mg/L, and adjusting the solution pH value to 6. Under these conditions, the photocatalytic degradation performance is the best, with a degradation efficiency of 81.28%. The photocatalytic degradation scheme of $ZnFe_2O_4/Ba_{0.7}Sr_{0.3}TiO_3/Ag$ is shown in Figure S19.

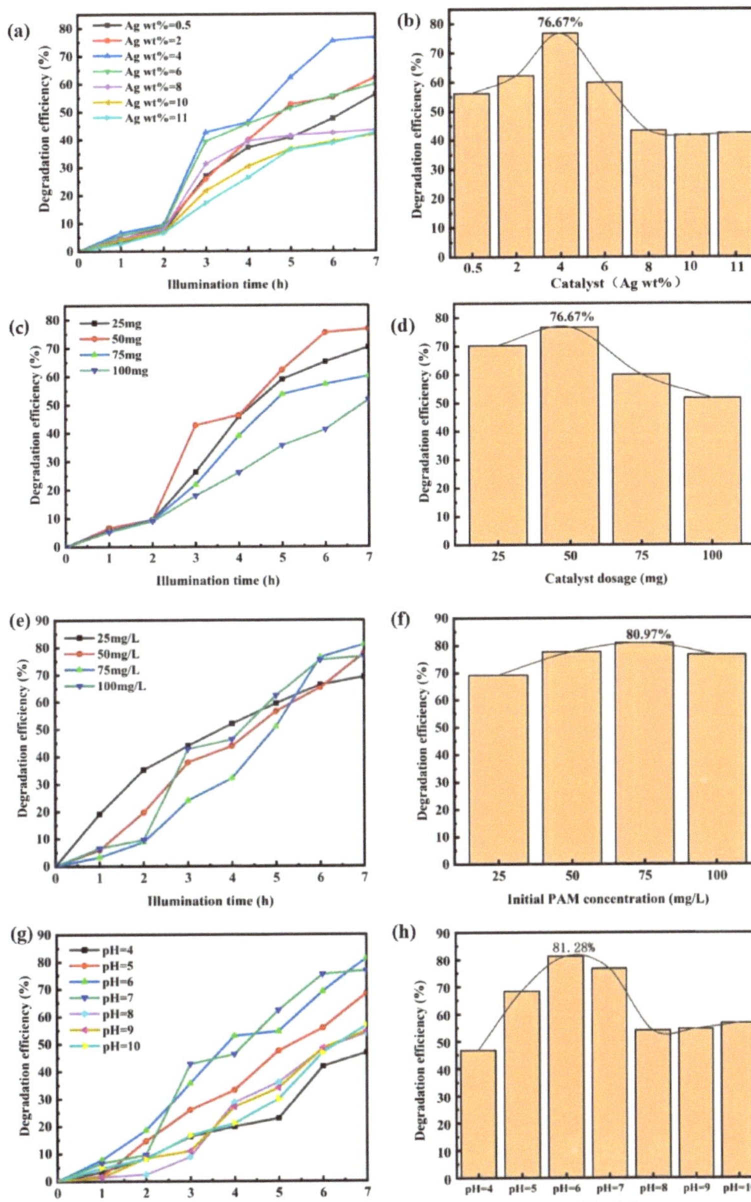

Figure 8. (**a**) The photocatalytic degradation performance of $ZnFe_2O_4/Ba_{0.7}Sr_{0.3}TiO_3/Ag$ catalyst with different Ag contents for PAM and (**b**) comparison of corresponding degradation efficiencies. (**c**) The photocatalytic degradation performance of $ZnFe_2O_4/Ba_{0.7}Sr_{0.3}TiO_3/Ag$ under different catalyst dosage for PAM and (**d**) comparison of corresponding degradation efficiencies. (**e**) The photocatalytic degradation performance of $ZnFe_2O_4/Ba_{0.7}Sr_{0.3}TiO_3/Ag$ catalyst under different initial concentration of PAM and (**f**) comparison of corresponding degradation efficiencies. (**g**) The photocatalytic degradation performance of $ZnFe_2O_4/Ba_{0.7}Sr_{0.3}TiO_3/Ag$ catalyst under different pH values and (**h**) comparison of corresponding degradation efficiencies.

4. Materials and Methods

4.1. Preparation of $ZnFe_2O_4$

Reagent A was prepared by adding 3.15 g of citric acid to 100 mL of deionized water. The solution was stirred with a magnetic stirrer until complete dissolution. Similarly, reagent B was prepared by dissolving 4.04 g of iron nitrate dihydrate and 1.48 g of zinc nitrate dihydrate in 100 mL of deionized water. The mixture was stirred with a magnetic stirrer until complete dissolution. Subsequently, reagent B was added dropwise to reagent A, with A serving as the complexing agent, while continuously stirring. The pH was then adjusted to 8 by the addition of ammonia solution. Upon heating and stirring the solution, it transformed into a viscous gel which was later removed from the drying oven after 12 h of drying at 90 °C. The resulting gel was ground into a powder. Next, the solid powder was calcined at 500 °C for 2 h, cooled to room temperature and ground to obtain $ZnFe_2O_4$ (ZFO) sample.

4.2. Preparation of $Ba_{0.7}Sr_{0.3}TiO_3$

Dissolve 6.5 mL of hydrochloric acid and 150 mL of anhydrous ethanol in 30 mL of deionized water with magnetic stirring to obtain a uniform solution, designated as reagent A. Reagent B was comprised of 1.7 g of tetra-*n*-butyl titanate and 5 mL of anhydrous ethanol, which were mixed evenly by stirring with a magnetic stirrer. A solution containing 0.7 g of barium carbonate and 0.225 g of strontium carbonate was added to reagent B and stirred with a magnetic stirrer until a uniform mixture was obtained. The resulting mixture was subjected to 540 W ultrasonic treatment for 90 min. Reagent A was then added dropwise to the well-sonicated reagent B and stirred until a sol was formed. The addition of reagent A was stopped at this point. The sol was placed in a dark environment for 3 h to promote aging and the formation of BST gel precursors. Remove the gel precursor from the drying oven after drying at 70 °C for 9 h and grind it into a powder. Take out the sample from the muffle furnace after calcining at 1100 °C for 3 h and cooling to room temperature and grind the sample to obtain a solid powder sample of $Ba_{0.7}Sr_{0.3}TiO_3$ (BST).

4.3. Preparation of $ZnFe_2O_4/Ba_{0.7}Sr_{0.3}TiO_3$

As shown in Figure S20, the preparation process of $ZnFe_2O_4/Ba_{0.7}Sr_{0.3}TiO_3$ composite material is as follows: A certain proportion of barium carbonate, strontium carbonate, and previously prepared different masses of zinc ferrite were added to reagent B. The mixture was stirred with a magnetic stirrer until it was thoroughly mixed. It was then subjected to 540 W ultrasonic treatment for 90 min. Reagent A was added dropwise to the well-sonicated reagent B and stirred until a sol was formed. The addition of reagent A was then stopped. The sol was placed in a dark environment for 3 h to promote aging and the formation of BST gel precursors. Remove the gel precursor from the drying oven after drying at 70 °C for 9 h and grind it into a powder. Take out the sample from the muffle furnace after calcining at 1100 °C for 3 h and cooling to room temperature and grind the sample to obtain solid powder samples of $ZnFe_2O_4/Ba_{0.7}Sr_{0.3}TiO_3$ (ZFO-BST) with different component content ratios.

4.4. Preparation of $ZnFe_2O_4/Ba_{0.7}Sr_{0.3}TiO_3/Ag$

Perform photocatalytic performance testing on the samples obtained in the previous step to select the sample with the best photocatalytic performance, namely (ZFO wt% = 31), and further carry out the loading of silver nanoparticles. The detailed steps are as follows: A certain amount of $ZnFe_2O_4/Ba_{0.7}Sr_{0.3}TiO_3$ sample was dispersed in deionized water and subjected to ultrasonic treatment using 540 W for 30 min. It was then stirred with a magnetic stirrer for 25 min. A certain amount of sodium borohydride was weighed and added to deionized water. The mixture was then stirred with a magnetic stirrer until it was completely dissolved to obtain a uniform solution. Polyvinylpyrrolidone (PVP), with an average molecular weight of 1,300,000, was used as a dispersant. PVP was weighed in a ratio of $mPVP/mAgNO_3 = 0.5$ and added to the sample solution. The mixture was stirred until it was completely dissolved. Different masses of silver nitrate were weighed and added to the sample solution, which was stirred until it was completely dissolved. The prepared sodium borohydride solution was added dropwise, and the mixture was stirred for 30 min to complete the reaction. The sample was washed three times with water using a high-speed centrifuge, dried in a drying oven at 90 °C for 12 h, and ground to obtain a solid powder sample of $ZnFe_2O_4/Ba_{0.7}Sr_{0.3}TiO_3/Ag$.

4.5. *Performance Testing of Photocatalytic Degradation of PAM*

4.5.1. Evaluation of Catalytic Performance of $ZnFe_2O_4/Ba_{0.7}Sr_{0.3}TiO_3$

For the photocatalytic performance study of $ZnFe_2O_4/Ba_{0.7}Sr_{0.3}TiO_3$ with different component contents, use PAM as the target degradation substance. Prepare a 100 mg/L PAM aqueous solution one day in advance to simulate aged wastewater. Take 50 mg of $ZnFe_2O_4/Ba_{0.7}Sr_{0.3}TiO_3$ samples with ZFO wt% of 15, 21, 27, 31, 35, and 39, and add them to transparent crucibles containing the PAM aqueous solution, and label them accordingly. Stir the sample solution in the dark for 30 min to mix the sample and PAM aqueous solution uniformly and reach the adsorption–desorption equilibrium state. A 300 W xenon lamp was used as the light source, and samples were taken at regular intervals of 1 h for absorbance detection through centrifugation. The solution concentration was calculated using the standard curve fitting formula, and the degradation efficiency was calculated based on the difference in concentration before and after. The total duration of the experiment was set to 7 h.

4.5.2. Evaluation of Catalytic Performance of $ZnFe_2O_4/Ba_{0.7}Sr_{0.3}TiO_3/Ag$

The photodegradation performance of $ZnFe_2O_4/Ba_{0.7}Sr_{0.3}TiO_3/Ag$ with PAM as the target degradable substance was studied. A 100 mg/L PAM aqueous solution was prepared one day in advance for simulated wastewater aging under dark conditions. Weigh 50 mg (Ag wt% of 0.5, 2, 4, 6, 8, 10, 11) of $ZnFe_2O_4/Ba_{0.7}Sr_{0.3}TiO_3/Ag$ sample and add to the transparent crucible containing the PAM solution. The sample solution was stirred in the dark for 30 min to achieve an adsorption–desorption equilibrium state. A 300 W xenon lamp was used as the light source, and samples were taken and centrifuged every 1 h. The absorbance was measured using the turbidity method, and the concentration values were calculated using a fitted formula based on a standard curve. The degradation efficiency was calculated based on the difference in concentration before and after treatment. The total experimental duration was set to 7 h, and the results are shown in Figure 8a,b. Under the same experimental conditions (100 mg/L PAM solution, Ag wt% = 4%), samples of $ZnFe_2O_4/Ba_{0.7}Sr_{0.3}TiO_3/Ag$ with masses of 25 mg, 50 mg, 75 mg, and 100 mg were tested as catalysts, and the results are shown in Figure 8c,d. Similarly, using PAM aqueous solutions with mass concentrations of 25 mg/L, 50 mg/L, 75 mg/L, and 100 mg/L as the target degradable substances, $ZnFe_2O_4/Ba_{0.7}Sr_{0.3}TiO_3/Ag$ samples with 50 mg Ag wt% = 4% were tested as catalysts, and the results are shown in Figure 8e,f. A 75 mg/L PAM aqueous solution was selected, and 50 mg of $ZnFe_2O_4/Ba_{0.7}Sr_{0.3}TiO_3/Ag$ sample with Ag wt% = 4 was used as the catalyst. Dilute hydrochloric acid and 0.1 mol/L sodium hydroxide solution were used to adjust the pH value to 4, 5, 6, 7, 8, 9, and 10 for testing; the results are shown in Figure 8g,h.

4.6. *Sample Characterization*

The characterization instruments of this article are as follows: The vertical X-ray diffractometer model TD-3500 produced by Dandong Tongda Science Technology Co., Ltd., Dandong, China, the test current is 25 mA, the voltage is 35 kV, and the scanning angle ranges from 10° to 90°; model S8 TIGER X-ray fluorescence spectrometer produced by Bruker Corporation, Bremen, Germany; the FTIR-8400S Fourier transform infrared spectrometer produced by SHIMADZU, Kyoto, Japan, tested using potassium bromide (KBr) pressure plates in the wavelength range of 4000 to 400 cm^{-1}, with a resolution of 0.85 cm^{-1}, a coaddition number of 16, and a detector of DLATGS; the fluorescence spectrophotometer model RF-5301PC produced by SHIMADZU, Kyoto, Japan; the SUPRA55 SAPPHIRE model field-emission scanning electron microscope energy spectrometer produced by Carl Zeiss AG using surface scanning, Shanghai, China; synchronous thermal analyzer with the model of Setsy Evolution produced by Setaram, Caluire, France; KuboX1000 specific surface area and pore size analyzer produced by Beijing Bi'aode Electronic Technology Co., Ltd., Beijing, China; Thermo Fisher ESCALAB XI + X-ray photoelectron spectrometer produced by Thermo Fisher Scientific, Waltham, MA, USA, the JEM2100F transmission electron microscope produced by JEOL, Tokyo, Japan; Shimadzu 3600 model UV visible spectrophotometer produced by SHIMADZU, Kyoto, Japan.

5. Conclusions

In this study, $ZnFe_2O_4/Ba_{0.7}Sr_{0.3}TiO_3$ heterostructures composites with different ratios were prepared using the sol–gel method, combined with a process of water bath heating, drying, and calcination. The crystal structure, morphology, surface state, photoelectric properties, and photocatalytic degradation performance of this series of heterostructure composites were systematically characterized and evaluated. It was found that the specific surface area of the $ZnFe_2O_4/Ba_{0.7}Sr_{0.3}TiO_3$ composite material was relatively small, but it exhibited a high efficiency in electron–hole separation, a narrow bandgap, and strong absorption of visible light. When the content of $ZnFe_2O_4$ was 31%, the $ZnFe_2O_4/Ba_{0.7}Sr_{0.3}TiO_3$ composite exhibited the lowest efficiency in electron–hole recombination, the highest separation efficiency, and achieved the highest degradation efficiency of PAM (46.54%). Subsequently, the introduction of Ag nanoparticles onto the surface of the $ZnFe_2O_4/Ba_{0.7}Sr_{0.3}TiO_3$ composite increased the specific surface area of the sample. Subsequently, Ag was loaded onto the surface of $ZnFe_2O_4/Ba_{0.7}Sr_{0.3}TiO_3$ and improved the experimental conditions, resulting in an increase in the specific surface area of the sample and an improvement in the efficiency of photo-generated electron migration to the surface. This led to an increase in the efficiency of electron–hole separation, a narrowing of the bandgap width, a stronger absorption of visible light, and an increase in photocatalytic activity to 81.28%. The main reason behind this improvement is that the introduction of Ag nanoparticles not only increased the number of active sites and enhanced light-absorption capacity, but also accelerated the separation of photo-generated charges. This work provides a new approach for the construction of spinel/perovskite heterogeneous composite materials and has reference significance for the application of photocatalytic degradation in the treatment of wastewater-containing polymers.

Supplementary Materials: The following supporting information can be downloaded at: https://www.mdpi.com/article/10.3390/catal13111424/s1. Figure S1. (a,b) SEM images of $ZnFe_2O_4$ samples, (c) XRD diffraction pattern, (d) FT-IR spectrum, (e) PL spectrum, (f) Nitrogen adsorption–desorption isotherm, (g) Pore-size distribution; Figure S2. (a,b) SEM images of $Ba_{0.7}Sr_{0.3}TiO_3$, (c) XRD diffraction pattern, (d) FT-IR spectrum, (e) PL spectrum; Figure S3. FT-IR spectra of $ZnFe_2O_4/Ba_{0.7}Sr_{0.3}TiO_3$ samples with different components; Figure S4. (a) TEM image and (b,c) HRTEM of $ZnFe_2O_4/Ba_{0.7}Sr_{0.3}TiO_3$ (ZFO wt% = 31); Figure S5. TG-DSC curve of $ZnFe_2O_4/Ba_{0.7}Sr_{0.3}TiO_3$ gel precursor (ZFO wt% = 31); Figure S6. Full XPS spectrum of $ZnFe_2O_4/Ba_{0.7}Sr_{0.3}TiO_3$ sample (ZFO wt% = 31); Figure S7. (a) Nitrogen adsorption–desorption isotherm and (b) Pore-size distribution of $ZnFe_2O_4/Ba_{0.7}Sr_{0.3}TiO_3$ (ZFO wt% = 31); Figure S8. (a) UV-visible absorption spectrum of $ZnFe_2O_4/Ba_{0.7}Sr_{0.3}TiO_3$ (ZFO wt% = 31), (b) corresponding Tauc plot curve of (a); Figure S9. Speculative pathways for photocatalytic degradation of PAM; Figure S10. Photodegradation of PAM aqueous solution by $ZnFe_2O_4$ (ZFO) and $Ba_{0.7}Sr_{0.3}TiO_3$ (BST); Figure S11. The recycling performance of $ZnFe_2O_4/Ba_{0.7}Sr_{0.3}TiO_3$ (ZFO wt% = 31); Figure S12. XRD patterns of $ZnFe_2O_4/Ba_{0.7}Sr_{0.3}TiO_3/Ag$ with different Ag contents; Figure S13. (a) TEM image and (b,c) HRTEM of $ZnFe_2O_4/Ba_{0.7}Sr_{0.3}TiO_3/Ag$ (Ag wt% = 4); Figure S14. XPS spectra of $ZnFe_2O_4/Ba_{0.7}Sr_{0.3}TiO_3/Ag$: (a) full spectrum, (b) Ba 3d, (c) O 1s, (d) Fe 2p, (e) Sr 3d, (f) Ti 2p, (g) Zn 2p and (h)Ag 3d; Figure S15. EDS spectra of $ZnFe_2O_4/Ba_{0.7}Sr_{0.3}TiO_3/Ag$ (Ag wt% = 4); Figure S16. (a) PL spectra of $ZnFe_2O_4/Ba_{0.7}Sr_{0.3}TiO_3/Ag$ with different Ag content and (b) Partial enlarged view of (a); Figure S17. (a) UV-visible absorption spectra of $ZnFe_2O_4/Ba_{0.7}Sr_{0.3}TiO_3$ (ZFO-BST) and $ZnFe_2O_4/Ba_{0.7}Sr_{0.3}TiO_3/Ag$ (ZFO-BST-Ag, Ag wt% = 4), (b) Corresponding Tauc plot curves of (a); Figure S18. The irradiation spectrum of the xenon lamp; Figure S19. The photocatalytic degradation scheme of $ZnFe_2O_4/Ba_{0.7}Sr_{0.3}TiO_3/Ag$; Figure S20. Preparation process diagram of $ZnFe_2O_4/Ba_{0.7}Sr_{0.3}TiO_3$; Table S1. Comparison between XRF analysis results and theoretical values of $ZnFe_2O_4/Ba_{0.7}Sr_{0.3}TiO_3$; Table S2. Specific surface area, pore size, and pore volume of $ZnFe_2O_4/Ba_{0.7}Sr_{0.3}TiO_3$ (ZFO wt% = 31); Table S3. Specific surface area, pore size, and pore volume of $ZnFe_2O_4/Ba_{0.7}Sr_{0.3}TiO_3/Ag$ (Ag wt% = 4).

Author Contributions: Conceptualization: J.Q., H.W. and G.C.; methodology, validation, formal analysis and investigation: Q.Z., Y.L. and K.Y.; resources: H.W.; data curation: H.W.; writing—original draft preparation: Q.Z. and J.Q.; writing—review and editing: J.Q., H.W. and G.C.; visualization: H.W.; supervision: J.Q. and H.W.; project administration: J.Q. and H.W.; funding acquisition: J.Q. and H.W. All authors have read and agreed to the published version of the manuscript.

Funding: This work was supported by the Natural Science Foundation of China under Grant (Nos. 51972306, 21604007).

Data Availability Statement: Data is contained within the article or Supplementary Material.

Conflicts of Interest: The authors declare no conflict of interest.

References

1. Hu, H.; Liu, J.F.; Li, C.Y.; Yang, S.Z.; Gu, J.D.; Mu, B.Z. Anaerobic biodegradation of partially hydrolyzed polyacrylamide in long-term methanogenic enrichment cultures from production water of oil reservoirs. *Biodegradation* **2018**, *29*, 233–243. [CrossRef]
2. Zhang, D.; Wang, F.; Wang, T. Progress of research work on physical and chemical degradation of polyacrylamide. *Coal Prep. Technol.* **2020**, *2*, 1–5. [CrossRef]
3. Moussa, T.; Tiu, C. Factors affecting polymer degradation in turbulent pipe flow. *Chem. Eng. Sci.* **1994**, *49*, 1681–1692. [CrossRef]
4. van Wijngaarden, L. Mechanics of collapsing cavitation bubbles. *Ultrason. Sonochem.* **2016**, *29*, 524–527. [CrossRef] [PubMed]
5. Vijayalakshmi, S.P.; Madras, G. Effect of temperature on the ultrasonic degradation of polyacrylamide and poly(ethylene oxide). *Polym. Degrad. Stab.* **2004**, *84*, 341–344. [CrossRef]
6. Yen, H.Y.; Yang, M.H. The ultrasonic degradation of polyacrylamide solution. *Polym. Test.* **2003**, *22*, 129–131. [CrossRef]
7. Tsay, D.K.; Yang, M.H.; Wang, J.H. On the thermal degradation of polysulfones IX. The early stages of thermal degradation of poly(1-butene sulfone) and poly(2-methyl-1-pentene sulfone). *Polym. Degrad. Stab.* **2002**, *76*, 251–257. [CrossRef]
8. Silva, M.; Dutra, E.R.; Mano, V.; Machado, J.C. Preparation and thermal study of polymers derived from acrylamide. *Polym. Degrad. Stab.* **2000**, *67*, 491–495. [CrossRef]
9. Smith, E.A.; Prues, S.L.; Oehme, F.W. Environmental degradation of polyacrylamides.1. Effects of artificial environmental conditions: Temperature, light, and pH. *Ecotoxicol. Environ. Saf.* **1996**, *35*, 121–135. [CrossRef]
10. Keshipour, S.; Hadidi, M.; Gholipour, O. A Review on Hydrogen Generation by Photo-, Electro-, and Photoelectro-Catalysts Based on Chitosan, Chitin, Cellulose, and Carbon Materials Obtained from These Biopolymers. *Adv. Polym. Technol.* **2023**, *2023*, 8835940. [CrossRef]
11. Vijayalakshmi, S.P.; Senapati, D.; Madras, G. Pulsed laser degradation of polyethylene oxide and polyacrylamide in aqueous solution. *Polym. Degrad. Stab.* **2005**, *87*, 521–526. [CrossRef]
12. Wang, X.B.; Guan, F.W.; Huang, Z.G.; He, H.; Wang, L.; Li, K.F. Study on low temperature plasma combined with AC/Mn + TiO_2-Al_2O_3 catalytic treatment of sewage-containing polyacrylamide. *Water Sci. Technol.* **2023**, *87*, 879–891. [CrossRef] [PubMed]
13. Wang, X.B.; Feng, C.; Wang, L.; Liu, J.Q.; Jin, A.; Zhu, C.Y.; Guan, F.W.; Huang, Z.G.; He, H. Experimental study on reducing the viscosity of sewage containing PAM catalyzed by low temperature plasma synergistic AC/Mn + TiO_2. *Environ. Technol.* **2022**, 1–15. [CrossRef]
14. Xiong, B.Y.; Loss, R.D.; Shields, D.; Pawlik, T.; Hochreiter, R.; Zydney, A.L.; Kumar, M. Polyacrylamide degradation and its implications in environmental systems. *Npj Clean Water* **2018**, *1*, 17. [CrossRef]
15. Du, J.; Lv, C.H.; Lan, X.T.; Song, J.F.; Liu, P.L.; Chen, X.; Wang, Q.; Liu, J.M.; Guo, G.X. A review on viscosity retention of PAM solution for polymer flooding technology. *Pet. Sci. Technol.* **2022**, 1–34. [CrossRef]
16. Zhu, B.J.; Jiang, G.F.; Lv, Y.; Liu, F.; Sun, J. Photocatalytic degradation of polyacrylamide by rGO@Fe_3O_4/Cu_2O@ZnO magnetic recyclable composites. *Mater. Sci. Semicond. Process.* **2021**, *131*, 105841. [CrossRef]
17. Rong, X.S.; Qiu, F.X.; Zhang, C.; Fu, L.; Wang, Y.Y.; Yang, D.Y. Preparation of Ag-AgBr/TiO_2-graphene and its visible light photocatalytic activity enhancement for the degradation of polyacrylamide. *J. Alloys Compd.* **2015**, *639*, 153–161. [CrossRef]
18. Khordadpour Siahkal Mahalleh, M.; Ahour, F.; Keshipour, S. Development of copper electrochemical sensor using D-penicillamine functionalized graphene oxide modified electrode. *Appl. Chem.* **2023**, *18*, 71–90. [CrossRef]
19. Jabbar, Z.H.; Graimed, B.H. Recent developments in industrial organic degradation via semiconductor heterojunctions and the parameters affecting the photocatalytic process: A review study. *J. Water Process Eng.* **2022**, *47*, 102671. [CrossRef]
20. Wang, H.; Zhang, L.; Chen, Z.; Hu, J.; Li, S.; Wang, Z.; Liu, J.; Wang, X. Semiconductor heterojunction photocatalysts: Design, construction, and photocatalytic performances. *Chem. Soc. Rev.* **2014**, *43*, 5234–5244. [CrossRef]

21. Naiyun, L.; Yixian, L.; Yunliang, L.; Yaxi, L.; Yuanyuan, C.; Haitao, L. Modulation of photogenerated holes for enhanced photoelectrocatalytic performance. *Microstructures* **2023**, *3*, 2023001. [CrossRef]
22. Peng, C.; Jingwei, H.; Lianzhou, W. Metal-organic framework-tailored perovskite solar cells. *Microstructures* **2022**, *2*, 2022014. [CrossRef]
23. Wang, S.J.; Zhang, B.P. SPR propelled visible-active photocatalysis on Au-dispersed Co_3O_4 films. *Appl. Catal. A Gen.* **2013**, *467*, 585–592. [CrossRef]
24. Cho, Y.J.; Moon, G.H.; Kanazawa, T.; Maeda, K.; Choi, W. Selective dual-purpose photocatalysis for simultaneous H-2 evolution and mineralization of organic compounds enabled by a Cr_2O_3 barrier layer coated on Rh/$SrTiO_3$. *Chem. Commun.* **2016**, *52*, 9636–9639. [CrossRef]
25. Chandel, N.; Sharma, K.; Sudhaik, A.; Raizada, P.; Hosseini-Bandegharaei, A.; Thakur, V.K.; Singh, P. Magnetically separable ZnO/$ZnFe_2O_4$ and ZnO/$CoFe_2O_4$ photocatalysts supported onto nitrogen doped graphene for photocatalytic degradation of toxic dyes. *Arab. J. Chem.* **2020**, *13*, 4324–4340. [CrossRef]
26. Mousavi-Salehi, S.; Keshipour, S.; Ahour, F. Gold supported on graphene oxide/silica photocatalyst for hydrogen generation from formic acid. *J. Phys. Chem. Solids* **2023**, *176*, 111239. [CrossRef]
27. Tang, H.X.; Sodano, H.A. Ultra High Energy Density Nanocomposite Capacitors with Fast Discharge Using $Ba_{0.2}Sr_{0.8}TiO_3$ Nanowires. *Nano Lett.* **2013**, *13*, 1373–1379. [CrossRef]
28. Asghari, A.; Keshipour, S. Green and one-pot synthesis of Co(II) citrate complex nanoparticles/graphene oxide nanocomposites towards photocatalytic hydrogen evolution. *Int. J. Hydrogen Energy* **2023**, *48*, 34750–34765. [CrossRef]
29. Li, M.X.; Li, B.C.; Chen, J.S.; Shen, X.Y.; Cui, S.; He, X.C.; Liu, K.R.; Han, Q. Analysis of thermal decomposition of acidified sediments in gold plants and harmless disposal of it. *J. Hazard. Mater.* **2022**, *431*, 128472. [CrossRef]
30. Cigeroglu, Z.; Kazan-Kaya, E.S.; El Messaoudi, N.; Fernine, Y.; Americo-Pinheiro, J.H.P.; Jada, A. Remediation of tetracycline from aqueous solution through adsorption on g-C_3N_4-ZnO-$BaTiO_3$ nanocomposite: Optimization, modeling, and theoretical calculation. *J. Mol. Liq.* **2023**, *369*, 120866. [CrossRef]
31. Kar, P.; Shukla, K.; Jain, P.; Sathiyan, G.; Gupta, R.K. Semiconductor based photocatalysts for detoxification of emerging pharmaceutical pollutants from aquatic systems: A critical review. *Nano Mater. Sci.* **2021**, *3*, 25–46. [CrossRef]
32. Ghorpade, S.P.; Krishna, R.H.; Melavanki, R.M.; Dubey, V.; Patil, N. Effect of Eu^{3+} on optical and energy bandgap of SrY_2O_4 nanophosphors for FED applications. *Optik* **2020**, *208*, 164533. [CrossRef]
33. Zhang, L.; Liu, H.; Zhang, B. Mechanism of Enhancement Ozonation by Ultrasound for Degradation of Polyacrylamide in ASP Produced. *CIESC J.* **2015**, *6*, 2242–2247.
34. Su, J.; Shang, Q.K.; Guo, T.T.; Yang, S.; Wang, X.Y.; Ma, Q.; Guan, H.Y.; Xu, F.; Tsang, S.C. Construction of heterojunction $ZnFe_2O_4$/ZnO/Ag by using ZnO and Ag nanoparticles to modify $ZnFe_2O_4$ and its photocatalytic properties under visible light. *Mater. Chem. Phys.* **2018**, *219*, 22–29. [CrossRef]
35. Xu, X.; Chen, S.; Wu, Z.; Jia, Y.; Xiao, L.; Liu, Y. Strong pyro-electro-chemical coupling of $Ba_{0.7}Sr_{0.3}TiO_3$@Ag pyroelectric nanoparticles for room-temperature pyrocatalysis. *Nano Energy* **2018**, *50*, 581–588. [CrossRef]
36. Chen, P.; Zhang, Y.; Zhou, Y.; Dong, F. Photoelectrocatalytic carbon dioxide reduction: Fundamental, advances and challenges. *Nano Mater. Sci.* **2021**, *3*, 344–367. [CrossRef]
37. Abu Tariq, M.; Faisal, M.; Saquib, M.; Muneer, M. Heterogeneous photocatalytic degradation of an anthraquinone and a triphenylmethane dye derivative in aqueous suspensions of semiconductor. *Dye. Pigment.* **2008**, *76*, 358–365. [CrossRef]

catalysts

MDPI

Article

Propane Dehydrogenation over Cobalt Aluminates: Evaluation of Potential Catalytic Active Sites

Aleksey N. Chernov, Svetlana V. Cherepanova , Evgeny Yu. Gerasimov , Igor P. Prosvirin , Galina A. Zenkovets, Alexei A. Shutilov, Anna S. Gorbunova, Konstantin Yu. Koltunov * and Vladimir I. Sobolev *

Boreskov Institute of Catalysis, Russian Academy of Sciences, Pr. Akademika Lavrentieva, 5, Novosibirsk 630090, Russia; anch@catalysis.ru (A.N.C.); svch@catalysis.ru (S.V.C.); gerasimov@catalysis.ru (E.Y.G.); prosvirin@catalysis.ru (I.P.P.); zenk@catalysis.ru (G.A.Z.); alshut@catalysis.ru (A.A.S.); gorbunova@catalysis.ru (A.S.G.)

* Correspondence: koltunov@catalysis.ru (K.Y.K.); visobo@catalysis.ru (V.I.S.)

Abstract: Non-oxidative propane dehydrogenation (PDH) is becoming an increasingly important approach to propylene production, while cobalt-containing catalysts have recently demonstrated great potential for use in this reaction, providing efficiencies comparable to those of industrially employed Pt- and Cr-based catalytic systems. It is therefore essential to clarify the nature of their active sites, especially since contradictory opinions on this issue are expressed in the literature. In this study, efforts were made to determine the state of Co in cobalt aluminates ($CoAl_2O_4$-Al_2O_3) responsible for PDH under typical operating conditions (600 °C, 1 atm). It is shown that the catalyst with a low cobalt content (Co/Al = 0.1) ensured the highest selectivity to propylene, ca. 95%, while maintaining significant propylene conversion. The structural motifs such as cobalt oxide and metallic cobalt nanoparticles, in addition to tetrahedral Co^{2+} species in the $CoAl_2O_4$ spinel system, were evaluated as potential active-site ensembles based on the obtained catalytic performance data in combination with the XRD, H_2-TPR, TEM and XPS characteristics of as-synthesized, spent and spent–regenerated catalysts. It is revealed that the most likely catalytic sites linked to PDH are the Co-oxide forms tightly covering alumina or embedded in the spinel structure. However, additional in situ tuning is certainly needed, probably through the formation of surface oxygen vacancies rather than through a deeper reduction in Co^0 as previously thought.

Keywords: propylene; propane dehydrogenation; cobalt catalysts; heterogeneous catalysis; reaction mechanism

Citation: Chernov, A.N.; Cherepanova, S.V.; Gerasimov, E.Y.; Prosvirin, I.P.; Zenkovets, G.A.; Shutilov, A.A.; Gorbunova, A.S.; Koltunov, K.Y.; Sobolev, V.I. Propane Dehydrogenation over Cobalt Aluminates: Evaluation of Potential Catalytic Active Sites. *Catalysts* **2023**, *13*, 1419. https://doi.org/10.3390/catal13111419

Academic Editors: Leonarda Liotta, Narendra Kumar and Konstantin Ivanov Hadjiivanov

Received: 12 October 2023
Revised: 2 November 2023
Accepted: 3 November 2023
Published: 6 November 2023

1. Introduction

Propylene, one of the basic organic chemicals, is largely produced by non-selective and energy-intensive naphtha steam cracking and fluid catalytic processes [1]. However, given the increased availability of propane as a feedstock, non-oxidative propane dehydrogenation (PDH) is now growing in importance as the most targeted and atom-efficient propylene production route [1–10]. Nonetheless, it is noteworthy that Pt- and CrO_x-based catalysts commonly used in commercial PDH technologies have significant drawbacks, such as the high cost and difficult reactivation of the platinum catalyst and the negative environmental impact of chromium pollution [11].

Quite naturally, there is great interest in developing competitive PDH catalysts that do not contain precious metals and chromium. The most promising of them encompass a variety of transition metal oxides, namely CoO_x, VO_x, GaO_x, MoO_x, FeO_x, WO_x, InO_x, ZnO, ZrO_2 and TiO_2, deposited on silica, alumina, zeolites and other oxide carriers [2–10,12,13]. The cobalt-containing catalysts should be especially highlighted because of their significant activity, relative affordability and moderate toxicity. In addition, carbon-based materials enriched with cobalt have also shown considerable potential in PDH [14–20]. However, remarkably, it is still not clear what structural motif in the cobalt catalysts promotes the PDH reaction. According to major conflicting opinions, Co^{+2} oxide forms, rather than metallic cobalt (Co^0), and vice versa, may actually be the active sites [21–43]. As a compromise, some authors attribute catalytic activity to both oxidation states, Co^{+2} and Co^0 [33,35], or conclude that "the true active cobalt species for propane production are from tetrahedral Co^{+2} ions during the reaction" [21]. Additionally, apart from these possibilities, less attention has been drawn to the fact that low-valent first-row transition metal ions, including Co^{+1} in particular, can cause quite facile dehydrogenation of light alkanes [44–47]. Importantly, the oxidative addition/reductive elimination mechanism proposed for such reactions is independently well documented for a wide range of C–H bond functionalizations involving Co^{+1} organometallic compounds [48]. And lastly, atomically dispersed cobalt–nitrogen complexes deposited on carbon have also been suggested as possible catalytically active sites in the case of PDH over Co-N-C composites [14–20].

Given the strong interest in the activity of cobalt oxides in PDH, and cobalt–aluminum oxides in particular [21–23,26,28,29,33,34,41,43], we describe here a study on the behavior of $CoAl_2O_4$-Al_2O_3 catalysts in this reaction with the aim of elucidating the contributions of a variety of potential catalytic sites to catalytic performance.

2. Results

2.1. Catalyst Characterization

A number of textural characteristics of the $CoAl_2O_4$-Al_2O_3 catalysts, designated as CoAl-X (where X is the estimated molar ratio of Co/Al), are summarized in Table 1. The BET surface area was the highest for as-prepared CoAl-0.1 (291 $m^2\ g^{-1}$), which is comparable to that of CoAl-0.05 (274 $m^2\ g^{-1}$), despite the varying cobalt content, namely 6.7 and 4.3%, respectively. In addition, the total pore volume gradually increased as the cobalt content decreased. However, the BET surface areas and likewise the pore volumes of the spent–regenerated samples CoAl-0.25-R and CoAl-0.1-R were reduced markedly.

Table 1. Textural properties of the catalysts.

	Catalyst	Co/Al [1]	Co (wt%) [2]	S_{BET} ($m^2\ g^{-1}$)	Pore Size (nm)	V_{total} ($cm^3\ g^{-1}$)
1	CoAl-0.5	1:2	28.8	152	12.9	0.49
2	CoAl-0.25	1:4	17.0	219	16.5	0.90
3	CoAl-0.25-R [3]	1:4	13.2	137	21.7	0.74
4	CoAl-0.1	1:10	6.7	291	14.8	1.07
5	CoAl-0.1-R [3]	1:10	6.9	191	19.0	0.91
6	CoAl-0.05	1:20	4.3	274	16.3	1.12
7	Co/Al_2O_3	-	3.8	177	6.0	0.26
8	Al_2O_3	-	-	166	7.3	0.30

[1] The molar ratio of $Co(NO_3)_2$ and $Al(NO_3)_3$ taken for the catalyst preparation. [2] Determined by ICP. [3] The spent–regenerated (re-oxidized) catalyst.

The XRD patterns of as-synthesized and spent–regenerated catalysts allowed us to estimate the relative contributions of cobalt oxide and aluminum oxide crystalline phases (Figure 1). Table S1 summarizes the results of refining the structure of these samples. As expected, the peaks of Co_3O_4 form for CoAl-0.5, CoAl-0.25 and spent–regenerated CoAl-0.25-R catalysts were much stronger than for other catalysts. It is noteworthy that the crystallinity level of Co_3O_4 in the CoAl-0.25-R sample increased, whereas in the case of the CoAl-0.1-R catalyst the opposite occurred (Table S1). In general, the XRD data show that only CoAl-0.05, CoAl-0.1 and CoAl-0.1-R samples demonstrated the clear domination of a mixed Co−Al oxide crystalline form, which can be related to a separate $CoAl_2O_4$ spinel structure.

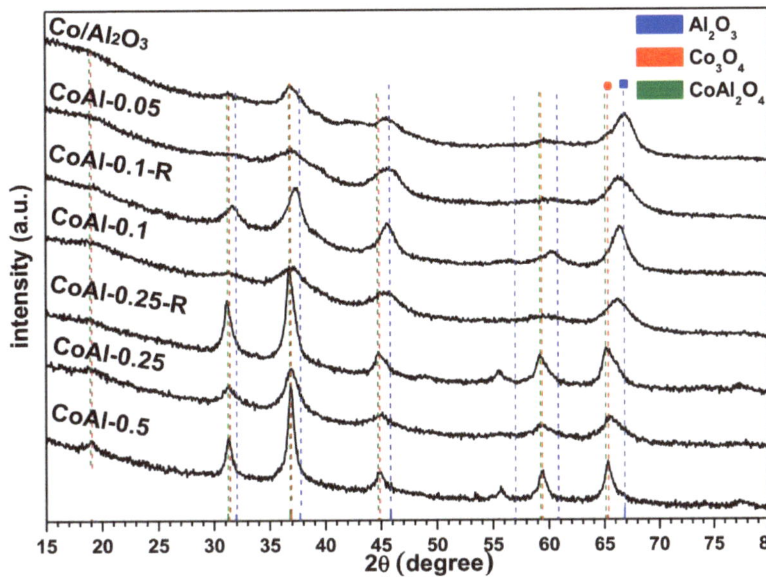

Figure 1. Powder XRD patterns of as-synthesized and spent–regenerated catalysts.

Figure 2 shows the H_2-TPR profiles of the catalysts, revealing that only CoAl-0.5, CoAl-0.25 and Co/Al_2O_3 exhibited a substantial H_2 uptake. For these catalysts, the reduction temperature of the first peak attributed to the reduction of Co_3O_4 to CoO was about 500 °C, which is notably higher than that of bulk Co_3O_4 (200–400 °C [37,38]) due to the strong interaction of Co_3O_4 with Al_2O_3, which prevents the catalyst reduction. The second reduction peak is located between 550 and 750 °C. Overall, the two-step reduction process corresponds to sequential CoO_x reduction into CoO and then into Co^0 states of cobalt. The increased reduction temperature indicates that CoO_x is tightly bound to the Al_2O_3 surface or forms a mixed oxide with it [37,38]. The weak hydrogen uptake for other catalysts (at least up to 600 °C and even higher) shows that their reduction is notably hindered, and the cobalt oxide phase remains largely resistant to reduction to a metal state.

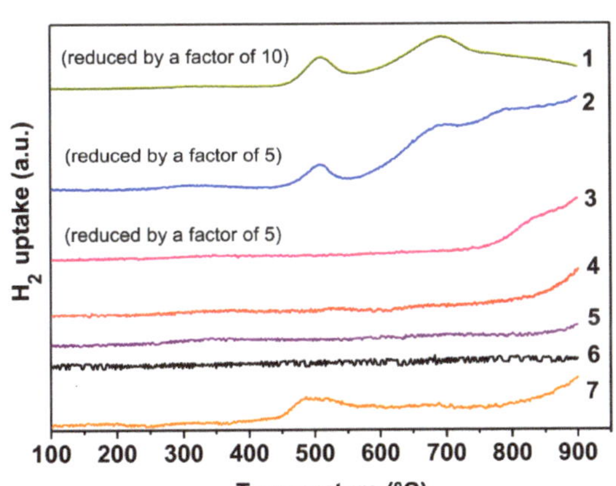

Figure 2. H_2-TPR profiles of (**1**) CoAl-0.5, (**2**) CoAl-0.25, (**3**) CoAl-0.25-R, (**4**) CoAl-0.1, (**5**) CoAl-0.1-R, (**6**) CoAl-0.05 and (**7**) Co/Al_2O_3 catalysts.

According to TEM data (Figure 3), the CoAl-0.1 sample consists mainly of a separate phase of cobalt aluminate $CoAl_2O_4$ (Figure 3b). In addition, the catalyst contains Al_2O_3 crystals, which are uniformly covered by CoO_x particles with a size of about 1–2 nm (Figure 3c). Notably, in Figure 3c the white particles are attributed to cobalt based on the relative mass (atomic number) of scattering Co and Al atoms, which affect the HAADF signal differently. In contrast, CoAl-0.25 contains, in addition to the above-mentioned species, well-distinguishable particles with an average grain size of 10–20 nm that can be assigned to both $CoAl_2O_4$ and Co_3O_4, taking into account their close lattice parameters (Figure 4).

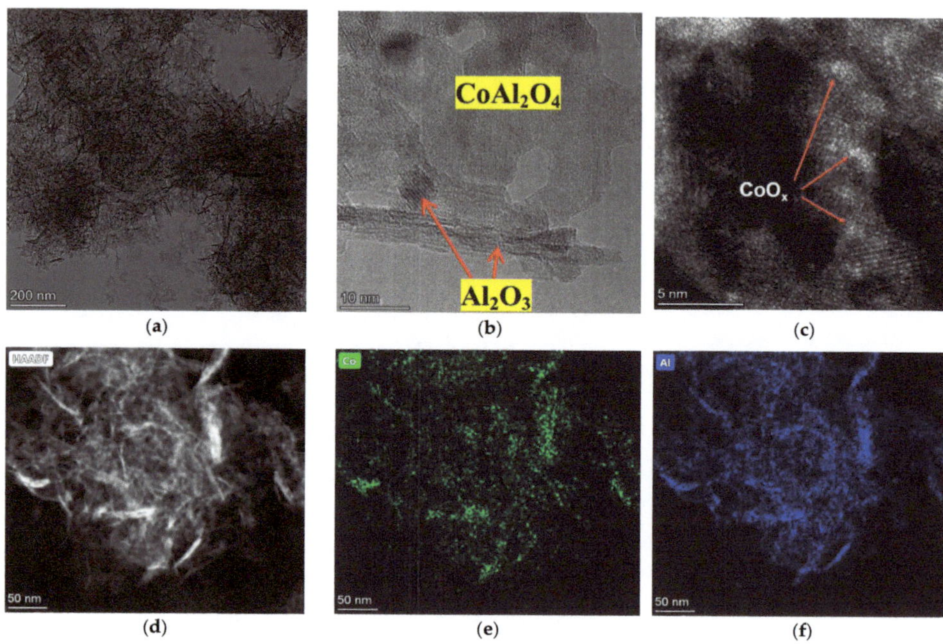

Figure 3. (**a**,**b**) TEM, (**c**,**d**) HAADF-STEM and (**e**,**f**) related EDX-STEM Co and Al elemental mapping images of as-synthesized CoAl-0.1 catalyst.

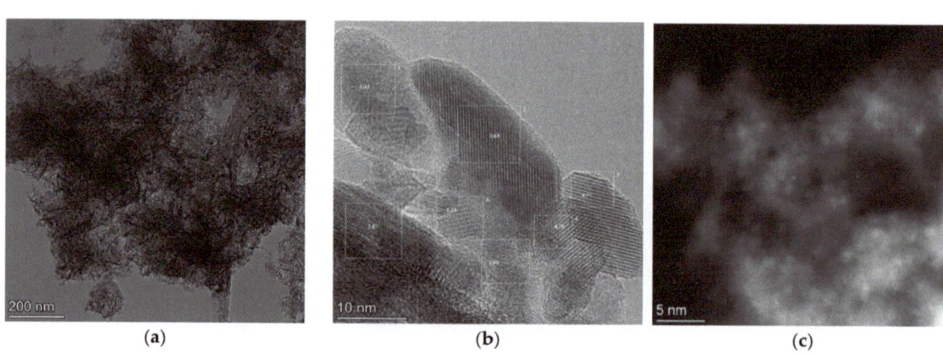

Figure 4. Low- (**a**) and higher- (**b**) magnification TEM and (**c**) HAADF-STEM images of as-synthesized CoAl-0.25 catalyst, representing a variety of phases, such as crystalline alumina, cobalt aluminate and cobalt oxides (including small particles on alumina).

To further characterize the nature of cobalt in the catalysts, the XPS spectra were obtained (Figure 5 and Table S2). In the high-resolution Co 2p spectra, the peaks centered at

781.3–781.8 (Co $2p_{3/2}$) and ~795 eV (Co $2p_{1/2}$) are related to Co^{2+} (Figure 5). An associated Co^{2+} satellite peak (shake-up peak) centered at ~787.0 eV is also present, confirming the predominant state of cobalt as tetrahedral Co^{2+} ions. As expected, the Al 2p and O 1s spectra of the catalysts are very similar, confirming the lack of differentiation in the electron state of aluminum and oxygen among the catalysts.

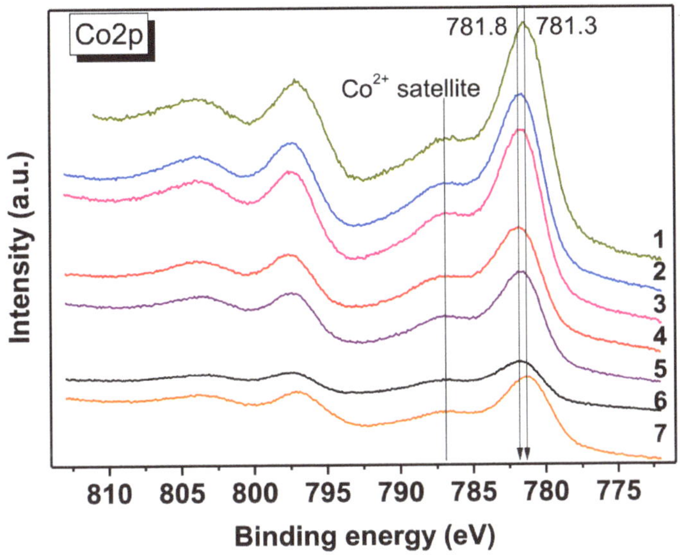

Figure 5. High-resolution Co 2p XPS spectra of (**1**) CoAl-0.5, (**2**) CoAl-0.25, (**3**) CoAl-0.25-R, (**4**) CoAl-0.1, (**5**) CoAl-0.1-R, (**6**) CoAl-0.05 and (**7**) Co/Al_2O_3 catalysts.

2.2. Gas-Phase Catalytic Reactions of Propane in a Flow Reactor

The activity of CoAl catalysts in PDH was estimated at 600 °C using a 10% propane/N_2 mixture. Unlike other samples, CoAl-0.5 proved unsuitable to catalyze PDH, as it was prone to rapid carbonization, clogging the reactor. This applies to both the fresh and spent–regenerated catalysts, and can be explained by the presence of the relatively high content of an easily reduced massive CoO_x form, which tends to induce cracking reactions [35,37–40]. In contrast, fresh CoAl-0.1 and CoAl-0.05 catalysts provided 47 and 36% of propane conversion after a 20 and 30 min induction period, respectively (Figure 6). Then, there was a gradual decrease in catalytic activity, while the selectivity to propylene remained at the initial level of ca. 95%. It is noteworthy that despite the two-and-a-half-times-higher Co content in the CoAl-0.25 catalyst (Table 1), its activity was nevertheless comparable to that of CoAl-0.1 (Figure 6). However, the induction period for CoAl-0.25 was twice as long, whereas selectivity to propylene was notably lower, reaching 90% only at 1.5 h. This result illustrates that the readily reducible extra-framework CoO_x species still present in the fresh CoAl-0.25 catalyst impairs the target catalytic performance.

It should be noticed that under the specified reaction conditions, especially during the induction period, propane underwent concurrent cracking reactions yielding mainly methane, along with carbonaceous material (see below), which is typical for high-temperature PDH [2–10].

The pre-treatment of the fresh CoAl-0.1 catalyst with hydrogen (as described in Section 4.5) shortened the induction period and led to a slight increase in the maximal propane conversion (Figure 7). However, regeneration of the spent CoAl-0.1 catalyst by high-temperature treatment with oxygen (Section 4.5) reduced its activity by about 20% compared to the fresh catalyst (Figure 7a). On the other hand, the selectivity to propylene increased to 96% (Figure 7b). The same is true of the spent–regenerated CoAl-0.25-R catalyst, which provided notably higher selectivity to propylene, a shorter induction period, and well-reproducible catalytic activity tested in several runs (Figure 8).

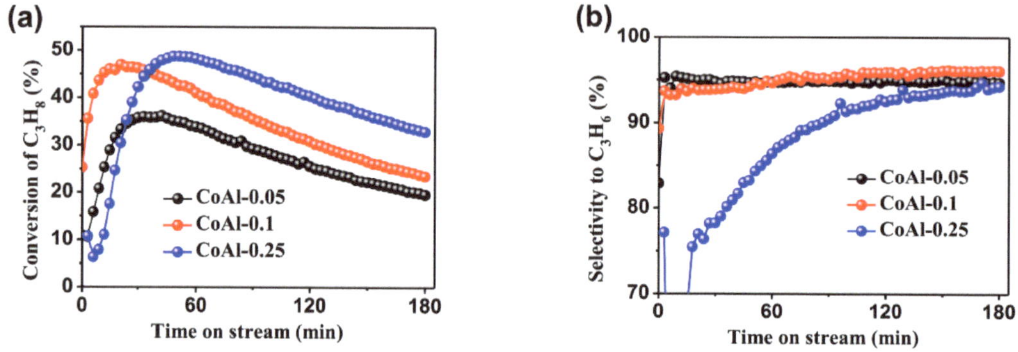

Figure 6. Propane conversion (**a**) and selectivity to propylene (**b**) as functions of time on stream over CoAl catalysts at T = 600 °C and P = 1 atm. Gas mixture: 10 vol.% propane, N_2 balance; GHSV = 7500 mL h^{-1} $g_{cat}{}^{-1}$.

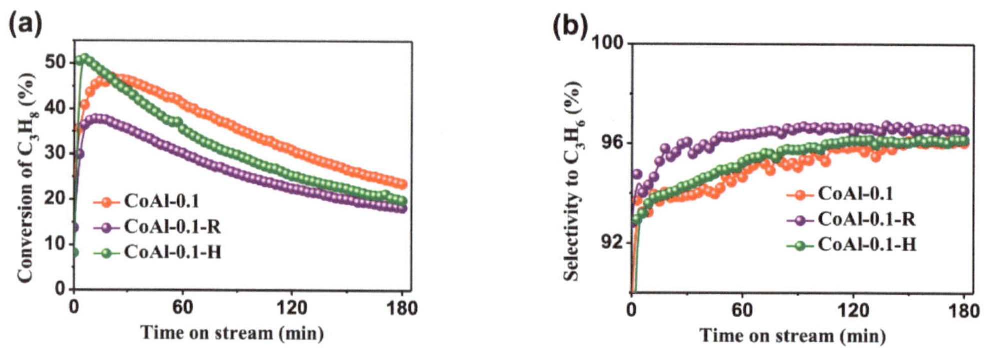

Figure 7. Propane conversion (**a**) and selectivity to propylene (**b**) as functions of time on stream over CoAl-0.1, CoAl-0.1-R (spent–regenerated) and CoAl-0.1-H (pre-treated with H_2) catalysts at T = 600 °C and P = 1 atm. Gas mixture: 10 vol.% propane, N_2 balance; GHSV = 7500 mL h^{-1} $g_{cat}{}^{-1}$.

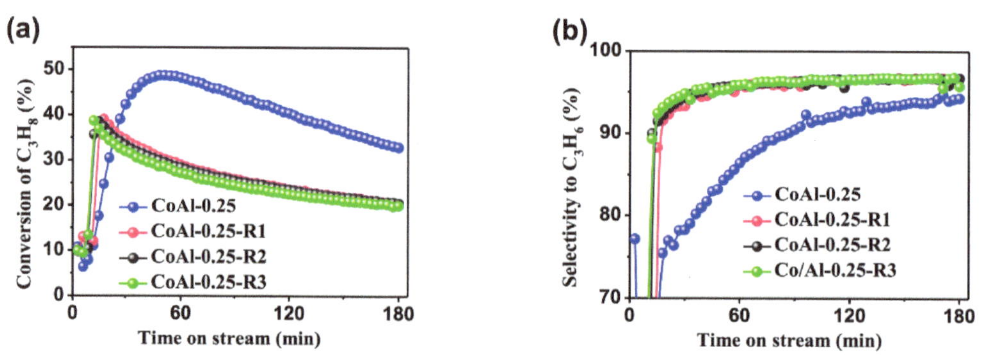

Figure 8. Propane conversion (**a**) and selectivity to propylene (**b**) as functions of time on stream over CoAl-0.25 and spent–regenerated CoAl-0.25-RN catalysts (where N is the execution number) at T = 600 °C and P = 1 atm. Gas mixture: 10 vol.% propane, N_2 balance; GHSV = 7500 mL h^{-1} $g_{cat}{}^{-1}$.

Furthermore, the catalytic properties of the "reference" Co/Al_2O_3 sample turned out to be comparable with that of CoAl-0.05 (Figure S1). This is not surprising, taking into account the close content of cobalt (Table 1), while the other characteristics of these catalysts are somewhat different. The selectivity to propylene over Co/Al_2O_3 was even the highest among the tested catalysts, reaching 97% and 98% for fresh and spent–regenerated samples, respectively (Figure S2). However, the relative drop in maximal propane conversion (from 36 to 25%) after the catalyst regeneration was the most significant (Figure S2).

2.3. Structural Changes in Catalysts during PDH and Subsequent Regeneration

According to the literature, the deactivation of cobalt-containing catalysts during PDH is commonly due to the deposition of coke and the restructuring of cobalt species under high-temperature reducing conditions [19,20,28,29]. Indeed, the microscopic images of the spent CoAl-0.1 catalyst show considerable growth of carbon nanotubes (Figure 9a,b). However, encapsulation of Co-Al oxides in carbon did not occur (cf. [19,20]), which means that carbonization by itself is not a cause of deactivation. More important is the fact of partial reduction of cobalt to the metallic state with the formation of Co nanoparticles of relatively large sizes about 10 nm (Figure 9b,d,e), although small particles of Co on alumina did not undergo substantial aggregation (Figure 9c).

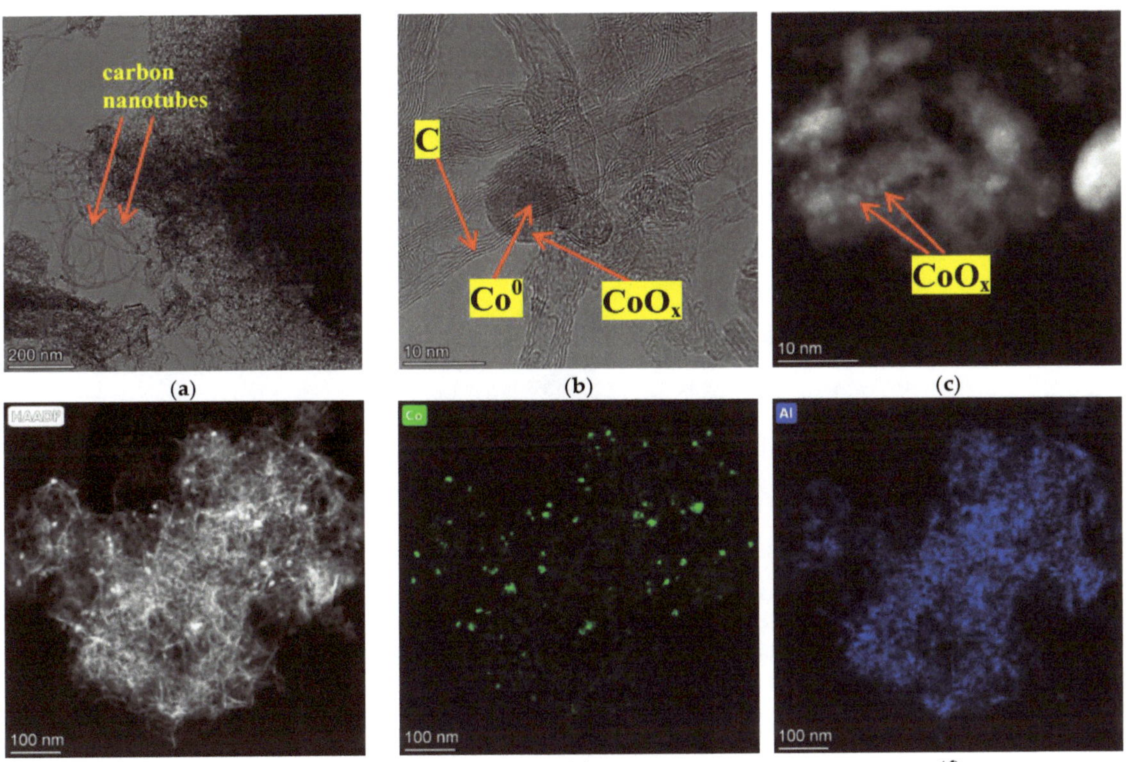

Figure 9. (**a**,**b**) TEM, (**c**,**d**) HAADF-STEM and (**e**,**f**) related EDX-STEM Co and Al elemental mapping images of the spent CoAl-0.1 catalyst (after 3 h of operation at T = 600 °C and P = 1 atm for the gas mixture of 10 vol% propane, N_2 balance, GHSV = 7500 mL h^{-1} $g_{cat}{}^{-1}$).

Regeneration of the CoAl-0.1 spent catalyst by high-temperature treatment with molecular oxygen results in the redispersion of aggregated cobalt nanoparticles, providing a catalyst quite similar in appearance to the original (Figure 10). The CoAl-0.25 catalyst also restores its structure after the use in PDH and subsequent heat treatment in oxygen, although more significant differentiation in CoAl-0.25-R morphology has occurred. Thus, separate phases of alumina (weakly enriched with cobalt), cobalt aluminate and rather large particles of Co_3O_4 can be observed via TEM (Figures S3 and S4). Remarkably, according to H_2-TPR data, the CoO_x species in CoAl-0.25-R are not so readily reducible with hydrogen, as was found for CoO_x in the fresh CoAl-0.25 catalyst (Figure 2).

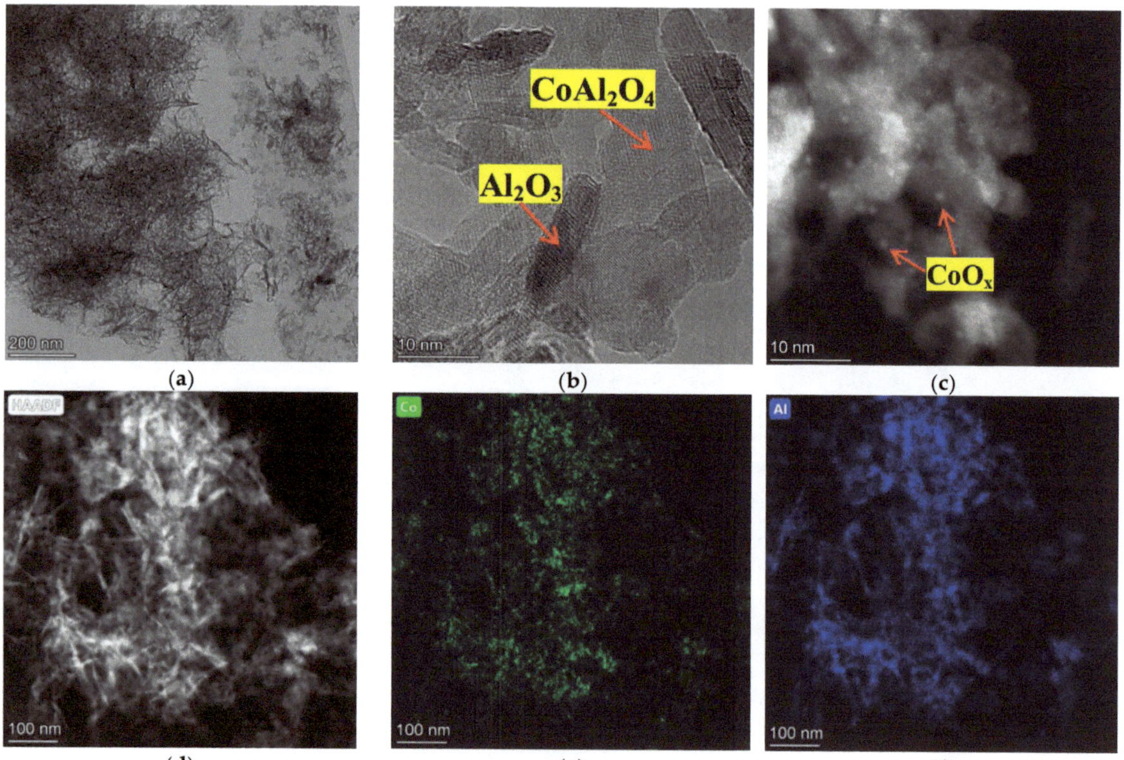

Figure 10. (**a**,**b**) TEM, (**c**,**d**) HAADF-STEM and (**e**,**f**) related EDX-STEM Co and Al elemental mapping images of the spent–regenerated CoAl-0.1-R catalyst.

3. Discussion

The CoAl-0.1 sample, which demonstrated the best catalytic performance in PDH, comprises three structural elements: Al_2O_3, small CoO_x particles (in the size range of 1–2 nm) on its surface and adjacent areas of crystalline cobalt aluminate ($CoAl_2O_4$). The CoAl-0.25 catalyst, exhibiting comparable catalytic activity, contains, in addition, Co_3O_4 particles of about 10 nm size and the Co_3O_4 phase of a comparatively low crystallinity (XRD data, Figure 1). The latter phase was easily reduced and could be responsible for the lower selectivity to propylene and longer induction period compared to the CoAl-0.1 catalyst. Indeed, CoAl-0.25-R, showing better selectivity to propylene and a shortened induction period (Figure 8), differed from CoAl-0.25 through its increased Co_3O_4 particle size (up to 100 nm, Figure S4) and their increased crystallinity (Figure 1, Table S1). Additionally, CoAl-0.25-R became practically non-reducible with hydrogen until reaching 700 °C (Figure 2). Furthermore, the high-temperature treatment with oxygen of both of the spent CoAl-0.1 and CoAl-0.25 catalysts resulted in a marked decrease in specific surface area (Table 1), which may be one of the reasons for the partial loss of their activity.

It is obvious that the target activity of CoAl catalysts in PDH may be associated with the small CoO_x particles on alumina, as well as with comparatively large and slowly reducible $CoAl_2O_4$ and Co_3O_4 crystalline phases, which can undergo a common evolutionary chain during PDH or pre-reduction with hydrogen: $Co^{+2/+3}$ → framework Co^{+2} → isolated Co^0 → small-size $(Co^0\text{-}Co^0)_n$ cluster → large cobalt particle (cf. [35–39]). Among these forms, as mentioned in Section 1, Co^{+2} oxides and/or small Co^0 species are widely suggested to be the active catalytic sites [21–43].

It is noteworthy that the basic rationale for the key contribution of tetrahedral Co^{+2} ions is their exclusive observation by physical–chemical techniques with no detection of Co^0 under PDH conditions [21–36]. The proposed reaction mechanism suggests a non-redox pathway involving the heterolytic C–H bond cleavage over four-coordinated Co^{+2} incorporated in the framework of oxide support [25,36]. In contrast, extra-framework Co_3O_4 weakly bonded to the support (if present) can easily be reduced to comparatively large metallic cobalt particles [35,37–40], leading to propane cracking until these zones are strongly carbonized [19–22,33].

However, equally convincing is the evidence that well-dispersed Co^0 species generated in situ by partial reduction of Co^{+2} ions are the alternative active catalytic sites [33,35,40–42]. The main argument here is based on the apparent relationship between the appearance of cobalt nanoparticles and the increase in catalytic activity. Furthermore, Kondratenko et al. have shown recently that propylene formation depends linearly on Co^0 accumulation in Co/silicalite-1 catalysts [43]. Though perhaps, the catalytic activity could simply correlate with the steady reduction of Co^{+2} to Co^0 due to a related but different process.

With regard to the CoAl catalytic system, it should be noted that the most effective PDH catalysts, such as CoAl-0.05, CoAl-0.1, CoAl-0.1-R and CoAl-0.25-R, are actually most resistant to reduction (metallization) by hydrogen (Figure 2). Moreover, a slow reduction of CoO_x to metallic cobalt followed by its aggregation to cobalt nanoparticles (Figure 9b) is accompanied by the catalyst's progressive deactivation, which may be related. On the other hand, the key catalytic activity of four-coordinated Co^{+2} ions only (whether they are part of CoO_x particles on aluminum or are included in $CoAl_2O_4$ and Co_3O_4 crystals) cannot explain the results of numerous case studies where a definite relationship between of the catalytic activity and Co^0 accumulation was found. Nevertheless, this apparent contradiction can be resolved by examining the missing link between the cobalt states Co^{+2} and Co^0 under the PDH reaction.

Indeed, a close literary analogy of the process of nickel(II) oxide reduction by hydrogen reveals that NiO(100) crystals exhibiting negligible reactivity to H_2 become active in splitting H-H bonds when surface defects, namely O vacancies (OVs), are created [49–51]. Remarkably, a correlation was found between the concentration of OVs and the rate of NiO reduction to Ni^0 [49]. It seems reasonable that a similar process of OV generation occurs on the surface of tetrahedral Co^{+2} oxide forms in the course of PDH. The lattice oxygen consumption leading to the formation of OVs under conditions of thermal gaseous reduction is quite typical for transition metal oxides, including CoO_x [52–55]. Moreover, the importance of OVs for efficient C−H bond activation in PDH has already been proven for a number of metal oxides, such as CeO_2 [56] and TiO_2 [57,58], and even for "non-reducible" oxides, such as ZrO_2 [59] and Al_2O_3 [60]. Likewise, the catalytic activity of WO_3 in PDH, provided by pre-treatment with hydrogen, may have an analogous origin [61].

The involvement of OVs could explain the induction period normally required for PDH catalyzed by cobalt oxides, the positive effect of pre-reduction with hydrogen (see, for example, Figure 7) and the observed correlation of catalytic activity with Co^0 accumulation, while an overly extensive reduction of Co^{+2} to Co^0 causes catalyst deactivation. Also notably, Co atoms surrounding OVs and carrying uncompensated positive charge can actually simulate Co^{+1} species able to catalyze PDH through the oxidative addition/reductive elimination pathway (Scheme 1), thus linking together the Co^{+2}, Co^{+1} and Co^0 forms, considered earlier as alternatives in the literature.

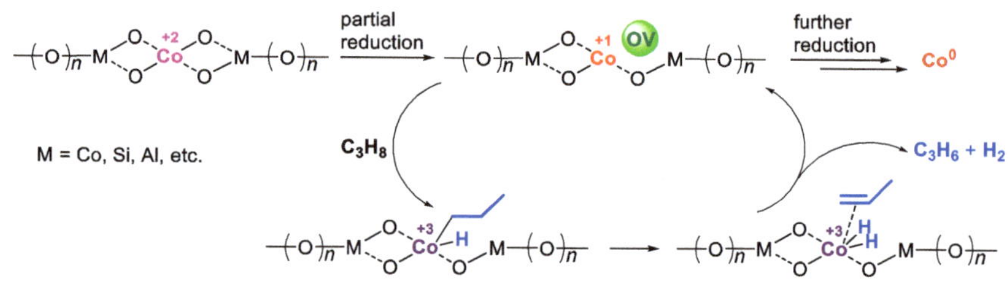

Scheme 1. Probable mechanism of PDH involving OVs generated on the surface of tetrahedral Co^{+2} oxide forms.

Summarizing the above considerations, nanocrystalline $CoAl_2O_4$ and Co_3O_4, as well as fine CoO_x particles on alumina (all present in CoAl active catalysts) may exhibit the same catalytic properties due to similar evolution under PDH conditions, namely partial reduction to metallic cobalt (ultimately) through the intermediacy of surface OVs. The small content of cobalt in CoAl catalysts ensures their strong interaction with alumina and embedding in the spinel structure, which generally prevents deep reduction of CoO_x to Co^0 under PDH conditions and provides the catalyst's relative stability. Based on these assumptions, progress in the design of cobalt-containing catalysts and similar transition metal oxide catalytic systems will require their optimal ability to generate surface Ovs, along with their resistance to subsequent reduction and metallization to be taken into account.

4. Materials and Methods

4.1. Materials

Cobalt(II) nitrate hexahydrate ($Co(NO_3)_2{\cdot}6H_2O$, ≥98%, Fisher Scientific UK, Loughborough, UK) and aluminum nitrate nonahydrate ($Al(NO_3)_3{\cdot}9H_2O$, ≥98%, Sigma-Aldrich, Merck Life Science LLC, Moscow, Russia) were used as received from the chemical suppliers.

4.2. Catalyst Preparation

The cobalt aluminate spinels ($CoAl_2O_4$-Al_2O_3) were obtained by mixing 2 M $Co(NO_3)_2$ and 2 M $Al(NO_3)_3$ aqueous solutions followed by co-precipitating with a 12.5% NH_4OH at constant pH 7.5 and temperature of 70 °C. The resulting precipitates were filtered and washed with distilled water until the ammonia was removed. The obtained materials were dried to an air-dry state at room temperature, then at 110 °C for 12 h, and calcined in air at 650 °C for 4 h in a muffle furnace. Accordingly, based on the molar ratio of cobalt and aluminum salts taken, the catalyst samples, denoted as CoAl-0.5, CoAl-0.25, CoAl-0.1 and CoAl-0.05, were prepared in the amount of 50 g each (Table 1).

The Co/Al_2O_3 catalyst was prepared via the incipient wetness impregnation of γ-Al_2O_3 with a cobalt(II) nitrate solution followed by drying at 110 °C for 12 h and thermal treatment in air at 650 °C for 4 h.

4.3. Catalyst Characterization Techniques

The X-ray diffraction (XRD) data were obtained on the diffractometer ARL X'tra (Thermo Fisher Scientific, Waltham, MA, USA) using monochromatic CuKα radiation (λ = 1.5418 Å) in a range of 2θ = 15–80° with a step of 0.02° (1°/min) using linear detector Mythen2R 1D (Dectris, Baden-Daettwil, Switzerland) with an angle of 7.0°. Refinement of the lattice parameters and determination of average crystallite sizes and weight ratios of Co_3O_4/Al_2O_3 were carried out by the Rietveld method applied for a range of 2θ = 63–70° using the TOPAS 4.2 program and structural data from the ICSD database (Table S1).

Temperature-programmed reduction (H_2-TPR) was performed in a flow system with a thermal conductivity detector. The samples (0.05 g, 0.25–0.5 mm fraction) were heated at rates of 10 °C min^{-1} to 900 °C; the flow rate of the reducing mixture (10% H_2 in argon) was 40 cm^3 min^{-1}.

The porous structure parameters were determined from N_2 adsorption isotherms at 77 K on an ASAP-2400 (Micromeritics, Norcross, GA, USA) analyzer after degassing the samples at 150 °C to a residual pressure of 30 mTorr (4 Pa).

The content of Co element in the samples was measured with an Optima 4300 DV inductively coupled plasma optical emission spectrometer (ICP-OES) (PerkinElmer, Wellesley, MA, USA).

High-resolution transmission electron microscopy (HRTEM) and scanning transmission electron microscopy (STEM) were conducted with a Themis Z (Thermo Fisher Scientific, Waltham, MA, USA) instrument with an accelerating voltage of 200 kV and lattice resolution of 0.07 nm. The TEM images were recorded with a Ceta 16 (Thermo Fisher Scientific, USA) CCD matrix. Elemental analysis was performed with a Super-X EDS detector (Thermo Fisher Scientific, USA). High-angle annular dark-field (HAADF) STEM images were recorded using a standard ThemisZ detector.

X-ray photoelectron spectra (XPS) were recorded on a SPECS (Berlin, Germany) photoelectron spectrometer using a hemispherical PHOIBOS-150-MCD-9 analyzer (Al Kα radiation, hν = 1486.6 eV, 150 W). The binding energy (BE) scale was pre-calibrated using the positions of the peaks of Au 4f$_{7/2}$ (BE = 84.0 eV) and Cu 2p$_{3/2}$ (BE = 932.67 eV) core levels. The BE values were corrected by reference to the C 1s internal standard (284.5 eV).

4.4. Catalytic Performance Tests

The PDH reactions were carried out at atmospheric pressure in a 7.5 mm quartz tube reactor loaded with a catalyst (0.2 g, 0.25–0.5 mm fraction), mixed with 1 mL quartz particles (d = 0.45 to 1 mm). The gas mixture (10 vol% propane balanced with nitrogen) was fed to the reactor at a gas hourly space velocity (GHSV) of 7500 mL h^{-1} $g_{cat}$$^{-1}$ (propane weight hourly space velocity of 1.4 h^{-1}). The reaction temperature, 600 °C, was controlled inside the reactor with an accuracy of ±0.5 °C. The reaction products were analyzed by an online gas chromatograph (Chromos GC-1000, Chromos Engineering, Dzerzhinsk, Russia) with a flame ionization detector, using a quartz capillary column of an inner diameter of 0.32 mm and a length of 7 m. The column was prepared by the static high-pressure method, yielding a monolithic layer of porous SiO_2 formed in the process of sol-gel synthesis directly inside the capillary [62]. Prior to the measurements, the catalysts were heated at 600 °C in a flow of N_2 for 1 h. The blank experiment (the Al_2O_3/ quartz particles only) showed propane conversion of ~1.2% and selectivity to propylene of about 70%.

The propane conversion (X) and propylene selectivity (S) were calculated according to the carbon atom balance method:

$$X_{C_3H_8} = \frac{C^0_{C_3H_8} - C_{C_3H_8}}{C^0_{C_3H_8}} \times 100\%;$$

$$S_{C_3H_6} = \frac{C_{C_3H_6}}{C^0_{C_3H_8} - C_{C_3H_8}} \times 100\%;$$

where $C_{C_3H_8}$ and $C_{C_3H_6}$ are the mole flow rates of propane and propylene, respectively. In view of the reaction-induced changes in the gas flow due to the conversion of one molecule of propane into two molecules (propylene and hydrogen) during the reaction, the values of $C_{C_3H_8}$ and $C_{C_3H_6}$ were normalized by carbon balance accordingly.

4.5. Catalyst Pre-Reduction with H_2 and Regeneration Procedures

The treatment of CoAl catalysts with hydrogen was carried out in the reactor at 600 °C in the H_2 flow (15 mL min^{-1}) for 1.5 h and then in the N_2 flow (30 mL min^{-1}) for 1 h.

After catalytic testing in PDH, the spent CoAl catalysts were treated in the reactor at 600 °C in the O_2 flow (15 mL min^{-1}) for 1.5 h and then in the N_2 flow (30 mL min^{-1}) for 1 h.

5. Conclusions

It was shown that cobalt aluminate containing 4–7 wt.% Co, and as a consequence, consisting mainly of alumina (with inclusions of small CoO_x particles on the surface) and $CoAl_2O_4$ spinel crystallites, can be an effective catalyst for PDH reactions under practical operating conditions (600 °C, 1 atm). Further increasing the Co content to 17 wt.% resulted in the formation of an additional Co_3O_4 phase without increasing the catalyst efficiency towards PDH.

The catalysts' loss of activity with time on stream is related to cobalt aggregation in the form of metallic Co nanoparticles; however, coke deposition, largely in the form of carbon nanotubes, does not appear to be the critical reason for the catalysts' deactivation. The high-temperature treatment of spent catalysts with molecular oxygen leads to their regeneration with a partial decrease in activity, but at the same time is accompanied by the stabilization of the catalysts' operation.

The structural motifs, such as tetrahedral Co^{2+} species located in the $CoAl_2O_4$ and Co_3O_4 spinel forms, as well as in the small CoO_x particles tightly covering alumina, can be regarded as the feasible primary active-site ensembles. However, their subsequent evolution under PDH reaction conditions, consisting of a partial reduction to metallic cobalt through the intermediacy of surface OVs, may provide further rationale for the origin of the catalytic activity of cobalt aluminates in PDH.

Supplementary Materials: The following supporting information can be downloaded at: https://www.mdpi.com/article/10.3390/catal13111419/s1, Table S1: Grid parameters (a), average crystallite sizes and weight ratios; Figure S1: Propane conversion and selectivity to propylene as functions of time on stream over CoAl and Co/Al_2O_3 catalysts; Figure S2: Propane conversion and selectivity to propylene as functions of time on stream over Co/Al_2O_3 and Co/Al_2O_3-R catalysts; Figure S3: Low- and higher-magnification TEM images of the spent–regenerated CoAl-0.25-R catalyst; Figure S4: HAADF-STEM and related EDX-STEM Co and Al elemental mapping images of different areas of CoAl-0.25-R; Table S2: High-resolution XPS spectra of Al 2p and O 1s of the samples; Table S3: The N_2 adsorption/desorption isotherms and the pore size distribution of the samples.

Author Contributions: Conceptualization, A.N.C.; investigation, A.N.C., S.V.C., E.Y.G., I.P.P., G.A.Z., A.S.G. and A.A.S.; data curation, K.Y.K. and V.I.S.; writing—original draft preparation, K.Y.K.; writing—review and editing, K.Y.K. and V.I.S.; supervision, V.I.S. All authors have read and agreed to the published version of the manuscript.

Funding: This work was supported by the Ministry of Science and Higher Education of the Russian Federation within the governmental order for the Boreskov Institute of Catalysis (project AAAA-A21-121011390008-4).

Data Availability Statement: Not applicable.

Acknowledgments: The studies were carried out using the facilities of the shared research center "National center of investigation of catalysts" at Boreskov Institute of Catalysis.

Conflicts of Interest: The authors declare no conflict of interest.

References

1. Zimmermann, H. Propene. In *Ullmann's Encyclopedia of Industrial Chemistry*; Wiley-VCH Verlag GmbH & Co., KGaA: Weinheim, Germany, 2013. [CrossRef]
2. Sattler, J.J.H.B.; Ruiz-Martinez, J.; Santillan-Jimenez, E.; Weckhuysen, B.M. Catalytic dehydrogenation of light alkanes on metals and metal oxides. *Chem. Rev.* **2014**, *114*, 10613–10653. [CrossRef] [PubMed]
3. Hu, Z.-P.; Yang, D.; Wang, Z.; Yuan, Z.-Y. State-of-the-art catalysts for direct dehydrogenation of propane to propylene. *Chin. J. Catal.* **2019**, *40*, 1233–1254. [CrossRef]
4. Martino, M.; Meloni, E.; Festa, G.; Palma, V. Propylene Synthesis: Recent Advances in the Use of Pt-Based Catalysts for Propane Dehydrogenation Reaction. *Catalysts* **2021**, *11*, 1070. [CrossRef]
5. Dong, S.; Altvater, N.R.; Mark, L.O.; Hermans, I. Assessment and comparison of ordered & non-ordered supported metal oxide catalysts for upgrading propane to propylene. *Appl. Catal. A-Gen.* **2021**, *617*, 118121. [CrossRef]
6. Chen, S.; Chang, X.; Sun, G.; Zhang, T.; Xu, Y.; Wang, Y.; Pei, C.; Gong, J. Propane dehydrogenation: Catalyst development, new chemistry, and emerging technologies. *Chem. Soc. Rev.* **2021**, *50*, 3315–3354. [CrossRef] [PubMed]

7. Otroshchenko, T.; Jiang, G.; Kondratenko, V.A.; Rodemerck, U.; Kondratenko, E.V. Current status and perspectives in oxidative, non-oxidative and CO_2-mediated dehydrogenation of propane and isobutane over metal oxide catalysts. *Chem. Soc. Rev.* **2021**, *50*, 473–527. [CrossRef]
8. Dai, Y.; Gao, X.; Wang, Q.; Wan, X.; Zhou, C.; Yang, Y. Recent progress in heterogeneous metal and metal oxide catalysts for direct dehydrogenation of ethane and propane. *Chem. Soc. Rev.* **2021**, *50*, 5590–5630. [CrossRef]
9. Wang, Y.; Hu, P.; Yang, J.; Zhu, Y.-A.; Chen, D. C–H bond activation in light alkanes: A theoretical perspective. *Chem. Soc. Rev.* **2021**, *50*, 4299–4358. [CrossRef]
10. Liu, S.; Zhang, B.; Liu, G. Metal-based catalysts for the non-oxidative dehydrogenation of light alkanes to light olefins. *React. Chem. Eng.* **2021**, *6*, 9–26. [CrossRef]
11. Phung, T.K.; Pham, T.L.M.; Vu, K.B.; Busca, G. (Bio)Propylene production processes: A critical review. *J. Environ. Chem. Eng.* **2021**, *9*, 105673. [CrossRef]
12. Yuan, Y.; Lee, J.S.; Lobo, R.F. Ga^+-Chabazite Zeolite: A Highly Selective Catalyst for Nonoxidative Propane Dehydrogenation. *J. Am. Chem. Soc.* **2022**, *144*, 15079–15092. [CrossRef]
13. Yuan, Y.; Lobo, R.F. Zinc Speciation and Propane Dehydrogenation in Zn/H-ZSM-5 Catalysts. *ACS Catal.* **2023**, *13*, 4971–4984. [CrossRef]
14. Xie, J.; Kammert, J.D.; Kaylor, N.; Zheng, J.W.; Choi, E.; Pham, H.N.; Sang, X.; Stavitski, E.; Attenkofer, K.; Unocic, R.R.; et al. Atomically Dispersed Co and Cu on N-Doped Carbon for Reactions Involving C–H Activation. *ACS Catal.* **2018**, *8*, 3875–3884. [CrossRef]
15. Cao, T.; Dai, X.; Li, F.; Liu, W.; Bai, Y.; Fu, Y.; Qi, W. Efficient Non-Precious Metal Catalyst for Propane Dehydrogenation: Atomically Dispersed Cobalt-nitrogen Compounds on Carbon Nanotubes. *ChemCatChem* **2021**, *13*, 3067–3073. [CrossRef]
16. Wang, Y.; Suo, Y.; Ren, J.-T.; Wang, Z.; Yuan, Z.-Y. Spatially isolated cobalt oxide sites derived from MOFs for direct propane dehydrogenation. *J. Colloid Interface Sci.* **2021**, *594*, 113–121. [CrossRef]
17. Wang, Q.; Xu, W.; Ma, Z.; Yu, F.; Chen, Y.; Liao, H.; Wang, X.; Zhou, J. Highly Effective Direct Dehydrogenation of Propane to Propylene by Microwave Catalysis at Low Temperature over Co–Sn/NC Microwave Catalyst. *ChemCatChem* **2021**, *13*, 1009–1022. [CrossRef]
18. Li, Y.-M.; Liu, Z.-Y.; Zhang, Q.-Y.; Wang, Y.-J.; Cui, G.-Q.; Zhao, Z.; Xu, C.-M.; Jiang, G.-Y. Influence of carbonization temperature on cobalt-based nitrogen-doped carbon nanopolyhedra derived from ZIF-67 for nonoxidative propane dehydrogenation. *Pet. Sci.* **2022**, *20*, 559–568. [CrossRef]
19. Chernov, A.N.; Sobolev, V.I.; Koltunov, K.Y. Propane dehydrogenation to propylene over Co@N-doped carbon: Structure-activity-selectivity relationships. *Catal. Commun.* **2022**, *170*, 106495. [CrossRef]
20. Chernov, A.N.; Sobolev, V.I.; Gerasimov, E.Y.; Koltunov, K.Y. Propane Dehydrogenation on Co-NC/SiO_2 Catalyst: The Role of Single-Atom Active Sites. *Catalysts* **2022**, *12*, 1262. [CrossRef]
21. Sun, Y.; Wu, Y.; Shan, H.; Li, C. Studies on the Nature of Active Cobalt Species for the Production of Methane and Propylene in Catalytic Dehydrogenation of Propane. *Catal. Lett.* **2015**, *145*, 1413–1419. [CrossRef]
22. Sun, Y.N.; Gao, Y.N.; Wu, Y.M.; Shan, H.H.; Wang, G.W.; Li, C.Y. Effect of sulfate addition on the performance of Co/Al_2O_3 catalysts in catalytic dehydrogenation of propane. *Catal. Commun.* **2015**, *60*, 42–45. [CrossRef]
23. Li, W.; Yu, S.Y.; Meitzner, G.D.; Iglesia, E. Structure and Properties of Cobalt-Exchanged H-ZSM5 Catalysts for Dehydrogenation and Dehydrocyclization of Alkanes. *J. Phys. Chem. B* **2001**, *105*, 1176–1184. [CrossRef]
24. Hu, B.; Getsoian, A.; Schweitzer, N.M.; Das, U.; Kim, H.; Niklas, J.; Poluektov, O.; Curtiss, L.A.; Stair, P.C.; Miller, J.T.; et al. Selective propane dehydrogenation with single-site Co-II on SiO_2 by a non-redox mechanism. *J. Catal.* **2015**, *322*, 24–37. [CrossRef]
25. Estes, D.P.; Siddiqi, G.; Allouche, F.; Kovtunov, K.V.; Safonova, O.V.; Trigub, A.L.; Koptyug, I.V.; Coperet, C. C–H Activation on Co,O Sites: Isolated Surface Sites versus Molecular Analogs. *J. Am. Chem. Soc.* **2016**, *138*, 14987–14997. [CrossRef] [PubMed]
26. Hu, B.; Kim, W.-G.; Sulmonetti, T.P.; Sarazen, M.L.; Tan, S.; So, J.; Liu, Y.; Dixit, R.S.; Nair, S.; Jones, C.W. A Mesoporous Cobalt Aluminate Spinel Catalyst for Nonoxidative Propane Dehydrogenation. *ChemCatChem* **2017**, *9*, 3330–3337. [CrossRef]
27. Song, S.J.; Li, J.; Wu, Z.J.; Zhang, P.; Sun, Y.Q.; Song, W.Y.; Li, Z.X.; Liu, J. In situ encapsulated subnanometric CoO clusters within silicalite-1 zeolite for efficient propane dehydrogenation. *AIChE J.* **2022**, *68*, e17451. [CrossRef]
28. Dewangan, N.; Ashok, J.; Sethia, M.; Das, S.; Pati, S.; Kus, H.; Kawi, S. Cobalt-Based Catalyst Supported on Different Morphologies of Alumina for Non-oxidative Propane Dehydrogenation: Effect of Metal Support Interaction and Lewis Acidic Sites. *ChemCatChem* **2019**, *11*, 4923–4934. [CrossRef]
29. Dai, Y.; Gu, J.; Tian, S.; Wu, Y.; Chen, J.; Li, F.; Du, Y.; Peng, L.; Ding, W.; Yang, Y. γ-Al_2O_3 sheet-stabilized isolate Co^{2+} for catalytic propane dehydrogenation. *J. Catal.* **2020**, *381*, 482–492. [CrossRef]
30. Bian, Z.F.; Dewangan, N.; Wang, Z.G.; Pati, S.; Xi, S.B.; Borgna, A.; Kus, H.; Kawi, S. Mesoporous-Silica-Stabilized Cobalt(II) Oxide Nanoclusters for Propane Dehydrogenation. *ACS Appl. Nano Mater.* **2021**, *4*, 1112–1125. [CrossRef]
31. Wu, L.Z.; Ren, Z.Z.; He, Y.S.; Yang, M.; Yu, Y.K.; Liu, Y.M.; Tan, L.; Tang, Y. Atomically Dispersed Co^{2+} Sites Incorporated into a Silicalite-1 Zeolite Framework as a High Performance and Coking-Resistant Catalyst for Propane Nonoxidative Dehydrogenation to Propylene. *ACS Appl. Mater. Interfaces* **2021**, *13*, 48934–48948. [CrossRef]
32. Ren, Z.; He, Y.; Yang, M.; Deng, H.; Zhang, Y.; Yang, H.; Tang, Z.; Tan, L.; Tang, Y.; Wu, L. The investigation into the different Co species over Silicalite-1 via modulating heat-treatment atmosphere for propane dehydrogenation. *Mol. Catal.* **2022**, *530*, 112580. [CrossRef]
33. Jeon, N.; Oh, J.; Tayal, A.; Jeong, B.; Seo, O.; Kim, S.; Chung, I.; Yun, Y.J. Effects of heat-treatment atmosphere and temperature on cobalt species in Co/Al_2O_3 catalyst for propane dehydrogenation. *J. Catal.* **2021**, *404*, 1007–1016. [CrossRef]
34. Ge, M.; Chen, X.; Li, Y.; Wang, J.; Xu, Y.; Zhang, L. Perovskite-derived cobalt-based catalyst for catalytic propane dehydrogenation. *React. Kinet. Mech. Catal.* **2020**, *130*, 241–256. [CrossRef]

35. Gao, Y.; Peng, L.; Long, J.; Wu, Y.; Dai, Y.; Yang, Y. Hydrogen pre-reduction determined Co–silica interaction and performance of cobalt catalysts for propane dehydrogenation. *Micropor. Mesopor. Mater.* **2021**, *323*, 111187. [CrossRef]
36. Hu, Z.-P.; Qin, G.; Han, J.; Zhang, W.; Wang, N.; Zheng, Y.; Jiang, Q.; Ji, T.; Yuan, Z.-Y.; Xiao, J.; et al. Atomic Insight into the Local Structure and Microenvironment of Isolated Co-Motifs in MFI Zeolite Frameworks for Propane Dehydrogenation. *J. Am. Chem. Soc.* **2022**, *144*, 12127–12137. [CrossRef]
37. Bulavchenko, O.A.; Cherepanova, S.V.; Malakhov, V.V.; Dovlitova, L.S.; Ishchenko, A.V.; Tsybulya, S.V. In situ XRD study of nanocrystalline cobalt oxide reduction. *Kinet. Catal.* **2009**, *50*, 192–198. [CrossRef]
38. Ji, Y.; Zhao, Z.; Duan, A.; Jiang, G.; Liu, J. Comparative Study on the Formation and Reduction of Bulk and Al_2O_3-Supported Cobalt Oxides by H_2-TPR Technique. *J. Phys. Chem. C* **2009**, *113*, 7186–7199. [CrossRef]
39. Tomic-Tucakovic, B.; Majstorovic, D.; Jelic, D.; Mentus, S. Thermogravimetric study of the kinetics of Co_3O_4 reduction by hydrogen. *Thermochim. Acta* **2012**, *541*, 15–24. [CrossRef]
40. Chen, C.; Zhang, S.M.; Wang, Z.; Yuan, Z.Y. Ultrasmall Co confined in the silanols of dealuminated beta zeolite: A highly active and selective catalyst for direct dehydrogenation of propane to propylene. *J. Catal.* **2020**, *383*, 77–87. [CrossRef]
41. Li, X.Y.; Wang, P.Z.; Wang, H.R.; Li, C.Y. Effects of the state of Co species in Co/Al_2O_3 catalysts on the catalytic performance of propane dehydrogenation. *Appl. Surf. Sci.* **2018**, *441*, 688–693. [CrossRef]
42. Wu, Y.; Long, J.; Wei, S.; Gao, Y.; Yang, D.; Dai, Y.; Yang, Y. Non-oxidative propane dehydrogenation over Co/Ti-ZSM-5 catalysts: Ti species-tuned Co state and surface acidity. *Micropor. Mesopor. Mater.* **2022**, *341*, 112115. [CrossRef]
43. Li, Y.; Zhang, Q.; Fu, S.; Kondratenko, V.A.; Otroshchenko, T.; Bartling, S.; Zhang, Y.; Zanina, A.; Wang, Y.; Cui, G.; et al. Active species and fundamentals of their creation in Co-containing catalysts for efficient propane dehydrogenation to propylene. *Chem. Eng. J.* **2023**, *460*, 14177. [CrossRef]
44. Jacobson, D.; Freiser, B. Studies of the Reactions of Group 8 Transition-Metal Ions Fe^+, Co^+, and Ni^+ with Linear Alkanes. Determination of Reaction Mechanisms and $MC_nH_{2n}{}^+$ Ion Structures Using Fourier Transform Mass Spectrometry Collision-Induced Dissociation. *J. Am. Chem. Soc.* **1983**, *105*, 5197–5206. [CrossRef]
45. Freas, R.B.; Campana, J.E. Dehydrogenation of isobutane by oxygen-deficient cobalt/oxygen cluster ions. *J. Am. Chem. Soc.* **1986**, *108*, 4659–4661. [CrossRef]
46. Van Koppen, P.; Brodbelt-Lustig, J.; Bowers, M.; Dearden, D.; Beauchamp, J.; Fisher, E.R.; Armentrout, P. Transition Metal Ion Mediated C–H and C–C Bond Activation of Alkanes: Dynamical Coupling between Entrance and Exit Channel Transition States. *J. Am. Chem. Soc.* **1991**, *113*, 2359–2369. [CrossRef]
47. Van Koppen, P.A.; Bowers, M.T.; Haynes, C.L.; Armentrout, P.B. Reactions of Ground-State Ti^+ and V^+ with Propane: Factors That Govern C–H and C–C Bond Cleavage Product Branching Ratios. *J. Am. Chem. Soc.* **1998**, *120*, 5704–5712. [CrossRef]
48. Gao, K.; Yoshikai, N. Low-Valent Cobalt Catalysis: New Opportunities for C−H Functionalization. *Acc. Chem. Res.* **2014**, *47*, 1208–1219. [CrossRef]
49. Rodriguez, J.A.; Hanson, J.C.; Frenkel, A.I.; Kim, J.Y.; Perez, M. Experimental and Theoretical Studies on the Reaction of H_2 with NiO: Role of O Vacancies and Mechanism for Oxide Reduction. *J. Am. Chem. Soc.* **2002**, *124*, 346–354. [CrossRef]
50. Jeangros, J.; Hansen, T.W.; Wagner, J.B.; Damsgaard, C.D.; Dunin-Borkowski, R.E.; Hebert, C.; Van herle, J.; Hessler-Wyser, A. Reduction of nickel oxide particles by hydrogen studied in an environmental TEM. *J. Mater. Sci.* **2012**, *48*, 2893–2907. [CrossRef]
51. Rasi, N.M.; Ponnurangam, S.; Mahinpey, N. First-principles investigations into the effect of oxygen vacancies on the enhanced reactivity of NiO via Bader charge and density of states analysis. *Catal. Today* **2023**, *407*, 172–181. [CrossRef]
52. Li, Y.; Chen, T.; Zhao, S.; Wu, P.; Chong, Y.; Li, A.; Zhao, Y.; Chen, G.; Jin, X.; Qiu, Y.; et al. Engineering Cobalt Oxide with Coexisting Cobalt Defects and Oxygen Vacancies for Enhanced Catalytic Oxidation of Toluene. *ACS Catal.* **2022**, *12*, 4906–4917. [CrossRef]
53. Kumar, J.; Jung, H.J.; Neiber, R.R.; Soomro, R.A.; Kwon, Y.J.; Hassan, N.U.; Shon, M.; Lee, J.H.; Baek, K.-Y.; Cho, K.Y. Recent advances in oxygen deficient metal oxides: Opportunities as supercapacitor electrodes. *Int. J. Energy Res.* **2022**, *46*, 7055–7081. [CrossRef]
54. Liu, H.; Fu, H.; Liu, Y.; Chen, X.; Yu, K.; Wang, L. Synthesis, characterization and utilization of oxygen vacancy contained metal oxide semiconductors for energy and environmental catalysis. *Chemosphere* **2021**, *272*, 129534. [CrossRef] [PubMed]
55. Zhang, X.; Liu, X.; Zeng, Y.; Tong, Y.; Lu, X. Oxygen Defects in Promoting the Electrochemical Performance of Metal Oxides for Supercapacitors: Recent Advances and Challenges. *Small Methods* **2020**, *4*, 1900823. [CrossRef]
56. Zeeshan, M.; Chang, Q.-Y.; Zhang, J.; Hu, P.; Sui, Z.-J.; Zhou, X.-G.; Chen, D.; Zhu, Y.-A. Effects of Oxygen Vacancy and Pt Doping on the Catalytic Performance of CeO_2 in Propane Dehydrogenation: A First-Principles Study. *Chin. J. Chem.* **2021**, *39*, 2391–2402. [CrossRef]
57. Xie, Z.; Yu, T.; Song, W.; Li, J.; Zhao, Z.; Liu, B.; Gao, Z.; Li, D. Highly active nanosized anatase TiO2–x oxide catalysts in situ formed through reduction and Ostwald ripening processes for propane dehydrogenation. *ACS Catal.* **2020**, *10*, 14678–14693. [CrossRef]
58. Li, C.-F.; Guo, X.; Shang, Q.-H.; Yan, X.I.; Ren, C.; Lang, W.-Z.; Guo, Y.-J. Defective TiO_2 for propane dehydrogenation. *Ind. Eng. Chem. Res.* **2020**, *59*, 4377–4387. [CrossRef]
59. Zhang, Y.; Zhao, Y.; Otroshchenko, T.; Lund, H.; Pohl, M.-M.; Rodemerck, U.; Linke, D.; Jiao, H.; Jiang, G.; Kondratenko, E.V. Control of coordinatively unsaturated Zr sites in ZrO_2 for efficient C–H bond activation. *Nat. Commun.* **2018**, *9*, 3794. [CrossRef]
60. Xie, Z.; Li, Z.; Tang, P.; Song, Y.; Zhao, Z.; Kong, L.; Fan, X.; Xiao, X. The effect of oxygen vacancies on the coordinatively unsaturated Al–O acid-base pairs for propane dehydrogenation. *J. Catal.* **2021**, *397*, 172–182. [CrossRef]

61. Yun, Y.; Araujo, J.R.; Melaet, G.; Baek, J.; Archanjo, B.S.; Oh, M.; Alivisatos, A.P.; Somorjai, G.A. Activation of tungsten oxide for propane dehydrogenation and its high catalytic activity and selectivity. *Catal. Lett.* **2017**, *147*, 622–632. [CrossRef]
62. Golovnya, R.V.; Samusenko, A.L.; Mistryukov, E.A. Analysis of polar compounds on PEG-40M/KF glass capillary columns. *J. High Resolut. Chromatogr.* **1979**, *2*, 609–612. [CrossRef]

catalysts

MDPI

Article

Cu-Doped $SrTiO_3$ Nanostructured Catalysts for CO_2 Conversion into Solar Fuels Using Localised Surface Plasmon Resonance

Lorenzo Rizzato [1,*], Jonathan Cavazzani [1], Andrea Osti [1], Marco Scavini [2] and Antonella Glisenti [1,*]

1 Department of Chemical Sciences, University of Padova, Via F. Marzolo, 1, 35131 Padova, Italy; jonathan.cavazzani@phd.unipd.it (J.C.); andrea.osti.1@phd.unipd.it (A.O.)
2 Department of Chemistry, University of Milano, Via Golgi, 19, 20133 Milano, Italy; marco.scavini@unimi.it
* Correspondence: lorenzo.rizzato.1@phd.unipd.it (L.R.); antonella.glisenti@unipd.it (A.G.)

Abstract: Carbon dioxide valorisation is one of the most discussed topics amongst researchers; indeed, finding a way to significantly reduce CO_2 concentration in the atmosphere is crucial in order to mitigate climate change effects in the next decades. In this study, $SrTiO_3$-supported Cu nanoparticles are exploited as Localised Surface Plasmon Resonance (LSPR)-mediated catalysts for CO_2 reduction. The materials were prepared via sol–gel citrate route methodology, inserting Cu as a dopant in the perovskite structure; reducing treatments at different temperatures were performed to promote copper atom exsolution, thus forming nanostructures upon the surface. The perovskitic structure was confirmed via ex situ and operando XRD analysis, while compositional analysis was carried out through XPS and EDS; SEM and TEM images revealed morphological changes with different reducing treatments, and bulk reducibility was analysed with H_2-TPR, revealing different Cu species in the material. Band gap analysis via DRS showed the successful incorporation of copper in the perovskite, affecting the light absorption properties. Finally, catalytic tests showed that copper nanoparticles play a role in CO_2 activation with sunlight, proving that LSPR could be exploited for catalytic means.

Keywords: photocatalysis; $Sr_{0.9}TiO_3$ perovskites; CO_2 valorisation; Cu doping

Citation: Rizzato, L.; Cavazzani, J.; Osti, A.; Scavini, M.; Glisenti, A. Cu-Doped $SrTiO_3$ Nanostructured Catalysts for CO_2 Conversion into Solar Fuels Using Localised Surface Plasmon Resonance. *Catalysts* **2023**, *13*, 1377. https://doi.org/10.3390/catal13101377

Academic Editors: Leonarda Liotta, Bo Weng, Konstantin Ivanov Hadjiivanov and Narendra Kumar

Received: 31 August 2023
Revised: 9 October 2023
Accepted: 17 October 2023
Published: 19 October 2023

1. Introduction

As one of the most discussed topics in the last decade, carbon dioxide has gained much attention due to its high environmental impact, endangering ecosystems and human society survival. CO_2 emissions have been raising annually for more than 100 years [1], contributing 70% of the overall GHG (greenhouse gas) concentration [2] and accelerating global warming [3], thus increasing extreme climate events such as hurricanes, floods and droughts [4]. As an IR-active molecule [5], CO_2 can trap radiation from sunlight and Earth, contributing to the problem of global warming [6]. From the work of the Intergovernmental Panel on Climate Change (IPCC), it is estimated that about 730 additional trees should be grown to contain the anthropogenic CO_2 emissions of one single person [7]. For this reason, science is making a great effort to find alternative ways to convert carbon dioxide into useful and valuable chemical products; however, the unfavourable thermodynamics of CO_2 make the activation difficult to achieve in normal conditions [8]. Indeed, high C-O bond energy and high stability raise the activation energy required to adsorb molecules onto the catalysts surface and trigger bond cleavage [9].

To address the thermodynamic problem of CO_2 activation, different approaches have been explored from the early 1970s; different methodologies have been studied so far, including different energy vectors such as electricity, heat and photons [10–12]. Moreover, different classes of catalysts, including metal oxides, perovskites, metal–organic frameworks (MOFs), MXenes, metal complexes and many others [13–16], are currently under investigation. From the large plethora of catalytic processes studied so far, the photocatalytic pathway has produced interesting results: using solar light as an energy vector greatly reduces the environmental impact of the process, as the energy source that drives the reaction is sustainable and readily available [17]. Moreover, alternative catalytic methods, such as thermocatalysis or electrocatalysis, often require high temperatures and pressures, thus resulting in harsher conditions and less sustainable processes.

The difficult C–O bond cleavage and the short charge lifetime before recombination in the catalyst inhibit the photoactivation of carbon dioxide, worsening the conversion rates at ambient temperature and pressure [17,18]. For these reasons, catalyst design is crucial in order to ensure that a high number of electrons are generated via visible light and effective e^-/h^+ separation. Achieving these characteristics would provide efficient CO_2 activation and reduction [19]; however, catalytic active sites with optimal carbon dioxide adsorption and desorption properties must be present on the catalyst surface. To overcome these challenges, catalyst design can exploit nanocomposites such as surface-decorated nanostructures; indeed, Localised Surface Plasmon Resonance (LSPR) can be employed, given nanostructuresstrong interaction with the supporting material. Metallic nanoparticles (NPs) possess enhanced light harvesting properties, better light absorption and better electron/hole pair formation [20]. Moreover, the presence of an intense electromagnetic field in the plasmonic nanoparticle surroundings drives electrons to migrate towards the surface of the material, thus increasing the number of possible active sites in the catalyst [21]. Hot electrons, generated through LSPR excitation and injected into the photocatalytic supporting material, also enhance its catalytic performance; indeed, their high kinetic energy can trigger the photo-reductive process of CO_2 when transferred to the conduction band (CB) of the semiconductor. Plasmonic nanoparticles also possess the interesting possibility to change their light absorption properties with size, shape, composition and support interaction variations, making them suitable candidates for CO_2 reduction catalysts [22,23]. Indeed, finding the best combination of physical properties that match the reaction requirements is crucial for achieving good performance and effectively converting carbon dioxide into useful chemical products.

A different class of materials may be employed in the photocatalytic activation of carbon dioxide. MOFs, as an example, are widely studied for their tuneable porous structures that could preferentially coordinate reactants towards favourable interaction [24], but they are difficult to synthesise. Several simple metal oxides have revealed promising activities towards the formation of various products, such as CO, CH_4 or methanol [25,26]. An interesting class of inorganic compounds is polycationic oxides, in particular perovskites; this class of catalysts is widely studied for different catalytic applications and possesses good tunability regarding numerous properties [27,28]. Indeed, compositional changes affect the material structural and electronic behaviour, thus making these mixed oxides interesting for a great number of applications, including photocatalysis [29]. In general, oxide-based simple perovskites are composed of an ABO_3 structure, with A and B being 12- and 6-coordinated cations, respectively; the correct selection of these cations allows them to develop active surface sites with respect to CO_2 reduction.

As one of the most studied materials for photocatalytic applications, $SrTiO_3$ could be a feasible choice for the photoreduction of CO_2, given its good stability in reducing atmospheres and in a wide range of temperatures [30]. However, its large band gap of 3.2 eV and its short charge lifetime makes it difficult to be used with visible light and for catalytic applications [31], as the catalyst–gas interaction process is relatively slow (>100 ps), while charge recombination occurs with shorter times (100 fs–100 ps) [32]; thus, optimisation of the material is crucial for slowing the recombination rate and shifting the absorption range towards a higher wavelength. Doping the B-site of the perovskite with an aliovalent cation results in band structure modification, introducing defect levels that narrow the band gap; additionally, defect centres can act as trap sites, extending charge lifetime [33,34]. Different kinds of metals can be taken into consideration for $SrTiO_3$ doping. As for its catalytic properties and enhanced electron conductivity, Cu doping seemed to be a feasible choice for the scope of this work [35,36]. Moreover, the formation of small and dispersed nanoparticles can beeasily achieved: it is the so-called exsolution [37,38], i.e., the formation of socketed nanoparticles on the surface of the oxide upon a thermal treatment in a reducing atmosphere. The exploitation of such a phenomenon can lead to the creation of a nanocomposite system composed of a supporting $SrTiO_3$-based material and surface-dispersed Cu nanoparticles (NPs); while the semiconductor possesses a good interaction with CO_2 and H_2, NPs can harvest light efficiently and provide hot electrons to the active site. Indeed, unlike other deposition methods, e.g., wet impregnation, the exsolved NPs interact more strongly with the supporting material and are characterised by smaller dimensions, higher dispersion and enhanced stability. Thus, an increased number of active sites and better charge injection properties in the perovskite are expected.

In this work, a plasmonic-assisted photocatalytic approach to carbon dioxide activation, using an economic, scalable and low-energy-consumption reactor, is discussed. A perovskite-based semiconducting material, $Sr_{0.9}Ti_{0.95}Cu_{0.05}O_3$, was successfully synthetised using a sol–gel method, while surface decoration with small metallic nanoparticles was achieved through exsolution. The starting perovskite was chosen to be A-site-deficient to facilitate exsolution under a hydrogen atmosphere. Exsolution was studied as a function of the reduction temperature up to 700 °C; indeed, the exsolution environment deeply affects the type and surface distribution of copper NPs and thus the photocatalytic behaviour. The sample compositions and treatment temperatures are summarised in Table 1. The prepared Cu NP/titanate nanocomposite catalysts were characterised from structural, electronic and functional points of view and compared with the corresponding undoped material ($Sr_{0.9}TiO_3$), showing promising activity towards CO_2 reduction.

Table 1. Summary of specific surface area, grain size and pore size of STO and STCuO samples.

Sample	Composition	SSA (m^2/g)	Grain Size (nm) [1]	Mean Pore Size (nm) [2]
STO	$Sr_{0.9}TiO_3$	5.9	74.6	25.5
STCuO	$Sr_{0.9}Ti_{0.95}Cu_{0.05}O_3$	2.5	94.6	17.4
STCuO-r300	$Sr_{0.9}Ti_{0.95}Cu_{0.05}O_3$	-	100.9	-
STCuO-r500	$Sr_{0.9}Ti_{0.95}Cu_{0.05}O_3$	-	80.5	-
STCuO-r700	$Sr_{0.9}Ti_{0.95}Cu_{0.05}O_3$	-	67.9	-

[1] Calculated via Scherrer's equation using 40.0° reflection. [2] Calculated using BJH analysis.

2. Results and Discussion

2.1. Structural and Morphological Properties

From a structural investigation of $Sr_{0.9}TiO_3$ (STO) and $Sr_{0.9}Ti_{0.95}Cu_{0.05}O_3$ (STCuO) (Figure 1) performed using X-ray diffraction, the pattern of the as-prepared reference STO sample reveals the presence of cubic strontium titanate as the main phase (JCPDS no. 35-0734) [39] with little impurities of TiO_2 in anatase form (JCPDS no. 21-1272) [40]. Thus, the perovskitic structure is expected to be less A-site-deficient, as observed in other perovskitic structures [41]. Likewise, the STCuO sample shows a main phase, identified as cubic strontium titanate; as in STO, the STCuO sample shows reflections related to the presence of titanium dioxide impurities [42]. Other small reflections can be ascribed to the formation of mixed oxides phases, including copper-containing impurities, that suggest the incomplete incorporation of Cu in the perovskitic structure. From Rietveld refinements of the two as-prepared perovskites, a unit cell parameter of 3.9 Å could be calculated, consistent with the data found in the literature [43]. To optimise the exsolution process, operando XRD analysis was performed from RT to 800 °C in a vacuum to ensure low-oxygen partial pressure, thus ensuring the required driving force to promote oxygen vacancies (OVs) formation and metal atom diffusion. Rietveld analysis on the diffraction patterns obtained from STO and STCuO revealed different behaviour in the unit cell expansion, as can be seen in Figure 2. Indeed, the STO sample shows a linear expansion when the temperature is increased, as observed in the literature; conversely, Cu-doped strontium titanate does not follow the same expansion rate, showing two different linear regions, one up to 400 °C and the other at higher temperatures. The first expansion process is slower in STCuO compared with STO, and it can be correlated with the partial exsolution of Cu from the oxide structure; indeed, the creation of a metal ion vacancy due to its diffusion causes unit cell shrinkage and a slowdown of its thermal expansion. At higher temperatures, the unit cell thermal expansion is not limited by the Cu diffusion, and the expansion rate becomes similar to the STO sample.

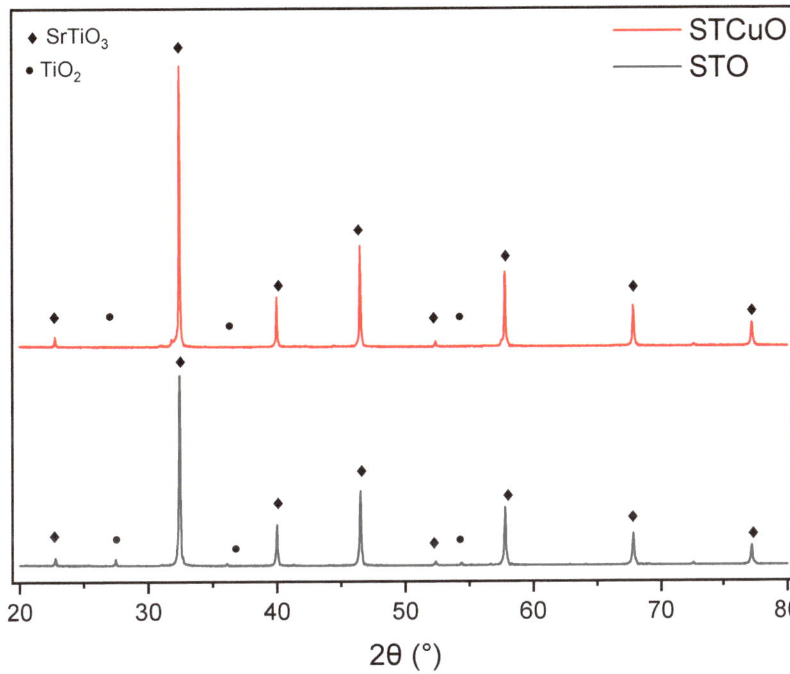

Figure 1. XRD patterns of STO and STCuO; the main perovskitic phase can be observed with squared dots, while the TiO_2 impurities are marked with circle dots.

Comparing the main peak positions of STCuO after the thermal treatment and reference STO [43], a slight shift towards lower angles can be observed; the aforementioned displacement can be related to a partial incorporation of Cu^{2+} ions into the perovskitic structure after the exsolution treatment, thus causing unit cell expansion as a consequence of the larger ionic radius of copper with respect to Ti^{4+} (Cu^{2+} = 0.73 Å > Ti^{4+} = 0.61 Å) [38].

Impurities due to mixed oxide reflections almost disappear after reduction (Figure 3). STCuO-r300 presents an XRD pattern similar to STCuO, confirming that the reduction treatment does not significantly affect the perovskitic structure. As found in the literature, a reducing atmosphere and adequate temperature trigger exsolution promote partial Cu diffusion from the first layers of the material and form Cu NPs on the surface [44]; however, exsolved metallic nanoparticles are generally small, making their detection hard in the XRD pattern. Samples STCuO-r500 and STCuO-r700 present a weak reflection at 43.5°, ascribed to metallic Cu. Atomic diffusion is a thermally activated phenomenon: simultaneously exsolved Cu cations can move more easily towards the surface. The reducing atmosphere allows the diffusion of elemental copper, while high temperatures promote particle aggregation; therefore, larger-dimension particles are expected at higher temperatures, significantly contributing to the diffraction pattern. When treating an STCuO sample at 700 °C, a new shift towards low angles can be observed (Figure S1), suggesting further expansion of the unit cell of the material; similar behaviour was described by Brennow et al. [45], who suggested the presence of Ti^{3+} ions after treating $SrTiO_3$ at a high temperature in reducing atmospheres. Thus, the cell expansion can be related to the reduction of Ti^{4+} ions to Ti^{3+}, and to the subsequent incorporation of a certain amount of Cu^{2+} ions in the structure. Indeed, copper ion diffusion in the bulk could be enhanced by the lower ionic radius mismatch between Cu^{2+} and Ti^{3+} with respect to Ti^{4+}, thus promoting the formation of nanoparticles on the surface of the material.

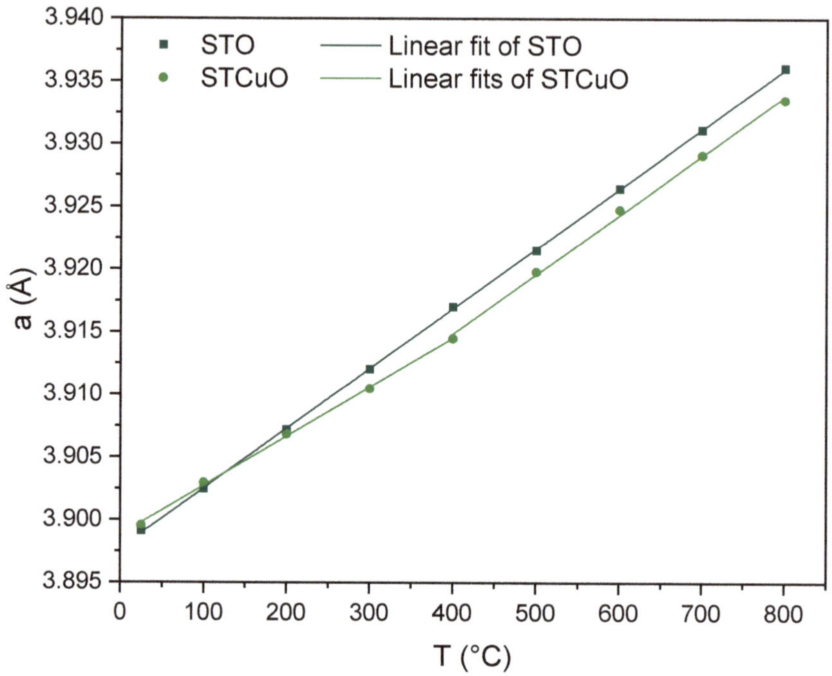

Figure 2. Cell parameter variation in STO and STCuO with temperature in low-oxygen partial pressure conditions. Linear cell expansion is observed for the undoped strontium titanate, while two regions can be distinguished, with two different expansion rates.

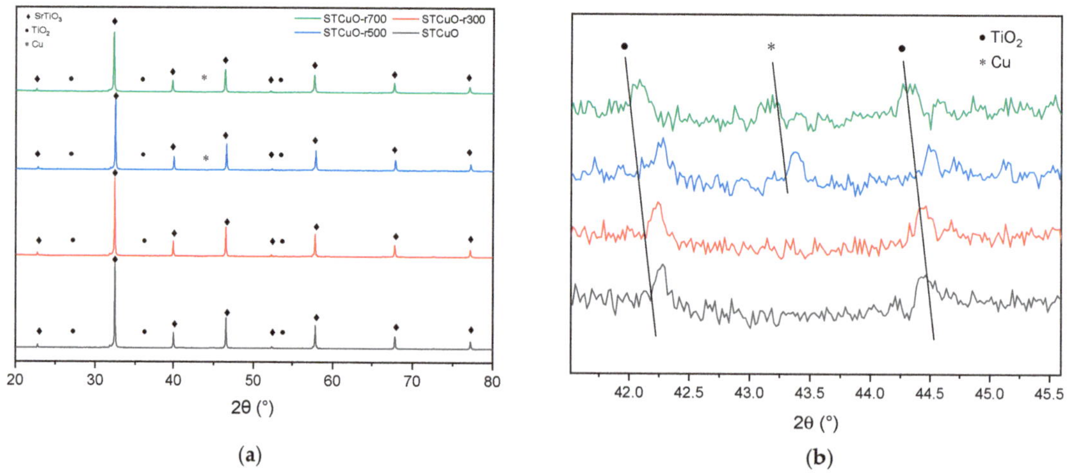

Figure 3. XRD pattern of as-prepared STCuO and of reduced STCuO at different temperatures (**a**); an enlargement in the metallic Cu region is presented (**b**).

From the BET analysis of N_2 adsorption–desorption isotherms, a type II isotherm together with an H3 hysteresis is observed for STO and for the Cu-doped counterpart (Figure S2); therefore, a macroporous material with little mesopores is expected from these observations. The specific surface area (SSA) of the strontium titanate is 5.9 $m^2 \cdot g^{-1}$, while the analysis of the doped perovskite results in a 2.5 $m^2 \cdot g^{-1}$ surface area, consistent with the literature data for a similar structure and synthetic procedure [46].

When comparing these results with the BJH pore distribution (Figure S3a,b), a higher density of mesopores, ranging from 2 to 20 $cm^3 \cdot g^{-1} \cdot nm^{-1}$, can be observed in the undoped sample with respect to STCuO; this evidence is consistent with the lower surface area and the more compact morphology revealed via SEM analysis in STCuO.

The atomic composition of the catalysts (Table 2) was determined using both energy-dispersive X-ray spectroscopy (EDS) and X-ray photoelectron spectroscopy (XPS) to gain information from different depths (with XPS being much more surface-sensitive). Therefore, compositional analysis and a comparison between sub-surface (EDS) and surface (XPS) measurements were performed, only considering metallic cations, as the EDS analysis on oxygen did not match the desired accuracy and reliability; the full XPS quantitative analyses are presented in Table S1. Composition-wise, the pure STO sample presents a slight excess of Sr in the first bulk and on the surface, probably also related to the formation of TiO_2 impurities during the synthetic process; comparing the EDS and the XPS compositions, a significant difference can be observed, consistent with the surface segregation of Sr. The presence of XPS peaks at 132.4 and 288.9 eV in the high-resolution spectra of the Sr 3d and C 1s regions, respectively, is symptomatic of Sr carbonate formation. In STCuO, the surface segregation of Sr is less evident, while the Ti content slightly changes between samples, generally showing a small deficiency with respect to the nominal composition.

Table 2. Elemental composition calculated from theoretical chemical formula; XPS and EDS data of STO, STCuO and STCuO reduced at different temperatures.

Element	Sample	Element Concentration		
		Nominal	EDS	XPS
Sr	STO	0.47	0.52	0.55
	STCuO	0.47	0.42	0.48
	STCuO-r300	0.47	0.41	0.47
	STCuO-r500	0.47	0.42	0.48
	STCuO-r700	0.47	0.47	0.47
Ti	STO	0.53	0.48	0.45
	STCuO	0.50	0.46	0.49
	STCuO-r300	0.50	0.46	0.49
	STCuO-r500	0.50	0.48	0.49
	STCuO-r700	0.50	0.46	0.50
Cu	STO	-	-	-
	STCuO	0.03	0.12	0.03
	STCuO-r300	0.03	0.13	0.04
	STCuO-r500	0.03	0.10	0.03
	STCuO-r700	0.03	0.07	0.03

Comparing sample composition between the as-prepared STCuO and the reduced samples, Cu content observed with EDS is higher with respect to the XPS signal; this could be related to the presence of Cu in the sub-surface of the catalyst. When reducing the doped perovskite, however, the EDS Cu content varies as a function of the reducing temperature: after the treatment at 300 °C, the Cu content is slightly increased, probably due to bulk Cu^{2+} ions that migrate towards the surface, a phenomenon already observed in perovskites [44], while at a higher temperature, the copper content decreases, approaching the nominal one. This behaviour could be ascribed as a new Cu incorporation process in the bulk due to the partial reduction of Ti^{4+}, as described above.

A comparison among the high-resolution XPS spectra of the various STCuO samples allows interesting insights; as mentioned before, the surface segregation of Sr is consistent with the formation of carbonate species on the surface. Besides the contribution due to carbonates ($3d_{5/2}$ around 132.4 eV), another doublet ($3d_{5/2}$ around 132.8 eV) can be ascribed to Sr–O bonding in the perovskite structure [47]. It is worth noting that no significant change in the peaks' position was found when comparing the fresh STCuO sample with the reduced materials, while a change in the carbonate/perovskite intensity ratio could be observed: increasing the reducing temperature, the carbonate-related signal decreased, thus showing that carbonates decompose and Sr is either embedded in the structure or can be converted to SrO.

As regards the Cu $2p_{3/2}$ region (Figure S6), the signal can be well fitted with multiple components; the main peak, at 934.5 eV for STCuO, is related to the presence of Cu^{2+} ions [48], while the second peak, at 943 eV, can be ascribed to the shake-up peak of Cu(II). When reducing the material, the position of the main component slightly changes, shifting to 935 eV in STCuO-r300 and then returning to its original position at 934.5 eV in STCuO-r700. No direct evidence of metallic Cu is present in the XPS spectra; however, the slight shift of the main peak observed in STCuO-r300 compared with STCuO can be ascribed to partial reduction of the Cu species [49,50].

Scanning electron microscopy images of STO (Figure 4) show a porous structure characterised by globular-shaped agglomerates with a high dimensional dispersion, ranging from 50 nm to 200 nm; fresh STCuO (Figure S4a) images indicate a similar morphology with a slightly more compact and less porous structure. Perovskite grains tend to merge after thermal treatment in a reducing atmosphere, reducing the porosity of the sample (Figure S4b–d).

(a) (b)

Figure 4. SEM images of STO (**a**) and STCuO (**b**). A different morphology can be observed: smaller grains appear on STO, while STCuO seems to possess a more compact structure.

To further confirm the presence of Cu nanostructures, TEM analysis was performed on the Cu-doped samples (Figure 5). The images revealed the same morphology, with dimensionally dispersed grains; on STCuO-r300, 3–5 nm particles were visible on the perovskitic agglomerates (Figure S5, circled). On the other samples, no visible signs of different nanostructures were clearly seen, therefore supporting the partial dissolution of the metallic structures at higher temperatures.

A two-step mechanism could explain the behaviour of STCuO: an initial exsolution step, occurring at low temperatures, involves Cu atoms in the bulk of the material diffusing towards the surface. In the second step, taking place at higher temperatures, the reducing atmosphere leads to the formation of Ti^{3+} ions and the subsequent diffusion of copper species towards the bulk. Moreover, the higher temperature reduction seems to enhance the growth of the nanoparticles, causing more dimensional dispersion and less stability of the system. As for the purpose of the study, the best reduction temperature was found to be the lowest possible, as the Cu nanoparticle formation was enhanced and the agglomeration or diffusion to the bulk steps were limited.

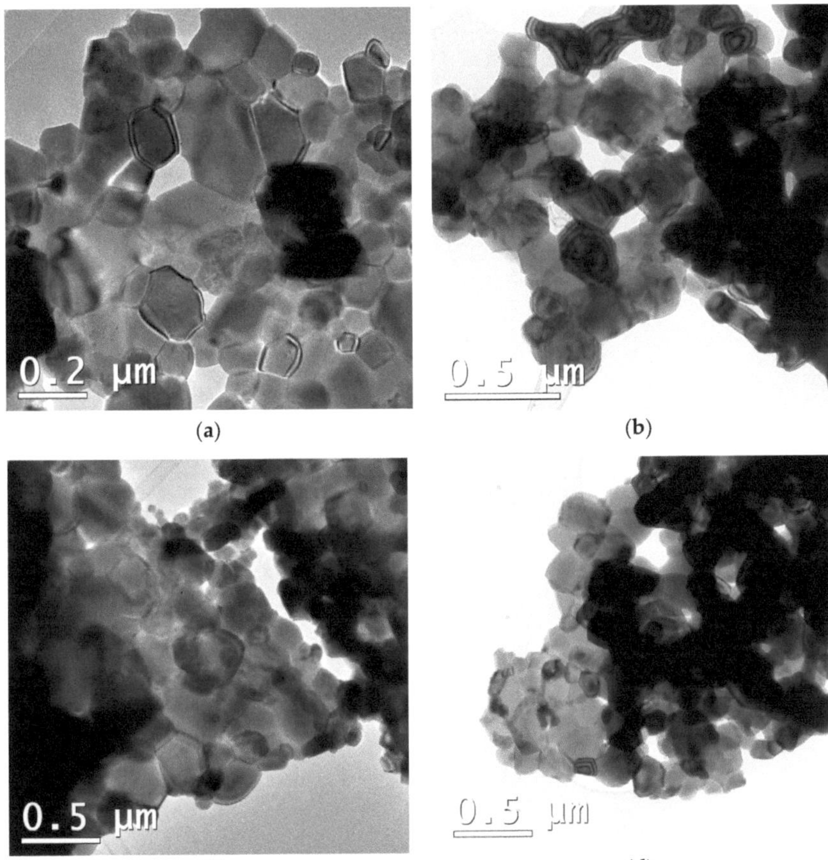

(a) (b) (c) (d)

Figure 5. TEM images of as-prepared STCuO (**a**) and reduced samples: STCuO-r300 (**b**), STCuO-r500 (**c**) and STCuO-r700 (**d**).

2.2. Functional Properties

2.2.1. Bulk Reducibility

To gain information about the bulk reducibility of STO and STCuO, temperature-programmed reduction was performed (Figure 6). As-prepared strontium titanate TPR resulted in the absence of significant peaks, thus suggesting that no reductions occurred; however, Ti^{4+} to Ti^{3+} conversion was expected between 400 °C and 500 °C [51]. The absence of signals in this region could be ascribed to the small quantity of Ti species that were reduced in the process, which contributed to the signal for a fraction smaller than the instrument detection limit.

As regards STCuO, an intense signal could be observed at around 250 °C, which can be assigned to Cu^{2+} ions reduction; three contributions were derived from the peak fit, as can be seen in Figure S7. Different interpretations of this signal are found in the literature [52,53], assigning the components to different reduction processes: the subsequent $Cu^{2+} \rightarrow Cu^{+}$ and $Cu^{+} \rightarrow Cu^{0}$ reactions, or the presence of Cu ions with different chemical surroundings. Comparing the TPR signal with the XRD pattern, both interpretations are feasible. A two-step partial reduction of shallow copper together with various reduction processes of Cu^{2+} species, present in grain boundaries, impurities and in the first layers, can occur, resulting in a reduction temperature shift to higher values with respect to pure CuO [54].

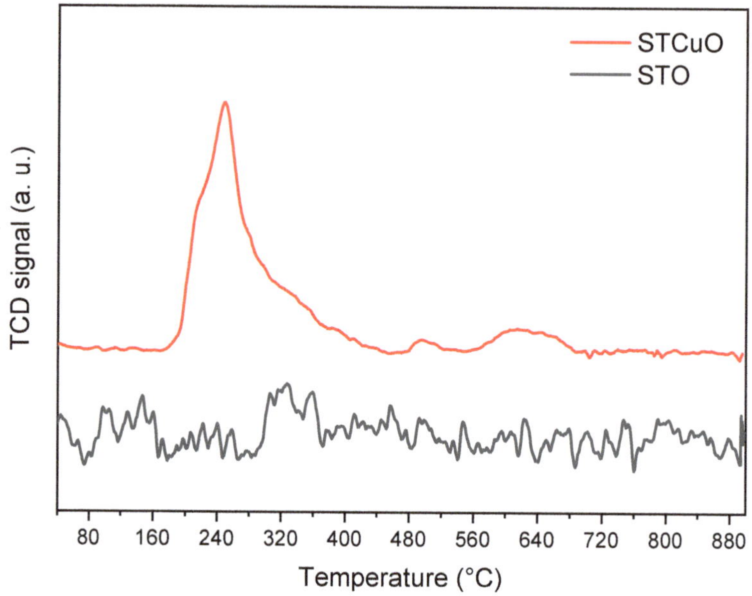

Figure 6. TPR signals of STO (black) and STCuO (red). No evident peaks were observed for undoped strontium titanate, while for STCuO, a set of low-temperature peaks, together with two higher temperature peaks, could be assigned to different reduction processes of Cu ions.

Two less intense signals are observed at around 500 °C and 620 °C; from the work of Gwóźdź et al. [54], these peaks could be ascribed to the reduction of Cu ions embedded in the material bulk, thus being more bound to the structure and therefore less reducible.

2.2.2. Electronic Properties

To further understand the properties of STCuO, diffuse reflectance UV-Vis spectra (DRS) were collected for STO and Cu-doped samples; subsequently, the Tauc plot was derived following Equations (2) and (3) described in Section 3.

From the Tauc plot (Figure 7) analysis, a band gap of 3.2 eV was found for the STO sample, which is in good agreement with the literature data. When doping the perovskite with Cu, the band gap decreases at 2.9 eV. From computational studies, a value of 2.2 eV [55] was expected; this suggests that the Cu atoms are only partially embedded in the perovskitic structure, thus further confirming the hypothesis previously made. The DRS spectra of reduced STCuO produce no significant changes in the band gap; this could be due to the extraction of Cu atoms from the sub-surface layers only, thus not significantly changing the perovskitic structure and Cu content in the bulk.

Comparing the two samples with respect to the minimum absorption wavelength, a redshift of 40 nm (from 385 nm to 425 nm) was found when inserting copper; this result indicates that better visible light absorption is expected from STCuO with respect to STO. The literature suggests that the insertion of copper into STO creates additional donor levels below the Fermi level, thus lowering the conduction band edge and narrowing the band gap [55]; this could be beneficial not only from a light absorption point of view, but also from a catalytic perspective: a higher energy valence band results in a better alignment with CO_2 reduction potentials [56]. To further confirm the position of the valence and the conduction band, studies are underway to investigate this more thoroughly.

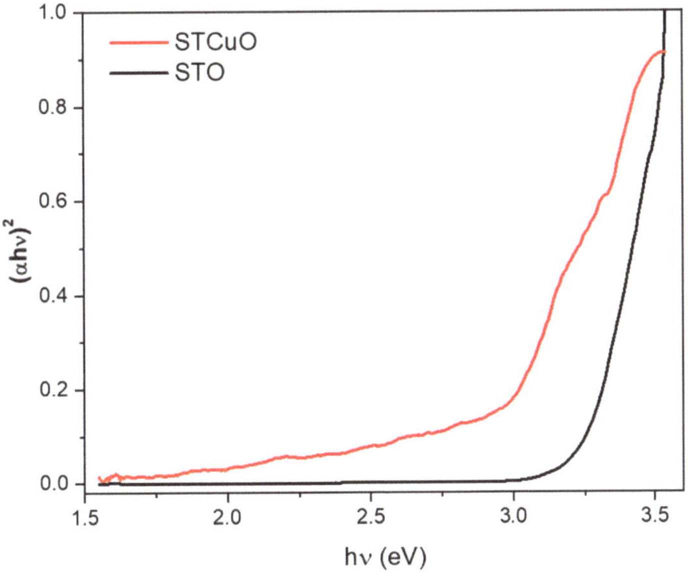

Figure 7. Tauc plot derived from DRS spectra of STO (black) and STCuO (red). A lower band gap can be calculated from the STCuO sample, thus confirming the Cu incorporation in the strontium titanate structure.

2.3. Catalytic Activity

Samples of STO and STCuO-r300 were tested for their ability to activate the CO_2 molecule with hydrogen in different conditions; a better understanding of which active sites are present in the materials and what type of interactions co-exist during the catalytic processes (Figure 8a) was expected. Additionally, the role of Cu nanoparticles was investigated to further evaluate the plasmonic resonance contribution to the reaction and confirm the results previously found [57].

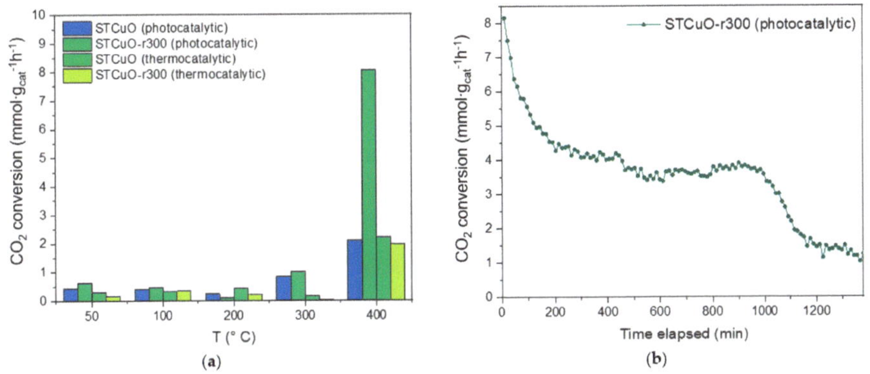

Figure 8. (**a**) Catalytic tests performed in dark and light conditions for STCuO and STCuO-r300 samples. Five temperatures (50, 100, 200, 300 and 400 °C) were chosen to analyse the behaviour of the materials in different thermodynamic conditions; (**b**) stability test conducted for 24 h in photo-assisted catalysis conditions with the STCuO-r300 sample.

The STO sample did not convert any significant amount of CO_2 during the reaction, either with only the thermal contribution or with the photo-assisted process. The absence of activity in the presence of a light source was caused by the high band gap, which limited the light absorption to the UV part of the spectrum, and by the low number of active sites in which the CO_2 molecule could effectively adsorb.

When adding Cu in the material structure, the catalytic tests in dark conditions showed a significant CO_2 conversion of about 2 mmol·$g_{cat}{}^{-1}$·h^{-1} at 400 °C; the conversion is ascribed to the presence of Cu sites that were able to activate the carbon dioxide molecule. Therefore, a significant contribution to the reaction is derived from the material structure and composition, meaning it is able to adsorb the reactant molecule and activate it during the reaction process. In the photo-assisted catalytic test, the presence of light further enhanced the onset shift, as 1 mmol·$g_{cat}{}^{-1}$·h^{-1} CO_2 was converted at 300 °C. Indeed, the more efficient light absorption in the visible region leads to higher charge carrier production with respect to the thermocatalytic experiments, thus increasing the probability of reaction with the adsorbed CO_2 [58].

After reduction, the STCuO-r300 sample behaviour was similar to that of the fresh STCuO sample in light-off conditions: at 400 °C, the conversion difference was minimal, thus allowing us to conclude that Cu nanoparticles do not play a particular role in CO_2 reduction under dark conditions. However, when the material was exposed to visible light, the results differed significantly to the fresh STCuO: CO_2 conversion could already be observed at 50 °C; moreover, a four-times-higher conversion was determined at 400 °C compared to the fresh STCuO sample or to the same material not exposed to light. The conversion at 50 °C could be ascribed to the Cu nanoparticles on the surface that acted as active sites for CO_2 reduction and were photo-catalytically active. As the copper nanoparticle number was limited by the low doping content and the exsolution of the metal atoms embedded in the perovskitic structure was only partial, the number of active sites was consequently low, thus limiting the reaction rate. Moreover, the presence of Cu NPs can lead to the activation of their plasmonic behaviour, thus further enhancing catalytic activity; indeed, low temperatures favour the LSPR effect, limiting charge recombination and thus prolonging their lifetime, as well as enhancing the probability to reach surface active sites. By increasing temperature, catalytic activity in the presence of light decreases; this suggests the intervention of side-processes in the material, as phononic or electronic scattering, that prevent the LSPR from effectively activating the CO_2 molecule. As the temperature rises (300 °C), photo-assisted STCuO-mediated catalysis can still be observed; thus, a different catalytic mechanism can be hypothesised: at a higher temperature, the transport properties of STCuO change [59], allowing the generated SPR hot electrons to be injected more efficiently into the supporting material and therefore activating a larger number of active sites, enhancing the catalytic performance of the system.

As regards the products obtained from the catalytic tests, CO formation was observed for all of the active catalysts, as can be seen in Figure S8; however, the amount of carbon monoxide in the overall mixture was <5.3 mmol·$g_{cat}{}^{-1}$·h^{-1}, that is, the maximum sensibility of the instrument. Thus, the selectivity of the catalysts towards CO was found to be <66%. Comparing the CO TCD signals of STCuO and STCuO-r300, a similar trend to CO_2 conversion was observed: the catalysts in dark conditions exhibited reaction onset at about 320 °C, while the photo-assisted measurements indicated a lower onset at about 250 °C. Additionally, STCuO-r300 showed a 2-fold enhancement in CO production in photo-assisted conversion, confirming the hypothesis made above: light can significantly contribute to c performance, activating the catalytic sites and triggering CO_2-RR.

To address the stability of the STCuO material in working conditions, the catalyst was put under photo-assisted reaction conditions for 24 h; from the graph in Figure 8b, an initial significative decrease in the catalytic performance can be observed, followed by a plateau region and a second decrease between 17 h and 20 h after the start of the reaction. The evidence of two different deactivation regions can be ascribed to the presence of two different species of active sites, thus further confirming the hypothesis made above. From the literature, it seems that the inactivation of Cu active sites may occur due to partial oxidation of the metal [60], thus causing one of the conversion decrease curves [58]; the other deactivation process could be related to the oxygen vacancies (OVs) that are filled during the process, as stated by Ma et al. [61]. Indeed, the presence of OVs in the supporting material is crucial for enhancing both catalytic and transport properties [62].

Comparing the performances of the catalysts with the state-of-the-art materials for CO_2 reduction, a significant difference can be spotted, showing lower performance of STCuO with respect to group VIII nanoparticles supported on Al_2O_3. Indeed, Ru, Rh and Ni nanostructures showed a much higher conversion rate, up to three orders of magnitude better than the present; indeed, a CO_2 reduction rate > 10 mol·${g_{met}}^{-1}$·h^{-1} was observed in batch conditions for Ru-based materials. In spite of this, the STCuO-r300 catalyst showed better catalytic activity than conventional oxide-based perovskites, resulting in a 10-times-higher CO_2 conversion rate [63]. In addition, STCuO materials do not contain noble metals and are synthesised using only aqueous solutions, and the CO_2-RR is conducted using a cost-effective reactor; thus, the results obtained are interesting for the increased sustainability and scalability of the overall process.

To summarise, the presence of Cu nanoparticles on the surface of the perovskite results in better catalytic activity, due to the plasmonic effect that induces the excitation of hot electrons that either trigger a catalytic reaction on the nanoparticle surface or migrate to the supporting material active sites, as found in similar systems [64]. Copper atoms diffusion towards the surface can result in new defect points in the strontium titanate structure, thus making them possible active sites for effective CO_2 adsorption, making the catalyst performance better with respect to the undoped perovskite. Plasmonic effects can be visible at low temperatures (50 °C), in which charge carrier recombination is lower and the electron lifetimes are prolonged; additionally, a further contribution can be found at higher temperatures (400 °C), when the transport properties of strontium titanate allow effective hot electron injection through the active sites.

3. Materials and Methods

3.1. Synthetic Procedure

$Sr_{0.9}TiO_3$ (STO) was synthetised by means of a sol–gel/citrate route method, starting from $SrCO_3$ (98% Sigma-Aldrich, Hamburg, Germany) and $C_{12}H_{28}O_4Ti$ (97% Sigma-Aldrich, Gillingham, UK); the precursors were dissolved in an aqueous solution of nitric acid (≥65% Sigma-Aldrich, Schnelldorf, Germany) containing citric acid monohydrate as a complexing agent (≥99.0% Sigma-Aldrich, Vienna, Austria) in a citric acid/sum of metal cations equal to 1.9 and then neutralised with NH_3 (32%, Sigma-Aldrich, Darmstadt, Germany). The solution was aged at 80 °C for 24 h to form a wet gel; subsequently, the temperature was risen to 400 °C to ignite the self-combustion and therefore eliminate the organic framework and nitrates. $Sr_{0.9}Ti_{0.9}Cu_{0.05}O_3$ (STCuO) was synthetised using the same procedure, adding CuO (≥99% Sigma-Aldrich, St. Louis, MO, USA) as a precursor to the mother solution. The powders obtained from the combustion process were grinded and subsequently calcined at 1000 °C for 12 h with a temperature ramp of 6 °C/min.

The calcined samples containing Cu were treated in a 5% H_2/Ar atmosphere at different temperatures (300 °C, 500 °C and 700 °C) for 2 h with a temperature ramp of 6 °C/min, according to the observed TPR signal, to promote exsolution and gain information about the formation of nanoparticles and the structural and morphological changes in the sample.

3.2. Materials' Characterisation

XRD measurements were performed using a BRUKER D8 ADVANCE diffractometer with Bragg–Brentano geometry and using Cu K_α radiation (λ = 0.1548 nm). Data were collected with 2θ ranging from 20° to 80° and with a rate of 0.02 °/step and 0.35 s/step, while 40 kV and 40 mA were set as working conditions. Operando XRD analysis was performed by heating the samples from RT to 800 °C at 10 °C/min and collecting the diffraction pattern every 100 °C after equilibrating the system for 10 min; a constant oxygen partial pressure of $p_{O2} = 10^{-7}$ atm was used in the chamber to ensure the driving force needed to trigger exsolution. Scanning electron microscopy (SEM) images were taken using a 5 kV (for STCuO) and 15 kV (for STO) electron beam and secondary electron detection was conducted using a Zeiss Supra 40VP; energy-dispersive X-ray spectroscopy (EDX) was carried out at 20 kV using the same instrument. Specific surface area (SSA) evaluation was possible by means of N_2 physisorption isotherm analysis using a Micromeritics Asap 2020 Plus. The samples were loaded in a quartz reactor and degassed at 200 °C for 16 h; subsequently, N_2–sorption curves were obtained at liquid nitrogen temperature (T = 77 K) to evaluate the SSA from the BET equation. Temperature-programmed reduction (TPR) measurements were performed using the Micromeritics AutoChem II 2920 coupled with a TCD detector and a quartz tube reactor loaded with 50 mg of the sample. TPR measures were carried out by gassing the samples with 50 mL/min He and subsequently heating from RT to 900 °C at 10 °C/min in a 5% H_2/Ar atmosphere. XPS measurements were performed using a Thermo Scientific ESCALAB QXi. Extended spectra of the samples were acquired assuming the BE of the Au $4f_{7/2}$ to be 84.0 eV with respect to the Fermi level. Extended XPS spectra were acquired using a monochromatic Al source, with 100 eV pass energy and 0.5 eV/step, while detailed spectra were acquired with 20 eV pass energy and 0.1 eV/step. Surface compositions were determined after a Shirley-type background subtraction and using the Thermo Scientific sensitivity factors. Elemental composition analysis for EDS was performed considering the total metal concentration as the reference, as seen in Equation (1), where %M is the metal concentration and M^{sgn} is the signal of the same metal:

$$\%M = M^{sgn} / \sum {M_i}^{sgn} \quad (1)$$

To gain information about the electronic structure of the samples, Diffuse Reflectance Spectroscopy (DRS) was performed using an Oxford FT-1000 fluorimeter coupled with an integrating sphere. The spectra were acquired from 300 nm to 850 nm at 1 nm/step and 0.5 s/step. The collected data were then processed using the Tauc plot method to obtain the sample band gap; a direct allowed electronic transition was assumed based on the literature data [55], resulting in the following equation:

$$(KM \cdot h\nu)^{1/2} = A \cdot (h\nu - E_g) \quad (2)$$

where KM is the absorbance calculated using the Kubelka–Munk equation from the reflectance (R) spectra:

$$KM = (1 - R)^2 \cdot (2R)^{-1} \quad (3)$$

Catalytic tests were performed by loading 50 mg of the sample in a quartz tube reactor with a 6 mm internal diameter; the powder was compressed between two quartz wool flocks to ensure gas diffusion through the catalyst. Light-on/light-off tests were performed at different temperatures to evaluate the photo-thermocatalytic response of the materials. All photocatalytic tests were performed using an 85 W LED visible lamp with 200 mW/cm^2 power density measured using a Thorlabs S405C power meter coupled with a PM101 interface. A 100 mL/min stoichiometric gas mixture of 8% H2, 2% CO_2 and 90% Ar as a carrier gas was used for all of the tests; the reaction products were then analysed through gas chromatography, using an Agilent 7890 GC. To prevent water injection in the GC, the gas mixture was cooled with a chiller; for this reason, condensable products such as CH_3OH could not be detected. Thus, only the CO_2 conversion was quantified and not the products obtained from the reaction.

4. Conclusions

To summarise, $Sr_{0.9}TiO_3$ and $Sr_{0.9}Ti_{0.95}Cu_{0.05}O_3$ were successfully synthesised using a sol–gel citrate route, with minor impurities of TiO_2 for STO and spurious phase for STCuO. A cubic $SrTiO_3$ phase was observed for STO, while a distortion of the lattice was detected for STCuO, probably due to the insertion of copper in the structure. Thermal treatments at different temperatures in 5% H_2/Ar pointed out the general stability of the perovskitic phase in the reducing conditions, while evidence of copper nanoparticles was spotted in STCuO-r300. The compositional analysis revealed a segregation of Cu in the fresh sample, which slightly increased at a low temperature and significantly decreased at higher reduction temperatures. Operando XRD analysis showed that the presence of Cu limits the cell thermal expansion below 400 °C, proving its diffusion outside of the perovskite cell. Copper insertion in the structure resulted in band gap narrowing and better visible light absorption; better valence band positioning was supposed, but more thorough studies have to be conducted to confirm this assumption. Comparing the catalytic performance of different samples, an increased CO_2 conversion rate of 8 mmol·g_{cat}^{-1}·h^{-1} was obtained with the reduced sample in photo-assisted conditions at 400 °C. A two-step reaction mechanism was proposed: firstly, plasmonic activation of the catalyst occurs, in which the Cu nanoparticles act as active sites and LSPR provides electrons for the reaction; secondly, LSPR-excited hot electrons are injected into the supporting material, activating the catalytic sites. To further understand the reaction process however, kinetics measurements are needed; specific in situ and operando techniques are crucial for observing the active sites evolution, the adsorption processes and the intermediates' formation, thus unravelling the reaction mechanism.

Supplementary Materials: The following supporting information can be downloaded at https://www.mdpi.com/article/10.3390/catal13101377/s1: Figure S1: XRD enlargement of Cu-doped samples; Figure S2: N_2 adsorption–desorption isotherms; Figure S3a: BJH adsorption curve; Figure S3b: BJH adsorption curve (enlargement); Figure S4a–d: SEM images of STCuO fresh and reduced samples; Figure S5: TEM enlargement of STCuO-r300; Figure S6: $Cu2p_{3/2}$ high-resolution XPS spectra for STCuO, STCuO-r300, STCuO-r500 and STCuO-r700; Figure S7: Peak fit of STCuO H_2-TPR; Figure S8: CO production curves; Table S1: XPS complete compositional analysis.

Author Contributions: Conceptualisation, L.R. and A.G.; methodology, L.R. and A.O.; validation, L.R., A.O. and J.C.; formal analysis, L.R., M.S., A.O. and J.C.; investigation, L.R., M.S., A.O. and J.C.; resources, A.G.; writing—original draft preparation, L.R.; writing—review and editing, A.O., J.C. and A.G.; visualisation, L.R., A.O. and J.C.; supervision, A.G.; project administration, L.R.; funding acquisition, A.G. All authors have read and agreed to the published version of the manuscript.

Funding: This research was supported by European Union HORIZON EUROPE under grant agreement number 101091534 KNOWSKITE-X.

Data Availability Statement: All data are available upon request to the main author: lorenzo.rizzato.1@phd.unipd.it.

Acknowledgments: The authors acknowledge the KNOWSKITE-X project for providing funding. The authors would like to thank Ilaria Fortunati for her support in Diffuse Reflectance Spectroscopy measurements. Special thanks go to Stefano Mercanzin, Lorenzo Dainese and Alberto Doimo from the University of Padova and Green Pool s. r. l. for the technical support with the reactor design and construction.

Conflicts of Interest: The authors declare no conflict of interest. The funders had no role in the design of the study; in the collection, analyses or interpretation of data; in the writing of the manuscript or in the decision to publish the results.

References

1. Etheridge, D.M.; Steele, L.P.; Langenfelds, R.L.; Francey, R.J.; Barnola, J.M.; Morgan, V.I. Natural and Anthropogenic Changes in Atmospheric CO_2 over the Last 1000 Years from Air in Antarctic Ice and Firn. *J. Geophys. Res. Atmos.* **1996**, *101*, 4115–4128. [CrossRef]
2. Yoro, K.O.; Daramola, M.O. CO_2 Emission Sources, Greenhouse Gases, and the Global Warming Effect. *Adv. Carbon Capture Methods Technol. Appl.* **2020**, 3–28. [CrossRef]
3. Cox, P.M.; Betts, R.A.; Jones, C.D.; Spall, S.A.; Totterdell, I.J. Acceleration of Global Warming Due to Carbon-Cycle Feedbacks in a Coupled Climate Model. *Nature* **2000**, *408*, 184–187. [CrossRef] [PubMed]
4. Perera, A.T.D.; Nik, V.M.; Chen, D.; Scartezzini, J.L.; Hong, T. Quantifying the Impacts of Climate Change and Extreme Climate Events on Energy Systems. *Nat. Energy* **2020**, *5*, 150–159. [CrossRef]
5. Taylor, J.H.; Benedict, W.S.; Strong, J. Infrared Spectra of H_2O and CO_2 at 500 °C. *J. Chem. Phys.* **1952**, *20*, 1884–1898. [CrossRef]
6. Anderson, T.R.; Hawkins, E.; Jones, P.D. CO_2, the Greenhouse Effect and Global Warming: From the Pioneering Work of Arrhenius and Callendar to Today's Earth System Models. *Endeavour* **2016**, *40*, 178–187. [CrossRef] [PubMed]

7. Shukla, P.R.; Skea, J.; Calvo Buendia, E.; Masson-Delmotte, V.; Pörtner, H.-O.; Roberts, D.C.; Zhai, P.; Slade, R.; Connors, S.; van Diemen, R.; et al. *IPCC, 2019: Climate Change and Land: An IPCC Special Report on Climate Change, Desertification, Land Degradation, Sustainable Land Management, Food Security, and Greenhouse Gas Fluxes in Terrestrial Ecosystems;* Cambridge University Press: Cambridge, UK, 2019.
8. Schneider, J.; Jia, H.; Muckerman, J.T.; Fujita, E. Thermodynamics and Kinetics of CO_2, CO, and H^+ Binding to the Metal Centre of CO_2 reductioncatalysts. *Chem. Soc. Rev.* **2012**, *41*, 2036–2051. [CrossRef]
9. Zheng, Y.; Zhang, W.; Li, Y.; Chen, J.; Yu, B.; Wang, J.; Zhang, L.; Zhang, J. Energy Related CO_2 Conversion and Utilization: Advanced Materials/Nanomaterials, Reaction Mechanisms and Technologies. *Nano Energy* **2017**, *40*, 512–539. [CrossRef]
10. Nahar, S.; Zain, M.F.M.; Kadhum, A.A.H.; Hasan, H.A.; Hasan, M.R. Advances in Photocatalytic CO_2 Reduction with Water: A Review. *Materials* **2017**, *10*, 629. [CrossRef]
11. Resasco, J.; Bell, A.T. Electrocatalytic CO_2 Reduction to Fuels: Progress and Opportunities. *Trends Chem.* **2020**, *2*, 825–836. [CrossRef]
12. Alam, M.I.; Cheula, R.; Moroni, G.; Nardi, L.; Maestri, M. Mechanistic and Multiscale Aspects of Thermo-Catalytic CO_2 Conversion to C 1 Products. *Catal. Sci. Technol.* **2021**, *11*, 6601–6629. [CrossRef]
13. Li, X.; Xiong, J.; Tang, Z.; He, W.; Wang, Y.; Wang, X.; Zhao, Z.; Wei, Y. Recent Progress in Metal Oxide-Based Photocatalysts for CO_2 Reduction to Solar Fuels: A Review. *Molecules* **2023**, *28*, 1653. [CrossRef] [PubMed]
14. Li, D.; Kassymova, M.; Cai, X.; Zang, S.Q.; Jiang, H.L. Photocatalytic CO_2 Reduction over Metal-Organic Framework-Based Materials. *Coord. Chem. Rev.* **2020**, *412*, 213262. [CrossRef]
15. Zeng, S.; Kar, P.; Thakur, U.K.; Shankar, K. A Review on Photocatalytic CO_2 Reduction Using Perovskite Oxide Nanomaterials. *Nanotechnology* **2018**, *29*, 052001. [CrossRef] [PubMed]
16. Nguyen, T.P.; Tuan Nguyen, D.M.; Tran, D.L.; Le, H.K.; Vo, D.V.N.; Lam, S.S.; Varma, R.S.; Shokouhimehr, M.; Nguyen, C.C.; Le, Q. Van MXenes: Applications in Electrocatalytic, Photocatalytic Hydrogen Evolution Reaction and CO_2 Reduction. *Mol. Catal.* **2020**, *486*, 110850. [CrossRef]
17. Qian, R.; Zong, H.; Schneider, J.; Zhou, G.; Zhao, T.; Li, Y.; Yang, J.; Bahnemann, D.W.; Pan, J.H. Charge Carrier Trapping, Recombination and Transfer during TiO_2 Photocatalysis: An Overview. *Catal. Today* **2019**, *335*, 78–90. [CrossRef]
18. Vu, N.-N.; Kaliaguine, S.; Do, T.-O.; Vu, N.; Kaliaguine, S.; Do, T. Critical Aspects and Recent Advances in Structural Engineering of Photocatalysts for Sunlight-Driven Photocatalytic Reduction of CO_2 into Fuels. *Adv. Funct. Mater.* **2019**, *29*, 1901825. [CrossRef]
19. Neaţu, Ş.; Maciá-Agulló, J.A.; Garcia, H. Solar Light Photocatalytic CO_2 Reduction: General Considerations and Selected Bench-Mark Photocatalysts. *Int. J. Mol. Sci.* **2014**, *15*, 5246–5262. [CrossRef] [PubMed]
20. Hou, W.; Cronin, S.B. A Review of Surface Plasmon Resonance-Enhanced Photocatalysis. *Adv. Funct. Mater.* **2013**, *23*, 1612–1619. [CrossRef]
21. Hutter, E.; Fendler, J.H. Exploitation of Localized Surface Plasmon Resonance. *Adv. Mater.* **2004**, *16*, 1685–1706. [CrossRef]
22. Chen, K.; Leong, E.S.P.; Rukavina, M.; Nagao, T.; Liu, Y.J.; Zheng, Y. Active Molecular Plasmonics: Tuning Surface Plasmon Resonances by Exploiting Molecular Dimensions. *Nanophotonics* **2015**, *4*, 186–197. [CrossRef]
23. Xu, G.; Tazawa, M.; Jin, P.; Nakao, S.; Yoshimura, K. Wavelength Tuning of Surface Plasmon Resonance Using Dielectric Layers on Silver Island Films. *Appl. Phys. Lett.* **2003**, *82*, 3811–3813. [CrossRef]
24. Klein, N.; Senkovska, I.; Gedrich, K.; Stoeck, U.; Henschel, A.; Mueller, U.; Kaskel, S. A Mesoporous Metal-Organic Framework. *Angew. Chem. Int. Ed.* **2009**, *48*, 9954–9957. [CrossRef] [PubMed]
25. Kohno, Y.; Tanaka, T.; Funabiki, T.; Yoshida, S. Photoreduction of CO_2 with H_2 over ZrO_2. A Study on Interaction of Hydrogen with Photoexcited CO_2. *Phys. Chem. Chem. Phys.* **2000**, *2*, 2635–2639. [CrossRef]
26. Inoue, T.; Fujishima, A.; Konishi, S.; Honda, K. Photoelectrocatalytic Reduction of Carbon Dioxide in Aqueous Suspensions of Semiconductor Powders. *Nature* **1979**, *277*, 637–638. [CrossRef]
27. Zhang, G.; Liu, G.; Wang, L.; Irvine, J.T.S. Inorganic Perovskite Photocatalysts for Solar Energy Utilization. *Chem. Soc. Rev.* **2016**, *45*, 5951–5984. [CrossRef]
28. Schanze, K.S.; Kamat, P.V.; Yang, P.; Bisquert, J. Progress in Perovskite Photocatalysis. *ACS Energy Lett.* **2020**, *5*, 2602–2604. [CrossRef]
29. Tanaka, H.; Misono, M. Advances in Designing Perovskite Catalysts. *Curr. Opin. Solid State Mater. Sci.* **2001**, *5*, 381–387. [CrossRef]
30. Luo, C.; Zhao, J.; Li, Y.; Zhao, W.; Zeng, Y.; Wang, C. Photocatalytic CO_2 Reduction over $SrTiO_3$: Correlation between Surface Structure and Activity. *Appl. Surf. Sci.* **2018**, *447*, 627–635. [CrossRef]
31. Cheng, C.; Long, R. Charge-Compensated Doping Extends Carrier Lifetimes in $SrTiO_3$ by Passivating Oxygen Vacancy Defects. *J. Phys. Chem. Lett.* **2021**, *12*, 12040–12047. [CrossRef]
32. Ghoussoub, M.; Xia, M.; Duchesne, P.N.; Segal, D.; Ozin, G. Principles of Photothermal Gas-Phase Heterogeneous CO_2 Catalysis. *Energy Env. Sci* **2019**, *12*, 1122–1142. [CrossRef]
33. Konta, R.; Ishii, T.; Kato, H.; Kudo, A. Photocatalytic Activities of Noble Metal Ion Doped $SrTiO_3$ under Visible Light Irradiation. *J. Phys. Chem. B* **2004**, *108*, 8992–8995. [CrossRef]
34. Xu, Y.; Liang, Y.; He, Q.; Xu, R.; Chen, D.; Xu, X.; Hu, H. Review of Doping $SrTiO_3$ for Photocatalytic Applications. *Bull. Mater. Sci.* **2022**, *46*, 6. [CrossRef]
35. Rahman, Q.I.; Ahmad, M.; Misra, S.K.; Lohani, M. Efficient Degradation of Methylene Blue Dye over Highly Reactive Cu Doped Strontium Titanate ($SrTiO_3$) Nanoparticles Photocatalyst under Visible Light. *J. Nanosci. Nanotechnol.* **2012**, *12*, 7181–7186. [CrossRef]

36. Gawande, M.B.; Goswami, A.; Felpin, F.X.; Asefa, T.; Huang, X.; Silva, R.; Zou, X.; Zboril, R.; Varma, R.S. Cu and Cu-Based Nanoparticles: Synthesis and Applications in Catalysis. *Chem. Rev.* **2016**, *116*, 3722–3811. [CrossRef]
37. Neagu, D.; Tsekouras, G.; Miller, D.N.; Ménard, H.; Irvine, J.T.S. In Situ Growth of Nanoparticles through Control of Non-Stoichiometry. *Nat. Chem.* **2013**, *5*, 916–923. [CrossRef]
38. Cavazzani, J.; Squizzato, E.; Brusamarello, E.; Glisenti, A. Exsolution in Ni-Doped Lanthanum Strontium Titanate: A Perovskite-Based Material for Anode Application in Ammonia-Fed Solid Oxide Fuel Cell. *Int. J. Hydrogen Energy* **2022**, *47*, 13921–13932. [CrossRef]
39. Yeshodamma, S.; Hareesh, K.; Nagabhushana, H.; Dwivedi, J.; Petwal, V.C.; Sunitha, D.V. Influence of Electron Beam Irradiation on Structural, Thermo and Photoluminescence Properties of $SrTiO_3$:Sm^{3+} Nanophosphor. *Optik* **2020**, *223*, 165483. [CrossRef]
40. Mohamed, I.M.A.; Dao, V.D.; Yasin, A.S.; Mousa, H.M.; Mohamed, H.O.; Choi, H.S.; Hassan, M.K.; Barakat, N.A.M. Nitrogen-Doped&SnO_2-Incoportaed TiO_2 Nanofibers as Novel and Effective Photoanode for Enhanced Efficiency Dye-Sensitized Solar Cells. *Chem. Eng. J.* **2016**, *304*, 48–60. [CrossRef]
41. Osti, A.; Rizzato, L.; Cavazzani, J.; Glisenti, A. Optimizing Citrate Combustion Synthesis of A-Site-Deficient La,Mn-Based Perovskites: Application for Catalytic CH_4 Combustion in Stoichiometric Conditions. *Catalysts* **2023**, *13*, 1177. [CrossRef]
42. Zhang, J.Y.; Boyd, I.W.; O'Sullivan, B.J.; Hurley, P.K.; Kelly, P.V.; Sénateur, J.P. Nanocrystalline TiO_2 Films Studied by Optical, XRD and FTIR Spectroscopy. *J. Non-Cryst. Solids* **2002**, *303*, 134–138. [CrossRef]
43. Wu, C.; Li, J.; Fan, Y.; Xing, J.; Gu, H.; Zhou, Z.; Lu, X.; Zhang, Q.; Wang, L.; Jiang, W. The Effect of Reduced Graphene Oxide on Microstructure and Thermoelectric Properties of Nb-Doped A-Site-Deficient $SrTiO_3$ Ceramics. *J. Alloy. Compd.* **2019**, *786*, 884–893. [CrossRef]
44. Zhang, Z.; Chen, R.; Wang, L.; Chen, X.; Ding, J.; Zhang, J.; Wan, H.; Guan, G. Synergistic Effect of Cu^{2+} and Cu^{+} in $SrTiO_3$ Nanofibers Promotes the Photocatalytic Reduction of CO_2 to Methanol. *Appl. Surf. Sci.* **2023**, *609*, 155297. [CrossRef]
45. Blennow, P.; Hagen, A.; Hansen, K.K.; Wallenberg, L.R.; Mogensen, M. Defect and Electrical Transport Properties of Nb-Doped $SrTiO_3$. *Solid State Ion.* **2008**, *179*, 2047–2058. [CrossRef]
46. Zhang, W.; Du, L.; Bi, F.; He, H. A Novel $SrTiO_3$/HZSM-5 Photocatalyst Prepared by Sol–Gel Method. *Mater. Lett.* **2015**, *157*, 103–105. [CrossRef]
47. Divya, A.; Mathavan, T.; Harish, S.; Archana, J.; Benial, A.M.F.; Hayakawa, Y.; Navaneethan, M. Synthesis and Characterization of Branchlet-like $SrCO_3$ Nanorods Using Triethylamine as a Capping Agent by Wet Chemical Method. *Appl. Surf. Sci.* **2019**, *487*, 1271–1278. [CrossRef]
48. Li, B.; Hong, J.; Ai, Y.; Hu, Y.; Shen, Z.; Li, S.; Zou, Y.; Zhang, S.; Wang, X.; Zhao, G.; et al. Visible-near-Infrared-Light-Driven Selective Oxidation of Alcohols over Nanostructured Cu Doped $SrTiO_3$ in Water under Mild Condition. *J. Catal.* **2021**, *399*, 142–149. [CrossRef]
49. Ghodselahi, T.; Vesaghi, M.A.; Shafiekhani, A.; Baghizadeh, A.; Lameii, M. XPS Study of the Cu@Cu_2O Core-Shell Nanoparticles. *Appl. Surf. Sci.* **2008**, *255*, 2730–2734. [CrossRef]
50. Gaudin, P.; Fioux, P.; Dorge, S.; Nouali, H.; Vierling, M.; Fiani, E.; Molière, M.; Brilhac, J.F.; Patarin, J. Formation and Role of Cu^{+} Species on Highly Dispersed CuO/SBA-15 Mesoporous Materials for SO_x Removal: An XPS Study. *Fuel Process. Technol.* **2016**, *153*, 129–136. [CrossRef]
51. Bai, L.; Polo-Garzon, F.; Bao, Z.; Luo, S.; Moskowitz, B.M.; Tian, H.; Wu, Z. Impact of Surface Composition of $SrTiO_3$ Catalysts for Oxidative Coupling of Methane. *ChemCatChem* **2019**, *11*, 2107–2117. [CrossRef]
52. López-Suárez, F.E.; Parres-Esclapez, S.; Bueno-López, A.; Illán-Gómez, M.J.; Ura, B.; Trawczynski, J. Role of Surface and Lattice Copper Species in Copper-Containing (Mg/Sr)TiO_3 Perovskite Catalysts for Soot Combustion. *Appl. Catal. B* **2009**, *93*, 82–89. [CrossRef]
53. Rodriguez, J.A.; Kim, J.Y.; Hanson, J.C.; Pérez, M.; Frenkel, A.I. Reduction of CuO in H_2: In Situ Time-Resolved XRD Studies. *Catal. Lett.* **2003**, *85*, 247–254. [CrossRef]
54. Gwóźdź, P.; Łącz, A.; Mizera, A.; Drożdż, E. Some Aspects of Cu Incorporation into $SrTiO_3$ Structure. *J. Therm. Anal. Calorim.* **2022**, *147*, 9949–9958. [CrossRef]
55. Rizwan, M.; Ali, A.; Usman, Z.; Khalid, N.R.; Jin, H.B.; Cao, C.B. Structural, Electronic and Optical Properties of Copper-Doped $SrTiO_3$ Perovskite: A DFT Study. *Phys. B Condens. Matter* **2019**, *552*, 52–57. [CrossRef]
56. Huang, H.; Song, H.; Kou, J.; Lu, C.; Ye, J. Atomic-Level Insights into Surface Engineering of Semiconductors for Photocatalytic CO_2 Reduction. *J. Energy Chem.* **2022**, *67*, 309–341. [CrossRef]
57. Huang, H.B.; Yu, K.; Zhang, N.; Xu, J.Y.; Yu, X.T.; Liu, H.X.; Cao, H.L.; Lü, J.; Cao, R. Localized Surface Plasmon Resonance Enhanced Visible-Light-Driven CO_2 Photoreduction in Cu Nanoparticle Loaded ZnInS Solid Solutions. *Nanoscale* **2020**, *12*, 15169–15174. [CrossRef]
58. Li, Z.; Zhu, G.; Zhang, W.; Zhu, L.; Cao, B.; Gao, J.; Shi, X.; Huang, Y.; Liu, P.; Hojamberdiev, M. Dual-Functional Copper (Cu0/Cu2+)-Modified SrTiO3-δ Nanosheets with Enhanced Photothermal Catalytic Performance for CO_2 Reduction and H_2 Evolution. *Chem. Eng. J.* **2023**, *452*, 139378. [CrossRef]
59. Shang, D.S.; Sun, J.R.; Shi, L.; Wang, J.; Wang, Z.H.; Shen, B.G. Electronic Transport and Colossal Electroresistance in $SrTiO_3$:Nb -Based Schottky Junctions. *Appl. Phys. Lett.* **2009**, *94*, 052105. [CrossRef]
60. Kang, Q.; Wang, T.; Li, P.; Liu, L.; Chang, K.; Li, M.; Ye, J. Photocatalytic Reduction of Carbon Dioxide by Hydrous Hydrazine over Au–Cu Alloy Nanoparticles Supported on $SrTiO_3$/TiO_2 Coaxial Nanotube Arrays. *Angew. Chem.* **2015**, *127*, 855–859. [CrossRef]

61. Ma, H.; Yang, W.; Tang, H.; Pan, Y.; Li, W.; Fang, R.; Shen, Y.; Dong, F. Enhance the Stability of Oxygen Vacancies in $SrTiO_3$ via Metallic Ag Modification for Efficient and Durable Photocatalytic NO Abatement. *J. Hazard. Mater.* **2023**, *452*, 131269. [CrossRef]
62. Rusevich, L.L.; Tyunina, M.; Kotomin, E.A.; Nepomniashchaia, N.; Dejneka, A. The Electronic Properties of $SrTiO_{3-\delta}$ with Oxygen Vacancies or Substitutions. *Sci. Rep.* **2021**, *11*, 23341. [CrossRef] [PubMed]
63. Teh, Y.W.; Chee, M.K.T.; Kong, X.Y.; Yong, S.T.; Chai, S.P. An Insight into Perovskite-Based Photocatalysts for Artificial Photosynthesis. *Sustain. Energy Fuels* **2020**, *4*, 973–984. [CrossRef]
64. Luo, S.; Ren, X.; Lin, H.; Song, H.; Ye, J. Plasmonic Photothermal Catalysis for Solar-to-Fuel Conversion: Current Status and Prospects. *Chem. Sci.* **2021**, *12*, 5701–5719. [CrossRef] [PubMed]

catalysts

MDPI

Article

The Role of Carbon Nanotube Deposit in Catalytic Activity of FeO_X-Based PECVD Thin Films Tested in RWGS Reaction

Bartosz Panek [1], Hanna Kierzkowska-Pawlak [1], Paweł Uznański [2], Stefan Nagy [3], Veronika Nagy-Trembošová [3] and Jacek Tyczkowski [1,*]

1 Department of Molecular Engineering, Faculty of Process and Environmental Engineering, Lodz University of Technology, Wolczanska 213, 93-005 Lodz, Poland; bartosz.panek.1@p.lodz.pl (B.P.); hanna.kierzkowska-pawlak@p.lodz.pl (H.K.-P.)
2 Centre of Molecular and Macromolecular Studies, Polish Academy of Sciences, Sienkiewicza 112, 90-363 Lodz, Poland; pawel.uznanski@cbmm.lodz.pl
3 Institute of Materials and Machine Mechanics, Slovak Academy of Sciences, Dúbravská cesta 9, 845-11 Bratislava, Slovakia; nagy.stefan@savba.sk (S.N.); veronika.trembosova@savba.sk (V.N.-T.)
* Correspondence: jacek.tyczkowski@p.lodz.pl

Abstract: While the reverse water-gas shift (RWGS) reaction holds great promise as a method of converting CO_2 to CO and subsequently into valuable fuels, achieving its commercial viability requires the development of highly efficient, selective, durable, and low-cost catalysts. Recently, thin-film nanocatalysts produced through plasma deposition (PECVD) have garnered significant attention in this domain. Among them, FeOx-based catalytic films deposited using $Fe(CO)_5$ as a precursor, under reduced pressure (4–5 Pa) and a 13.56 MHz glow discharge, have demonstrated particular interest. Our study shows that by appropriately tuning the parameters of the plasma deposition process, it is feasible to generate nanocatalyst films exhibiting exceptional CO_2 conversion (38% at 673 K) and CO selectivity (97%). Moreover, the study has revealed the formation of a carbon deposit containing carbon nanotubes (CNTs) during the RWGS reaction, significantly increasing the catalytic activity of the films. Through an analysis involving X-ray photoelectron spectroscopy (XPS), X-ray diffraction (XRD), and electron microscopy techniques (SEM and HRTEM), we have determined that CNTs not only serve as carriers for highly catalytically active Fe nanoparticles but also create nanoscale heterojunctions (*p-n*) with Fe_2O_3 nanoparticles, thereby enhancing their catalytic effect. This paper attempts to elucidate the differences and changes in the surface structure of FeOx-based films dictating the catalytic activity, which stems from both the conditions of plasma deposition and the environmental impact during the catalytic process.

Keywords: thin-film nanocatalysts; plasma deposition; RWGS reaction; CO_2 hydrogenation; iron nanoparticles; iron oxides; nanoscale heterojunctions

Citation: Panek, B.; Kierzkowska-Pawlak, H.; Uznański, P.; Nagy, S.; Nagy-Trembošová, V.; Tyczkowski, J. The Role of Carbon Nanotube Deposit in Catalytic Activity of FeO_X-Based PECVD Thin Films Tested in RWGS Reaction. *Catalysts* **2023**, *13*, 1302. https://doi.org/10.3390/catal13091302

Academic Editors: Leonarda Liotta, Narendra Kumar and Konstantin Ivanov Hadjiivanov

Received: 19 August 2023
Revised: 9 September 2023
Accepted: 14 September 2023
Published: 17 September 2023

1. Introduction

Undoubtedly, thin-film nanocatalysts produced through cold plasma deposition (PECVD) have been attracting increasing interest due to their unique properties and wide range of applications in various catalytic structured reactors [1–4]. Building on our previous research on FeOx-based PECVD thin-film catalysts tested for the CO_2 hydrogenation process [5,6], our current focus is investigating the impact of carbon deposits on their catalytic activity, which turned out to be formed as carbon nanotubes during this process.

Beyond the imperative to significantly reduce atmospheric CO_2 levels, this substrate offers a valuable source of pristine carbon that can be utilized for large-scale production of various important chemicals, including liquid fuels. It is, therefore, not surprising that substantial efforts have recently been dedicated to developing existing and exploring new approaches for CO_2 utilization. Among them, the reverse water-gas shift (RWGS) reaction is particularly interesting [7,8]:

$$CO_2 + H_2 \rightarrow CO + H_2O \qquad \Delta H^0{}_{298} = 42.1 \text{ kJ/mol} \tag{1}$$

This mildly endothermic and competitive reaction with CO_2 methanation (Sabatier reaction) holds great potential and is considered an essential intermediate stage in various key CO_2 hydrogenation reactions [9], and it also serves as a route to produce syngas compatible with existing industrial infrastructure [10]. However, although the RWGS reaction shows immense promise, achieving commercial viability and efficiency still necessitates the development of highly efficient, selective, durable, and low-cost catalysts [11,12].

A lot of attention in this area is devoted to iron-based catalysts, which have been intensively studied for a long time, not only in the RWGS reaction [13] but also in the Fischer-Tropsch (FT) synthesis [14,15] and the water-gas shift (WGS) reaction [16]. However, in all these cases, there is still an intense debate about the mechanisms of these processes and the role that the individual phases of iron-based catalysts, such as oxides, carbides, and elemental metal (Fe), play in them [17,18]. Furthermore, the problem is complicated by the complex transformation between these phases, which occurs during the catalytic processes. Nevertheless, based on current findings, iron oxides are frequently considered to be notably active phases of the RWGS reaction, where oxygen vacancies act as catalytic sites for CO_2 activation and CO release [19]. On the other hand, there are also reports indicating metallic iron nanoparticles as the active phase in this reaction [20].

Another challenge in finding suitable catalysts for the RWGS process is the already well-known accumulation of carbonaceous deposits (also called coke) on their surfaces during catalytic reactions [21,22]. In the vast majority of cases, such a deposit deactivates the catalyst by blocking pores of various sizes or directly blocking active sites. In the RWGS process in the relatively lower temperature range (<873 K), two reactions of carbon deposit formation should be considered [23]:

$$2CO \rightarrow C + CO_2, \tag{2}$$

$$CO + H_2 \rightarrow C + H_2O \tag{3}$$

The first is the Boudouard reaction, often attributed to the formation of carbon deposits on iron oxides in the presence of carbon monoxide [24]. In the second case, it is assumed that the initial step is the adsorption of CO and H_2 on the catalyst surface. H_2 molecules directly dissociate, and hydrogen atoms can assist in the dissociation of CO and the formation of, among other species, a carbon deposit [25].

The nature of carbonaceous deposits formed on the catalyst depends on the catalyst's composition, textural properties, structure, feed composition, and reaction conditions; they can take various forms, such as amorphous and graphitic carbon islands on the catalyst surface, encapsulating layers, carbon filaments, whiskers, nanofibers, and even creating chemical connections as carbides [26]. A special place among them is occupied by carbon nanotubes (CNTs), which, unlike most other forms of carbon deposit, are able to enhance catalytic activity [27].

The ability to form carbon nanotubes on fine Fe particles or its compounds (like iron oxides or carbides) as effective catalysts for this process was established quite a long time ago in the reactions of thermal decomposition of carbon monoxide or hydrocarbons [28], before the concept of carbon nanotubes was known. They were called filamentous carbons at that time. Interestingly, the formation of carbon nanotubes with the participation of magnetite (Fe_3O_4) was also found in interplanetary space by studying micrometeorites. It is assumed that carbon was formed there in the catalytic reaction of CO disproportionation (Equation (2)), i.e., analogous to the one observed during the processes of Fischer-Tropsch and RWGS (Equation (1)) [29].

In some cases, studies of the catalytic properties of nanohybrids formed from FeOx and CNTs revealed better catalytic activity than that of pure FeOx. This has been shown, for example, in the case of Fe_2O_3/CNT (then reduced to Fe/CNT, where Fe nanoparticles were inside or outside of carbon nanotubes) tested in the Fischer-Tropsch reaction [30], Fe_3O_4/CNT nanohybrids used in catalytic ozonation [31], or α-Fe_2O_3/CNT in photo-catalytic oxidation [32]. It should be emphasized, however, that all these nanohybrids were produced in a planned manner and were not the result of uncontrolled carbon deposit formation.

Although the explanation of the effect of CNTs on the improvement of catalytic activity is still debatable, some interesting suggestions point to the possible influence of electronic interaction between the particles of the catalytic material and the walls of the CNTs, which may entail the transfer of electrons and, consequently, facilitate the activation of the reaction substrates [33].

In this work, we also demonstrate the improving effect of the deposition of carbon nanotubes formed during the RWGS reaction (Equation (1)) on the catalytic activity of the FeOx-based PECVD thin-film catalyst in this reaction. The paper attempts to explain the mechanism of this phenomenon.

2. Results and Discussion

2.1. General Characteristics of Catalytic Films

The studies were carried out on catalytic FeOx-based thin films produced by cold plasma deposition (PECVD) from $Fe(CO)_5$ as a precursor. Preliminary catalytic tests performed for the CO_2 hydrogenation showed that the catalytic activity of the films was significantly affected by the parameters of the plasma deposition process and, above all, by the partial pressure of the precursor, which led to the distinction of two basic groups of these films—with low and high catalytic activity. Representatives for each of these groups were selected for further analysis, namely two nanocatalysts produced at different partial pressures of $Fe(CO)_5$: 0.1 Pa (low activity) and 0.3 Pa (high activity), while maintaining other parameters constant, such as discharge power P = 80 W, partial pressure of argon (carrier gas) p_{Ar} = 4.0 Pa, and deposition temperature T = 298 K. Before the catalytic tests, the produced films were calcined at 673 K in an argon atmosphere. In the next part of the work, the films produced at the precursor partial pressure of 0.1 Pa will be called 0.1FeOx, and those obtained at the pressure of 0.3 Pa will be called 0.3FeOx. Their thicknesses were about 200 nm and 650 nm for 0.1FeOx and 0.3FeOx, respectively.

Investigations of the basic elemental composition of the film surfaces performed by XPS spectroscopy (Table 1) exhibit a significant similarity between both types of fresh films (as prepared, before their use in the catalytic tests). A deeper look at the molecular structure, however, revealed quite significant differences. Figure 1a,b shows the core level of the Fe 3p spectrum for both types of films. The spectra were numerically deconvoluted, and the calculated bands were assigned to appropriate oxidation states of Fe [6,34]. On this basis, it can be concluded that while metallic iron (Fe^0) is present in the 0.3FeOx film, it is completely absent in the 0.1FeOx film. In turn, the presence of Fe^{2+} and Fe^{3+} in both films indicates the possibility of the existence of iron oxides, such as FeO, Fe_2O_3, and Fe_3O_4. However, in the case of 0.1FeOx, Fe_2O_3 (Fe^{3+}) is dominant on the surface, while all three of these oxides may be present on the surface of 0.3FeOx, although the Fe^{3+}/Fe^{2+} content ratio = 2.02 may indicate Fe_3O_4 as the main component.

Table 1. Elemental atomic composition of the fresh and spent 0.1FeOx and 0.3FeOx films.

Catalyst →	0.1FeOx Fresh	0.1FeOx Spent	0.3FeOx Fresh	0.3FeOx Spent	0.3FeOx Spent (CNTs Partially Removed)
Element ↓			**Content [at%]**		
Fe	23 ± 2	24.0 ± 0.5	25.0 ±0.4	5.0 ± 0.3	10 ± 1
O	45 ± 2	49.7 ±0.4	48 ± 1	11 ± 1	23 ± 1
C	32.0 ± 0.1	26.3 ± 0.8	27 ± 2	84 ± 1	67 ± 2

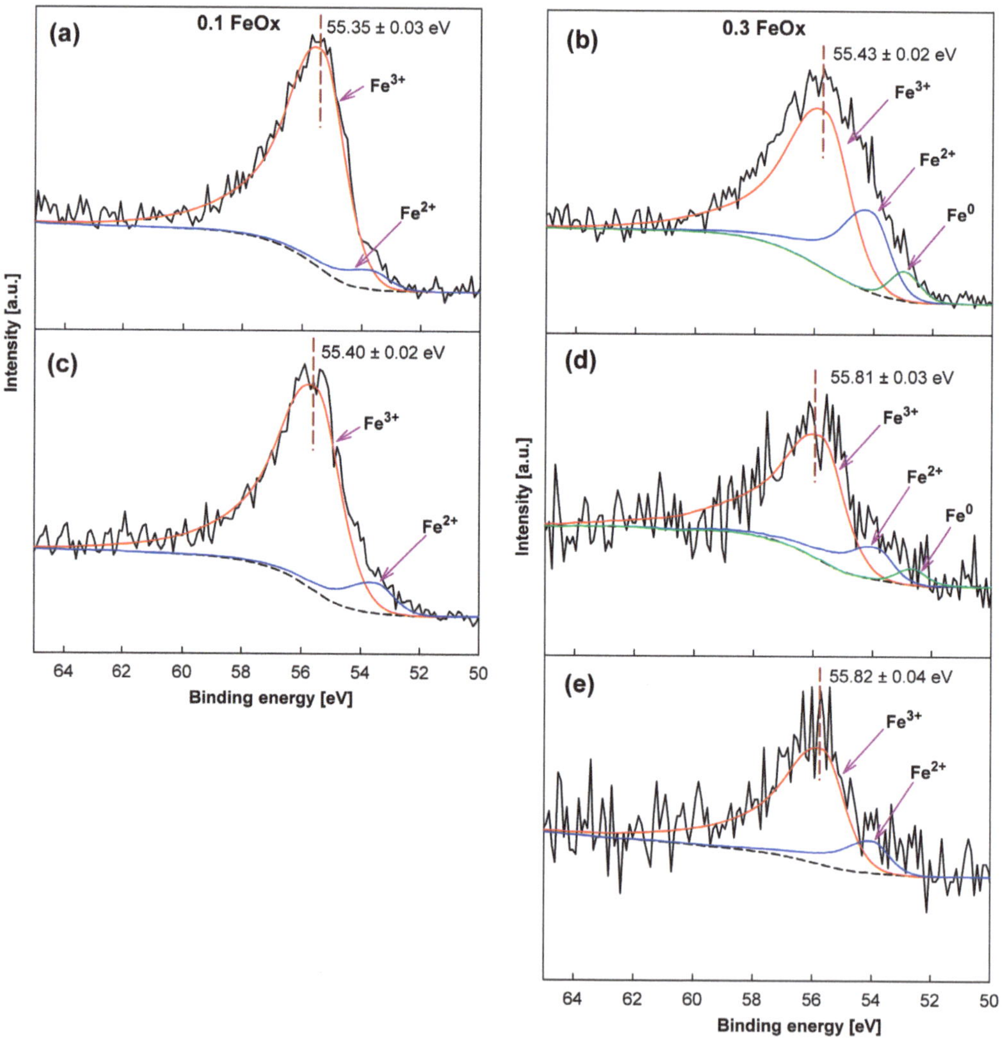

Figure 1. Typical XPS Fe 3p spectra for: (**a**,**c**) 0.1FeOx film; (**b**,**d**,**e**) 0.3FeOx film. (**a**,**b**) Fresh (as-prepared) films; (**c**,**d**) Spent films; (**e**) 0.3FeOx spent film after removing part of the moss-like carbon deposit.

To supplement the above studies, XRD measurements were performed. Raw XRD spectra (Supplement Figure S1) in both films reveal the likely presence of an amorphous fraction while also containing clear nanocrystalline structures. Figure 2 shows "normalized" XRD patterns by interpreting individual diffraction peaks based on the ICDD PDF database [35], using the following ICDD-PDF cards: FeO (card no. 01-089-0687), Fe_3O_4 (card no. 01-072-2303), and Fe (card no. 04-002-3692). However, it should be remembered that the XRD results refer to the entire thickness of the film, while XPS covers only its very thin surface layer (5–10 nm), the structure of which may differ from the bulk structure. This is the case with 0.1FeOx, where XRD (Figure 2a) indicates the presence of Fe_3O_4 (approx. 25 nm) and FeO (approx. 10 nm) nanocrystallites in the film, while XPS on the film surface basically shows only the presence of iron at the oxidation state (Fe^{3+}). However, a common result for both measurement techniques is the complete lack of metallic iron (Fe^0) in the 0.1FeOx film. In turn, in the 0.3FeOx film, the XRD pattern (Figure 2b) indicates the presence of metallic iron nanocrystallites with a cubic structure of bcc (approx. 23 nm) and Fe_3O_4 (approx. 25 nm), confirming in this case compliance with the XPS result.

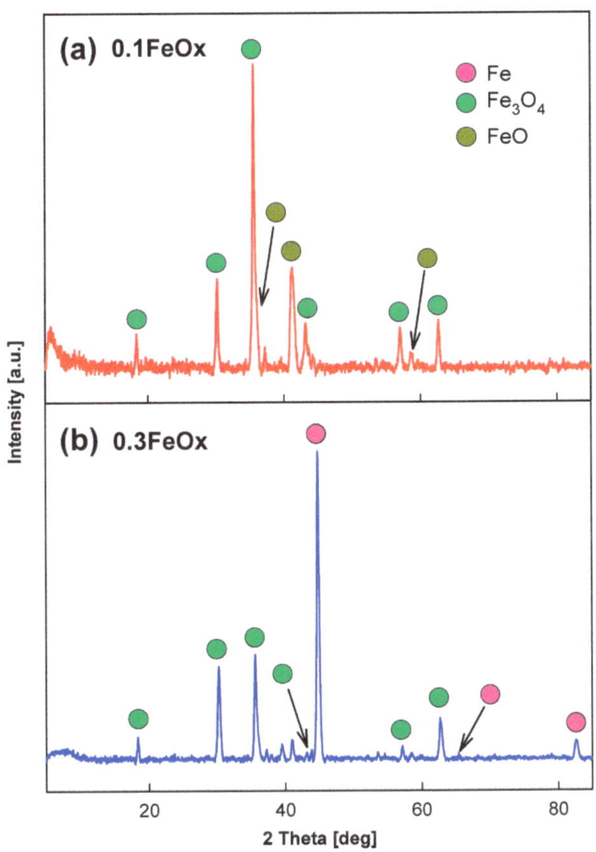

Figure 2. XRD patterns with interpreted diffraction peaks for: (**a**) 0.1FeOx film; (**b**) 0.3FeOx film.

2.2. Catalytic Activity

The catalytic tests were carried out for the CO_2 hydrogenation process (in the reaction mixture, $H_2/CO_2 = 4:1$). Figure 3 shows the dependencies of CO_2 conversion and selectivity to CO and CH_4 as a function of the catalytic process temperature for the two types of nanocatalysts tested.

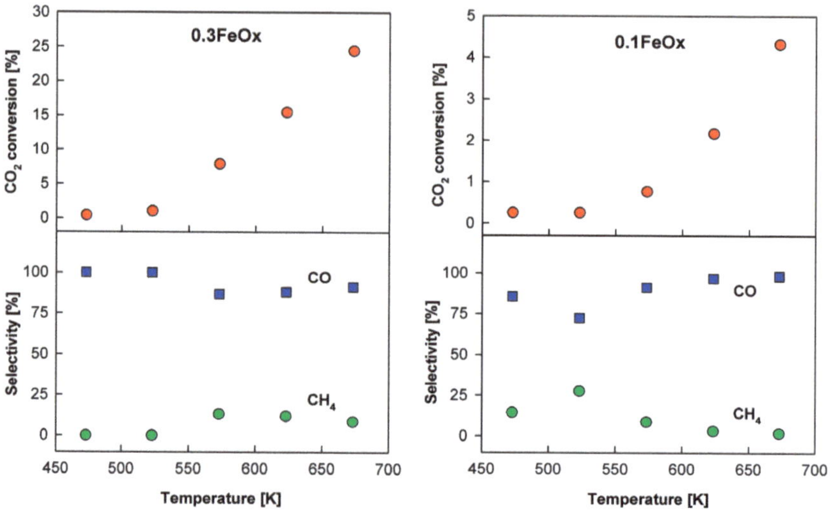

Figure 3. CO_2 conversion and selectivity to CO and CH_4 as a function of temperature for 0.3FeOx and 0.1FeOx films.

As can be seen, the CO_2 hydrogenation reaction proceeds much more efficiently on the 0.3FeOx than on the 0.1FeOx film. The next step in the analysis of the catalytic process was to determine the conversion as a function of time (time-on-stream). Figure 4 shows such dependencies for both nanocatalysts after they reach (in the reaction mixture) the temperature of 673 K. In the case of 0.3FeOx, a clear increase in CO_2 conversion is visible over time, until reaching a high stable value after about 4–6 h (Test A), close to the equilibrium value (Supplement Material Figure S2 which under the reaction conditions is 41.5%. Meanwhile, for the 0.1FeOx film, such an induction period is not observed (Test B), and the conversion remains at a very low level.

Interestingly, the end of the catalytic process carried out on 0.3FeOx after 20 h (Test A), cooling the catalyst to room temperature in the reaction mixture, leaving it in an air atmosphere for 12 h, and then carrying out another catalytic test in the same way as the first test, reveals a similar increase in conversion over time after reaching the temperature of 673 K (Test C). In the next, third, analogous test, we obtain a similar, as in the previous, course of the CO_2 conversion curve versus time (Test D). This type of behavior indicates evident periodic changes in the molecular structure of the 0.3FeOx nanocatalyst, which may result from changes in temperature as well as the environment in which the catalyst is located.

To obtain more complete information about changes in the structure of 0.3FeOx, two further catalytic tests were carried out. In the first of them, the catalyst, after the previously performed test (Test A) and prepared as for Test C, was heated from room temperature to 673 K, but not in the reaction mixture and only in a helium atmosphere. After 2 h of keeping the catalyst at the final temperature, the reaction mixture (H_2/CO_2 = 4:1) was introduced. The time dependence of the CO_2 conversion is shown in Figure 4 (Test E). The observed period of the conversion increase is much longer in this case (approx. 8–10 h) than in the tests where the catalyst was heated to 673 K immediately in the reaction mixture (Tests A, C, and D). This difference in the induction period is indicated by the dotted lines in Figure 4.

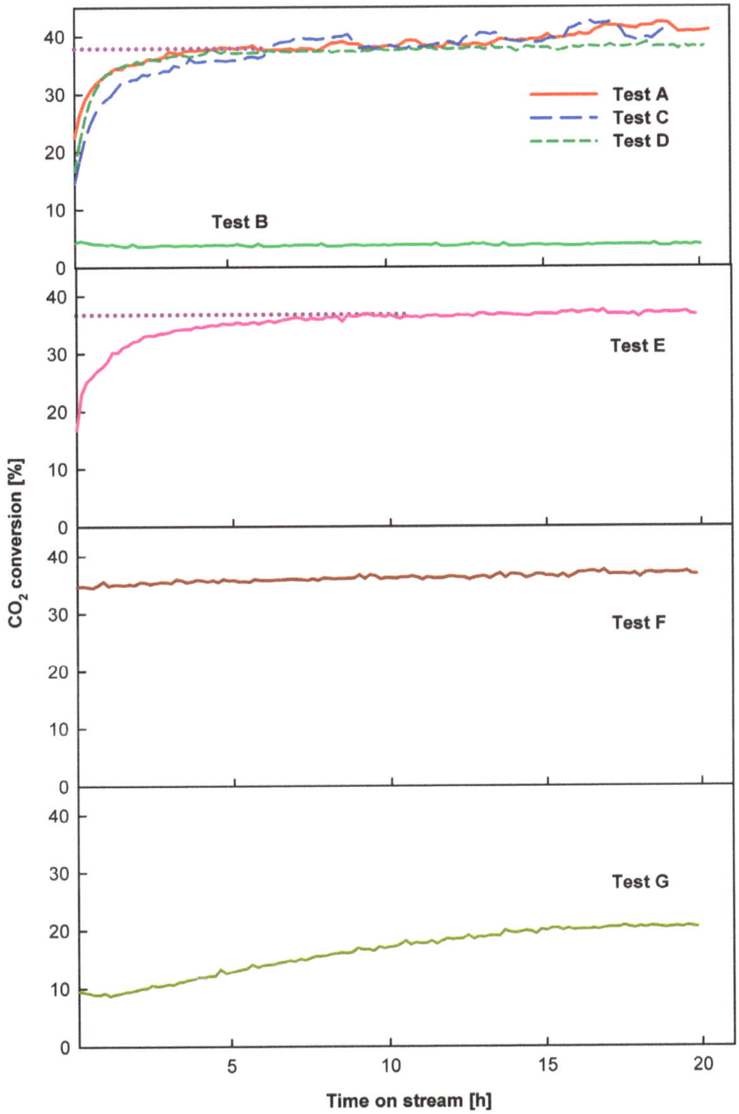

Figure 4. CO_2 conversion at 673 K as a function of time-on-stream: (Tests A, C, D) Successive runs for 0.3FeOx film starting in reaction mixture (H_2/CO_2 = 4:1) from room temperature; (Test B) Run for 0.1FeOx film starting in reaction mixture (H_2/CO_2 = 4:1) from room temperature; (Test E) Run for the 0.3FeOx film after Test A and heated from room temperature to 673 K in He; (Test F) Run for 0.3FeOx film after Test A and heated from room temperature to 673 K in H_2; (Test G) Run for 0.3FeOx film after Test A and removing part of the moss-like carbon deposit. (More details in the text.).

The second test was carried out analogously to the first one, except that the heating and holding at a temperature of 673 K for 2 h were carried out in a hydrogen atmosphere. The curve (Test F) in Figure 4 illustrates the dependence of the CO_2 conversion on time, which in this case is stable and reaches a high value from the beginning (as in Tests A, C, and D). The tests performed indicate that the reducing environment (hydrogen) plays an important role in the catalytic activity of 0.3FeOx. The reduced structure of the catalyst is oxidized upon contact with air and needs to be reduced again to achieve a stabilized CO_2 conversion value.

An interesting observation was provided by studies performed using scanning electron microscopy (SEM). Figure 5 shows SEM images of the surface of the 0.1FeOx and 0.3FeOx as-prepared and spent catalysts (after Tests B and A in Figure 4, respectively). In the case of 0.3FeOx, unlike 0.1FeOx, a fine moss-like deposit appears on the catalyst surface after the catalytic test and persists during subsequent tests. However, this deposit can be removed (at least partially) by blowing it off with compressed air. A stream of synthetic air at a pressure of 0.35 MPa for 3 min was used. The elimination of the deposit causes a drastic, irreversible change in the catalytic activity of 0.3FeOx, which is shown by the curve of CO_2 conversion versus time at 673 K (Test G, carried out in the same way as Test A). Performing further catalytic tests gives a very similar result. SEM observations also do not reveal the presence of a moss-like deposit in these cases.

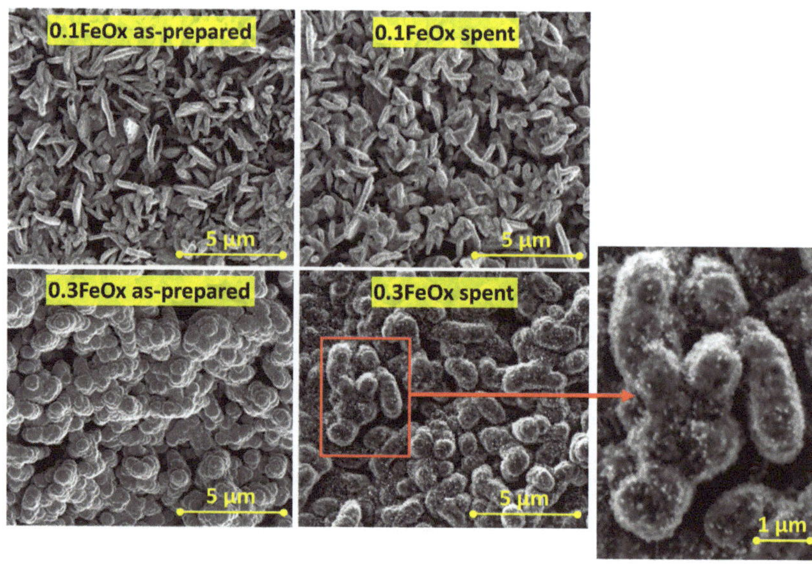

Figure 5. SEM images of 0.1FeOx and 0.3FeOx films before (as-prepared) and after 20 h (spent) of the RWGS reaction. The moss-like deposit is clearly visible on the surface of the spent 0.3FeOx nanocatalyst.

The composition of the mixture consisting of substrates (CO_2 and H_2) and products (CO, H_2O, and a small amount of CH_4) suggests with high probability that the emerging deposit is made of carbon. This is confirmed by XPS investigations carried out on 0.1FeOx and 0.3FeOx films after catalytic Tests B and A, respectively. Table 1 shows the elemental compositions of the surface of these films, which, compared with the composition of the films before the catalytic process, indicates a drastic increase in the carbon content on the 0.3FeOx film and practically no change in the composition for the 0.1FeOx film. The core-level analysis of the Fe 3p spectrum also shows no significant differences for the 0.1FeOx film before and after the catalytic test (Figure 1a,c). Similarly, no significant differences in the Fe 3p spectrum, apart from signal reduction and noise increase caused by a significant increase in carbon content, are visible for the 0.3FeOx film before and after the catalytic test (Figure 1b,d). Changes, however, occur after removing the superficial moss-like layer of carbon deposit with compressed air (as was achieved before the catalytic Test G), where we do not now record the presence of metallic iron (Figure 1e). Analysis of the elemental composition, however, indicates that a significant part of the carbon deposit still remains on the surface (Table 1).

Attention should also be paid to the location of the maximum of the Fe^{3+} band, which in the 0.3FeOx film is shifted by about 0.38 eV towards the higher binding energy after the catalytic process, while in the 0.1FeOx film, it remains practically unchanged. This effect will be discussed in Section 2.4.

2.3. HRTEM Analysis

The finding of the formation of a carbon deposit on the 0.3FeOx films in the catalytic process and its moss-like structure shown in the SEM micrographs (Figure 5) suggests the possibility of building this deposit from carbon nanotubes. To delve deeper into the structure of the deposit, studies were carried out using high-resolution transmission electron microscopy (HRTEM).

Indeed, HRTEM micrographs of the moss-like structure taken from the surface of the 0.3FeOx film confirmed the presence of multi-walled carbon nanotubes (CNTs) formed during the CO_2 hydrogenation process (Figure 6).

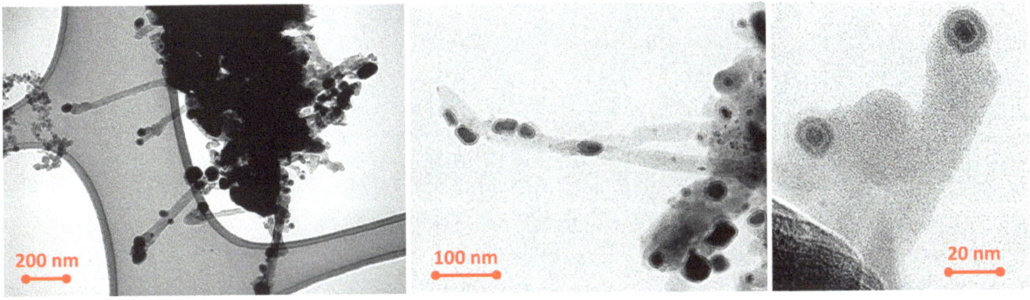

Figure 6. Examples of HRTEM micrographs of the moss-like deposit taken from the surface of the spent 0.3FeOx nanocatalyst.

As can be seen, the nanotubes are accompanied by nanoparticles with an average size of about 20–35 nm, the location of which proves their participation in the formation of CNTs and indicates the tip-growth mechanism. Nano-EDX measurements allowed for a more precise determination of the nature of the nanoparticles. Figure 7 shows the atomic intensity maps of C, Fe, and O for a typical nanoparticle located on top of the nanotube. It is easy to see that the nanoparticles are composed of an iron-rich core and an oxygen-rich shell (about 5 nm thick). A more detailed analysis of these results based on the atomic intensity profile across the diameter of the nanoparticle (Supplement Figure S3) indicates that the core is composed of pure iron, while the shell is composed of iron oxide, most likely Fe_2O_3.

The above conclusions are confirmed by the analysis of the crystallographic structure of the nanoparticles (Figure 8). The clearly visible interplanar spacing in the core of the nanoparticle is 0.211 nm, which is attributed to the plane (111) in the Fe cubic (fcc) structure. In contrast to the core, the shell shows a much more amorphous structure. However, small, ordered areas composed of several planes can be found there. The interplanar spacing determined for them is 0.274 nm, which can be attributed to the plane (104) in the rhombohedral Fe_2O_3, thus the previously suggested shell structure.

Crystallographic structure analysis was also performed for CNTs. Figure 9 shows an exemplary micrograph of a nanotube along with a fast Fourier transformation (FFT) diffraction pattern that reveals two main typical reflections corresponding to crystallographic planes (002) and (100), characteristic of CNTs [36].

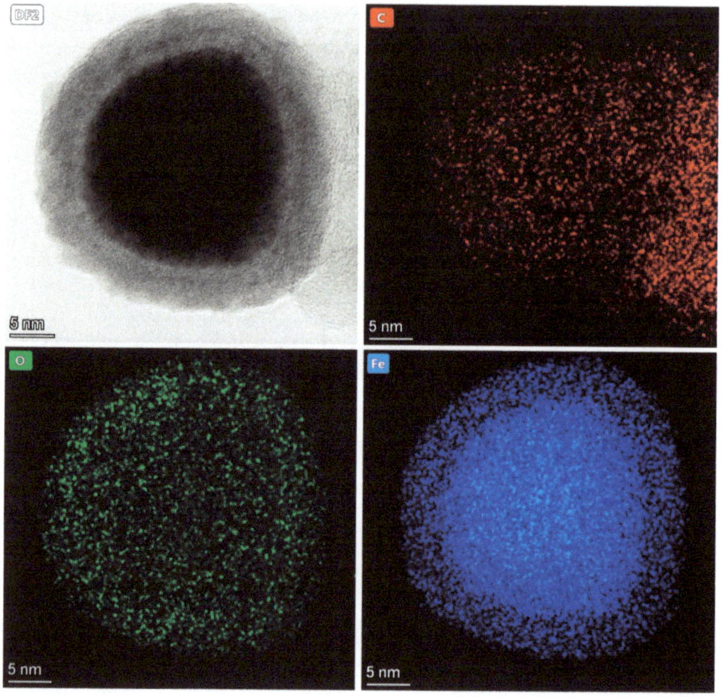

Figure 7. DF2 HRTEM micrograph of a typical Fe-based nanoparticle (spent 0.3FeOx) and its atomic intensity maps for carbon (C), oxygen (O) and iron (Fe) determined from nano-EDX measurements.

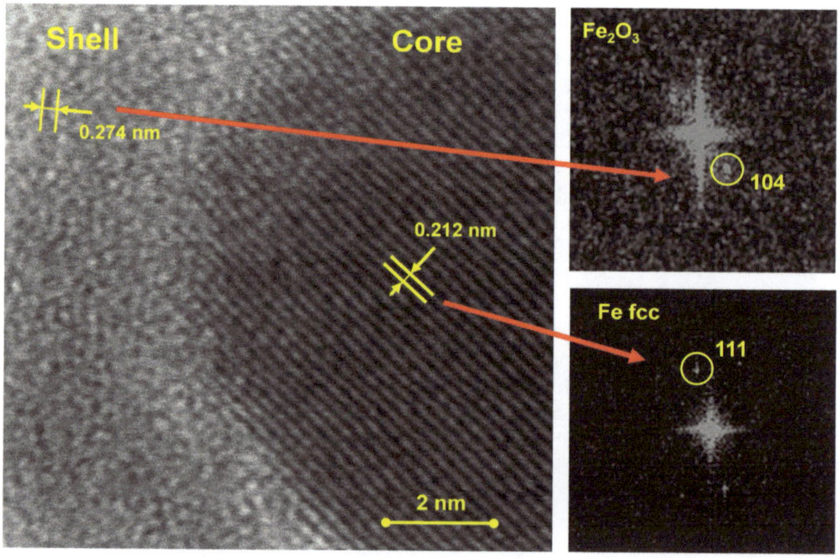

Figure 8. ABF HRTEM micrograph and the FFT diffraction patterns analyzing core and shell structures for a Fe-based nanoparticle.

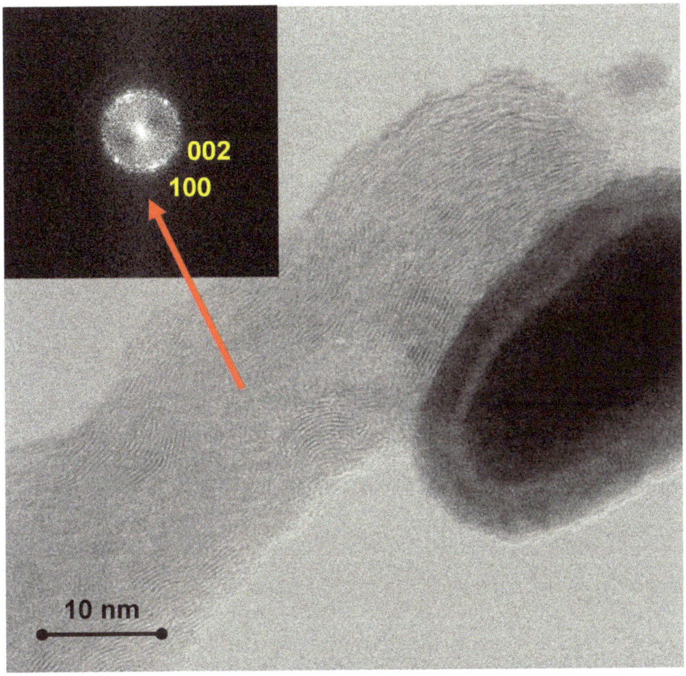

Figure 9. DF2 HRTEM micrograph of a typical Fe-based nanoparticle and its atomic intensity maps for carbon (C), oxygen (O) and iron (Fe) determined from nano-EDX measurements.

2.4. What Is the Role of Carbon Nanotubes in the Catalytic Activity?—A Proposed Explanation

There are two key findings to build upon in starting this discussion. Firstly, the much better catalytic activity of 0.3FeOx films, on which the CNT deposit is formed, than of 0.1FeOx films, on which such a deposit is not formed (Figure 4, tests A, B, C, and D). Secondly, the significant decrease in catalytic activity after removing part of the CNT deposit with associated iron nanoparticles from the surface of the 0.3FeOx films, although still higher than for the 0.1FeOx films. Both of these results clearly indicate the positive role of the CNT deposit in the RWGS reaction on thin-film FeOx-based nanocatalysts deposited by PECVD.

Comparing the CO_2 conversion for fresh 0.3FeOx and 0.1FeOx films (Figure 4, Test A and Test B, respectively) and the molecular structure of their surfaces (Figure 1a,b), it can be assumed that a much higher conversion value for the 0.3FeOx films is associated with the presence of metallic iron nanoparticles, which are the active phase on the surface of such a film. As a result of the CO_2 hydrogenation process, CO is produced in an amount enabling the formation of a carbon deposit (as a result of the reactions presented by Equation (2) or Equation (3)), which is deposited in the form of CNTs using Fe nanoparticles, widely known as the archetypal material that catalyzes the efficient growth of CNTs [37,38]. Growing nanotubes, in accordance with the tip-growth mechanism, lift up Fe nanoparticles, where they continue to act catalytically in the CO_2 hydrogenation process. It is worth noting that, although a difference was observed in the crystallographic structure of Fe nanoparticles between those carried by nanotubes, for which the structure of fcc was determined in HRTEM studies (Section 2.3), and those present in the fresh film, whose structure was defined by XRD to be bcc (Section 2.1), it is not known at the present stage of research whether this affects the catalytic activity. It is also difficult to determine whether the change in the structure results from changes occurring during the transfer of Fe nanoparticles by CNTs in the catalytic process, which is not completely excluded [39], or whether it results from the difference between the surface and bulk structure of the catalytic film that was fixed in the plasma deposition process, which is also probable, for example in the light of XPS and XRD investigations for the 0.1FeOx film, where the difference between the surface and bulk structure is clearly visible (Section 2.1).

If the 0.3FeOx film comes into contact with air before initiating the catalytic test (e.g., Tests A, C, and D (Figure 4)), the surface of the Fe nanoparticles oxidizes, forming a Fe_2O_3 shell with a thickness of approx. 5 nm (Figure 7). Upon starting the test, a certain induction period is required to achieve full catalytic activity. This effect is attributed to the reduction of the shell to metallic Fe in the presence of hydrogen, which is a component of the reaction mixture [40]. This assumption is strongly supported by the results of Test E described in Section 2.2, and most notably by Test F, where the presence of the hydrogen atmosphere eliminates the induction period. Based on the above discussion, an important conclusion can be drawn: in the investigated nanocatalyst, metallic iron nanoparticles are primarily responsible for the high catalytic activity. Although Fe_2O_3 shells also exhibit catalytic activity, it is at a significantly lower level.

After removing the CNT fragments with the Fe nanoparticles they carry, the CO_2 conversion clearly and irreversibly decreases (Figure 4, Test G), although it does not fall to zero. This finding proves that some catalytic activity is still retained by the surface of the 0.3FeOx film. However, this surface (Figure 1e and Table 1) differs from that of the fresh film (Figure 1b and Table 1) in two aspects. Firstly, it lacks iron nanoparticles, implying that they are not the sole contributors to the catalytic activity. Secondly, it contains the remaining part of the CNTs deposit, which may enhance this activity, as evidenced by comparison with the activity of the 0.1FeOx film. In the former case, the catalytic activity is clearly greater than in the latter (Tests B and G, respectively (Figure 4)). Meanwhile, the XPS spectra of Fe 3p for the spent film of 0.3FeOx devoid of Fe nanoparticles (Figure 1e) and 0.1FeOx (Figure 1c) are analogous, indicating the dominant presence of Fe^{3+} in both cases, most likely Fe_2O_3. We have already attributed some catalytic activity to this structure in the case of Fe nanoparticles carried by CNTs.

The enhanced catalytic activity of 0.3FeOx films can be attributed to the creation of nanohybrids from carbon nanotubes and Fe_2O_3 nanoparticles present on the film surface [32]. This leads to the formation of nanoscale heterojunctions between them. Assuming that Fe_2O_3 is an n-type semiconductor [41] and CNTs are a p-type semiconductor [42], the junction formation results in the appearance of space charges, positive on Fe_2O_3 and negative on CNTs, in the resulting depletion regions.

The small size of Fe_2O_3 nanoparticles (approx. 25 nm, determined analogously to the size of Fe_3O_4 nanoparticles (Figure 2b), which are oxidized to Fe_2O_3 on the film surface) is comparable to the average size of the depletion region [43]. This means that the entire volume of Fe_2O_3 nanoparticles becomes filled with a positive charge. On the other hand, the CNTs deposit constitutes a much larger volume, and the negatively charged depletion region occupies only a small space in the vicinity of the junctions.

This concept is confirmed by the aforementioned shift of the XPS bands for Fe^{3+} by approx. 0.38 eV towards higher binding energies after the catalytic process, i.e., after the formation of CNTs (Figure 1b,d). The presence of a positive charge causes a stronger binding of electrons, hence shifting the band maximum towards higher binding energy [44]. Such an effect is not observed in the case of the 0.1FeOx film, where no CNTs are formed in the catalytic process (Figure 1a,c).

The presence of a positive charge in Fe_2O_3 nanoparticles can generally be interpreted as an enhancement of the acidic nature of the surface of the 0.3FeOx film, leading, for example, to an increase in the density of oxygen vacancies. This change results in an elevated CO_2 conversion towards CO [45], as evident when compared with the 0.1FeOx film. Another aspect to consider is the possibility of reducing Fe_2O_3 with hydrogen from the reaction mixture, similar to the case with the Fe_2O_3 shell on Fe nanoparticles. However, it is reasonable to assume that Fe_2O_3 nanoparticles, with a size of about 25 nm and present in much larger quantities on the film surface compared with the thin Fe_2O_3 shells (of a much smaller size, about 5 nm) on Fe nanoparticles, may undergo reduction, if at all, more reluctantly [46]. The slow increase in CO_2 conversion observed during Test G could possibly be attributed to a partial, slow reduction of Fe_2O_3 to Fe.

Despite several questions that arise when attempting to explain the mechanism of action of Fe_2O_3-based nanocatalysts produced by the PECVD method, the positive effect of carbon nanotubes on their catalytic activity is evident. However, the pursuit of a comprehensive understanding and potential application in the design of nanocatalysts encourages further in-depth research and analysis.

3. Materials and Methods

3.1. Thin-Film Nanocatalyst Preparation

The FeOx-based thin films were produced by plasma-enhanced chemical vapor deposition (PECVD) in a parallel-plate radio-frequency (RF 13.56 MHz) reactor, the construction of which is described in more detail elsewhere [47]. The film fabrication procedure was similar to that presented in our previous work [6]. Iron pentacarbonyl ($Fe(CO)_5$, Stream Chemicals, Newburyport, MA, USA) was used as a precursor, which was fed into the reactor chamber as a vapor in a mixture with a carrier gas (Ar, 99.999%, Linde Gas, Cracow, Poland). Argon had a constant flow rate of 1.0 sccm, corresponding to a partial pressure of 4.0 Pa, while the precursor was fed at two different partial pressures: 0.1Pa (0.1FeOx films) or 0.3 Pa (0.3FeOx films). The same discharge power of 80 W and the same deposition time of 30 min were used in all experiments. For catalytic tests, the films were deposited on a precalcined wire-mesh support [6], while for XPS and XRD studies, zero-background Si wafers (Si 510, Institute of Electronic Materials Technology—Łukasiewicz Research Network, Warsaw, Poland) were used as substrates. After deposition, all films were thermally treated at 673 K for 30 min under a continuous flow of argon at a rate of 2 L/min. Samples for the HRTEM study were collected from the surface of the 0.3FeOx film after the catalytic test.

3.2. Characteristics of the Film Structure

The elemental composition and molecular structure of the films were investigated by X-ray photoelectron spectroscopy (XPS) using a Kratos AXIS Ultra spectrometer (Kratos Analytical Ltd., Manchester, UK) with a monochromatic Al-Kα X-ray source (1486.6 eV). The power of the anode was set at 180 W, and the hemispherical electron energy analyzer was operated at a pass energy of 20 eV for all high-resolution measurements. The measurements were performed with a charge neutralizer. XPS spectra were analyzed using Kratos Vision 2.2.10 software, calibrated by setting the C1s carbon peak assigned to sp^2 carbon at 284.6 eV. The background subtraction was performed with Shirley's algorithm.

The nanocrystalline structure of the films was determined by X-ray diffraction. The measurements were performed in a reflection Bragg–Brentano mode using an Empyrean diffractometer (Malvern Panalytical Ltd., Malvern, UK) with Cu K_α radiation (λ = 1.541874 Å), operating at 45 kV and 40 mA. The diffractometer was equipped with a PIXcel3D detector with 255 active channels. Soller slits of 0.04 rad, a fixed mask of 20 mm, and a divergence slit of 1/4 deg were used on an incident beam path, while the diffracted beam path was also equipped with Soller slits of 0.04 rad. The diffractograms were obtained in the range of 2Θ = 5–85 deg using a continuous scan mode with a step size of 0.0263 deg and a counting time of 176 s. The samples were spined with a rotation time of 8 s.

The morphology of the films was examined using both scanning electron microscopy (SEM) and high-resolution transmission electron microscopy (HRTEM). SEM micrographs were obtained with an Apreo 2S microscope (Thermo Fisher Scientific, Waltham, MA, USA) in high vacuum mode at low voltage (1.0 kV) with a T2 detector. The tested samples were without sputtered coatings. In turn, the HRTEM investigations were performed using a Titan Themis 300 S/TEM (Thermo Fisher Scientific, Waltham, MA, USA) in scanning mode (STEM), equipped with a Super-X EDS system. Scanning transmission micrographs were acquired simultaneously by four detectors: a high-angle annular dark field (HAADF) detector, annular dark-field detectors (DF2, DF4), and an annular bright-field (ABF) detector.

Copper grids with lacey carbon film were used to observe the deposits taken from the surface of the catalytic films.

3.3. Catalytic Tests

The catalytic tests were carried out in a quartz tube-in-tube reactor at atmospheric pressure with an internal diameter of 2.3 cm and an overall length of 30 cm. The reactor was operated under continuous flow conditions and placed horizontally in an electric furnace equipped with a programmable temperature controller. Further details on the experimental setup can be found in our previous publication [6]. Here we present a concise description supplemented with details on testing the performance of the catalyst under the conditions used in this work.

The catalyst bed was formed by six circular wire mesh discs (OD = 2.3 cm) with a hole (ID = 1.0 cm), which were fixed on the inner tube of the tubular reactor. To ensure a fair comparison between the different catalysts, the same number of mesh discs were always placed in the reactor in a uniform manner. The total geometrical area of the catalyst bed was 38 cm^2, and its measured length was 10 cm.

The system was purged with 99.999% He before starting a typical catalytic test. Thereafter, a gas mixture consisting of CO_2 99.99% and H_2 99.999% (all gases supplied by Linde Gas, Cracow, Poland) flowed into the system at a total rate of 25 sccm with a 4:1 H_2/CO_2 feed composition. Once helium was flushed out, the system was preheated to 473 K. Kinetic tests were performed in the range of 473–673 K with a temperature ramp of 50 K and a heating rate of 5 K/min. The test was then continued at 673 K, and the time-on-stream performance was evaluated over a period of up to 20 h.

Additionally, several tests were conducted to evaluate the performance and repeatability of the catalysts. Following the 20 h time-on-stream run, the system was cooled to room temperature under the flow of the reaction mixture (25 sccm). The catalyst was then held in the air for 12 h before the next catalytic test was restarted. Tests were also carried out in which the catalyst reached a temperature of 673 K, in accordance with the procedure described above, in a flow of helium (25 sccm) or hydrogen (25 sccm) instead of the reaction mixture, and it was maintained in this way for 2 h. The flow of He or H_2 was then stopped, and the feed of the reaction mixture was started. After approx. 30 min, the progress of CO_2 hydrogenation was monitored over time at 673 K.

The outlet gas mixture from the reactor was passed through a cold trap to condense the water vapor, and then the dry gas was analyzed. The GC analysis was conducted in triplicate at 10 min intervals at each temperature. Carbon-containing components, including CO_2, CO, and CH_4, were analyzed on a HayeSep D column (Agilent) using a gas chromatograph (SRI 8610C, SRI Instruments, Torrance, CA, USA) with a thermal conductivity detector (TCD). Based on the calibration curves for CO_2, CO, and CH_4, the conversion of CO_2 and selectivity to CO and CH_4 were determined.

We attempted to determine the mass of carbon deposited on the surface during the catalytic tests. This was achieved by calculating the difference in mass between the fresh and spent catalysts. The measured mass (after 20 h of the process) remained consistent within an uncertainty of 0.1 mg, corresponding to the measurement sensitivity of an analytical balance. As a result, the mass of deposited carbon per unit of time is orders of magnitude lower than the flow rate of carbon in the tests performed (2.7 mg/min), and this value does not impact the accuracy of the carbon balance, which is solely based on the gaseous reactants (CO_2, CO, and CH_4) and was calculated as presented in Ref. [5].

4. Conclusions

This paper presents two significant aspects of research related to the search for new solutions in the area of catalytic materials for CO_2 conversion into syngas. The first aspect confirms the broad potential offered by the PECVD method in producing thin-film nanocatalytic materials with unique properties. These materials are particularly useful in the construction of structured packings for catalytic reactors. Through plasma deposition using $Fe(CO)_5$ as a precursor, conveniently carried out by means of PECVD, thin-film nanocatalysts based on FeOx were obtained. The structure and catalytic activity of these nanocatalysts can be precisely controlled by selecting appropriate deposition process parameters. This approach enabled the production of films (referred to as 0.3FeOx in this work) that achieved CO_2 conversion in the reverse water-gas shift (RWGS) reaction very close to the equilibrium value under reaction conditions (X_{CO2} = approx. 38% at 673 K for H_2/CO_2 = 4:1, with an equilibrium value of 41.5%), and with a selectivity to CO of approx. 96%. Additionally, these films exhibited no signs of nanocatalyst destabilization over long periods of time.

The second important aspect of the research involved the finding of a carbon deposit containing carbon nanotubes (CNTs) on the surface of highly catalytically active films (0.3FeOx). This deposit was formed during the RWGS reaction. Contrary to the commonly held opinion, the presence of this deposit turned out to be highly beneficial in this case, significantly enhancing the catalytic activity of the films. According to the proposed mechanism, carbon nanotubes play a dual role in this context. Firstly, CNTs formed through the tip-growth mechanism lift up the iron nanoparticles present on the surface of the fresh 0.3FeOx film. These nanoparticles serve as both the nuclei for CNT growth and an active catalyst in the RWGS reaction. The removal of these iron nanoparticles leads to a significant reduction in catalytic activity. Secondly, CNTs simultaneously form nanoscale heterojunctions with Fe_2O_3 nanoparticles present on the film surface. The resulting positive charge on the Fe_2O_3 nanoparticles significantly enhances the conversion of CO_2 into CO.

The presented results undoubtedly contribute to a better understanding of the nature of the active form of plasma-deposited thin films based on FeOx as catalysts in the CO_2 hydrogenation process, thus opening up new possibilities for designing active and selective nanocatalysts in this area.

Supplementary Materials: The following supporting information can be downloaded at: https://www.mdpi.com/article/10.3390/catal13091302/s1, Supplementary Figure S1. X-ray diffraction (XRD); Supplementary Figure S2. RWGS reaction—thermodynamic analysis; Supplementary Figure S3. Calculation of composition of core-shell iron nanoparticle.

Author Contributions: Conceptualization and methodology, J.T. and H.K.-P.; catalytic tests and analysis, B.P. and H.K.-P.; XPS analysis, B.P. and J.T.; XRD analysis, P.U.; SEM analysis, B.P.; XRTEM analysis, S.N., V.N.-T. and J.T.; data curation, B.P., H.K.-P., P.U., S.N. and J.T; writing (draft, review, editing), J.T. and H.K.-P.; visualization, J.T., H.K.-P., P.U. and S.N.; supervision, J.T. and H.K.-P. All authors have read and agreed to the published version of the manuscript.

Funding: This research received no external funding and was financed as part of the statutory activity of the Lodz University of Technology.

Data Availability Statement: The data presented in this study are available on reasonable request from the corresponding author.

Conflicts of Interest: The authors declare no conflict of interest.

References

1. Tyczkowski, J. Cold Plasma produced catalytic materials. In *Plasma Science and Technology—Progress in Physical States and Chemical Reactions*; Mieno, T., Ed.; InTech Open: Vienna, Austria, 2016; pp. 53–93.
2. Wang, Z.; Zhang, Y.; Neyts, E.C.; Cao, X.; Zhang, X.; Jang, B.W.L.; Liu, C.J. Catalyst preparation with plasmas: How does it work? *ACS Catal.* **2018**, *8*, 2093–2110. [CrossRef]
3. Kołodziej, A.; Łojewska, J.; Tyczkowski, J.; Jodłowski, P.; Redzynia, W.; Iwaniszyn, M.; Zapotoczny, S.; Kustrowski, P. Coupled engineering and chemical approach to the design of a catalytic structured reactor for combustion of VOCs: Cobalt oxide catalyst on knitted wire gauzes. *Chem. Eng. J.* **2012**, *200–202*, 329–337. [CrossRef]
4. Kierzkowska-Pawlak, H.; Kruszczak, E.; Tyczkowski, J. Catalytic activity of plasma-deposited Co_3O_4-based thin films for CO_2 hydration—A new approach to carbon capture applications. *Appl. Catal. B Environ.* **2022**, *304*, 120961. [CrossRef]
5. Kierzkowska-Pawlak, H.; Tracz, P.; Redzynia, W.; Tyczkowski, J. Plasma deposited novel nanocatalysts for CO_2 hydrogenation to methane. *J. CO2 Util.* **2017**, *17*, 312–319. [CrossRef]

6. Kierzkowska-Pawlak, H.; Ryba, M.; Fronczak, M.; Kapica, R.; Sielski, J.; Sitarz, M.; Zając, P.; Łyszczarz, K.; Tyczkowski, J. Enhancing CO_2 conversion to CO over plasma-deposited composites based on mixed Co and Fe oxides. *Catalysts* **2021**, *11*, 883. [CrossRef]
7. Daza, Y.A.; Kuhn, J.N. CO_2 conversion by reverse water gas shift catalysis: Comparison of catalysts, mechanisms and their consequences for CO_2 conversion to liquid fuels. *RSC Adv.* **2016**, *6*, 49675–49691. [CrossRef]
8. González-Castaño, M.; Dorneanu, B.; Arellano-García, H. The reverse water gas shift reaction: A process systems engineering perspective. *React. Chem. Eng.* **2021**, *6*, 954. [CrossRef]
9. Mota, F.M.; Kim, D.H. From CO_2 methanation to ambitious long-chain hydrocarbons: Alternative fuels paving the path to sustainability. *Chem. Soc. Rev.* **2019**, *48*, 205–259. [CrossRef]
10. Wismann, S.Y.; Larsen, K.E.; Mortensen, P.M. Electrical reverse shift: Sustainable CO_2 valorization for industrial scale. *Angew. Chem. Int. Ed.* **2022**, *61*, e202109696. [CrossRef]
11. Su, X.; Yang, X.; Zhao, B.; Huang, Y. Designing of highly selective and high-temperature endurable RWGS heterogeneous catalysts: Recent advances and the future directions. *J. Energy Chem.* **2017**, *26*, 854–867. [CrossRef]
12. Chen, X.; Chen, Y.; Song, C.; Ji, P.; Wang, N.; Wang, W.; Cui, L. Recent advances in supported metal catalysts and oxide catalysts for the reverse water-gas shift reaction. *Front. Chem.* **2020**, *8*, 709. [CrossRef]
13. Chou, C.Y.; Loiland, J.A.; Lobo, R.F. Reverse water-gas shift iron catalyst derived from magnetite. *Catalysts* **2019**, *9*, 773. [CrossRef]
14. Ding, M.; Yang, Y.; Wu, B.; Wang, T.; Ma, L.; Xiang, H.; Li, Y. Transformation of carbonaceous species and its influence on catalytic performance for iron-based Fischer–Tropsch synthesis catalyst. *J. Mol. Catal. A Chem.* **2011**, *351*, 165–173. [CrossRef]
15. Otun, K.O.; Yao, Y.; Liu, X.; Hildenbrandt, D. Synthesis, structure, and performance of carbide phases in Fischer-Tropsch synthesis: A critical review. *Fuel* **2021**, *296*, 120689. [CrossRef]
16. Zhu, M.; Wachs, I.E. Iron-based catalysts for the high-temperature water–gas shift (HTWGS) reaction: A review. *ACS Catal.* **2016**, *6*, 722–732. [CrossRef]
17. Landau, M.V.; Meiri, N.; Utsis, N.; Nehemya, R.V.; Herskowitz, M. Conversion of CO_2, CO, and H_2 in CO_2 hydrogenation to fungible liquid fuels on Fe-based catalysts. *Ind. Eng. Chem. Res.* **2017**, *56*, 13334–13355. [CrossRef]
18. Zhang, Y.; Cao, C.; Zhang, C.; Zhang, Z.; Liu, X.; Yang, Z.; Zhu, M.; Meng, B.; Xu, J.; Han, Y.F. The study of structure-performance relationship of iron catalyst during a full life cycle for CO_2 hydrogenation. *J. Catal.* **2019**, *378*, 51–62. [CrossRef]
19. Puga, A.F. On the nature of active phases and sites in CO and CO_2 hydrogenation catalysts. *Catal. Sci. Technol.* **2018**, *8*, 5681–5707. [CrossRef]
20. Kim, D.H.; Han, S.W.; Yoon, H.S.; Kim, Y.D. Reverse water gas shift reaction catalyzed by Fe nanoparticles with high catalytic activity and stability. *J. Ind. Eng. Chem.* **2015**, *25*, 67–71. [CrossRef]
21. Saeidi, S.; Amin, N.A.S.; Rahimpour, M.R. Hydrogenation of CO_2 to value-added products—A review and potential future developments. *J. CO2 Util.* **2014**, *5*, 66–81. [CrossRef]
22. Zhou, J.; Zhao, J.; Zhang, J.; Zhang, T.; Ye, M.; Liu, Z. Regeneration of catalysts deactivated by coke deposition: A review. *Chin. J. Catal.* **2020**, *41*, 1048–1061. [CrossRef]
23. le Sache, E.; Reina, T.R. Analysis of dry reforming as direct route for gas phase CO_2 conversion. The past, the present and future of catalytic DRM technologies. *Prog. Energy Combust. Sci.* **2022**, *89*, 100970. [CrossRef]
24. Bost, N.; Ammar, M.R.; Bouchetou, M.L.; Poirier, J. The catalytic effect of iron oxides on the formation of nano-carbon by the Boudouard reaction in refractories. *J. Eur. Ceram. Soc.* **2016**, *36*, 2133–2142. [CrossRef]
25. Vogt, E.T.C.; Fu, D.; Weckhuysen, B.M. Carbon deposit analysis in catalyst deactivation, regeneration, and rejuvenation. *Angew. Chem. Int. Ed.* **2023**, *62*, e202300319. [CrossRef]
26. Papadopoulou, C.; Matralis, H.; Verykios, X. Deactivation due to coking. In *Catalysis for Alternative Energy Generation*; Guczi, L., Erdőhelyi, A., Eds.; Springer: New York, NY, USA, 2012; pp. 80–84.
27. Lu, W.; Li, N.; Chen, W.; Yao, Y. The role of multiwalled carbon nanotubes in enhancing the catalytic activity of cobalt tetraaminophthalocyanine for oxidation of conjugated dyes. *Carbon* **2009**, *47*, 3337–3345. [CrossRef]
28. Oberlin, A.; Endo, M.; Koyama, T. Filamentous growth of carbon through benzene decomposition. *J. Cryst. Growth* **1976**, *32*, 335–349. [CrossRef]
29. Bradley, J.P.; Brownlee, D.E. Carbon compounds in interplanetary dust: Evidence for formation by heterogeneous catalysis. *Science* **1984**, *223*, 56–57. [CrossRef]
30. Chen, W.; Fan, Z.; Pan, X.; Bao, X. Effect of confinement in carbon nanotubes on the activity of Fischer–Tropsch iron catalyst. *J. Am. Chem. Soc.* **2008**, *130*, 9414–9419. [CrossRef]
31. Bai, Z.Y.; Yang, Q.; Wang, J.I. Fe_3O_4/multi-walled carbon nanotubes as an efficient catalyst for catalytic ozonation of p-hydroxybenzoic acid. *Int. J. Environ. Sci. Technol.* **2016**, *13*, 483–492. [CrossRef]
32. Ismail, A.A.; Ali, A.M.; Harraz, F.A.; Faisal, M.; Shoukry, H.; Al-Salami, A.E. A facile synthesis of α-Fe_2O_3/carbon nanotubes and their photocatalytic and electrochemical sensing performances. *Int. J. Electrochem. Sci.* **2019**, *14*, 15–32. [CrossRef]
33. Chen, Y.; Wei, J.; Duyar, M.S.; Ordomsky, V.V.; Khodakov, A.Y.; Liu, J. Carbon-based catalysts for Fischer–Tropsch synthesis. *Chem. Soc. Rev.* **2021**, *50*, 2337–2366. [CrossRef] [PubMed]
34. Yamashita, T.; Hayes, P. Analysis of XPS spectra of Fe^{2+} and Fe^{3+} ions in oxide materials. *Appl. Surf. Sci.* **2008**, *254*, 2441–2449. [CrossRef]

35. Gates-Rector, S.; Blanton, T. The powder diffraction file: A quality materials characterization database. *Powder Diffr.* **2019**, *34*, 352–360. [CrossRef]
36. Das, R.; Hamid, S.B.A.; Ali, M.E.; Ramakrishna, S.; Yongzhi, W. Carbon nanotubes characterization by X-ray powder diffraction—A review. *Curr. Nanosci.* **2015**, *11*, 23–35. [CrossRef]
37. Wirth, C.T.; Bayer, B.C.; Gamalski, A.D.; Esconjauregui, S.; Weatherup, R.S.; Ducati, C.; Baehtz, C.; Robertson, J.; Hofmann, S. The phase of iron catalyst nanoparticles during carbon nanotube growth. *Chem. Mater.* **2012**, *24*, 4633–4640. [CrossRef]
38. McCafferty, L.; Stolojan, V.; King, S.G.; Zhang, W.; Haq, S.; Silva, S.R.P. Decoration of multiwalled carbon nanotubes with protected iron nanoparticles. *Carbon* **2015**, *84*, 47–55. [CrossRef]
39. Ou, X. Molecular dynamics simulations of fcc-to-bcc transformation in pure iron: A review. *Mater. Sci. Technol.* **2017**, *33*, 822–835. [CrossRef]
40. Jozwiak, W.K.; Kaczmarek, E.; Maniecki, T.P.; Ignaczak, W.; Maniukiewicz, W. Reduction behavior of iron oxides in hydrogen and carbon monoxide atmospheres. *Appl. Catal. A Gen.* **2007**, *326*, 17–27. [CrossRef]
41. Liao, P.; Toroker, M.C.; Carter, E.A. Electron transport in pure and doped hematite. *Nano Lett.* **2011**, *11*, 1775–1781. [CrossRef]
42. Hu, Q.; Lu, Z.; Wang, Y.; Wang, J.; Wang, H.; Wu, Z.; Lu, G.; Zhang, H.L.; Yu, C. Double doping approach for unusually stable and large n-type thermoelectric voltage from p-type multi-walled carbon nanotube mats. *J. Mater. Chem. A* **2020**, *8*, 13095–13105. [CrossRef]
43. Sze, S.M.; Ng, K.K. *Physics of Semiconductor Devices*, 3rd ed.; John Wiley & Sons, Inc.: Hoboken, NJ, USA, 2007; pp. 80–90.
44. Wang, L.; Chen, B.; Zhang, L.; Yu, J. In situ irradiated XPS investigation on S-scheme TiO_2@$ZnIn_2S_4$ photocatalyst for efficient photocatalytic CO_2 reduction. *Small* **2021**, *17*, 2103447. [CrossRef]
45. Gao, X.; Cai, P.; Wang, Z.; Lv, X.; Kawi, S. Surface acidity/basicity and oxygen defects of metal oxide: Impacts on catalytic performances of CO_2 reforming and hydrogenation reactions. *Top. Catal.* **2023**, *66*, 299–325. [CrossRef]
46. Pour, A.N.; Housaindokht, M.R.; Babakhani, E.G.; Irani, M.; Shahri, S.M.K. Size dependence on reduction kinetic of iron based Fischer–Tropsch catalyst. *J. Ind. Eng. Chem.* **2011**, *17*, 596–602. [CrossRef]
47. Tyczkowski, J.; Kierzkowska-Pawlak, H.; Kapica, R.; Balcerzak, J.; Sielski, J. Cold plasma – a promising tool for the production of thin-film nanocatalysts. Catal. *Today* **2019**, *337*, 44–54. [CrossRef]

catalysts

MDPI

Article

Optimizing Citrate Combustion Synthesis of A-Site-Deficient La,Mn-Based Perovskites: Application for Catalytic CH_4 Combustion in Stoichiometric Conditions

Andrea Osti *, Lorenzo Rizzato, Jonathan Cavazzani and Antonella Glisenti *

Department of Chemical Sciences, University of Padova, Via F. Marzolo, 1, 35131 Padova, Italy; lorenzo.rizzato.1@phd.unipd.it (L.R.); jonathan.cavazzani@phd.unipd.it (J.C.)
* Correspondence: andrea.osti.1@phd.unipd.it (A.O.); antonella.glisenti@unipd.it (A.G.)

Abstract: $LaMnO_3$-based perovskites are widely recognized as promising catalysts for several oxidation reactions, but the final physicochemical and catalytic properties can be greatly influenced by the adopted synthesis procedure. In this work, a series of A-site-deficient perovskites of composition $La_{0.8}MnO_3$ and $La_{0.8}Mn_{0.9}B_{0.1}O_3$ (B = Ni, Cu) were prepared through the citrate combustion route with variations in two synthesis parameters: a citric acid/metal cations molar ratio (CA/M) of either 1.1 or 1.5 and either acidic (given by HNO_3 + citric acid) or neutral (after NH_3 addition) pH of the precursor solution. The obtained samples were characterized by XRD, H_2-TPR, O_2-TPD, N_2 physisorption, SEM-EDX and XPS. Acidic pH coupled with a CA/M ratio of 1.1 clearly emerged superior among all the other combinations of the two parameters, resulting in smaller crystallite size, higher surface area and porosity, enhanced Mn^{4+} reducibility and the ability to release oxygen species; these features were even further improved by B-site substitution with 10 mol% Ni and Cu cations. The synthesized catalysts were tested in CH_4 oxidation to CO_2 under stoichiometric O_2, confirming the great superiority of samples prepared in acidic pH with a CA/M ratio of 1.1. Ni and Cu doping had a beneficial effect on catalytic activity, which, however, was more evident for less optimized perovskites (acidic pH and CA/M ratio of 1.5), without significance differences among the two dopants.

Keywords: $La_{0.8}MnO_3$ perovskite; Ni and Cu doping; citrate combustion synthesis; heterogeneous catalysis; CH_4 combustion

Citation: Osti, A.; Rizzato, L.; Cavazzani, J.; Glisenti, A. Optimizing Citrate Combustion Synthesis of A-Site-Deficient La,Mn-Based Perovskites: Application for Catalytic CH_4 Combustion in Stoichiometric Conditions. *Catalysts* **2023**, *13*, 1177. https://doi.org/10.3390/catal13081177

Academic Editors: Leonarda Liotta, Narendra Kumar and Konstantin Ivanov Hadjiivanov

Received: 23 June 2023
Revised: 25 July 2023
Accepted: 29 July 2023
Published: 1 August 2023

1. Introduction

Perovskite oxides with the general formula ABO_3 have attracted much attention in the last decades as catalysts in many reactions because of their stable crystal structure, hydrothermal stability, redox activity, great composition tunability and defect engineering [1,2]. Among them, $LaMnO_3$-based perovskites have been widely studied as promising catalysts for several oxidation (to CO_2) reactions, of, e.g., CO, volatile organic compounds (VOCs) and hydrocarbons (HCs) [3–5]. Factors like exposed surface area, cation reducibility, the content of highly valent cations (e.g., Mn^{4+} compared to the main Mn^{3+}) and the presence of mobile surface and lattice oxygen species have been reported to be decisive in perovskites to improve their catalytic activity in oxidation reactions [6–11]. However, such physicochemical properties can be greatly influenced by the adopted synthesis protocol, which must be properly optimized. The citrate combustion (or citrate sol-gel) route [12] has been extensively employed for the preparation of high-purity and crystalline oxide powders at a relatively low calcination temperatures (e.g., 600–900 °C), benefiting from the ability of citrate anions to complex and homogenously disperse metal cations, with the formation of a gel, which is then ignited at a suitable temperature through an in situ redox reaction between citric acid (fuel) and nitrates (oxidizer) used as precursors. The composition of the obtained gel, its combustion rate and heat generated upon burning depend on the fuel/oxidizer molar ratio, also affecting the final perovskite's textural properties. For instance, Taguchi et al. prepared $LaMnO_3$ with a citric acid/sum of metal cations (CA/M) molar ratio of between 0.016 and 1.07 (moles of citric acid from 0.0023 to 0.015 in the paper), showing that the crystallite size reached a minimum and the surface area a maximum at CA/M of about 0.5–0.6 (0.007 and 0.008 moles of citric acid), above which the combustion heat increased, negatively affecting the surface area [13]. Ghiasi et al. prepared $LaMnO_3$ with a citric acid/sum of nitrates ratio from 1.0 to 5.0 (corresponding to CA/M of 2.5 to 12.5), highlighting that the stoichiometric amount of citric acid (1.0 vs. nitrates, 2.5 vs. metal cations) was the best for lowering crystallite size and particle agglomeration, whereas too high an amount of citric acid degraded the perovskite structure because of excessive heat of combustion [14]. Sihaib et al. prepared $LaMnO_3$ with a CA/M ratio from 0.5 to 2.0, evidencing that intermediate CA/M values of 1.2 to 1.5 led to the lowest crystallite size, enhanced surface area, greater Mn^{4+} reducibility, the highest Mn/La surface ratio and, accordingly, better catalytic activity in toluene and propene oxidation [15]. When NH_3 was added to the $LaMnO_3$ precursor solution to neutralize the pH, as in the work of Li et al. with citric acid/nitrates ratio ranging from 0.5 to 3.0 (corresponding to CA/M of 1.25 to 7.5), the lowest particle size was obtained with a lower amount of citric acid (0.5 and 1.0 vs. nitrates, 1.25 and 2.5 vs. metal cations), although the phase purity was the maximum at a citric acid/nitrate ratio of 1.5 (CA/M = 3.75) [16]. These works point out the necessity of finding the right fuel/oxidizer molar ratio, which can vary depending on the other synthesis parameters (like precursor solution pH and calcination temperature). Furthermore, charge imbalances at the A-site, achieved by aliovalent doping (e.g., Sr^{2+} partially replacing La^{3+}) or by means of A-site cation deficiency, are known to increase the number of oxygen vacancies, which is useful for improving reactant adsorption and enhancing oxygen mobility [1,5,17]. A-site deficiency can also be desirable to counteract the usually observed La surface enrichment (non-redox active) and to increase the B-site cation oxidation state and reducibility [9,18,19]. Accordingly, $LaMn_{1.2}O_{3+\delta}$ (or $La_{0.83}MnO_3$) was found to be more active than stoichiometric $LaMnO_3$ towards CO and C_3H_8 oxidation [19]. A partial replacement of Mn at the B-site by either Ni or Cu cations was also shown to improve HCs' oxidation capability, thanks to improved surface area, reducibility and surface oxygen mobility [20,21]

Based on the above considerations, in the present work, a series of A-site-deficient $La_{0.8}MnO_3$ perovskites were synthesized by the citrate combustion route, with variations in terms of two synthesis parameters: a CA/M ratio of either 1.1 or 1.5 and either an acidic (given by HNO_3 and citric acid) or neutral (upon NH_3 addition) pH of the precursor solution. B-site doping with Ni and Cu cations was performed as well, preparing perovskites with the composition $La_{0.8}Mn_{0.9}Ni_{0.1}O_3$ and $La_{0.8}Mn_{0.9}Cu_{0.1}O_3$. The precursor gel decomposition process was analyzed by TGA, and the calcined samples were deeply characterized by XRD, N_2 physisorption, SEM-EDX, XPS, H_2-TPR and O_2-TPD techniques. The influence of the synthesis parameters (pH, CA/M ratio) and of B-site doping on physicochemical properties and on catalytic activity towards CH_4 combustion with stoichiometric O_2 was evaluated.

2. Results and Discussion

2.1. Physicochemical Characterizations

2.1.1. Structural and Morphological Features

A list of the eight prepared perovskites and their synthesis conditions are reported in Table 1. The subscript 'H^+' is used to refer to samples synthesized in acidic precursor solution before the gel formation, as determined by HNO_3 and citric acid, whereas the subscript 'NH_3' indicates samples prepared through pH neutralization by NH_3 addition before the gel formation (see Section 3.1 Synthesis protocol). The molar ratio citric acid/sum of metal cations (either 1.1 or 1.5) is indicated by the notation 'CA1.1' and 'CA1.5'. Let us note that the variation in the amount of citric acid (CA1.1 or CA1.5) in the acidic precursor solution did not substantially change the final pH, which was always <1 and probably even negative. This was primarily governed by the strong acid HNO_3. The ignition temperature was varied depending on whether or not NH_3 was introduced into the precursor solution, as further explained in Section 2.1.2. The chosen calcination temperature (750 °C) was a compromise between the necessity of approaching the conditions employed in the catalytic test (up to 800 °C) and the need to avoid excessive sintering induced by high temperatures.

Table 1. Prepared samples and synthesis conditions. CA/M* = citric acid/metal cations molar ratio.

Sample	CA/M* Molar Ratio	pH of Synthesis	Ignition T (°C)	Calcination T (°C)
$La_{0.8}MnO_3$-NH_3-CA1.1	1.1	7	350	750
$La_{0.8}MnO_3$-NH_3-CA1.5	1.5	7	350	750
$La_{0.8}MnO_3$-H^+-CA1.1	1.1	<1	200–230	750
$La_{0.8}MnO_3$-H^+-CA1.5	1.5	<1	200–230	750
$La_{0.8}Mn_{0.9}Ni_{0.1}O_3$-$H^+$-CA1.1	1.1	<1	200–230	750
$La_{0.88}Mn_{0.9}Ni_{0.1}O_3$-$H^+$-CA1.5	1.5	<1	200–230	750
$La_{0.88}Mn_{0.9}Cu_{0.1}O_3$-$H^+$-CA1.1	1.1	<1	200–230	750
$La_{0.88}Mn_{0.9}Cu_{0.1}O_3$-$H^+$-CA1.5	1.5	<1	200–230	750

XRD patterns of the synthesized perovskites are reported in Figure 1. For all of them, the recorded diffractograms match with the hexagonal perovskite phase $La_{0.887}MnO_3$ (space group R-3c:H, COD reference 1531294), with the characteristic reflection splitting of hexagonal $LaMnO_3$-type structures (see Figure S1). Additionally, for all $La_{0.8}MnO_3$ samples except H^+-CA1.1, weak reflections at $2\theta = 36.1°$ and $2\theta = 59.9°$ are probably ascribed to the presence of small Mn_3O_4 impurities, making these compositions less A-site-deficient than expected by stoichiometry. In the case of Ni-doped samples, no evident secondary phases emerged from the XRD patterns, which suggests the complete Ni^{2+} (or possibly Ni^{3+}) incorporation inside the perovskite lattice at 10 mol% doping. Instead, Cu-doped samples seem to display a weak reflection at $2\theta = 36.0°$ ascribed to the $CuMn_2O_4$ spinel phase. Thus, Cu^{2+} cations are not completely incorporated in the perovskite framework, probably because of a larger size difference with respect to Mn^{3+} main cations than in the case of Ni doping (0.72 Å Mn^{3+}, 0.87 Å Cu^{2+}, 0.83 Å Ni^{2+}, 0.74 Å Ni^{3+} [22]). It is interesting to note that samples prepared in acidic pH, and especially those with CA1.1, exhibit broader reflections and a less defined splitting, suggesting the presence of smaller crystallites with respect to samples prepared with NH_3 neutralization. Indeed, average crystallite size values reported in Table 2, calculated by means of the Williamson–Hall method, support this observation: the grain size tends to decrease for acidic pH synthesis and especially when there is less citric acid (CA1.1 samples), and the size tends to be even smaller after Ni and Cu doping compared to undoped $La_{0.8}MnO_3$ prepared with the same synthesis protocol, as already found in the literature [20,21]. The lowest crystallite size, 23 nm, was obtained for $La_{0.8}Mn_{0.9}Ni_{0.1}O_3$-$H^+$-CA1.1; the highest, 82 nm, for $La_{0.8}MnO_3$-NH_3-CA1.5. For the sake of comparison, Ghiasi et al. obtained $LaMnO_3$ with an average grain size ranging from 30 to 70 nm depending on the amount of citric acid and the calcination temperature [14]; Gholizadeh prepared $LaMnO_3$ calcined at 900 °C with an average size of 53 nm [23]; and Frozandeh-Mehr et al. obtained a crystallite size of just 14 nm for $LaMnO_3$ with a low calcination temperature of 600 °C [24]. Thus, crystallite sizes obtained in the present work are in reasonable agreement with those reported in literature for similar compositions and synthesis methods.

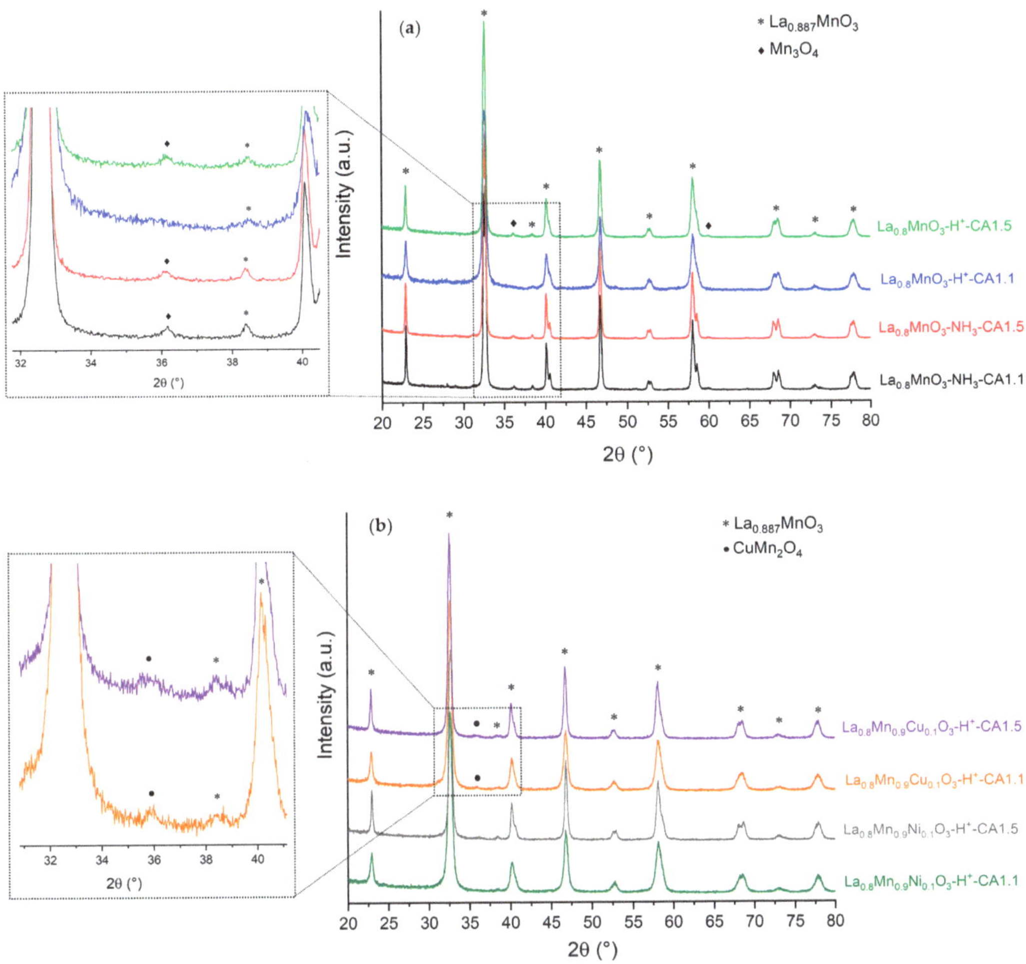

Figure 1. XRD patterns of (**a**) $La_{0.8}MnO_3$ samples, with zoom highlighting the presence of a weak Mn_3O_4 reflection at 2θ = 36.1°; (**b**) Ni- and Cu-doped samples, with zoom highlighting the presence of a weak $CuMn_2O_4$ reflection at 2θ = 36.0° on samples containing Cu.

Table 2. Specific surface area (SSA), total pore volume and average crystallite size (d_{XRD}) of the eight prepared perovskites.

Sample	SSA (m^2/g) [1]	Pore Vol. (cm^3/g) [2]	d_{XRD} (nm) [3]
$La_{0.8}MnO_3$-NH_3-CA1.1	4.9	0.013	69.3
$La_{0.8}MnO_3$-NH_3-CA1.5	4.0	0.0070	81.5
$La_{0.8}MnO_3$-H^+-CA1.1	25.0	0.081	33.1
$La_{0.8}MnO_3$-H^+-CA1.5	4.2	0.016	64.8
$La_{0.8}Mn_{0.9}Ni_{0.1}O_3$-$H^+$-CA1.1	43.9	0.152	22.7
$La_{0.88}Mn_{0.9}Ni_{0.1}O_3$-$H^+$-CA1.5	7.6	0.039	46.8
$La_{0.88}Mn_{0.9}Cu_{0.1}O_3$-$H^+$-CA1.1	37.1	0.145	33.0
$La_{0.88}Mn_{0.9}Cu_{0.1}O_3$-$H^+$-CA1.5	11.7	0.037	45.1

[1] Determined by BET method; [2] determined by BJH method; [3] determined by Williamson–Hall method, with K = 0.90 and λ = 0.154 nm.

In Figure 2, N_2 adsorption–desorption isotherms of the eight samples are reported, and the corresponding pore size distributions calculated from the desorption branches are shown in Figure S2. Specific surface area (SSA) and total pore volume, determined by BET and BJH methods, respectively, are listed in Table 2. Adsorption isotherms are similar to type III, with no evident "point B", indicating a weak N_2-adsorbent interaction; however, a type H3 hysteresis also appears, suggesting the presence of macro/mesopores probably formed by loose aggregates of particles [25]. Despite the similar shape of isotherms and hysteresis loops regardless of the synthesis protocol, SSA and pore volume can vary to a relatively large extent. Among the four $La_{0.8}MnO_3$ samples, H^+-CA1.1 clearly emerges superior for its much higher SSA (25 m^2/g) and pore volume, both values being at least five times higher than those of the other three samples (4–5 m^2/g), which is related to a larger amount of mesopores, as shown in Figure S2. In general, the synthesis in acidic pH seems to be preferable compared to synthesis with NH_3 neutralization, although a higher amount of citric acid (1.5) reduces the benefits in terms of SSA and porosity. Similar values ranging from 5 to 26 m^2/g, depending on CA/M ratio, were obtained by Sihaib et al. for $LaMnO_3$ calcined at 750 °C [15], and Taguchi et al. obtained intermediate surface areas of 13–16 m^2/g (again as a function of the amount of CA) by calcining $LaMnO_3$ at 700 °C [13]. In some cases, La-deficient compositions were reported to have higher surface area values compared to stoichiometric ones. For instance, Chen et al. obtained a value of 23 m^2/g for $La_{0.9}MnO_3$, compared to 20 m^2/g for $LaMnO_3$, both calcined at 700 °C; a marked increase in surface area was obtained for $La_{0.7}FeO_3$ compared to $LaFeO_3$ as well [26]. The SSA values obtained in this work are therefore comparable to the findings in the literature. At a parity of CA, B-site doping with Ni or Cu cations is even more beneficial in enhancing SSA and mesoporosity compared to benchmark La and Mn perovskites, pushing SSA values close to or even higher than 40 m^2/g in the case of less CA (1.1), whereas the enhancement is less marked with CA of 1.5 (SSA of 7–11 m^2/g). All these observations correlate well with the crystallite size determined from XRD patterns: a decrease in grain size, especially for H^+-CA1.1 samples, is accompanied by a marked increase in surface area values and pore volume. Increases in SSA upon Ni or Cu doping were observed in the literature as well, e.g., 26 m^2/g for $LaMn_{0.9}Ni_{0.1}O_3$ compared to 20 m^2/g for $LaMnO_3$, both calcined at 750 °C [20], and 15 m^2/g for $LaMnCu_{0.1}O_3$ compared to 8 m^2/g for $LaMnO_3$, calcined at 700 °C [21]. The SSA values of 44 and 37 m^2/g obtained in this work for $La_{0.8}Mn_{0.9}Ni_{0.1}O_3$-$H^+$-CA1.1 and $La_{0.8}Mn_{0.9}Cu_{0.1}O_3$-$H^+$-CA1.1, respectively, are therefore even higher than those in many works in the literature.

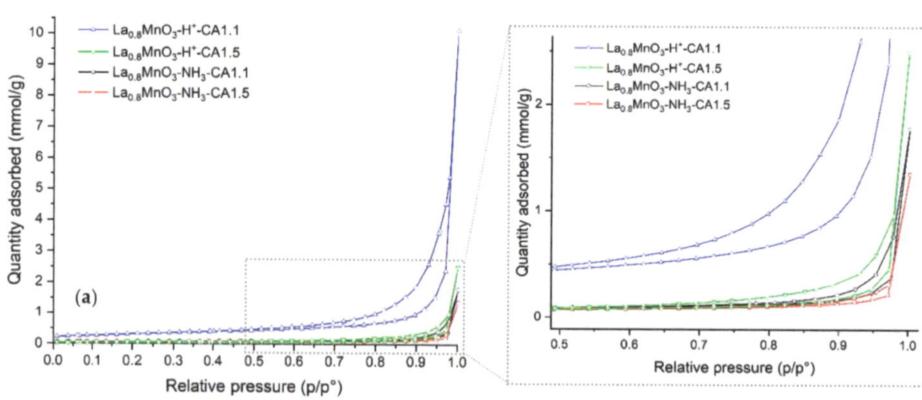

Figure 2. *Cont.*

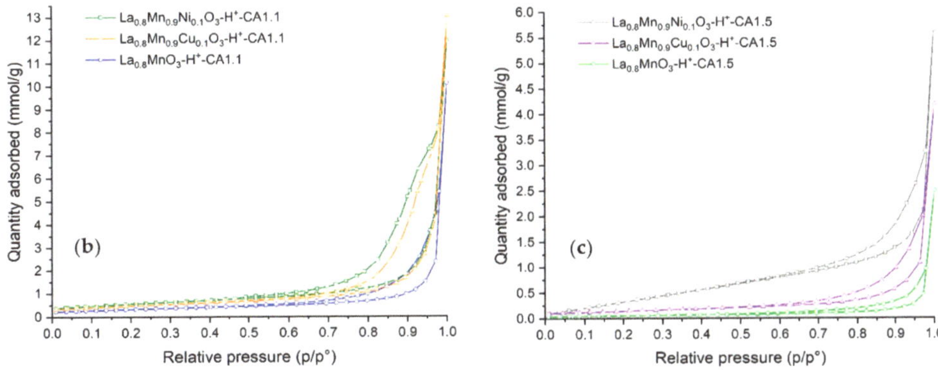

Figure 2. N_2 adsorption isotherm of the prepared catalysts: (**a**) the four $La_{0.8}MnO_3$ samples, with zoom on the high-pressure range of the low SSA samples; (**b**) Ni- and Cu-doped samples with CA1.1, compared to $La_{0.8}MnO_3$-H^+-CA1.1; (**c**) Ni- and Cu-doped samples with CA1.5, compared to $La_{0.8}MnO_3$-H^+-CA1.5.

Representative SEM micrographs of the eight prepared perovskites are shown in Figures 3 and S3. The morphology of NH_3 samples (Figures 3a and S3a) appears to be composed of particles with a size distribution from tens to hundreds of nm, which leave pores of variable dimension upon aggregation. In the case of H^+-CA1.1 samples (Figures 3b,c and S3c), individual grains are not clearly discernible, in agreement with their lower size calculated from XRD patterns, and many small pores in the meso-range are evident, especially for B-site-doped perovskites (Figures 3c and S3c), in agreement with N_2 physisorption results. H^+-CA1.5 samples (Figures 3d and S3b,d) have somewhat intermediate characteristics, with a more compact morphology compared to H^+-CA1.1 samples, as evidenced by the presence of several-μm-wide plate-like aggregates (Figure S3b,d) and of discernible grains (Figures 3d and S3b), although a certain mesoporosity can still be appreciated, especially for Ni- and Cu-doped samples (Figures 3d and S3d). This is again in agreement with both XRD and N_2 physisorption results (e.g., intermediate SSA values and pore volume).

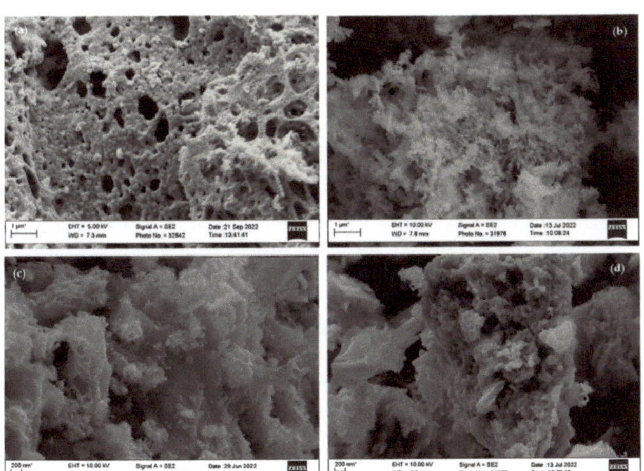

Figure 3. SEM micrographs of (**a**) $La_{0.8}MnO_3$-NH_3-CA1.1; (**b**) $La_{0.8}MnO_3$-H^+-CA1.1; (**c**) $La_{0.8}Mn_{0.9}Ni_{0.1}O_3$-$H^+$-CA1.1; (**d**) $La_{0.8}Mn_{0.9}Ni_{0.1}O_3$-$H^+$-CA1.5. Images (**a**,**b**) are taken at 25,000× magnification, images (**c**,**d**) at 50,000× magnification.

2.1.2. Precursor Decomposition

Since the calcination temperature was kept equal for all catalysts, the differences in morphological properties at varying synthesis parameters, as evidenced by XRD, N_2 physisorption and SEM, are clearly dependent on how the exothermic reaction occurs between nitrate oxidizer and citrate fuel during the perovskite synthesis process. By empirical observation, we generally noted that gels of samples prepared in acidic pH required temperatures lower than 250 °C (sometimes just above 150 °C) to start decomposing, with the evolution of gases and a change in color from yellowish/greenish to dark brown, as if carbonized. On the other hand, gels obtained after the addition of NH_3 seemed to require a temperature above 250 °C for this process to occur, at first with the evolution of gases, then with a clearly visible flame, which was not observed in the former case, producing a grey/black ash-like solid. To try to better understand the process of gel decomposition, thermogravimetric analysis (TGA) was carried out on four selected gel precursors after being dried at 120 °C; the resulting weight loss and corresponding first derivative profiles as a function of temperature are reported in Figure 4. As a general trend, a small weight loss of 5–10 % at up to 120–150 °C is ascribed to the removal of residual water of hydration. The main weight loss of up to 50–70 %, occurring between about 150 and 350 °C, is due to the exothermic nitrate–citrate redox reaction, and further weak weight losses can be ascribed to the decomposition of residual carbonates up to ≈500 °C, as well as to the formation of the final perovskite structure with some oxygen release by further increasing the temperature [15]. The gel of the $La_{0.8}MnO_3$-H^+-CA1.1 sample shows a very sharp and one-step weight loss at 195 °C because of the nitrate–citrate reaction. For $La_{0.8}MnO_3$-NH_3-CA1.1, and especially for $La_{0.8}MnO_3$-H^+-CA1.5, such decomposition occurs in more than one step, extending through a wider temperature range (the former up to 270 °C, the latter up to 350 °C). The delayed onset of decomposition when increasing the CA/M ratio (with acidic pH syntheses) was observed in previous works as well [13,15,16], explained by the lower amount of nitrate oxidizer in proportion to the citrate fuel; the same authors reported that either the longer reaction time or the higher amount of citrate fuel resulted in higher local temperatures, which produced bigger crystallites and decreased the final surface area. The values of crystallite size and SSA that we obtained for $La_{0.8}MnO_3$-H^+-CA1.1 and $La_{0.8}MnO_3$-H^+-CA1.5 are in line with these observations. In the case of the $La_{0.8}MnO_3$-NH_3-CA1.1 sample, the TGA profile suggests that the decomposition of the gel's citrate matrix starts to occur before the expected temperature of 250 °C, although the sharpest weight loss occurs right after this temperature, which we postulate might be due to the exothermic decomposition of the NH_4NO_3 formed in situ [27], further triggering the complete decomposition of the gel. The highest total weight loss observed for this sample is likely due to the initial presence of NH_3. The higher crystallite size and poorer SSA values observed for NH_3 samples could be again due to the longer decomposition time as observed in TGA plots, or perhaps to the occurrence of the flame, which would be characterized by a very high local temperature, resulting in particle sintering and a decrease in SSA, with only a weak dependency on the CA/M ratio. Even for the gel of the $La_{0.8}Mn_{0.9}Ni_{0.1}O_3$-$H^+$-CA1.1 sample, a multiple-step weight loss extending to almost 350 °C was noticed, which is thus above the temperature at which the gel carbonization was observed (as it was for $La_{0.8}MnO_3$-H^+-CA1.5); in fact, such carbonization might occur with the first weight loss centered at ~190°C. The reason for the lower final crystallite size and higher SSA achieved through Ni doping (as well as Cu doping), compared to $La_{0.8}MnO_3$-H^+-CA1.1, seems not to be clarified by TGA alone; it could be more related to the perovskite unit cell distortion induced by dopant cations rather than to the gel decomposition process.

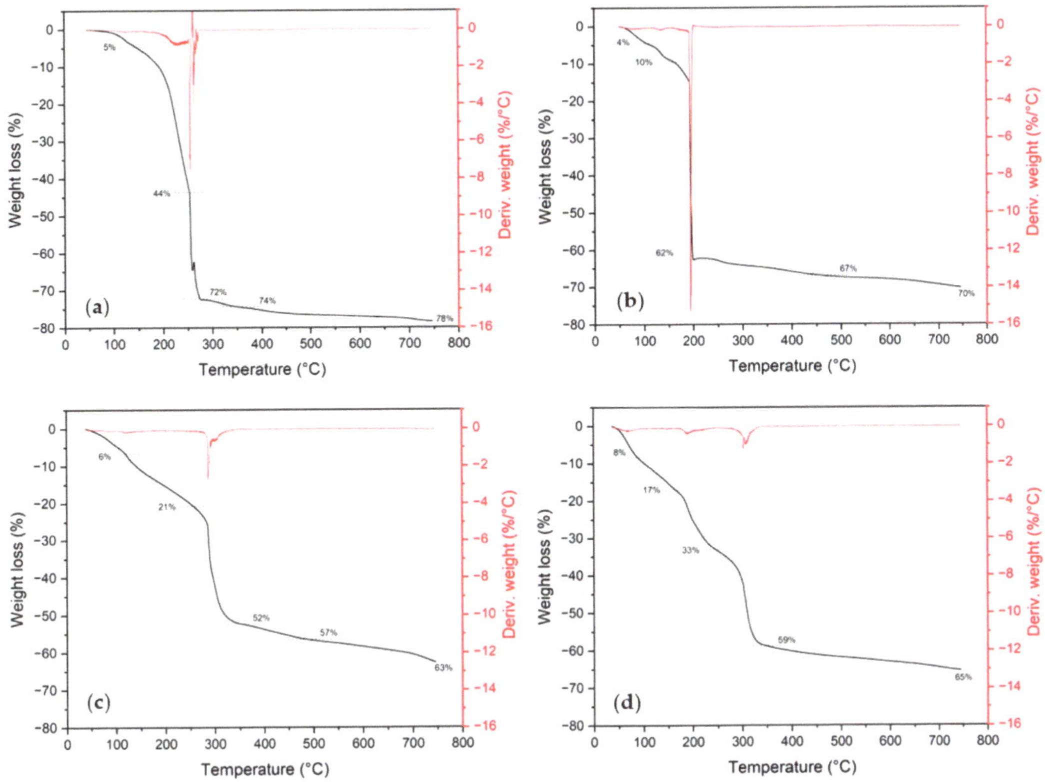

Figure 4. TGA weight loss as a function of temperature (black) and corresponding first derivative (red) of: (**a**) $La_{0.8}MnO_3$-NH_3-CA1.1; (**b**) $La_{0.8}MnO_3$-H^+-CA1.1; (**c**) $La_{0.8}MnO_3$-H^+-CA1.5; (**d**) $La_{0.8}Mn_{0.9}Ni_{0.1}O_3$-$H^+$-CA1.1.

2.1.3. Bulk and Surface Composition

The average perovskite composition until a depth of ~1 μm was explored by energy-dispersive X-ray analysis (EDX), and the average surface composition (depth of 5–10 nm) was explored by X-ray photoelectron spectroscopy (XPS). Element percentages are reported in Table S1 and Table 3, with and without considering the oxygen content, respectively, which is always excessive at the surface as well as in bulk, as usually occurs for perovskite oxides [28,29]. For all samples, all the expected elements (O, La, Mn, Ni/Cu, if any) were detected by both EDX and XPS, without evidence of a significant amount of impurities.

Table 3. Bulk (EDX) and surface (XPS) composition of the prepared perovskites, compared to the respective nominal amount according to sample stoichiometry.

Sample		La (at%) *	Mn (at%) *	Ni or Cu (at%) *
$La_{0.8}MnO_3$-NH_3-CA1.1	EDX	48.1	51.9	-
	XPS	45.0	55.0	-
	(nominal)	(44.5)	(55.5)	-
$La_{0.8}MnO_3$-NH_3-CA1.5	EDX	46.2	53.8	-
	XPS	42.9	57.1	-
	(nominal)	(44.5)	(55.5)	-

Table 3. *Cont.*

Sample		La (at%) *	Mn (at%) *	Ni or Cu (at%) *
$La_{0.8}MnO_3$-H^+-CA1.1	EDX	50.1	49.9	-
	XPS	41.7	58.3	-
	(nominal)	(44.5)	(55.5)	-
$La_{0.8}MnO_3$-H^+-CA1.5	EDX	46.5	53.5	-
	XPS	41.9	58.1	-
	(nominal)	(44.5)	(55.5)	-
$La_{0.8}Mn_{0.9}Ni_{0.1}O_3$-$H^+$-CA1.1	EDX	48.7	46.3	5.0
	XPS	37.9	60.1	2.0
	(nominal)	(44.5)	(50.0)	(5.5)
$La_{0.8}Mn_{0.9}Ni_{0.1}O_3$-$H^+$-CA1.5	EDX	48.3	45.7	6.0
	XPS	39.7	56.2	4.1
	(nominal)	(44.5)	(50.0)	(5.5)
$La_{0.8}Mn_{0.9}Cu_{0.1}O_3$-$H^+$-CA1.1	EDX	48.1	48.3	3.6
	XPS	37.2	51.5	11.3
	(nominal)	(44.5)	(50.0)	(5.5)
$La_{0.8}Mn_{0.9}Cu_{0.1}O_3$-$H^+$-CA1.5	EDX	48.2	45.7	6.1
	XPS	35.7	53.7	10.6
	(nominal)	(44.5)	(50.0)	(5.5)

* Metal percentage without considering the oxygen content.

In general, EDX reveals an enrichment in La and a deployment of Mn with respect to the nominal amounts, almost irrespective of the synthesis protocol, although this phenomenon seems more prominent in samples prepared in acidic pH. On the contrary, XPS reveals an opposite Mn enrichment at the expense of La, which is an effect of the A-site understoichiometry and could in principle be desirable for catalysis, with the surface being enriched with the redox active species (Mn^{n+}) [18]. Regarding Ni and Cu dopants, with EDX, there seems to be a slightly deficient amount with H^+-CA1.1 synthesis and a slightly excess amount in the case of H^+-CA1.5. At the surface, Ni is always deficient and Cu always in excess, likely confirming the complete Ni^{n+} incorporation inside the perovskite framework and the partial Cu^{2+} segregation as $CuMn_2O_4$ spinel, as revealed by XRD. Generally speaking, there are no evident correlations between synthesis parameters (CA/M ratio, pH) and the resulting surface and bulk composition.

2.1.4. Bulk Reducibility and Oxygen Mobility

Samples' reducibility was tested by H_2-TPR. The recorded profiles are reported in Figure 5, and reduction temperatures and quantification of H_2 consumption are listed in Table 4. Starting from the benchmark $La_{0.8}MnO_3$ composition (Figure 4a), two main reduction processes are discernible: $Mn^{4+} \rightarrow Mn^{3+}$ reduction in the lower temperature range (~150–550 °C), in parallel with the removal of the excessive lattice oxygen with respect to metal cations, and $Mn^{3+} \rightarrow Mn^{2+}$ reduction in the higher temperature range (~600–900 °C), leading to perovskite decomposition into single oxides/hydroxides (La_2O_3 or $La(OH)_3$ + MnO) [9,19]. Although the described reduction events are the same for all samples, peaks' characteristics (shape, area and position) are quite variable depending on the adopted synthesis protocol. In particular, the peaks tend to shift towards lower temperatures when decreasing the amount of CA (1.1) in the precursor solution and with acidic pH (H^+) versus neutral pH (NH_3), indicating a more ready reducibility. Indeed, $La_{0.8}MnO_3$-H^+-CA1.1 is by far the most reducible catalyst among the four, with the $Mn^{4+} \rightarrow Mn^{3+}$ reduction anticipated by at least 40 °C compared to the other three; in parallel, its low-temperature H_2 uptake is much enhanced (at the expense of the high-temperature uptake), indicating a greater lattice stabilization of Mn^{4+} at the expense of Mn^{3+}, whereas the three samples of H^+-CA1.5, NH_3-CA1.1 and NH_3-CA1.5 show approximately the same H_2 uptake in both T ranges. These results can be well interpreted in terms of the lower crystallite size and much greater surface area and porosity of the H^+-CA1.1 samples [15,19], leading to easier adsorption and activation of H_2 during TPR experiments in comparison to the other three samples, which displayed a very similar surface area and a weak Mn_3O_4 phase segregation, likely decreasing the amount of Mn^{4+} in the perovskite lattice. Furthermore, the low-temperature shoulder at 200–250 °C, evidently appearing only in H^+-CA1.1 samples, could be attributed to the reduction in highly reducible Mn^{4+} species, likely located at the abundant surface or pore sites of this sample; a similar result was obtained for nanocrystalline high-surface area CeO_2, which was shown to display an additional TPR peak at low temperatures compared to bulk low-surface area CeO_2 [30].

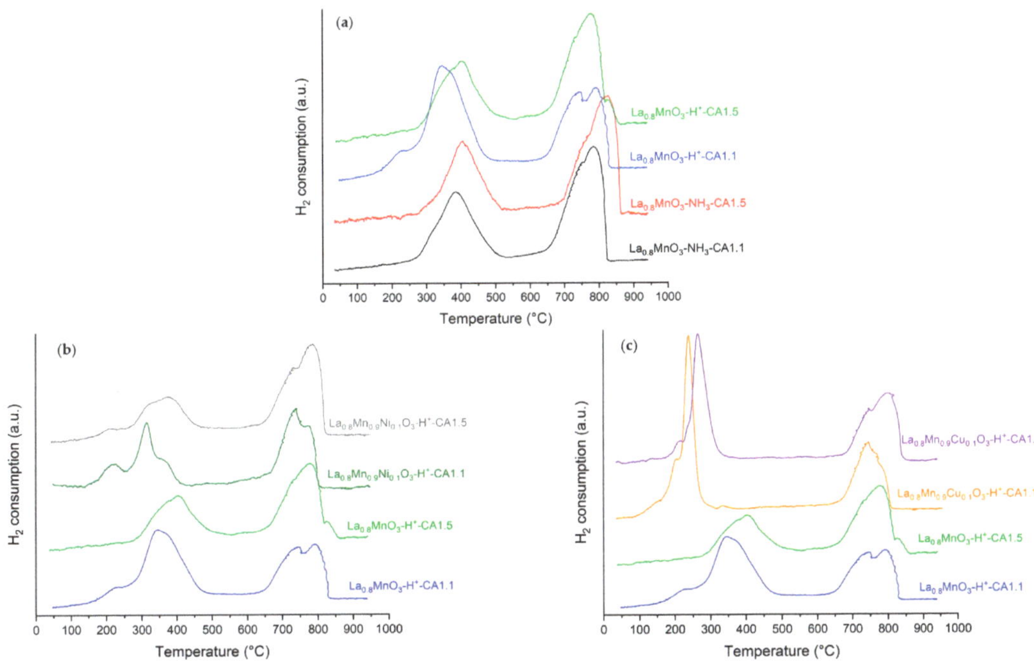

Figure 5. H_2-TPR profiles of: (**a**) the four $La_{0.8}MnO_3$ samples; (**b**) Ni-doped perovskites, compared to undoped ones prepared with the same protocol; (**c**) Cu-doped perovskites, compared to undoped ones prepared with the same protocol.

Table 4. Peak temperatures and quantitative analysis of H_2-TPR experiments.

Sample	Low Temperature Range [1]			High Temperature Range [2]		
	T_{max} (°C)	H_2 Uptake (mmol/g)	H/Mn + Cu	T_{max} (°C)	H_2 Uptake (mmol/g)	H/Mn + Ni
$La_{0.8}MnO_3$-NH_3-CA1.1	387	1.18	0.50	785	1.66	0.71
$La_{0.8}MnO_3$-NH_3-CA1.5	406	1.20	0.51	827	1.69	0.72
$La_{0.8}MnO_3$-H^+-CA1.1	346	1.85	0.79	794	1.32	0.56
$La_{0.8}MnO_3$-H^+-CA1.5	403	1.16	0.49	777	1.83	0.78
$La_{0.8}Mn_{0.9}Ni_{0.1}O_3$-$H^+$-CA1.1	315	1.35	0.58	740	1.73	0.74

Table 4. *Cont.*

Sample	Low Temperature Range [1]			High Temperature Range [2]		
	T_{max} (°C)	H_2 Uptake (mmol/g)	H/Mn + Cu	T_{max} (°C)	H_2 Uptake (mmol/g)	H/Mn + Ni
$La_{0.8}Mn_{0.9}Ni_{0.1}O_3$-$H^+$-CA1.5	378	1.02	0.44	785	1.96	0.84
$La_{0.8}Mn_{0.9}Cu_{0.1}O_3$-$H^+$-CA1.1	239	1.98	0.85	745	1.45	0.62
$La_{0.8}Mn_{0.9}Cu_{0.1}O_3$-$H^+$-CA1.5	265	1.59	0.68	793	1.68	0.72

[1] Corresponding to reduction steps $Mn^{4+} \rightarrow Mn^{3+}$ and $Cu^{2+} \rightarrow Cu^{+} \rightarrow Cu^{0}$; [2] corresponding to reduction steps $Mn^{3+} \rightarrow Mn^{2+}$ and $Ni^{2+} \rightarrow Ni^{0}$.

Considering, now, the Ni- and Cu-doped samples, the above written considerations about the influence of textural properties on reducibility are confirmed: at a parity of composition, H^+-CA1.1 samples are more reducible and display a greater lattice Mn^{4+} stabilization with respect to H^+-CA1.5 samples, thanks to lower crystallite size, abundant surface area and porosity, as obtained by XRD and N_2 physisorption. The effect of Cu B-site doping (Figure 3c) is quite evident: Cu^{2+} reduction to Cu^0 in two steps, with Cu^+ as an intermediary, appears on the TPR profile with two shoulders at very low temperatures (range 100–220 °C), confirmed by an increase in the total H_2 uptake at low temperatures as compared to un-doped perovskites. Furthermore, Mn^{4+} reduction is greatly anticipated (at more than 100 °C) compared to benchmark $La_{0.8}MnO_3$ and starts right after the complete reduction of Cu, thanks to the phenomenon of H_2 spillover from pre-formed Cu^0 particles to the perovskite support [21]. In contrast, Ni doping (Figure 3b) has a less marked effect: there are no evident peaks ascribed to Ni^{n+} reduction; based on previous literature with similar compositions [31], $Ni^{2+} \rightarrow Ni^0$ reduction might be located at around 600–650 °C, possibly hidden by the rising part in the Mn^{3+} reduction peak. In any case, Ni doping seems to render Mn^{4+} species slightly more reducible by anticipating their peak temperature of 20–30 °C compared to benchmark $La_{0.8}MnO_3$, although it is not clear whether this is an effect of Ni^{n+} cations themselves, or of the improved surface area and porosity arising from Ni^{n+} insertion in the lattice, as could be inferred by the enhancement of the low-temperature shoulder previously ascribed to highly reducible Mn^{4+} species located at surface/pore sites. Despite such improvement in Mn^{4+} reducibility, the total H_2 uptake in the low T region decreases, and that in the high T region increases upon Ni substitution, suggesting the presence of a greater amount of Mn^{3+} in the lattice at the expense of Mn^{4+}, as found elsewhere in the literature [32].

Apart from cation reducibility, oxygen mobility was further investigated by O_2-TPD on five selected samples. The O_2 desorption profiles are shown in Figure 6, and the calculation of evolved O_2 amounts at low and high temperatures is reported in Table 5. Let us note that the latter calculation cannot be considered very accurate because of the uncomplete desorption process at the maximum tested temperature of 900 °C, but it has been added for the sake of comparison among the different catalysts. Below 500 °C, the release of weakly bound surface oxygen species (α_1-O_2), likely chemisorbed on oxygen vacancies, can be detected; at intermediate temperatures of approximately 550 to 700 °C, lattice oxygen still coming from the surface or near the surface (α_2-O_2) is desorbed; finally, above 700 °C, the release of bulk lattice oxygen (β-O_2) can be appreciated [3], even occurring in more than one step [20]. Notably, H^+-CA1.1 samples display the highest amount of α_1-O_2 desorption, especially the Ni-doped sample, in agreement with their easy reducibility at low temperatures and superior surface area values. $La_{0.8}MnO_3$-H^+-CA1.1 has the most prominent α_2- and β-O_2 release as well, whereas these decrease upon Ni doping; this suggests that the release of bulk lattice oxygen is accompanied by bulk $Mn^{4+} \rightarrow Mn^{3+}$ reduction, whose H_2-TPR peak was less marked on $La_{0.8}Mn_{0.9}Ni_{0.1}O_3$ compared to $La_{0.8}MnO_3$. $La_{0.8}MnO_3$-NH_3-CA1.1 and $La_{0.8}MnO_3$-H^+-CA1.5 samples do not show an evident α_1-O_2 peak; it is likely very weak and hidden by experimental noise due to low surface area and delayed Mn^{4+} reducibility compared to H^+-CA1.1 samples, whereas $La_{0.8}Mn_{0.9}Cu_{0.1}O_3$-$H^+$-CA1.5

samples display a higher amount of oxygen desorption in the whole range compared to undoped $La_{0.8}MnO_3$-H^+-CA1.5, thanks to the somewhat improved surface area and Mn^{4+} reducibility arising from B-site doping.

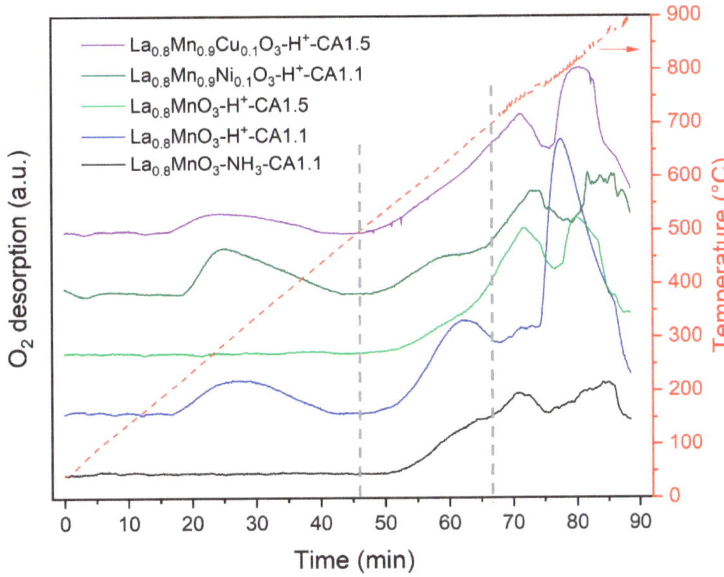

Figure 6. O_2-TPD profiles of five selected perovskites. Grey dotted vertical lines, at about 500 °C and 700 °C, roughly separate the three O_2 desorption stages: α_1, α_2 and β, in order of increasing T.

Table 5. Quantitative analysis of O_2-TPD experiments.

Sample	O_2 Desorption (μmol/g)	
	Low T Range (<500 °C)	**High T Range (>500 °C)**
$La_{0.8}MnO_3$-NH_3-CA1.1	-	86
$La_{0.8}MnO_3$-H^+-CA1.1	28	296
$La_{0.8}MnO_3$-H^+-CA1.5	-	149
$La_{0.8}Mn_{0.9}Ni_{0.1}O_3$-$H^+$-CA1.1	46	145
$La_{0.8}Mn_{0.9}Cu_{0.1}O_3$-$H^+$-CA1.5	8.3	208

2.2. Catalytic Activity

The prepared A-site-deficient perovskites were tested as catalysts for the complete combustion of CH_4 to CO_2 under a stoichiometric amount of O_2, as a model reaction occurring on methane-fed vehicles' three-way catalytic converters (TWC) [33]. No pre-treatment of the catalyst was performed, but further tests with a pre-reduction under diluted H_2 before exposure to the reaction mixture are currently ongoing, possibly modifying the final catalytic performances. CH_4 light-off conversion curves are reported in Figure 7, and the values of T_x (temperature necessary to achieve a certain value x of CH_4 conversion) and the reaction rates at 600 °C are listed in Table 6.

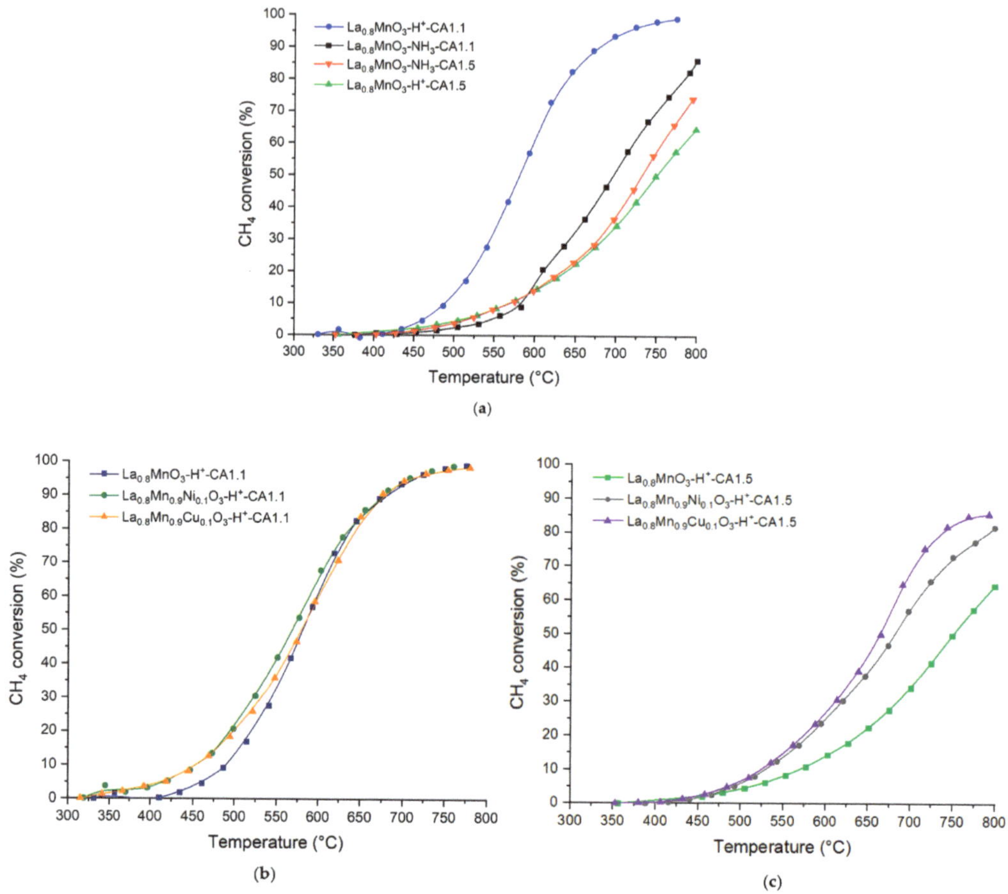

Figure 7. CH_4 conversion light-off curves of: (**a**) the four $La_{0.8}MnO_3$ samples; (**b**) Ni- and Cu-doped samples with CA1.1, compared to $La_{0.8}MnO_3$-H^+-CA1.1; (**c**) Ni- and Cu-doped samples with CA1.5, compared to $La_{0.8}MnO_3$-H^+-CA1.5.

Table 6. CH_4 oxidation performance of the prepared catalysts: temperature of 10%, 50% and 80% CH_4 conversion, specific and intrinsic reaction rates at 600 °C.

Sample	T_{10} (°C)	T_{50} (°C)	T_{80} (°C)	$r_{specific}$ (mol/s/g)	$r_{intrinsic}$ (mol/s/m^2)
$La_{0.8}MnO_3$-NH_3-CA1.1	583	696	783	3.7×10^{-6}	7.5×10^{-7}
$La_{0.8}MnO_3$-NH_3-CA1.5	570	732	>800	3.5×10^{-6}	8.8×10^{-7}
$La_{0.8}MnO_3$-H^+-CA1.1	489	581	639	9.9×10^{-6}	4.0×10^{-7}
$La_{0.8}MnO_3$-H^+-CA1.5	569	750	>800	3.6×10^{-6}	8.7×10^{-7}
$La_{0.8}Mn_{0.9}Ni_{0.1}O_3$-$H^+$-CA1.1	455	568	637	1.0×10^{-5}	2.3×10^{-7}
$La_{0.88}Mn_{0.9}Ni_{0.1}O_3$-$H^+$-CA1.5	528	681	790	5.6×10^{-6}	7.4×10^{-7}
$La_{0.88}Mn_{0.9}Cu_{0.1}O_3$-$H^+$-CA1.1	455	580	644	1.0×10^{-5}	2.7×10^{-7}
$La_{0.88}Mn_{0.9}Cu_{0.1}O_3$-$H^+$-CA1.5	524	664	736	6.2×10^{-6}	5.3×10^{-7}

Before examining the activity of the different samples, it is useful to introduce the mechanism of CH_4 (and other HCs) combustion on perovskite oxides. The reaction is reported to follow the Mars–Van Krevelen (MvK) mechanism [34], in which oxygen species on the perovskite surface are the primary active species for the reactant activation and complete oxidation to CO_2; the release of oxygen species induces the formation of oxygen vacancies (V_O) on the catalyst surface and the parallel reduction in B-site cations (in this case, $Mn^{4+} \rightarrow Mn^{3+}$); finally, gas-phase oxygen is dissociatively adsorbed on the perovskite surface, thus replenishing V_O and reoxidizing the reduced cations. However, such oxygen species can be different depending on the temperature range: at low temperatures, the mechanism is suprafacial, meaning that the reaction involves weakly bound oxygen species directly adsorbed on the catalyst surface, particularly on V_O, which is easily exchangeable with gaseous O_2; at high temperatures, the diffusion of lattice oxygen becomes thermally activated, and thus lattice oxygen species present in the sub-surface layers can migrate to the reaction sites to oxidize the reactant and/or replenish the generated V_O. Here, the mechanism is referred to as intrafacial [6–8,19,35]. Therefore, in addition to the amount of exposed surface active sites, other important factors for catalytic activity are the content of highly valent Mn^{4+} ions and their ease of reducibility, in connection with a facile release of weakly adsorbed oxygen species at low temperatures, and with a fast mobility of lattice oxygen at high temperatures.

All catalysts produce CO_2 with selectivity close to 100% as a product of CH_4 oxidation (a very low and hardly quantifiable amount of CO was detected in a few cases), since the stoichiometric amount of O_2 in the gaseous mixture and the ability of perovskites to release active oxygen in general prevent uncomplete combustion. Among undoped $La_{0.8}MnO_3$ compositions, the H^+-CA1.1 sample is by far the most active, having an anticipated T_{50} value of more than 100 °C compared to the other three samples; it is also the only sample that is able to achieve full CH_4 conversion at 800 °C. This is evidently an effect of the superior surface area, porosity, Mn^{4+} reducibility and surface and bulk oxygen mobility and is also possibly due to the absence of Mn_3O_4 segregated phases, as revealed by physicochemical characterizations. The other three samples, NH_3-CA1.1, NH_3-CA1.5 and H^+-CA1.5, have quite similar activities, since their physicochemical properties are not much different (e.g., SSA = 4–5 m^2/g for all), although the NH_3-CA1.1 sample seems to perform slightly better than the other two above 600 °C, possibly thanks to a slightly easier Mn^{4+} reducibility (T_{max} at TPR of 387 °C compared to 403–406 °C of H^+-CA1.5 and NH_3-CA1.5). The calculation of specific reaction rates (i.e., at parity of the catalyst weight) at 600 °C confirm the superiority of the H^+-CA1.1 sample and the similar performances of the other three samples. However, the trend in the intrinsic rate values (normalized by surface area and thus linked to the single active sites' activity) is not in agreement with a specific rate trend, with H^+-CA1.1 having the lowest intrinsic rate. This likely supports the thesis that specific surface area, implying a greater exposure of active sites, is one of the main factors in improving catalytic activity, since it also influences active Mn^{4+} ions' reducibility and the ability to release active oxygen species.

Concerning Cu- and Ni-doped perovskites, analogous considerations about the influence of different synthesis protocols on catalytic activities can be formulated. Samples prepared with a lower CA/M ratio (1.1 vs. 1.5) perform much better, thanks to more favorable morphological features and Mn^{4+} reducibility. H^+-CA1.5 samples are indeed not able to reach complete CH_4 conversion in the tested temperature range and have much greater T_{50} values (of at least 85 °C) compared to H^+-CA1.1 samples. The comparison with undoped $La_{0.8}MnO_3$ compositions reveals some activity gain upon Ni or Cu doping, which is more evident in the case of H^+-CA1.5 samples: the T_{50} values are 681 and 664 °C for $La_{0.8}Mn_{0.9}Ni_{0.1}O_3$ and $La_{0.8}Mn_{0.9}Cu_{0.1}O_3$, respectively (vs. 750 °C for undoped composition), with Cu doping performing slightly better than Ni doping in the higher temperature region. The better performances are always related to better surface area and porosity resulting from B-site doping, confirmed by the calculation of reaction rates at 600 °C: specific rates (per unit of mass) are greater for doped samples compared to undoped ones, and the reverse is true for intrinsic rates (per unit of exposed surface). O_2-TPD further proved that $La_{0.8}Mn_{0.9}Cu_{0.1}O_3$-$H^+$-CA1.5 can release a higher amount of active surface and bulk lattice oxygen species compared to $La_{0.8}MnO_3$-H^+-CA1.5. The activity enhancement achieved through B-site doping is instead almost negligible for H^+-CA1.1 samples, appearing only at low temperatures and to a slight extent: T_{10} = 455 °C for both $La_{0.8}Mn_{0.9}Ni_{0.1}O_3$ and $La_{0.8}Mn_{0.9}Cu_{0.1}O_3$, with respect to 489 °C for $La_{0.8}MnO_3$, whereas T_{50} and T_{80} are very similar for the three samples. In this case, the improved surface area only brings benefits to the low-temperature activity by enhancing the amount of surface oxygen release, as shown by O_2-TPD, thus facilitating the suprafacial mechanism; conversely, the mobility of surface and lattice oxygen is not enhanced by B-site doping (at least by Ni doping), thus not accelerating the intrafacial mechanism. A temperature value of around 600 °C might be the threshold for the change in the CH_4 oxidation mechanism under the tested conditions (stoichiometric amount of O_2), roughly corresponding to the temperature needed to activate surface lattice oxygen (α_2-O_2) at TPD. Overall, the most active catalyst is $La_{0.8}Mn_{0.9}Ni_{0.1}O_3$-$H^+$-CA1.1, which is the catalyst with the highest specific surface area.

3. Materials and Methods

3.1. Synthesis Protocol

$La_{0.8}MnO_3$ perovskites were prepared according to the citrate combustion route. The selected amount of La_2O_3 (≥99.9% Sigma-Aldrich, Beijing, China) and Mn(acetate)$_2$·4H_2O (≥99% Sigma-Aldrich, Schnelldorf, Germany) were dissolved in deionized water containing the stoichiometric amount of HNO_3 (≥65% Sigma-Aldrich, Schnelldorf, Germany), as if the corresponding $La(NO_3)_3$ and $Mn(NO_3)_2$ salts were used as precursors. Nitrate salts were generated indirectly in the precursor solution by HNO_3 addition because of the high hygroscopic nature of these salts, making the precise amount of cations more controllable with the chosen precursors La_2O_3 and $Mn(CH_3COO)_2$. Citric acid monohydrate (CA, ≥99.0% Sigma-Aldrich, Vienna, Austria) was added according to a molar ratio citric acid/sum of metal cations (CA/M) equal to 1.1 or 1.5. For the synthesis under acidic pH, the solution was directly evaporated overnight at ~120 °C, and a gel was formed, which was then decomposed in an oven at 200–230 °C for 1 h, obtaining a dark solid. For the synthesis under neutral pH, the necessary amount of NH_3 (32%, Sigma-Aldrich, Darmstadt, Germany) was added prior to solvent evaporation as above, and the resulting gel was ignited at 350 °C directly on the hotplate for few minutes, obtaining a dark ash-like solid. In both cases, the obtained solid was grinded and ultimately calcined at 750 °C (ramp 6 °C/min) for 6 h under static air. The same protocol was adopted for the preparation of $La_{0.8}Mn_{0.9}Ni_{0.1}O_3$ and $La_{0.8}Mn_{0.9}Cu_{0.1}O_3$ by dissolving the appropriate amount of $Ni(NO_3)_2$·6H_2O (≥97.0% Sigma-Aldrich, Schnelldorf, Germany) or CuO (≥99.0% Sigma-Aldrich, Tokio, Japan), together with La- and Mn-precursors, in the starting solution. For Ni, the chosen precursor was the nitrate salt and not NiO, since the latter is hardly dissolvable in HNO_3. A synthesis with NiO was attempted, but it resulted in lower crystalline purity of the final perovskite. For doped samples, only the synthesis in acidic pH was carried out, with a CA/M molar ratio of either 1.1 or 1.5. In the Results and Discussion section, the subscript H^+ is used to indicate the synthesis under acidic pH, the subscript NH_3 to indicate the synthesis under neutral pH, and the molar ratio of citric acid/metal cations is indicated by CA1.1 and CA1.5 (see Table 1).

3.2. Characterizations

The decomposition of some selected dried gels was studied by thermogravimetric analysis (TGA) on a LINSESIS 1000 instrument (Linseis Messgeraete GmbH, Selb, Germany). About 10–15 mg of sample, placed in an alumina crucible, were heated under static air from RT to 750 °C at 10 °C/min.

Powder X-ray diffraction (XRD) patterns were acquired with a Bruker D8 Advance diffractometer (Billerica, MA, USA) in Bragg–Brentano geometry, employing a Cu Kα source (λ = 0.154 nm), powered at 40 kV and 40 mA, with steps of 0.02° and a dwell time of 0.35 s/step. Average perovskites' crystallite size was estimated with the Williamson–Hall method by fitting three reflections, i.e., (012), (110) or (104) and (024) (see Figure S1 and Equation (S1)).

N_2 adsorption isotherms were collected at −196 °C with a Micromeritics ASAP2020 Plus instrument (Norcross, GA, USA), in the relative pressure range (p/p^0) of 0.01–1.0, and vice versa; prior to experiments, the samples (~1 g) were outgassed in a vacuum ($p < 1.3$ P_a) at 300 °C for 3 h. Specific surface areas (SSA) were determined according to the Brunauer–Emmett–Teller (BET) model, and pore size distributions according to Barrett–Joyner–Halenda (BJH) model.

H_2 temperature-programmed reduction (TPR) was carried out to investigate the materials' reducibility on a Micromeritics Autochem II 2920 instrument (Norcross, GA, USA). 50 mg of sample were loaded on a U-shaped quartz reactor, above a layer of quartz wool, and heated from RT to 950 °C (ramp 10 °C/min) under a 50 scm^3/min flow of 5% H_2/Ar. The H_2 consumption was monitored with a TCD detector after H_2O absorption by a cold trap.

O_2 temperature-programmed desorption (TPD) was performed on some selected perovskites with the same set-up as for H_2-TPR experiments. About 200 mg of sample were pretreated at 500 °C for 30 min in pure He (50 scm^3/min) and subsequently for 30 min in 5 vol% O_2/He (20 scm^3/min). After cooling down to 40 °C in the same environment, O_2 was allowed to be adsorbed for 30 min. The atmosphere was switched to 30 scm^3/min of He and, after a purging at 40 °C for 15 min to remove physiosorbed oxygen, the sample was finally heated up to 900 °C at 10 °C/min.

Scanning electron microscopy (SEM) images were acquired with a Zeiss SUPRA 40 V P microscope (Zeiss, Oberkochen, Germany), setting the electron acceleration voltage at 5 or 10 kV. Energy-dispersive X-ray analysis (EDX) was coupled with SEM for elemental quantification at 20 kV electron acceleration voltage, probing a large area of sample (rectangle of hundreds of μm on each side); X-ray peaks considered for quantification were O Kα, Mn Kα, La Lα, Ni Kα and Cu Kα.

X-ray photoelectron spectroscopy (XPS) was performed with a Thermo Scientific ESCALAB QXi spectrometer (Waltham, MA, USA), employing a monochromatized Al Kα source (hν = 1486.68 eV) and a charge compensation gun. Survey spectra were acquired at 100 eV pass energy, 0.5 eV/step and 25 ms/step dwell time. Elemental quantification was carried by the integration of La $3d_{5/2}$, Mn 2p, O 1s, Ni 3p and Cu 2p photopeaks (Figure S4), after Shirley-type background subtraction.

3.3. Catalytic Tests

The prepared catalysts were tested for complete oxidation of CH_4 to CO_2 under stoichiometric O_2. A fixed-bed quartz reactor (ID 6 mm) was employed, loading 50 mg of sample between two layers of quartz wool, and a K-type thermocouple was placed in contact with the catalyst bed to monitor its temperature. The reaction mixture was composed of 2 vol% CH_4, 4 vol% O_2, 16 vol% N_2 and He balance, with a total flow rate of 100 scm^3/min (weight hourly space velocity WHSV = 120 L/h/g). The catalytic tests were conducted in the temperature range of 300–800 °C without any pre-treatment of the catalyst, heating up at 2 °C/min. The outlet gas mixture was analyzed by an online GC (Agilent 7890A, Santa Clara, CA, USA) equipped with a TCD detector and two packed columns, MolSieve 13X (60/80 mesh, 1.8 m) and Porapak Q (80/100 mesh, 1.8 m), using He as a carrier gas.

Reaction rates of CH_4 conversion were calculated assuming the mass balance of a plug flow reactor (PFR) and a first-order kinetic with respect to the methane concentration, leading to the following Equations (1) and (2) for the specific rate ($r_{specific}$, normalized by catalyst mass) and intrinsic rate ($r_{intrinsic}$, normalized by catalyst surface area), respectively [36]. X_{CH4} is the methane conversion, m the catalyst mass (g), $F_{CH4,0}$ the inlet molar flow rate of methane (mol/s) and SSA (m^2/g) the catalyst specific surface area as determined from the BET method.

$$r_{specific} = \frac{F_{CH4,0}}{m}(1 - X_{CH4})\ln\frac{1}{1 - X_{CH4}} \text{ (mol/s/g)} \tag{1}$$

$$r_{intrinsic} = r_{specific}/SSA \left(\mathrm{mol/s/m^2}\right) \quad (2)$$

4. Conclusions

In summary, the A-site-deficient perovskites $La_{0.8}MnO_3$ and $La_{0.8}Mn_{0.9}B_{0.1}O_3$ (B = Ni, Cu) were synthesized through the citrate–nitrate combustion route. The effect of the citric acid/metal cations (CA/M) molar ratio and of the precursor solution's pH on textural, physicochemical and catalytic properties was investigated. It was found that a relatively low CA/M ratio, i.e., 1.1, in conjunction with acidic pH synthesis, produced perovskites featuring a high phase purity, a low crystallite size, a high surface area and mesoporosity and an improved Mn^{4+} reducibility and surface oxygen mobility, thanks to a fast citrate–nitrate decomposition process that likely avoided excessive particle sintering. Instead, both an increase in the CA/M ratio to 1.5 in acidic pH and neutralization by the addition of NH_3 resulted in longer precursor decomposition time and/or higher heat of combustion, increasing the local temperature and producing bigger particles with less porosity, and also decreasing Mn^{4+} reducibility and oxygen mobility. B-site doping with both Ni and Cu cations had a beneficial effect, further improving textural properties and Mn^{4+} reducibility, especially at a CA/M ratio of 1.1. In accordance with the results of physicochemical characterizations, catalytic performances in CH_4 oxidation to CO_2 were much better for samples prepared in acidic pH with a CA/M ratio of 1.1, and the exposed surface area was found to be one of the main factors affecting the catalytic process, in conjunction with the higher Mn^{4+} reducibility and surface oxygen mobility. Cu and Ni doping further improved catalytic activity in the case of a CA/M ratio of 1.5, thanks to improved surface area, reducibility and lattice oxygen mobility, whereas the activity gain was modest and only at low temperatures for samples with a CA/M ratio of 1.1, since in this case, the high temperature activity was mostly dependent on the mobility of lattice oxygen and not improved by B-site doping in contrast to surface oxygen mobility.

Supplementary Materials: The following supporting information can be downloaded at: https://www.mdpi.com/article/10.3390/catal13081177/s1, Figure S1: Extract of XRD pattern with phase matching; Figure S2: BJH pore size distribution; Figure S3: Other SEM micrographs; Figure S4: Representative XPS survey spectra; Equation S1: Williamson–Hall equation; Table S1: EDX and XPS compositions including oxygen content.

Author Contributions: Conceptualization, A.O. and A.G.; methodology, A.G.; validation, A.O., L.R. and J.C.; formal analysis, A.O.; investigation, A.O., L.R. and J.C.; resources, A.G.; visualization: A.O., L.R. and J.C.; writing—original draft preparation, A.O.; writing—review and editing, L.R., J.C. and A.G.; supervision, A.G.; project administration, A.O. and A.G.; funding acquisition, A.G. All authors have read and agreed to the published version of the manuscript.

Funding: This research was supported by European Union HORIZON EUROPE under grant agreement number 101091534 KNOWSKITE-X.

Data Availability Statement: All data are available upon request by contacting the corresponding author: andrea.osti.1@phd.unipd.it.

Acknowledgments: The authors acknowledge the KNOWSKITE-X project for the provision of funding. A special thank goes to Nicola Tiso for TGA measurements.

Conflicts of Interest: The authors declare no conflict of interest. The funders had no role in the design of the study; in the collection, analyses, or interpretation of data; in the writing of the manuscript; or in the decision to publish the results.

References

1. Zhu, J.; Li, H.; Zhong, L.; Xiao, P.; Xu, X.; Yang, X.; Zhao, Z.; Li, J. Perovskite Oxides: Preparation, Characterizations, and Applications in Heterogeneous Catalysis. *ACS Catal.* **2014**, *4*, 2917–2940. [CrossRef]
2. Peña, M.A.; Fierro, J.L.G. Chemical Structures and Performance of Perovskite Oxides. *Chem. Rev.* **2001**, *101*, 1981–2017. [CrossRef] [PubMed]

3. Najjar, H.; Batis, H. Development of Mn-Based Perovskite Materials: Chemical Structure and Applications. *Catal. Rev. Sci. Eng.* **2016**, *58*, 371–438. [CrossRef]
4. Yang, L.; Li, Y.; Sun, Y.; Wang, W.; Shao, Z. Perovskite Oxides in Catalytic Combustion of Volatile Organic Compounds: Recent Advances and Future Prospects. *Energ. Environ. Mater.* **2022**, *5*, 751–776. [CrossRef]
5. Yang, J.; Guo, Y. Nanostructured Perovskite Oxides as Promising Substitutes of Noble Metals Catalysts for Catalytic Combustion of Methane. *Chinese Chem. Lett.* **2018**, *29*, 252–260. [CrossRef]
6. Alifanti, M.; Kirchnerova, J.; Delmon, B.; Klvana, D. Methane and Propane Combustion over Lanthanum Transition-Metal Perovskites: Role of Oxygen Mobility. *Appl. Catal. A Gen.* **2004**, *262*, 167–176. [CrossRef]
7. Spinicci, R.; Delmastro, A.; Ronchetti, S.; Tofanari, A. Catalytic Behaviour of Stoichiometric and Non-Stoichiometric $LaMnO_3$ Perovskite towards Methane Combustion. *Mater. Chem. Phys.* **2002**, *78*, 393–399. [CrossRef]
8. Royer, S.; Alamdari, H.; Duprez, D.; Kaliaguine, S. Oxygen Storage Capacity of $La_{1-x}A'_xBO_3$ Perovskites (with A′=Sr, Ce; B=Co, Mn)—Relation with Catalytic Activity in the CH_4 Oxidation Reaction. *Appl. Catal. B Environ.* **2005**, *58*, 273–288. [CrossRef]
9. Chen, J.; Shen, M.; Wang, X.; Qi, G.; Wang, J.; Li, W. The Influence of Nonstoichiometry on $LaMnO_3$ Perovskite for Catalytic NO Oxidation. *Appl. Catal. B* **2013**, *134–135*, 251–257. [CrossRef]
10. Pinto, D.; Glisenti, A. Pulsed Reactivity on $LaCoO_3$-Based Perovskites: A Comprehensive Approach to Elucidate the CO Oxidation Mechanism and the Effect of Dopants. *Catal. Sci. Technol.* **2019**, *9*, 2749–2757. [CrossRef]
11. Ponce, S.; Peña, M.A.; Fierro, J.L.G. Surface Properties and Catalytic Performance in Methane Combustion of Sr-Substituted Lanthanum Manganites. *Appl. Catal. B Environ.* **2000**, *24*, 193–205. [CrossRef]
12. Marcilly, C.; Courty, P.; Delmon, B. Preparation of Highly Dispersed Mixed Oxides and Oxide Solid Solutions by Pyrolysis of Amorphous Organic Precursors. *J. Am. Ceram. Soc.* **1970**, *53*, 56–57. [CrossRef]
13. Taguchi, H.; Matsu-ura, S.; Nagao, M.; Choso, T.; Tabata, K. Synthesis of LaMnO3 by Firing Gels Using Citric Acid. *J. Solid. State Chem.* **1997**, *129*, 60–65. [CrossRef]
14. Ghiasi, E.; Malekzadeh, A.; Ghiasi, M. Moderate Concentration of Citric Acid for the Formation of $LaMnO_3$ and $LaCoO_3$ Nano-Perovskites. *J. Rare Earths* **2013**, *31*, 997–1002. [CrossRef]
15. Sihaib, Z.; Puleo, F.; Pantaleo, G.; La Parola, V.; Valverde, J.L.; Gil, S.; Liotta, L.F.; Fendler, A.G. The Effect of Citric Acid Concentration on the Properties of $LaMnO_3$ as a Catalyst for Hydrocarbon Oxidation. *Catalysts* **2019**, *9*, 226. [CrossRef]
16. Li, Y.; Xue, L.; Fan, L.; Yan, Y. The Effect of Citric Acid to Metal Nitrates Molar Ratio on Sol-Gel Combustion Synthesis of Nanocrystalline $LaMnO_3$ Powders. *J. Alloys Compd.* **2009**, *478*, 493–497. [CrossRef]
17. Nakamura, T.; Misono, M.; Yoneda, Y. Catalytic Properties of Perovskite-Type Mixed Oxides, $La_{1-x}Sr_xCoO_3$. *Bull. Chem. Soc. Jpn.* **1982**, *55*, 394–399. [CrossRef]
18. Schön, A.; Dujardin, C.; Dacquin, J.-P.; Granger, P. Enhancing Catalytic Activity of Perovskite-Based Catalysts in Three-Way Catalysis by Surface Composition Optimisation. *Catal. Today* **2015**, *258*, 543–548. [CrossRef]
19. Esmaeilnejad-Ahranjani, P.; Khodadadi, A.; Ziaei-Azad, H.; Mortazavi, Y. Effects of Excess Manganese in Lanthanum Manganite Perovskite on Lowering Oxidation Light-off Temperature for Automotive Exhaust Gas Pollutants. *Chem. Eng. J.* **2011**, *169*, 282–289. [CrossRef]
20. Zhang, C.; Zeng, K.; Wang, C.; Liu, X.; Wu, G.; Wang, Z.; Wang, D. $LaMnO_3$ Perovskites via a Facile Nickel Substitution Strategy for Boosting Propane Combustion Performance. *Ceram. Int.* **2020**, *46*, 6652–6662. [CrossRef]
21. Esmaeilnejad-Ahranjani, P.; Khodadadi, A.A.; Mortazavi, Y. Self-Regenerative Function of Cu in $LaMnCu_{0.1}O_3$ Catalyst: Towards Noble Metal-Free Intelligent Perovskites for Automotive Exhaust Gas Treatment. *Appl. Catal. A Gen.* **2020**, *602*, 117702. [CrossRef]
22. Shannon, R.D. Revised Effective Ionic Radii and Systematic Studies of Interatomie Distances in Halides and Chaleogenides. *Acta Cryst.* **1976**, *32*, 751–767. [CrossRef]
23. Gholizadeh, A. X-Ray Peak Broadening Analysis in $LaMnO_{3+\delta}$ Nano-Particles with Rhombohedral Crystal Structure. *J. Adv. Mater. Process.* **2015**, *3*, 71–83.
24. Frozandeh-Mehr, E.; Malekzadeh, A.; Ghiasi, M.; Gholizadeh, A.; Mortazavi, Y.; Khodadadi, A. Effect of Partial Substitution of Lanthanum by Strontium or Bismuth on Structural Features of the Lanthanum Manganite Nanoparticles as a Catalyst for Carbon Monoxide Oxidation. *Catal. Commun.* **2012**, *28*, 32–37. [CrossRef]
25. Thommes, M.; Kaneko, K.; Neimark, A.V.; Olivier, J.P.; Rodriguez-Reinoso, F.; Rouquerol, J.; Sing, K.S.W. Physisorption of Gases, with Special Reference to the Evaluation of Surface Area and Pore Size Distribution (IUPAC Technical Report). *Pure Appl. Chem.* **2015**, *87*, 1051–1069. [CrossRef]
26. Wu, J.; Zheng, Y.; Dacquin, J.P.; Djelal, N.; Cordier, C.; Dujardin, C.; Granger, P. Impact of Dual Calcium and Manganese Substitution of La-Deficient Perovskites on Structural and Related Catalytic Properties: Future Opportunities in next Three-Way-Catalyst Generation? *Appl. Catal. A Gen.* **2021**, *619*, 118137. [CrossRef]
27. Feick, G.; Hainer, R.M. On the Thermal Decomposition of Ammonium Nitrate. Steady-State Reaction Temperatures and Reaction Rate. *J. Am. Chem. Soc.* **1954**, *76*, 5860–5863. [CrossRef]
28. Brusamarello, E.; Blonda, C.; Salazar-Castro, C.; Pascui, A.E.; Canu, P.; Glisenti, A. Industrially Produced Fe- And Mn-Based Perovskites: Effect of Synthesis on Reactivity in Three-Way Catalysis: Part 1. *ACS Omega* **2021**, *6*, 24325–24337. [CrossRef]
29. Brusamarello, E.; Blonda, C.; Salazar-Castro, C.; Canu, P.; Glisenti, A. Industrially Produced Fe- And Mn-Based Perovskites: Effect of Synthesis on Reactivity in Three-Way Catalysis: Part 2. *ACS Omega* **2021**, *6*, 24316–24324. [CrossRef]

30. Giordano, F.; Trovarelli, A.; De Leitenburg, C.; Giona, M. A Model for the Temperature-Programmed Reduction of Low and High Surface Area Ceria. *J. Catal.* **2000**, *193*, 273–282. [CrossRef]
31. Wei, T.; Jia, L.; Zheng, H.; Chi, B.; Pu, J.; Li, J. LaMnO3-Based Perovskite with in-Situ Exsolved Ni Nanoparticles: A Highly Active, Performance Stable and Coking Resistant Catalyst for CO_2 Dry Reforming of CH_4. *Appl. Catal. A Gen.* **2018**, *564*, 199–207. [CrossRef]
32. Patcas, F.; Buciuman, F.C.; Zsako, J. Oxygen Non-Stoichiometry and Reducibility of B-Site Substituted Lanthanum Manganites. *Termochimica Acta* **2000**, *360*, 71–76. [CrossRef]
33. Huang, C.; Shan, W.; Lian, Z.; Zhang, Y.; He, H. Recent Advances in Three-Way Catalysts of Natural Gas Vehicles. *Catal. Sci. Technol.* **2020**, *10*, 6407–6419. [CrossRef]
34. Mars, P.; van Krevelen, D.W. Oxidations Carried out by Means of Vanadium Oxide Catalysts. *Chem. Eng. Sci.* **1954**, *3*, 41–59. [CrossRef]
35. Hammami, R.; Aïssa, S.B.; Batis, H. Effects of Thermal Treatment on Physico-Chemical and Catalytic Properties of Lanthanum Manganite $LaMnO_{3+y}$. *Appl. Catal. A Gen.* **2009**, *353*, 145–153. [CrossRef]
36. Wu, J.; Dacquin, J.P.; Djelal, N.; Cordier, C.; Dujardin, C.; Granger, P. Calcium and Copper Substitution in Stoichiometric and La-Deficient $LaFeO_3$ Compositions: A Starting Point in next Generation of Three-Way-Catalysts for Gasoline Engines. *Appl. Catal. B* **2021**, *282*, 119621. [CrossRef]

catalysts

MDPI

Communication

Sulfonated Silica Coated $CoFe_2O_4$ Magnetic Nanoparticles for the Synthesis of 3,4-Dihydropyrimidin-2(1H)-One and Octahydroquinazoline Derivatives

Mozhgan Afshari [1,*], Sónia A. C. Carabineiro [2,*] and Maryam Gorjizadeh [1]

1 Department of Chemistry, Shoushtar Branch, Islamic Azad University, Shoushtar, Iran
2 LAQV-REQUIMTE, Department of Chemistry, NOVA School of Science and Technology, Universidade NOVA de Lisboa, 2829-516 Caparica, Portugal
* Correspondence: mozhgan.afshari@iau.ac.ir (M.A.); sonia.carabineiro@fct.unl.pt (S.A.C.C.); Tel.: +98-9166-161-157 (M.A.); Fax: +98-6126-231-300 (M.A.)

Abstract: Sulfonated-silica-coated cobalt ferrite ($CoFe_2O_4$) magnetic nanoparticles (MNPs-SiCoFe-SO_3H) are efficient heterogeneous catalysts for the synthesis of 3,4-dihydropyrimidin-2(1H)-one and octahydroquinazoline derivatives in the absence of solvent. The effects of solvent, temperature, and catalyst amount on the reaction are investigated. The easy separation, reusability of the catalyst, simplicity of the procedure, mild reaction conditions, and good yields (68–95%) within short reaction times (15–70 min) are the advantages of this method. The catalyst can be reused up to eight times with not much loss of activity. Scanning electron microscopy images, X-ray diffraction spectra, and elemental analysis of the recycled catalyst show that the catalyst is stable after the reaction.

Keywords: cobalt ferrite; 3,4-dihydropyrimidin-2(1H)-one derivative; octahydroquinazoline derivatives; magnetic nanoparticles; sulfonic acid

Citation: Afshari, M.; Carabineiro, S.A.C.; Gorjizadeh, M. Sulfonated Silica Coated $CoFe_2O_4$ Magnetic Nanoparticles for the Synthesis of 3,4-Dihydropyrimidin-2(1H)-One and Octahydroquinazoline Derivatives. *Catalysts* **2023**, *13*, 989. https://doi.org/10.3390/catal13060989

Academic Editors: Leonarda Liotta, Narendra Kumar and Konstantin Ivanov Hadjiivanov

Received: 30 April 2023
Revised: 2 June 2023
Accepted: 7 June 2023
Published: 9 June 2023

1. Introduction

In recent decades, nanoparticles have been reported in several versatile applications, due to their high surface area. They are employed in many fields, such as biology, optics, magnetic applications, and catalysis [1–4]. Nanomagnetic catalysts are easily separated from the reaction medium and can be recycled several times, maintaining the catalytic efficiency [5–7]. Therefore, nanoparticles are significant alternatives to homogeneous catalysts, as they are able to provide high product yields with elevated reaction rates [2,8].

Recently, magnetic nanoparticles are used in the laboratory and industry as supports for sulfonic acid and reported as effective catalysts for the synthesis of benzimidazoles [9], triazolo[4,3-a]pyrimidines [10,11], thiazolidin-4-one derivatives [12], xanthene derivatives [13], 1,4-dihydropyridines [14], octahydroquinazolinone, and 3,4-dihydropyrimidinone derivatives [15]. The Biginelli reaction is a common, yet very essential, type of reaction, which provides a simple method to prepare N-heterocyclic compounds, such as octahydroquinazolinone- and 3,4-dihydropyrimidinone-based materials [16]. Such molecules are well known for their several biological and pharmacological applications, including anticonvulsant [17], antibacterial [18,19], antidiabetic [20], and anticancer [21] activities.

Several materials, such as $Yb(OTf)_3$ [22], MnO_2-MWCNT (MWCNT—multi-walled carbon nanotubes) [23], and Nafion-Ga [24] are reported for the preparation quinazolinone and 3,4-dihydropyrimidinone compounds, yet with low efficiencies. The use of expensive reagents is associated with a long period of destruction and deactivation of the catalyst and difficulty in recycling. Sulfonic-acid-supported, magnetic-nanoparticles-based catalysts can help to overcome these limitations due to their high catalytic activities, easy recyclability, low cost, long durability, and environmentally safe and green preparation.

The growing demand for environmentally friendly and cost-effective chemical processes led to a search for mild methods that use reusable catalysts and operate under solvent-free conditions. In response to this need, we developed an alternative for synthesizing 3,4-dihydropyrimidin-2(1H)-one and octahydroquinazoline derivatives [25–27]. In continuation of our previous works [28,29], this paper explores a straightforward synthesis of 3,4-dihydropyrimidin-2(1H)-one and octahydroquinazoline derivatives with various substituents by a one-pot reaction under solvent-free conditions. Silica (Si)-coated cobalt ferrite ($CoFe_2O_4$) magnetic nanoparticles functionalized with sulfonic acid (MNPs-SiCoFe-SO_3H) are used as environmentally friendly, efficient, and reusable nanocatalysts with good yields (Scheme 1).

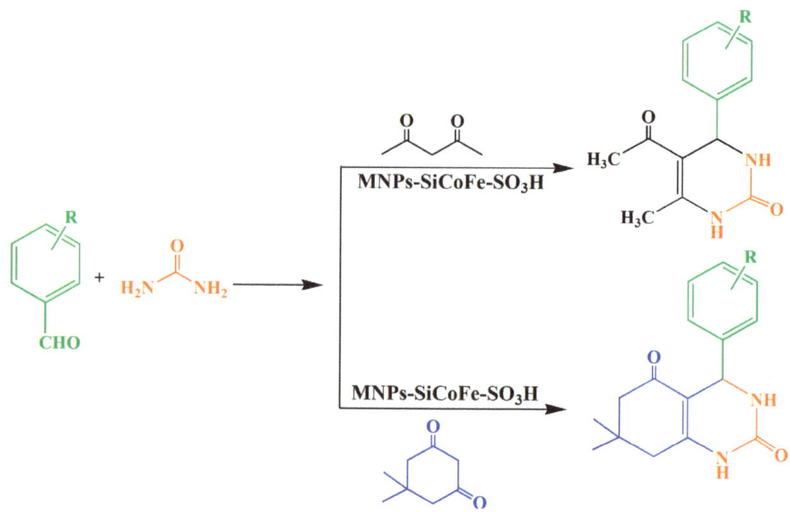

Scheme 1. MNPs-SiCoFe-SO_3H catalyzed synthesis of 3,4-dihydropyrimidin-2(1H)-one and octahydroquinazoline derivatives, under solvent-free conditions.

2. Results and Discussion

2.1. Characterization of Sulfonated Silica-Coated $CoFe_2O_4$ Magnetic Nanoparticles

The preparation of the catalysts used in this study can be found in the Experimental Section. The MNPs-SiCoFe-SO_3H catalyst was analyzed using various methods. The Fourier transform infrared spectroscopy (FT-IR) spectrum of MNPs-SiCoFe-SO_3H is shown in Figure 1 The Fe–O bonds were identified by the strong absorption bands at 589 cm^{-1} and 462 cm^{-1}. These peaks corresponded to the stretching vibrations of Fe–O bonds in the catalyst's structure. Furthermore, the FT-IR spectrum exhibited a broad peak within the range of 1000–1250 cm^{-1}, which could be attributed to the Si-O stretching vibrations present in the silica shell that surrounded the magnetic nanoparticles (MNPs) [30]. The presence of the alkyl groups in MNPs-SiCoFe-SO_3H was confirmed by the aliphatic weak C–H stretching vibrations appearing at 2943 cm^{-1}. However, detecting evidence of the SO_3H groups was challenging due to the low loadings of acidic components and the overlapping of peaks associated with Si–O–Si, Fe–O–Si, and symmetric SO_2 stretching [31].

The X-ray diffraction (XRD) patterns of MNPs-SiCoFe-SO_3H in the pure cobalt ferrite phase (JCPDS PDF #221086) showed the retention of the cubic reverse spinel structure of cobalt ferrite during coating and functionalization. An evident diffusion peak was observed around 20–28°, which could be attributed to the presence of amorphous silica in the shell of cobalt ferrite (Figure 2) [28].

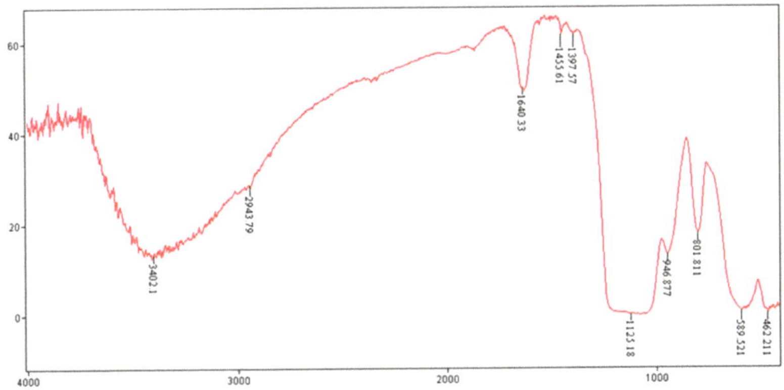

Figure 1. FT-IR spectrum of sulfonated silica-coated $CoFe_2O_4$ magnetic nanoparticles.

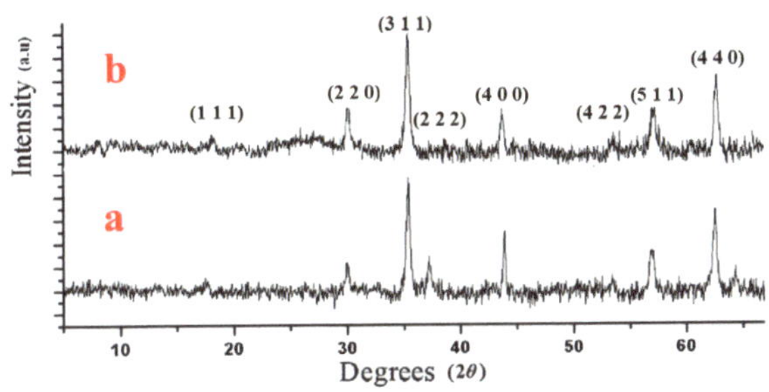

Figure 2. XRD pattern of (**a**) $CoFe_2O_4$, (**b**) sulfonated-silica-coated $CoFe_2O_4$ magnetic nanoparticles.

A transmission electron microscopy (TEM) image of MNPs-SiCoFe-SO_3H is presented in Figure 3, revealing the quasi-spherical shape of the particles. The average size of these nanoparticles is approximately 30 nm, and the magnetic core is visible as a dark spot inside the bright spherical SiO_2 thin shell.

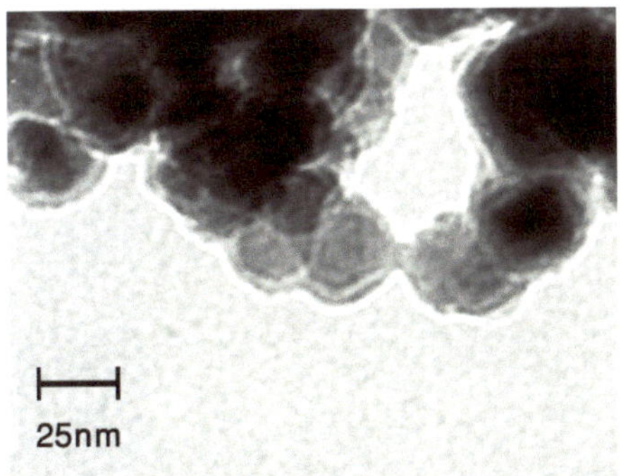

Figure 3. TEM image of the sulfonated-silica-coated $CoFe_2O_4$ magnetic nanoparticles.

Furthermore, sulfonic acid loadings of the catalyst (mmol of sulfonic acid per gram of MNPs-SiCoFe-SO_3H) could be determined using the back titration method. Initially, samples were allowed to react with 50 mL of distilled water containing 5 mL of 0.1 N NaOH overnight. Then, the remaining NaOH in the solution was titrated with a 0.1 M HCl solutions. The results of the back titration method showed that the value of loading of sulfonic acid was 0.56 mmol/g [32].

The magnetic properties are investigated by vibration sample magnetometry (VSM) at 27 °C, with a peak field of 10 kOe. The saturation magnetization (Ms), remanence magnetization (Mr), and coercivity (Hc) values of MNPs-CoFe, MNPs-SiCoFe, and MNPs-SiCoFe-SO_3H are shown in Table 1. The functionalization of cobalt ferrite nanoparticles led to a decrease in the mass saturation magnetization, which could be attributed to the presence of the nonmagnetic silica shell and functionalized groups. Despite the reduction in saturated magnetism from 59.49 emu/g to 31.16 emu/g, the synthesized nanocomposite exhibited the advantage of easy separation by an external magnet.

Table 1. Magnetic properties of the obtained nanoparticles: saturation magnetization (Ms), remanence magnetization (Mr), and coercivity (Hc).

Sample	Hc (Oe)	Mr (emu/g)	Ms (emu/g)
MNPs-CoFe	796.1	24.63	59.49
MNPs-SiCoFe	861.05	16.40	38.93
MNPs-SiCoFe-SO_3H	900	9.60	31.16

2.2. 3,4-Dihydropyrimidin-2(1H)-One Synthesis

In this work, the reaction of acetylacetone, benzaldehyde, and urea was used as a model reaction in different conditions. The results showed that the reaction proceeded slowly with low yields below 80 °C, and it did not improve above 80 °C. Additionally, the results of the model reaction, conducted with various amounts of MNPs-SiCoFe-SO_3H and urea, are presented. It was found that the amount of catalyst leading to the best results was 0.05 g and 5 mmol urea, which provided a high yield within 25 min (Table 2). In addition, we also reported the effect of the solvent on the model reaction and showed the results obtained with several solvents, such as 1,2 dichloroethane ($C_2H_4Cl_2$), acetonitrile (CH_3CN), and ethanol (EtOH), using MNPs-SiCoFe-SO_3H as catalysts, but the solvent-free conditions displayed the best result (Table 3). Moreover, we also tested the influence of the SO_3H group on the reaction by investigating silica magnetic nanoparticles (Si-MNPs) in the same reaction (Table 2, entry 5), showing that the catalysts displayed a weak activity in the 3,4-dihydropyrimidin-2(1H)-one synthesis.

Table 2. Optimization of synthesis conditions of 3,4-dihydropyrimidin-2(1H)-one [a].

Entry	Urea (mmol)	Amount of Catalyst (g)	Temperature (°C)	Time (min)	Yield (%) [b]
1	3	0.03	50	30	18
2	3	0.03	80	30	24
3	5	0.03	80	15	27
4	5	0.03	80	25	56
5	5	0.03	80	45	54
6	6	0.03	100	25	55
7	3	0.05	50	25	42
8	5	0.05	50	25	49
9	5	0.05	80	25	94
10	3	0.05	100	25	65
11	5	0.05	100	25	78
12	6	0.05	80	25	90
13	10	0.05	80	25	86
14	5	0.07	80	25	88
15	5	0.07	80	45	86
16	6	0.07	80	25	82
17	6	0.07	100	25	78

Highlighted values indicate the optimal conditions. [a] Reaction conditions: acetylacetone (3 mmol), benzaldehyde (3 mmol). [b] Isolated yield.

Table 3. Effect of different solvents on the synthesis of 3,4-dihydropyrimidin-2(1H)-one [a].

Entry	Catalyst	Solvent	Temperature (°C)	Time (min)	Yield [b] (%)
1	HSO_3-SiCoFe-MNPs	acetonitrile	85	25	80
2	HSO_3-SiCoFe-MNPs	1,2-dicholoroethane	80	25	85
3	HSO_3-SiCoFe-MNPs	ethanol	80	25	76
4	HSO_3-SiCoFe-MNPs	water	100	25	35
5	HSO_3-SiCoFe-MNPs	-	80	25	94
6	Si-MNPs	-	80	120	25

[a] Conditions: acetylacetone (3 mmol), benzaldehyde (3 mmol), urea (5 mmol), catalyst (0.05 g). [b] Isolated yield.

Additionally, different aldehydes with electron-donation or electron-withdrawing substituents are reported, under the above-mentioned optimal reaction conditions, with urea ethyl and acetoacetate, using MNPs-SiCoFe-SO_3H as catalysts. As shown in Table 4, the condensation is completed within 20–70 min and products are obtained in high yields (68–94%).

Table 4. Synthesis of several 3,4-dihydropyrimidin-2(1H)-one derivatives using MNPs-SiCoFe-SO_3H [a].

Entry	Aldehyde	Product	Time (min)	Yield [b] (%)
1	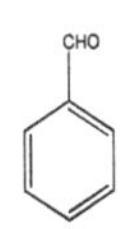	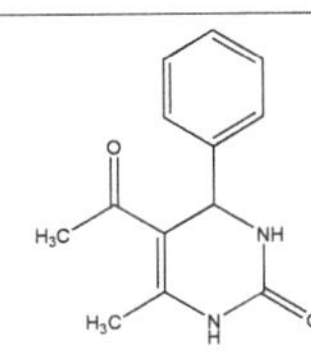	25	94
2	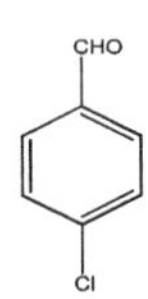	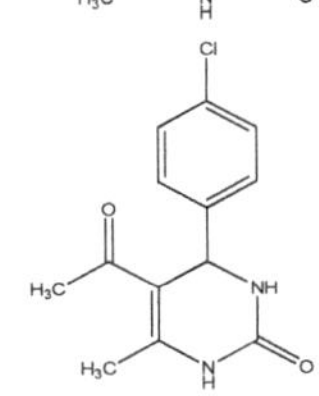	25	80
3	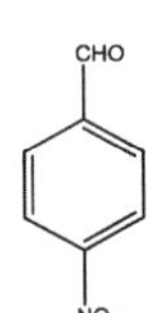	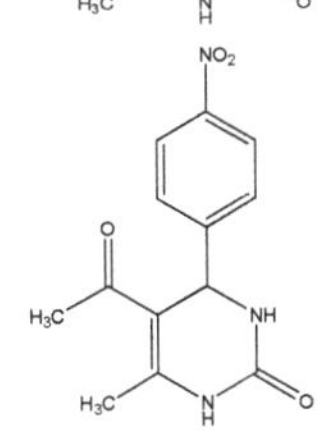	50	78
4	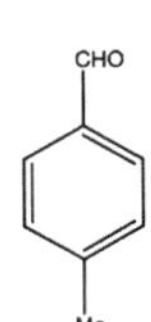	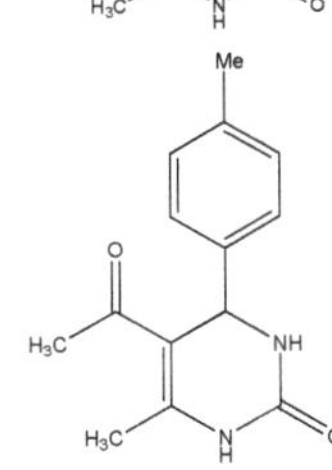	30	88

Table 4. *Cont.*

Entry	Aldehyde	Product	Time (min)	Yield [b] (%)
5	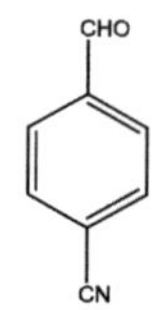	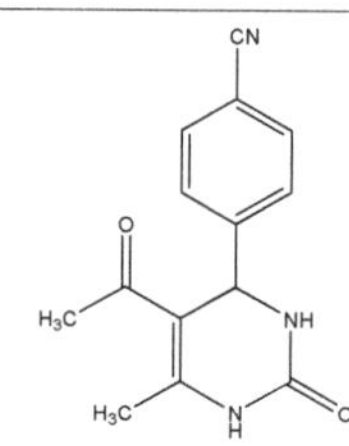	50	73
6	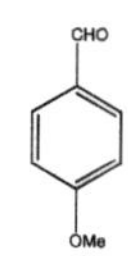	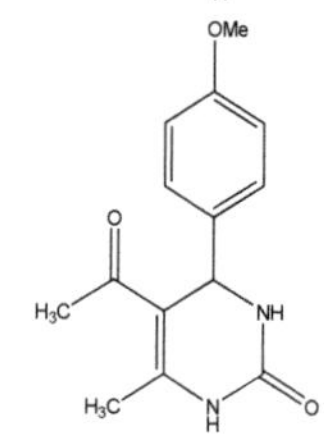	30	89
7	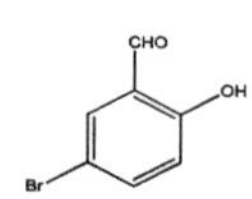	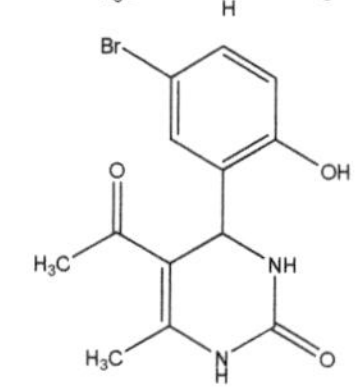	20	93
8	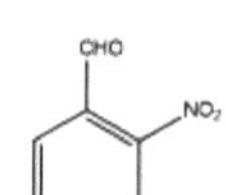	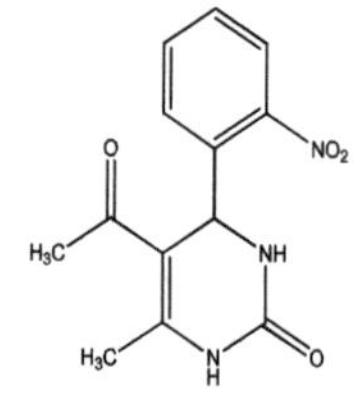	70	68

[a] Reaction conditions: benzaldehyde (3 mmol), acetylacetone (3 mmol), urea (5 mmol), catalyst (0.05 g), 80 °C.
[b] Isolated yield.

2.3. *Octahydroquinazoline Synthesis*

The efficiency of MNPs-SiCoFe-SO_3H in the condensation reactions of dimedone, benzaldehyde, and urea was also investigated. Using this reaction as a model, the temperature and amount of catalyst and solvent were optimized, similarly to what was mentioned for 3,4-dihydropyrimidin-2(1H)-one. The most effective combination was 80 °C 0.05 g MNPs-SiCoFe-SO_3H in solvent-free conditions after 15 min.

Using the optimized conditions for the model reaction, the efficiency of the MNPs-SiCoFe-SO_3H for several aromatic aldehydes with electron-withdrawing and donating groups was tested in order to obtain a series of octahydroquinazolinone derivatives (Table 5). The results showed that the substitution in the aromatic ring did not have a significant effect on the yield of the reaction (entries 3 and 8, Table 5). This method was compatible with aromatic aldehydes with different functional groups, such as –Br, –Cl, and –NO_2.

Table 5. MNPs-SiCoFe-SO_3H synthesis of diversified octahydroquinazoline derivatives [a].

Entry	Aldehyde	Product	Time (min)	Yield (%) [b]
1	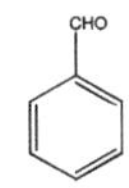	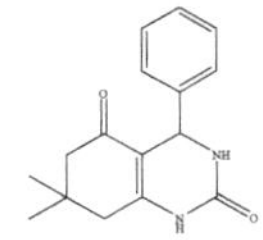	15	95
2	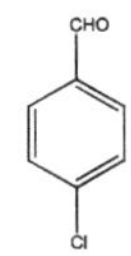	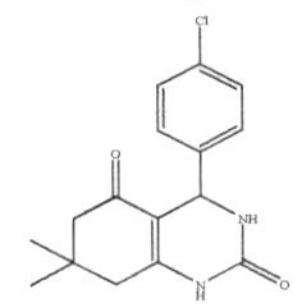	15	92
3	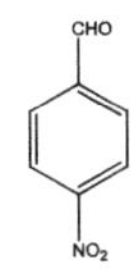	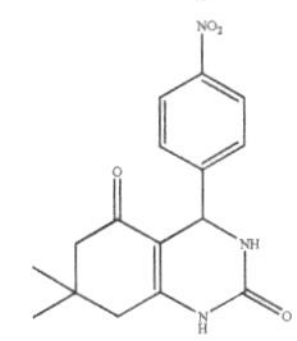	25	90
4	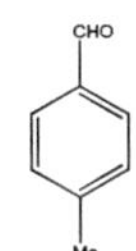	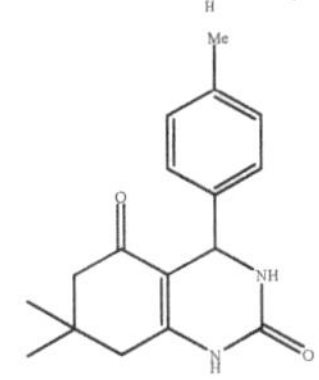	30	92
5	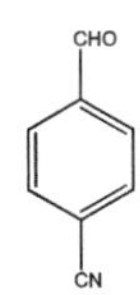	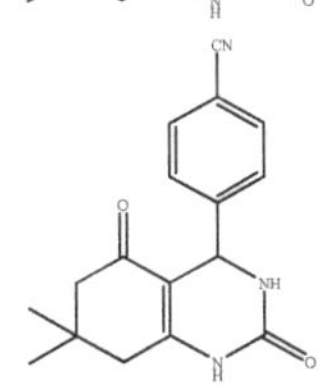	35	90
6	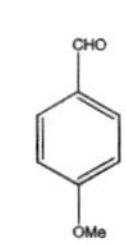	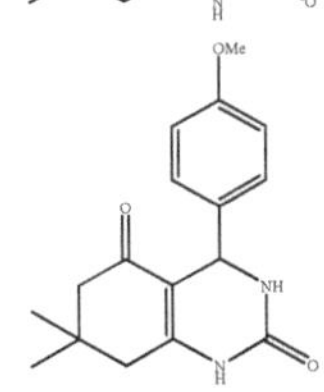	15	92
7	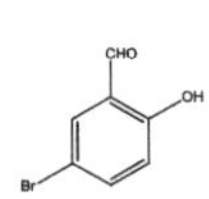	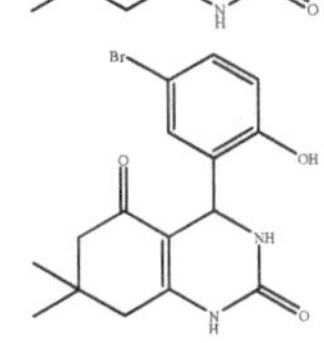	20	88
8	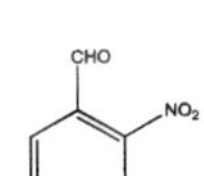	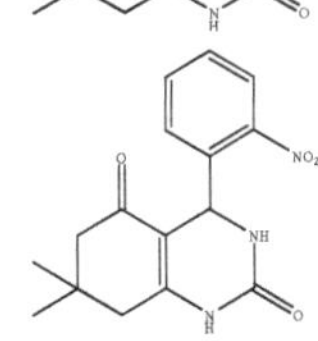	35	88

[a] Reaction conditions: aromatic aldehydes (1 mmol), dimedone (2 mmol), urea (3 mmol), catalyst (0.05 g), 80 °C.
[b] Isolated yield.

Table 6 compares the efficiency of MNPs-SiCoFe-SO_3H catalysts for the synthesis of the model compounds 3,4-dihydropyrimidin-2(1H)-one and octahydroquinazolinone, with some materials reported in earlier literature. From the collected data, it is clear that our catalytic conditions, in terms of the absence of solvent, cost-effective process, time, and high yield, are efficient in the fast synthesis of the products.

Table 6. Comparative performance of MNPs-SiCoFe-SO_3H with other reported catalysts from literature in the preparation of octahydroquinazolinone (1) and 3,4-dihydropyrimidin-2(1H)-one (2).

Entry	Catalyst	Product	Condition	Time (min)	Yield [a]	Reference
1	Fe_3O_4@SiO_2@OSO_3(0.01 g)	1	Ethanol/H_2O/Reflux	80	92	[33]
2	SiO_2-$NHSO_4$ (0.1 g)	1	H_2O/60–80 °C	90	95	[34]
3	γ-Al_2O_3/BF_3/Fe_3O_4 (0.04 g)	2	Solvent-free/80 °C	30	95	[35]
4	imidazole@Fe_3O_4(0.15 g)	2	Solvent-free/80 °C	30	91	[36]
5	[Btto] [p-TSA] (0.15 mmol)	2	Solvent-free/90 °C	30	96	[37]
8	Fe_3O_4@mesoporous SBA-15 (0.05 g)	2	EtOH/90 °C	360	85	[38]
9	cellulose sulfuric acid (0.05 g)	2	H_2O/100 °C	300	80	[39]
10	$[PVPP\text{-}SO_3H]^+Cl^-$ (0.06 g)	1	Solvent-free/70 °C	45	86	[40]
11	$[PVPP\text{-}SO_3H]^+Cl^-$ (0.06 g)	2	Solvent-free/70 °C	45	94	[40]
12	CuO@mTiO_2@$CoFe_2O_4$ (0.025 g)	2	Solvent-free/80 °C	60	92	[41]
13	n-TiO_2–NH_2 (0.024 g)	2	Solvent-free/100 °C	285	90	[42]
14	MNPs-SiCoFe-SO_3H (0.05 g)	1	Solvent-free/80 °C	15	95	This work
15	MNPs-SiCoFe-SO_3H (0.05 g)	2	Solvent-free/80 °C	25	94	This work

[a] Isolated yield.

A plausible pathway for the synthesis of 3,4-dihydropyrimidin-2(1H)-one and octahydroquinazolinone using MNPs-SiCoFe-SO_3H is shown in Scheme 2. The acid catalyst used in these reactions protonated the carbonyl group and produced an electrophilic center [32], which caused the nucleophilic addition of urea and the formation of an intermediate (1), which, subsequently, lost a water molecule. In addition, β-dicarbonyl compounds could easily convert into the enol form using MNPs-SiCoFe-SO_3H, which could easily react with the product of the previous reaction (2). Finally, cyclization and dehydration reactions occurred, and products were formed.

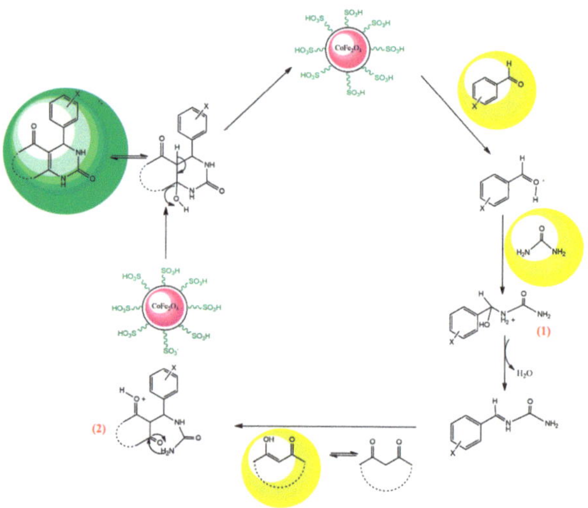

Scheme 2. Possible pathway of MNPs-SiCoFe-SO_3H catalyzed synthesis of dihydropyrimidinone and octahydroquinazolinone derivatives.

2.4. Catalyst Recycling

The catalyst recyclability and reusability of MNPs-SiCoFe-SO_3H were investigated using the reaction between dimedone, benzaldehyde, and urea under optimized conditions. The catalyst was removed by magnetic decantation, washed with methanol and $CHCl_3$, and dried. The recovered catalyst was further used for up to eight consecutive cycles. The obtained yields are shown in Table 7. There was negligible change in the catalytic efficiency. The slight decrease in activity could be attributed to the inevitable loss of material during the collection process.

Table 7. Catalyst recycling experiments for octahydroquinazolinone synthesis [a].

Cycle	0	1	2	3	4	5	6	7	8
Yield [b]	95	92	92	90	88	85	85	84	82

[a] Reaction conditions: aromatic aldehydes (1 mmol), dimedone (2 mmol), urea (3 mmol), catalyst (0.05 g), 80 °C.
[b] Isolated yield.

The XRD spectra (Figure 4) show the same peaks in both the fresh and recycled MNPs-SiCoFe-SO_3H, indicating retention of the crystalline spinel cobalt ferrite core structure (JCPDS PDF #221086), during the recycling process. Additionally, a comparison SEM of the fresh and separated catalysts (Figure 5) shows that reused MNPs-SiCoFe-SO_3H still keeps the nearly spherical morphological aspects, except for a slightly larger particle size than the fresh catalyst (about 30 nm), which may be due to the loss of some small particles during the recycling process. In addition, the elemental analysis shows that the amount of sulfur in the nanocomposite does not significantly change after recycling, confirming that the acidic component of the catalyst is firmly attached to it (Table 8).

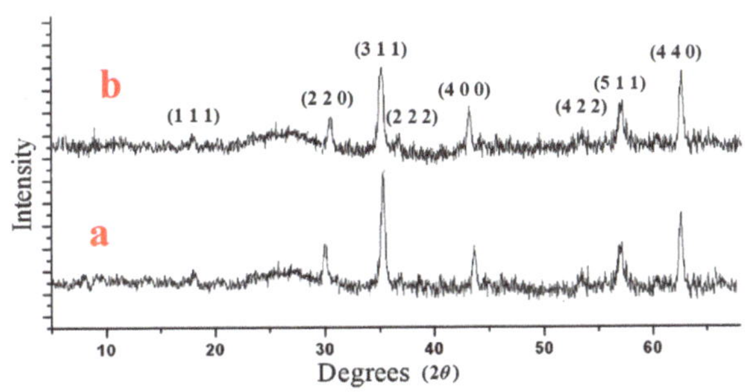

Figure 4. XRD diffractograms of (**a**) fresh MNPs-SiCoFe-SO_3H and (**b**) recycled MNPs-SiCoFe-SO_3H. The crystalline spinel cobalt ferrite core structure (JCPDS PDF #221086) is identified.

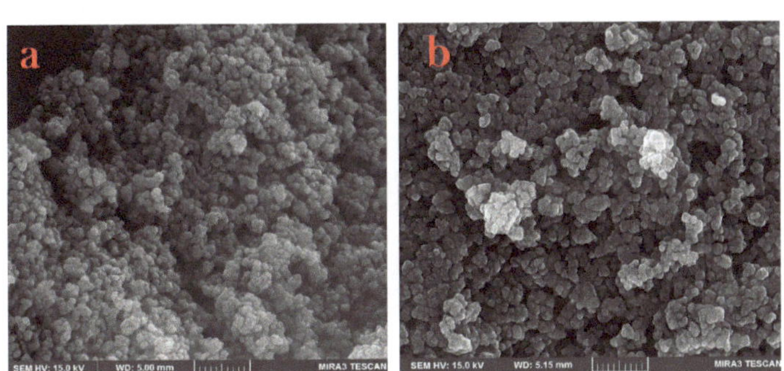

Figure 5. SEM images of (**a**) fresh MNPs-SiCoFe-SO_3H and (**b**) recycled MNPs-SiCoFe-SO_3H.

Table 8. Elemental analysis of fresh and recycled catalysts.

Compound	% C	% H	% S
Fresh MNPs-SiCoFe-SO_3H	3.65	0.66	3.25
MNPs-SiCoFe-SO_3H (1st recycle)	3.61	0.65	3.21
MNPs-SiCoFe-SO_3H (8th recycle)	3.50	0.63	3.11

3. Experimental Section

3.1. Materials Characterization

All the chemicals were supplied by Merck and used without further purification. X-ray diffraction (XRD) experiments were performed in a Philips X-ray diffractometer (PW1840 model) with Cu Kα radiation (λ = 1.54056 Å). Fourier transform infrared spectra (FT-IR) was recorded in a BOMEM MB-Series 1998 FT-IR spectrometer. Field emission scanning electron microscopy (FESEM) used a Hitachi Japan S4160 scanning electron microscope. The size and morphologies of sulfonated-silica-coated $CoFe_2O_4$ magnetic nanoparticles were recorded using a Leo 912 AB transmission electron microscope operated at a 100 keV. The vibration sample magnetometry (VSM) properties of the fabricated MNPs-SiCoFe-SO_3H composite and other samples were monitored by a Meghnatis Daghigh Kavir Company device. Reactions were followed by thin-layer chromatography (TLC). Nuclear magnetic resonance (NMR) spectra were obtained on a Bruker Advanced DPX 400 MHz spectrometer.

3.2. Synthesis of Sulfonated Silica-Coated $CoFe_2O_4$ Magnetic Nanoparticles

Cobalt ferrite ($CoFe_2O_4$) magnetic nanoparticles (CoFe-MNPs) were synthesized by the method described by Maaz et al. [43]. Briefly, a solution of iron (III) chloride (0.4 M, 25 mL) was mixed with a solution of cobalt (II) chloride (0.2 M, 25 mL), and sodium hydroxide solution (3 M) was added dropwise until the pH of the mixture reached approximately 11–12. To this mixture, oleic acid (0.6 mL) was added, and the solution was heated for 1 h at 80 °C. The resulting dark brown precipitate was separated using a centrifuge and washed multiple times with hot water and ethanol. Subsequently, it was dried overnight in an oven at 100 °C. The obtained sediment was then finely ground in a mortar and further subjected to heat treatment at 600 °C for 10 h.

Silica-coated $CoFe_2O_4$ nanoparticles (SiCoFe-MNPs) were obtained by the Stöber method [44–46]. To achieve this, cobalt ferrite (0.04 g) was added to a mixture of distilled water (20 mL) and ethanol (80 mL). The mixture was subjected to ultrasonic waves for 1 h to ensure proper dispersion. Then, ammonia (2.4 mL, 25%) was added, followed by the dropwise addition of 2 mL of tetraethylorthosilicate (TEOS) with stirring. The solution was left at room temperature for 24 h. The resulting black sediment was separated using an external magnet and washed thoroughly with water and ethanol. Finally, the sediment was dried in an oven at 100 °C for 2 h.

Sulfonic-acid-functionalized, silica-coated MNPs (HSO_3-SiCoFe-MNPs) were synthesized through a one-pot procedure (Scheme 3) [28]. Initially, a mixture of 3-mercapt opropyl trimethoxysilane (2.95 mL, 16 mmol) and 1,3-propanesultone (1.62 mL, 16 mmol) was heated for 24 h at 110 °C. The resulting product was dissolved in 25 mL of toluene, followed by the addition of Si-MNPs (0.55 g). The mixture was refluxed under a nitrogen atmosphere for 24 h. Finally, the silica-coated cobalt ferrite magnetic nanoparticles functionalized with sulfonic acid (MNPs-SiCoFe-SO_3H) were separated using an external magnet and washed thoroughly with toluene and ether.

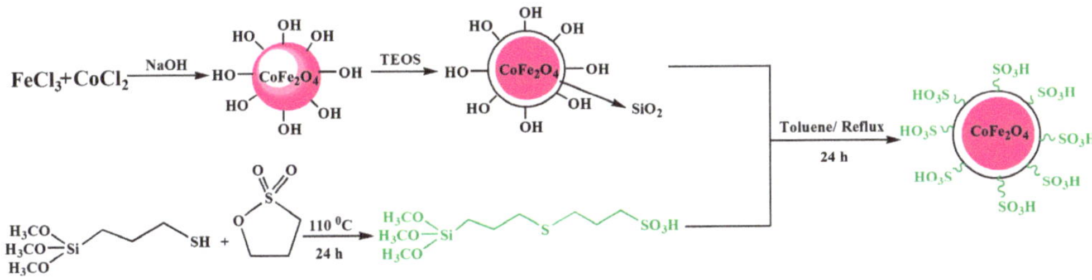

Scheme 3. Synthesis of sulfonated-silica-coated $CoFe_2O_4$ magnetic nanoparticles.

3.3. General Procedure for Catalytic 3,4-Dihydropyrimidin-2(1H)-One Synthesis

A mixture of acetylacetone (3 mmol), aromatic aldehyde (3 mmol), urea (5 mmol), and MNPs-SiCoFe-SO_3H (50 mg) was stirred at 80 °C, without any solvent, for 25–70 min. After the reaction was completed (see Table 4), the non-soluble material was dissolved in ethanol and decanted using a permanent magnet. The separated products were concentrated and recrystallized from hot ethanol and identified by Fourier transform infrared spectroscopy (FT-IR), proton nuclear magnetic resonance (^{1}H NMR), and melting point (mp) results (in comparison with the literature values).

Spectral Data for Selected Compounds

5-acetyl-4-(4-chlorophenyl)-6-methyl-3,4-dihydropyrimidin 2(1H)-one (Table 4, entry 2)

M.P.: 223–224 °C; FT-IR (KBr, cm^{-1}): 3293, 3098, 1700, 1619, 1235. ^{1}H NMR (DMSO, 400 MHz): δ = 9.21 (s, 1H), 7.84 (s, 1H), 7.37–7.40 (d, 2H), 7.25–7.27 (d, 2H) 5.2 (s, 1H), 2.29 (s, 3H), 2.145 (s, 3H). The NMR spectra is shown in Figure S1 of Supporting Information.

5-acetyl-6-methyl-4-(4-nitrophenyl)-3,4-dihydropyrimidin 2(1H)-one (Table 4,entry 3)

M.P.: 208–210 °C; FT-IR (KBr, cm^{-1}): 3243, 3119, 1708, 1674, 1232. ^{1}H NMR (DMSO, 400 MHz): δ = 9.33 (s, 1H), 7.98 (s, 1H,), 8.20–8.22 (d, 2H), 7.50–7.52 (d, 2H) 5.38 (s, 1H), 2.321 (s, 3H), 2.19 (s, 3H). The NMR spectra is shown in Figure S2 of Supporting Information.

5-acetyl-4-(4-methoxyphenyl)-6-methyl-3,4-dihydropyrimidin 2(1H)-one (Table 4, entry 6)

M.P.:172°C, FT-IR (KBR, cm^{-1}): 3309, 1700, 1617, 14669, 13647, 1239, 1180, 835; ^{1}H NMR (DMSO, 400 MHz): δ = 9.91 (s, 1H), 7.84 (s, 1H),7.21 (d, 2H), 6.86 (d, 2H), 5.40 (s, 1H), 3.79 (s, 3H), 2.35 (s, 3H), 2.12 (s, 3H). The NMR spectra is shown in Figure S3 of Supporting Information.

3.4. General Procedure for Catalytic Octahydroquinazoline Synthesis

A mixture of aromatic aldehyde (1 mmol), urea (3 mmol), dimedon (2 mmol), and MNPs-SiCoFe-SO_3H (50 mg) was stirred at a reflux of 80 °C. After the reaction was completed (see Table 5), the non-soluble product was dissolved in ethanol and decanted using an external magnet. The separated product was concentrated and purified from hot ethanol and identified by FT-IR, ^{1}H NMR, and melting point (M.P.) comparison with literature.

Spectral Data for Selected Compounds

4-(benzaldehyde)-7,7-dimethyl-1,2,3,4,5,6,7,8-octahydroquinazoline-2,5-dione (Table 5, entry 1)

M.P.:291–292 °C; FT-IR (KBR, cm^{-1}): 3328, 3259, 2964, 1712, 1679, 1614, 1449, 1374, 1238, 770, 693, 568, 489, 433. ^{1}H NMR: (DMSO, 400 MHz): δ = 0.95 (s, 3H); 1.11 (s, 3H); 2.21 (q, 2H); 2.39 (q, 2H); 5.26 (d,1H); 7.33–7.22 (m, 5H); 7.47 (s, 1H, NH); 9.39 (s, 1H). The NMR spectra is shown in Figure S4 of Supporting Information.

4-(4-methoxyphenyl)-7,7-dimethyl-1,2,3,4,5,6,7,8-octahydroquinazoline-2,5-dione (Table 5, entry 6)

M.P.: 274–275 °C; FT-IR (KBR, cm^{-1}): 3276 and 3025, 1638, 1251, 821, 778, 578, 523. ^{1}H NMR: (DMSO, 400 MHz): δ = 9.24 (s, 1H), 7.08 (s, 1H), 7.07 (d, 2H), 6.73 (d, 2H), 4.77 (s,

1H), 3.68 (s, 3H), 2.45 (d, 1H), 2.33 (d, 1H), 2.17 (d, 1H), 1.20 (d, 1H), 1.03 (s, 3H), 0.90 (s, 3H). The NMR spectra is shown in Figure S5 of Supporting Information.

4-(4-cholorobenzaldehyde)-7,7-dimethyl-1,2,3,4,5,6,7,8-octahydroquinazoline-2,5-dione (Table 5, entry 2)

M.P.: >300 °C; FT-IR (KBR, cm^{-1}): 3249, 2962,16979, 1615, 1489, 1379, 1243, 810, 767, 569, 512. ^1H NMR: (DMSO, 400 MHz): δ = 0.95 (s, 3H); 1.09 (s, 3H); 2.20 (q, 2H); 2.38 (q, 2H); 5.31 (d, 1H, CH); 7.20–7.31 (m, 4H); 7.55 (s, 1H); 9.37 (s, 1H). The NMR spectra is shown in Figure S6 of Supporting Information.

4. Conclusions

In this paper, we report a nano solid acid material as an environmentally safe catalyst for the synthesis of 3,4-dihydropyrimidin-2(1H)-one and octahydroquinazoline derivatives, in the absence of solvent, at 80 °C.

This method, using MNPs-SiCoFe-SO_3H, can promote reactions with a high range of substitutions in the reagents and offer benefits, such as short reaction times (15–70 min), good product yields (68–95%), solvent-free conditions, and a simple procedure. In addition, the catalyst shows high stability under reaction conditions and can be separated from the reaction mixture with a permanent magnet and can be reused up to eight times with not much loss in yield.

Supplementary Materials: The following supporting information can be downloaded at: https://www.mdpi.com/article/10.3390/catal13060989/s1, Figure S1: ^{1}H-NMR: 5-acetyl-4-(4-chlorophenyl)-6-methyl-3, 4-dihydropyrimidin 2(1H)-one. Figure S2: ^{1}HNMR: 5-acetyl-6-methyl-4-(4-nitrophenyl)-3,4-dihydropyrimidin 2(1H)-one. Figure S3: vHNMR: 5-acetyl-4-(4-methoxyphenyl)-6-methyl-3,4-dihydropyrimidin-2(1H)-one. Figure S4: ^{1}HNMR: 4-(benzaldehyde)-7,7-dimethyl-1,2,3,4,5,6,7,8-octahydroquinazoline-2,5-dione. Figure S5: ^{1}HNMR: 4-(4-methoxyphenyl)-7,7-dimethyl-1,2,3,4,5,6,7,8-octahydroquinazoline-2,5-dione. Figure S6: ^{1}HNMR: 4-(4-cholorobenzaldehyde)-7,7-dimethyl-1,2,3,4,5,6,7,8-octahydroquinazoline-2,5-dione.

Author Contributions: Conceptualization: M.A.; methodology: M.G.; validation: M.A. and S.A.C.C.; formal analysis: M.G.; investigation: M.A.; resources: M.A.; data curation: M.A.; writing—original draft preparation: M.A.; writing—review and editing: S.A.C.C.; visualization: M.G. and S.A.C.C.; supervision: M.A. and S.A.C.C.; project administration: M.A.; funding acquisition: M.A. All authors have read and agreed to the published version of the manuscript.

Funding: SACC is grateful to Fundação para a Ciência e a Tecnologia (FCT), Portugal for Scientific Employment Stimulus-Institutional Call (CEEC-INST/00102/2018) and to the Associate Laboratory for Green Chemistry-LAQV financed by national funds from FCT/MCTES (UIDB/50006/2020 and UIDP/5006/2020).

Institutional Review Board Statement: Not applicable.

Informed Consent Statement: Not applicable.

Data Availability Statement: Data will be made available upon request.

Acknowledgments: The authors wish to acknowledge the support of this work by the Research Council of Shoushtar Branch, Islamic Azad University, Shoushtar, Iran.

Conflicts of Interest: The authors declare no conflict of interest.

References

1. Somwanshi, S.B.; Somvanshi, S.B.; Kharat, P.B. Nanocatalyst: A brief review on synthesis to applications. *J. Phys. Conf. Ser.* **2020**, *1644*, 012046. [CrossRef]
2. Astruc, D. Introduction: Nanoparticles in catalysis. *Chem. Rev.* **2020**, *120*, 461–463. [CrossRef] [PubMed]
3. Mollazehi, F. Catalytic nanoparticles and magnetic nanocatalysts in organic reactions: A mini review. *Main Group Chem.* **2022**, *21*, 697–708. [CrossRef]
4. Terrasson, V.; Guenin, E. Nanomagnetic-Supported Catalysts. In *Novel Magnetic Nanostructures*; Elsevier: Amsterdam, The Netherlands, 2018; pp. 333–371.

5. Miceli, M.; Frontera, P.; Macario, A.; Malara, A. Recovery/reuse of heterogeneous supported spent catalysts. *Catalysts* **2021**, *11*, 591. [CrossRef]
6. Torok, B.; Schaefer, C.; Kokel, A. *Heterogeneous Catalysis in Sustainable Synthesis*; Elsevier: Amsterdam, The Netherlands, 2021.
7. Ghobadi, M.; Kargar Razi, M.; Javahershenas, R.; Kazemi, M. Nanomagnetic reusable catalysts in organic synthesis. *Synth. Commun.* **2021**, *51*, 647–669. [CrossRef]
8. Afshari, M.; Gorjizadeh, M.; Nazari, S.; Naseh, M. Cobalt salophen complex supported on magnetic nanoparticles as an efficient reusable catalyst for oxidation of benzylic alcohols. *Orient. J. Chem.* **2013**, *29*, 1523–1530. [CrossRef]
9. Yadav, P.; Kakati, P.; Singh, P.; Awasthi, S.K. Application of sulfonic acid fabricated cobalt ferrite nanoparticles as effective magnetic nanocatalyst for green and facile synthesis of benzimidazoles. *Appl. Catal. A Gen.* **2021**, *612*, 118005. [CrossRef]
10. Jahanbakhshi, A.; Farahi, M. Immobilized sulfonic acid functionalized ionic liquid on magnetic cellulose as a novel catalyst for the synthesis of triazolo [4, 3-a] pyrimidines. *Arab. J. Chem.* **2022**, *15*, 104311. [CrossRef]
11. Gebre, S.H. Recent developments in the fabrication of magnetic nanoparticles for the synthesis of trisubstituted pyridines and imidazoles: A green approach. *Synth. Commun.* **2021**, *51*, 1669–1699. [CrossRef]
12. Hajiarab, R.; Mohammad Shafiee, M.R.; Ghashang, M. Access to a Library of 3-(9-Methyl-9H-Carbazol-3-yl)-2-Arylthiazolidin-4-One Derivatives Using $NiFe_2O_4$@ SiO_2 Grafted Alkyl Sulfonic Acid as an Efficient Catalyst. *Polycycl. Aromat. Compd.* **2022**, *43*, 2032–2043. [CrossRef]
13. Nemati, F.; Sabaqian, S. Nano-Fe_3O_4 encapsulated-silica particles bearing sulfonic acid groups as an efficient, eco-friendly and magnetically recoverable catalyst for synthesis of various xanthene derivatives under solvent-free conditions. *J. Saudi Chem. Soc.* **2017**, *21*, S383–S393. [CrossRef]
14. Bakhtiarian, M.; Khodaei, M.M. Sonochemical synthesis of 1, 4-dihydropyridines using a bio-derived magnetic nanocomposite based on the pectin modified with the di-sulfonic acids under mild conditions. *Mater. Today Commun.* **2021**, *29*, 102791. [CrossRef]
15. Mozafari, R.; Gheisvandi, Z.; Ghadermazi, M. Covalently bonded sulfonic acid onto the surface of magnetic nanosilica obtained from rice husk: $CoFe_2O_4$@ RH-Pr-SO_3H as novel acid catalyst for synthesis of octahydroquinazolinone and 3, 4-dihydropyrimidinone. *J. Mol. Struct.* **2022**, *1265*, 133421. [CrossRef]
16. Kazemi, M. Magnetically reusable nanocatalysts in biginelli synthesis of dihydropyrimidinones (DHPMs). *Synth. Commun.* **2020**, *50*, 1409–1445. [CrossRef]
17. Narayanan, D.P.; Gopalakrishnan, A.; Yaakob, Z.; Sugunan, S.; Narayanan, B.N. A facile synthesis of clay–graphene oxide nanocomposite catalysts for solvent free multicomponent Biginelli reaction. *Arab. J. Chem.* **2020**, *13*, 318–334. [CrossRef]
18. Prakash, S.; Elavarasan, N.; Venkatesan, A.; Subashini, K.; Sowndharya, M.; Sujatha, V. Green synthesis of copper oxide nanoparticles and its effective applications in Biginelli reaction, BTB photodegradation and antibacterial activity. *Adv. Powder Technol.* **2018**, *29*, 3315–3326. [CrossRef]
19. Huseynzada, A.E.; Jelch, C.; Akhundzada, H.V.N.; Soudani, S.; Nasr, C.B.; Israyilova, A.; Doria, F.; Hasanova, U.A.; Khankishiyeva, R.F.; Freccero, M. Synthesis, crystal structure and antibacterial studies of dihydropyrimidines and their regioselectively oxidized products. *RSC Adv.* **2021**, *11*, 6312–6329. [CrossRef]
20. Dhumaskar, K.L.; Meena, S.N.; Ghadi, S.C.; Tilve, S.G. Graphite catalyzed solvent free synthesis of dihydropyrimidin-2 (1H)-ones/thiones and their antidiabetic activity. *Bioorganic Med. Chem. Lett.* **2014**, *24*, 2897–2899. [CrossRef]
21. Crnogorac, M.Đ.; Matić, I.Z.; Damjanović, A.; Janković, N.; Krivokuća, A.; Stanojković, T. 3D HeLa spheroids as a model for investigating the anticancer activity of Biginelli-hybrids. *Chem. Biol. Interact.* **2021**, *345*, 109565. [CrossRef]
22. Ma, Y.; Qian, C.; Wang, L.; Yang, M. Lanthanide triflate catalyzed Biginelli reaction. One-pot synthesis of dihydropyrimidinones under solvent-free conditions. *J. Org. Chem.* **2000**, *65*, 3864–3868. [CrossRef]
23. Safari, J.; Gandomi-Ravandi, S. MnO_2–MWCNT nanocomposites as efficient catalyst in the synthesis of Biginelli-type compounds under microwave radiation. *J. Mol. Catal. A Chem.* **2013**, *373*, 72–77. [CrossRef]
24. Prakash, G.; Lau, H.; Panja, C.; Bychinskaya, I.; Ganesh, S.K.; Zaro, B.; Mathew, T.; Olah, G.A. Synthesis of dihydropyrimidinones/thiopyrimidinones: Nafion-Ga, an efficient "green" lewis acid catalyst for the biginelli reaction. *Catal. Lett.* **2014**, *144*, 2012–2020. [CrossRef]
25. Tanaka, K.; Toda, F. Solvent-free organic synthesis. *Chem. Rev.* **2000**, *100*, 1025–1074. [CrossRef] [PubMed]
26. Kaur, M.; Bains, D.; Singh, G.; Kaur, N.; Singh, N. The solvent-free one-pot multicomponent tandem polymerization of 3, 4-dihydropyrimidin-2 (1 H)-ones (DHPMs) catalyzed by ionic-liquid@ Fe_3O_4 NPs: The development of polyamide gels. *Polym. Chem.* **2021**, *12*, 1165–1175. [CrossRef]
27. Yahyazadeh, A.; Nafei-Kohi, M.; Abbaspour-Gilandeh, E.; Aghaei-Hashjin, M. A New Procedure for the Preparation of 3, 4-dihydropyrimidin-2 (1H)-one and Octahydroquinazolinone Derivatives catalyzed by SCMNPs@ CA-EASO3H under Solvent-free Conditions. *Lett. Org. Chem.* **2022**, *19*, 867–883. [CrossRef]
28. Afshari, M.; Gorjizadeh, M.; Naseh, M. Supported sulfonic acid on magnetic nanoparticles used as a reusable catalyst for rapid synthesis of α-aminophosphonates. *Inorg. Nano-Met. Chem.* **2017**, *47*, 591–596. [CrossRef]
29. Afshari, M.; Gorjizadeh, M. Synthesis of 1-amidoalkyl-2-naphthol derivatives using Supported sulfonic acid on silica coated cobalt ferrite nanoparticles as a catalyst under solvent free conditions. *J. Chem. React. Synth.* **2022**, *12*, 34–41.
30. Ma, M.; Zhang, Y.; Yu, W.; Shen, H.-Y.; Zhang, H.-q.; Gu, N. Preparation and characterization of magnetite nanoparticles coated by amino silane. *Colloids Surf. A Physicochem. Eng. Asp.* **2003**, *212*, 219–226. [CrossRef]

31. Safari, J.; Zarnegar, Z. A magnetic nanoparticle-supported sulfuric acid as a highly efficient and reusable catalyst for rapid synthesis of amidoalkyl naphthols. *J. Mol. Catal. A Chem.* **2013**, *379*, 269–276. [CrossRef]
32. Kim, Y.; Choi, Y.; Kim, H.K.; Lee, J.S. New sulfonic acid moiety grafted on montmorillonite as filler of organic–inorganic composite membrane for non-humidified proton-exchange membrane fuel cells. *J. Power Sources* **2010**, *195*, 4653–4659. [CrossRef]
33. Ghasemzadeh, M.A. Preparation and Application of Fe_3O_4@ SiO_2@ OSO_3H Nanocomposite as a Green Catalyst for the Synthesis of Octahydroquinazolinones. *J. Appl. Chem. Res.* **2019**, *13*, 8–23.
34. Azzam, S.H.S.; Siddekha, A.; Nizam, A.; Pasha, M.A. SiO_2-$NaHSO_4$ as an efficient reusable heterogeneous catalyst for the one-pot three-component synthesis of octahydro-quinazolin-2, 5-diones in water. *Chin. J. Catal.* **2012**, *33*, 677–680. [CrossRef]
35. Bamoniri, A.; Mirjalili, B.B.F.; Mahmoodi Fard Chegeni, M. Synthesis of 3, 4-dihydropyrimidinones using nano γ-$Al_2O_3/BF_3/Fe_3O_4$ as an efficient magnetic nanocatalyst under solvent-free conditions. *J. Nanostructures* **2020**, *10*, 751–759.
36. Nazari, S.; Saadat, S.; Fard, P.K.; Gorjizadeh, M.; Nezhad, E.R.; Afshari, M. Imidazole functionalized magnetic Fe_3O_4 nanoparticles as a novel heterogeneous and efficient catalyst for synthesis of dihydropyrimidinones by Biginelli reaction. *Mon. Für Chem.-Chem. Mon.* **2013**, *144*, 1877–1882. [CrossRef]
37. Zhang, Y.; Wang, B.; Zhang, X.; Huang, J.; Liu, C. An efficient synthesis of 3, 4-dihydropyrimidin-2 (1*H*)-ones and thiones catalyzed by a novel brønsted acidic ionic liquid under solvent-free conditions. *Molecules* **2015**, *20*, 3811–3820. [CrossRef]
38. Mondal, J.; Sen, T.; Bhaumik, A. Fe_3O_4@ mesoporous SBA-15: A robust and magnetically recoverable catalyst for one-pot synthesis of 3, 4-dihydropyrimidin-2 (1 H)-ones via the Biginelli reaction. *Dalton Trans.* **2012**, *41*, 6173–6181. [CrossRef]
39. Rajack, A.; Yuvaraju, K.; Praveen, C.; Murthy, Y. A facile synthesis of 3, 4-dihydropyrimidinones/thiones and novel N-dihydro pyrimidinone-decahydroacridine-1, 8-diones catalyzed by cellulose sulfuric acid. *J. Mol. Catal. A Chem.* **2013**, *370*, 197–204. [CrossRef]
40. Chegini, M.G.; Mokhtary, M. Polyvinylpolypyrrolidone-supported chlorosulfonic acid: An efficient catalyst for one-pot synthesis of dihydropyrimidinones and octahydroquinazolin-2, 5-diones. *Polycycl. Aromat. Compd.* **2017**, *37*, 63–72. [CrossRef]
41. Ghosh, B.K.; Moitra, D.; Chandel, M.; Patra, M.K.; Vadera, S.R.; Ghosh, N.N. CuO nanoparticle immobilised mesoporous TiO_2–cobalt ferrite nanocatalyst: A versatile, magnetically separable and reusable catalyst. *Catal. Lett.* **2017**, *147*, 1061–1076. [CrossRef]
42. Tabrizian, E.; Amoozadeh, A.; Shamsi, T. A novel class of heterogeneous catalysts based on toluene diisocyanate: The first amine-functionalized nano-titanium dioxide as a mild and highly recyclable solid nanocatalyst for the Biginelli reaction. *React. Kinet. Mech. Catal.* **2016**, *119*, 245–258. [CrossRef]
43. Maaz, K.; Mumtaz, A.; Hasanain, S.; Ceylan, A. Synthesis and magnetic properties of cobalt ferrite ($CoFe_2O_4$) nanoparticles prepared by wet chemical route. *J. Magn. Magn. Mater.* **2007**, *308*, 289–295. [CrossRef]
44. Stöber, W.; Fink, A.; Bohn, E. Controlled growth of monodisperse silica spheres in the micron size range. *J. Colloid Interface Sci.* **1968**, *26*, 62–69. [CrossRef]
45. Deng, Y.-H.; Wang, C.-C.; Hu, J.-H.; Yang, W.-L.; Fu, S.-K. Investigation of formation of silica-coated magnetite nanoparticles via sol–gel approach. *Colloids Surf. A Physicochem. Eng. Asp.* **2005**, *262*, 87–93. [CrossRef]
46. Kooti, M.; Afshari, M. Molybdenum Schiff base complex covalently anchored to silica-coated cobalt ferrite nanoparticles as a novel heterogeneous catalyst for the oxidation of alkenes. *Catal. Lett.* **2012**, *142*, 319–325. [CrossRef]

catalysts

Article

Fabrication of New TiO_2 Photocatalyst for Removing Organic Dyes and Hazardous VOCs in Air Purifier System

Ji Won Lee [1,2], Rak Hyun Jeong [1,2], Ikjo Shin [3] and Jin-Hyo Boo [1,2,*]

[1] Department of Chemistry, Sungkyunkwan University, Suwon 16419, Republic of Korea; ljw9917@naver.com (J.W.L.); jrh1015@naver.com (R.H.J.)
[2] Institute of Basic Science, Sungkyunkwan University, Suwon 16419, Republic of Korea
[3] Dokyeung21 Co., Ltd., Suwon 13631, Republic of Korea; ikjoshin@hanmail.net
* Correspondence: jhboo@skku.edu; Tel.: +82-31-290-7072

Abstract: We synthesized an amorphous Ti-based hydroperoxo complex (ATPC) using a facile method involvingonly titanium hydride (TiH_2) and H_2O_2 under mild conditions. We chose TiH_2 as the precursor because it has more reactive sites than metal oxides such as TiO_2. Qualitative and quantitative optical measurements showed that our synthesized ATPC photocatalysts contained many hydroperoxo groups and various oxidation states of Ti (Ti^{2+}, Ti^{3+}, and Ti^{4+}). Thus, the synthesized ATPC exhibits excellent photocatalytic properties with very fast rates of organic decolorization compared to other conventional visiblelight catalysts. The presence of many hydroperoxo complexes increases the formation of active radicals, which can degrade VOCs such as acetaldehyde in a gas phase. To test the application of the synthesized ATPC, we fabricated a filter system in an air purifier using ATPC coating layers and successfully removed the VOCs. We also proposed a possible photocatalytic oxidation mechanism with ATPC based on this study. It is important to conduct application tests as well as commercialization in photocatalytic experiments.

Keywords: visiblelight photocatalyst; Ti-based hydroperoxo complex; coating layer; organic dye and hazardous VOC; air purifier system

Citation: Lee, J.W.; Jeong, R.H.; Shin, I.; Boo, J.-H. Fabrication of New TiO_2 Photocatalyst for Removing Organic Dyes and Hazardous VOCs in Air Purifier System. *Catalysts* **2023**, *13*, 935. https://doi.org/10.3390/catal13060935

Academic Editors: Leonarda Liotta, Narendra Kumar and Konstantin Ivanov Hadjiivanov

Received: 3 May 2023
Revised: 22 May 2023
Accepted: 24 May 2023
Published: 25 May 2023

1. Introduction

Recently, industrialization-oriented societies and the reckless use of chemical fuels have caused various environmental problems such as air and water pollution. To solve these issues, ecofriendly studies are being conducted. Among many environmental problems, air pollutants contained in industrial and car emissions, which are the major cause of global warming, are being addressed first. Volatile organic compounds (VOCs) are oneof the representative air pollutants that are harmful to both human health and the environment [1,2]. VOCs areharmful compoundsand can also cause photochemical reactions with other air pollutants, producing secondary pollutants such as ozone [3,4]. Formaldehyde, acetaldehyde (ACT), and benzene, for example, can cause headaches, vomiting, and dizziness if they come into contact with the skin or enter the respiratory tract [5,6]. These environmental problems need to be urgently resolved because they can cause damage to the nervous system when exposed to high concentrations. Recently, many studies have been conducted to effectively and rapidly remove VOCs, such as adsorption, thermal oxidation, photocatalytic oxidation, and biofiltration [7–9]. Among them, photocatalysts have been widely studied because they are environmentally friendly and produce water (H_2O) and carbon dioxide (CO_2) as final byproducts [10–12]. In addition, photocatalytic materials have the advantage of being activated with only solar energy. Resolving energy shortages and environmental pollution problems has become the most urgent challenge nowadays. Using a catalyst as one of the many solutions can lower the activation energy of the reaction, producing more energy for us to use. Among the various catalysts, photocatalysts, which are more environmentally friendly and utilize light sources, have been extensively studied.

There are three main reasons why many studies on photocatalysts have been conducted. First, photocatalysts can be active even at room temperature, unlike other catalysts that require high temperatures. Second, the photocatalytic reaction can be induced by utilizing sunlight, which is an infinite resource. Third, the photocatalytic reaction can be stopped at a desired point in time by blocking the supply of sunlight. Therefore, photocatalysts can play an important role in solving future environmental issues such as energy conservation and global warming.

Titanium dioxide (TiO_2) is a well-known semiconductor material and is widely used as a photocatalytic material owing to its strong oxidizing power and excellent durability [13]. TiO_2 is also utilized for solar energy conversion applications such as solar cells by utilizing its inherent bandgap [14]. However, TiO_2 has a bandgap of ~3.0–3.2 eV, which only allows it to absorb UVlight, making bandgap engineering necessary for the effective utilization of light sources. The development of a photocatalyst with visible lightactivity is crucial as most of the sunlight energy is in the visiblelight region. Many studies have been performed to overcome this problem, and a typical method of bandgap engineering is achieved by either metal doping or nonmetal doping to TiO_2 [15] as well as by coupling with a semiconductor with a narrow bandgap [16]. Although such photocatalytic materials have visiblelight activities, they also have several disadvantages. For example, the process of making narrow bandgaps by doping or coupling is complicated and requires high temperature and pressure [17], and these materials have low photocatalytic efficiency in visiblelight. Recently, extensive effort has been devoted to forming peroxo groups (Ti–OOH, --Ti–O–O–Ti–, and –O–Ti–O–) by oxidizing metal oxide surfaces with H_2O_2 [18]. The hydroperoxo photocatalysts have visiblelight activity and can cause electron transfer into the conduction band (CB) of metal oxide, forming a superoxide anion. Generating metal oxides with a peroxo (or hydroperoxo) surface is a relatively simple process, and such metal oxides with a hydroperoxo surface have excellent visiblelight photocatalytic efficiency that can be applied to various fields. Therefore, another study was conducted to synthesize photocatalysts that have visiblelight activity by intentionally creating oxygen vacancies in the TiO_2 lattice. For example, hydrogenated TiO_2 with an oxygen vacancy can be synthesized by adding a strong reducing agent such as $NaBH_4$ to TiO_2 [19]. Studies have also been conducted on the synthesis of Ti^{3+} self-doped TiO_2through the hydrothermal method using $TiCl_3$ and $(NH_4)_2TiF_6$ as the precursor of Ti^{3+} and Ti^{4+}, respectively [20]. These photocatalysts achieve visiblelight activity owing to the oxygen defects in the synthesized titania-based photocatalysts, ultimately allowingthe absorption of visiblelight by forming a new donor level in the band structure.

Herein, we synthesized an amorphous Ti-based hydroperoxo complex (ATPC) using only titanium hydride (metallic Ti) and H_2O_2undermild conditions through a facile method. Polymerization was used to generate the ATPC without annealing through an exothermic reaction. To investigate how the peroxo groups and oxygen vacancy of ATPC affect the visiblelight catalyst, the decolorization of Rhodamine B (RhB) dye was performed under visiblelight irradiation. Next, we tested the abilityof the ATPC to remove acetaldehyde (ACT), one of the VOCs in the gas phase. Finally, we tested the application of ATPC as a filter system in an air purifier.

2. Results and Discussion

2.1. Photocatalytic Activity Test with Amorphous Ti-Based Hydroperoxo Complex (ATPC)

To test the photocatalytic activity of the synthesized ATPC, we dissolved the ATPC powder in water and performed spin coating on a glass substrate. Two samples were prepared with different coating numbers to determine the thickness of the thin film showing appropriate catalytic characteristics. Figure 1c,d showed that a one-time coated layer yielded a thin film with a thickness of ~180 nm, while atwo-time coated layer yieldeda thin film with ~320 nm thickness. The surface shown in Figure 1b is rougher than that shown in Figure 1a, indicating different surface area. To measure the specific surface area of two different films, BET analysis was performed with a nitrogen gas adsorption–desorption isotherm. The thin films with thickness of 180 and 320 nm had the specific surface area of 129.9 and 166.5 m^2/g, respectively. This suggests thatthicker film will show more high catalytic activity than thinner film owing to a larger surface area. Therefore, the surface area strongly depends on the coating number. However, in our case if the film was thicker than 320 nm, the catalytic activity decreased, indicating that the catalytic reaction mainly occurred on the surface region rather than in the bulk area. Therefore, an optimized thickness of a photocatalyst is required to obtainthe maximum catalytic activity.

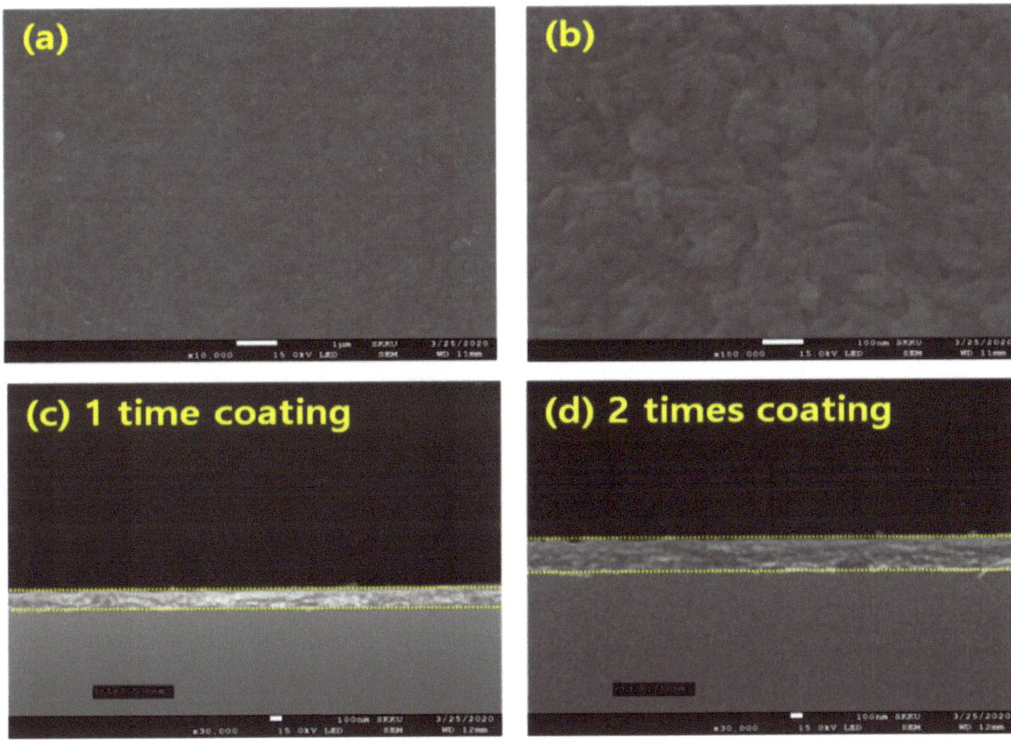

Figure 1. FE-SEM images of (**a**) one-time ATPC coated sample and (**b**) two-time APTC coated sample. (**c**,**d**) show their cross-sectional images, respectively.

We performed the decolorization of the RhB dye under visiblelight irradiation to confirm how the peroxo group and oxygen vacancy of ATPC surface affected the visiblelight catalyst (see also Section 3). Our synthesized ATPC photocatalysts showed very fast organic decolorization ratesowing to the formation of many active radicals such as OH radicals. We added 20 mg of samples to a concentration of RhB solution and stirred it in a dark room for 30 min to establish an adsorption–desorption equilibrium. We used a 200 W blue LED lamp as the light source, and the total irradiation time of each sample was 2 h. After 2 h, we separated the RhB mixture from the sample usingcentrifugation and measured the absorbance usinga UV–Visible spectrophotometer. Figure 2 shows that the two-time coated sample (Figure 2b) decomposed more RhB dye in the same time than the one-time coated sample (Figure 2a), reflecting thickness and surface area differences as well as peroxo groups on the ATPC surface. We also found that there was no considerable difference when coating two or more times. Therefore, we used the same sample of two-time coating with a thickness of ~320 nm to check the photocatalytic activities for other VOCs such as ACT in a gas phase.

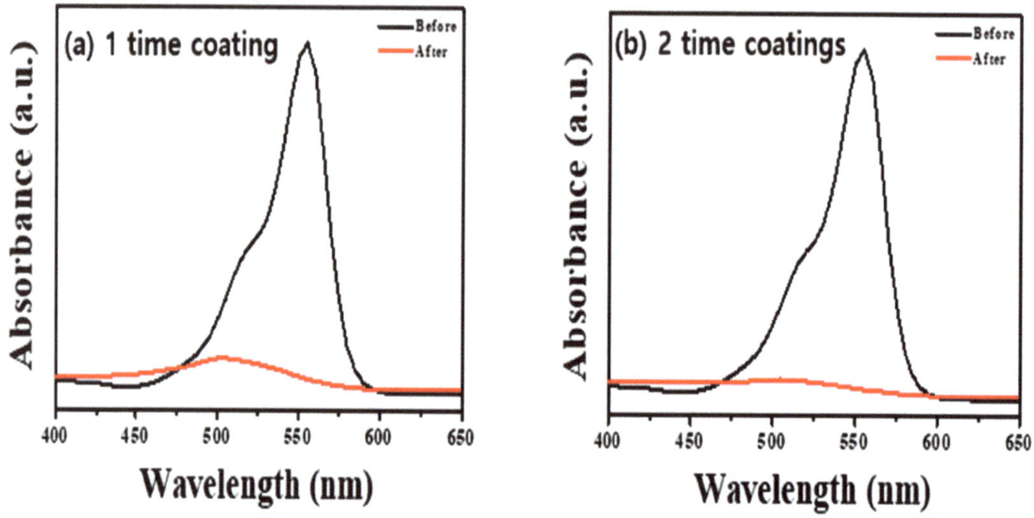

Figure 2. UV–Visible absorption spectra obtained before and after decolorization of RhB dye with (**a**) one-time ATPC coated sample and (**b**) two-time APTC coated sample.

The experiment involved the degradation of acetaldehyde (ACT) in a gas phase using a flow-type reactor equipped with an online gas chromatography (GC, HP 6890, HEWLETT PACKARD). The stationary phase and detector of the GC were the HP-PLOT Q column and the flame ionization detector, respectively. The synthesized ATPC powder (500 mg) in a quartz boat was placed at the center of the quartz tube. A gas mixture containing 107.7 ppm of ACT and dry air was flowed into a photocatalytic reactor using mass flow controllers, while blue light (10 W, KWLD B1207W, Kwangwon light) was irradiated on the sample at a distance of ~10 cm. The gas mixture that passed through the sample was injected into the GC every 10 min, and the amount of ACT and CO_2 were monitored with the lightirradiation time. Figure 3a shows that CO_2 is generated as ACT is removed, and it can be seen that ~15% of the ACT can be removed within 5 h. The power (10 W) used was relatively weak compared to 200 W for the decolorization of the RhB dye; hence, more time was required to remove the ACT. A 10 W small powered blue LED was used as the light source to consider its installation inside an air purifier system. Therefore, only 15% of the ACT was removed in 5 h. However, if a stronger power source, such as a 200 W blue LED lamp, was used, the percentage of removal could be increased to >60% within only 2 h, suggesting an increased functionality of photocatalytic activity.

Based on this study, we suggested a possible photocatalytic oxidation mechanism of ATPC as follows [21]. When the hydroperoxo group (Ti-OOH) absorbs visiblelight, radicals are formed and an electron (e^-) is also formed by combining with an unshared electron pair of oxygen. The formed e^- reacts with oxygen in the air to form superoxide anions, providing strong oxidation. The remaining plus-charged hydroperoxo groups react with water molecules on the photocatalyst's surface to form OH radicals, which degrade the contaminants (i.e., ACT) through oxidation [21–23]. To prove and clarify these mechanisms utilizing various analytical instruments, we have to carry out both in-situ and ex-situ experiments to detect intermediates, radicals, holes, and electrons by time-resolved photoluminescence (TRPL), time-dependent light irradiated UV–Vis spectra, transient absorption spectroscopy, electron spin resonance (ESR), X-ray photoelectron spectroscopy (XPS), Diffuse Reflectance Fourier Translation Infrared Spectroscopy (DRIFTS), etc. Recently, Shu et al. proposed similar photocatalytic oxidation of ammonia (NH_3) on TiO_2-based catalysts such as P25 and nanosheet TF4 [23]. For systematic investigation into the NH_3 adsorption and conversion behaviors, they utilized in-situ DRIFTS and obtainedaprogressed mechanism of NH_3 photocatalytic oxidation. Additionally, Karmakar et al. proposed the CO_2 photocatalytic reduction mechanism based on time-dependent in-situ DRIFTS experiments on CO_2 reduction in a mixture of CO_2 and H_2O vapor on Ce-MOF-RuII-bpy surface [24]. Based on our data and suggested mechanisms, therefore, our ATPC photocatalytic efficiency shows the best performance because it can form more electrons than conventional photocatalytic TiO_2 materials. For these reasons, as seen in Figure 3b, the adsorption and degradation of ACT proceeded rapidly on the surface of ATPC particles. Air pollutants such as ACT are initially adsorbed on the surface of the ATPC photocatalyst, and subsequentlyphoto-oxidative decomposition occurs through light-induced electron–hole pairs. After the reaction, the finally formed water (H_2O) and carbon dioxide (CO_2) are desorbed from the adsorption site of the ATPC photocatalyst surface.

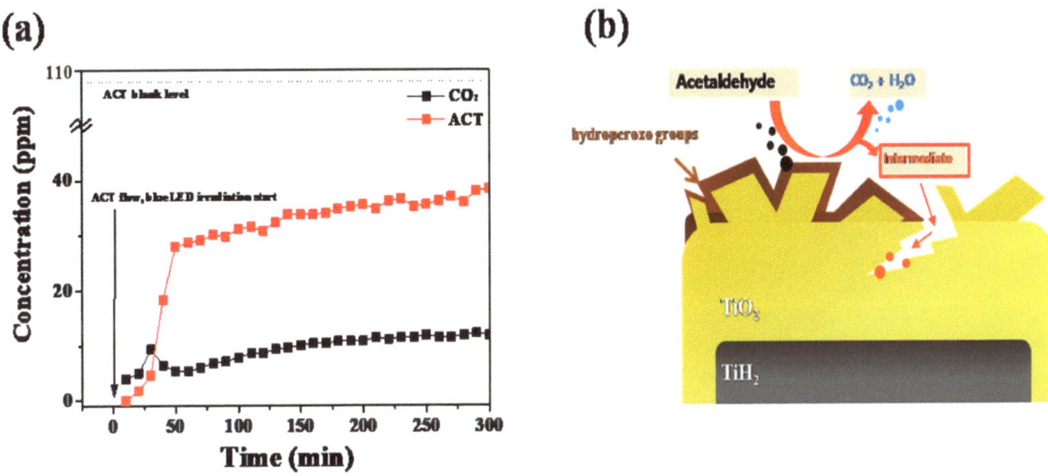

Figure 3. (**a**) Photocatalytic degradation test of acetaldehyde with two-time ATPC coated sample, and (**b**) photocatalytic oxidation mechanism of ACT on ATPC surface.

2.2. Application Test for Air Purifier System

As people spend more time indoors, there is a growing interest in creating a comfortable and healthy indoor air environment. This has led to increased research into ecofriendly airpurification mechanisms [25,26]. Based on the ecofriendly characteristics of our ATPC samples, they have potential for use in future applications that prioritizeefficiency and durability. We successfully incorporated our synthesized ATPC coating layers into a filter system in an air purifier and were able to effectively remove VOCs. To test the commercial viability of our photocatalyst, we applied it to the filter of a small car air purifier. Figure 4 provides a schematic of two different coating methods: (a) direct dip coating of the photocatalyst onto the ABS plastic filter substrate, and (b) dip coating with the addition of the adhesion-promoting leveling agent, EBECRYL 350 (EB-350), to improve the bond between the coating layer and the ABS plastic substrate surface. As shown in Figure 5a, it was very tough for us to obtaina uniform and thick ATPC film on the ABS plastic substrate with good adhesion. Therefore, we introduced an adhesion-promoting leveling agent, EBECRYL 350 (EB-350), to improve the adhesion between the coating layer and the plastic substrate surface. With EB-350, we could modify the plastic surface first, and then successfully grow uniform and thick ATPC film on the ABS plastic substrate, resulting in good adhesion and high photocatalytic activity.

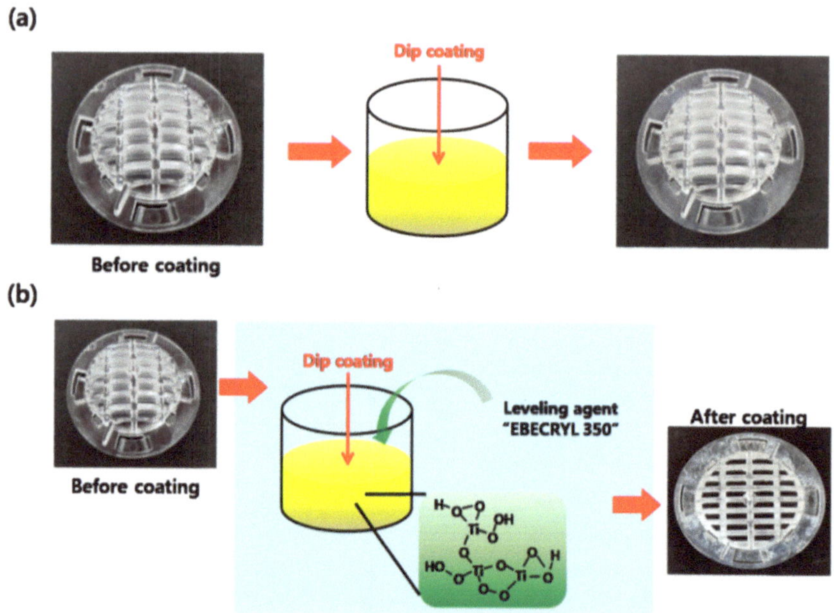

Figure 4. Schematicfor dip coatings of the APTC layer on the ABS plastic filter substrates (**a**) without EB-350 and (**b**) with EB-350.

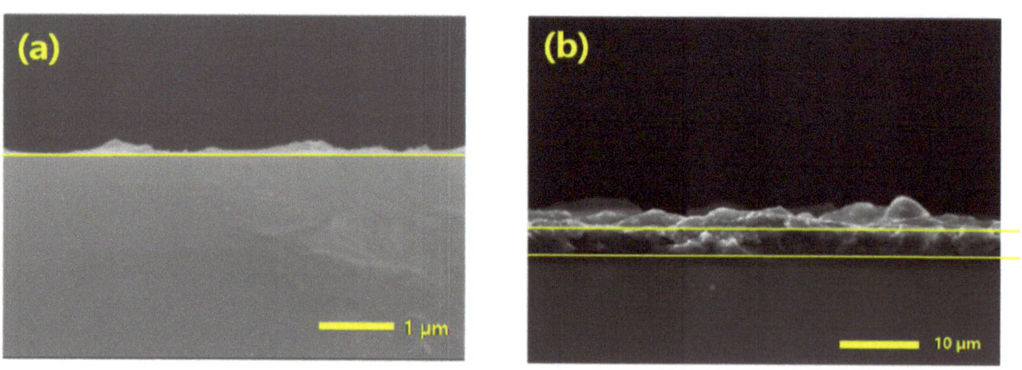

Figure 5. SEM cross-sectional images of the APTC coating layers on the ABS plastic filter substrates (**a**) without EB-350 and (**b**) with EB-350.

The cross-sectional images of the coatings are presented in Figure 5. Figure 5a shows the rough surface and thin film thickness of the photocatalyst coated by dip coating without EB-350, while Figure 5b shows a relatively uniform surface and a greaterfilm thickness with the addition of EB-350. A coating film was obtained in Figure 5b that exhibited photocatalytic performance similar to that obtained by spin coating twice on a glass substrate.

Figure 6 displays the results of the RhB removal experiment using the coatings shown in Figure 5. In Figure 6a, only ~40% of the RhB was removed in 2 h owing to the very small thickness and poor surface uniformity, whereas in Figure 6b, >90% of the RhB was removed in 2 h due to a thick film witha high surface area andgood uniformity, as well as better adhesion between the ATPC coating layer and the substrate. These results demonstrate that our synthesized ATPC photocatalyst can be used commercially and can be adopted as an airpurifier filter to remove more VOCs in the future.

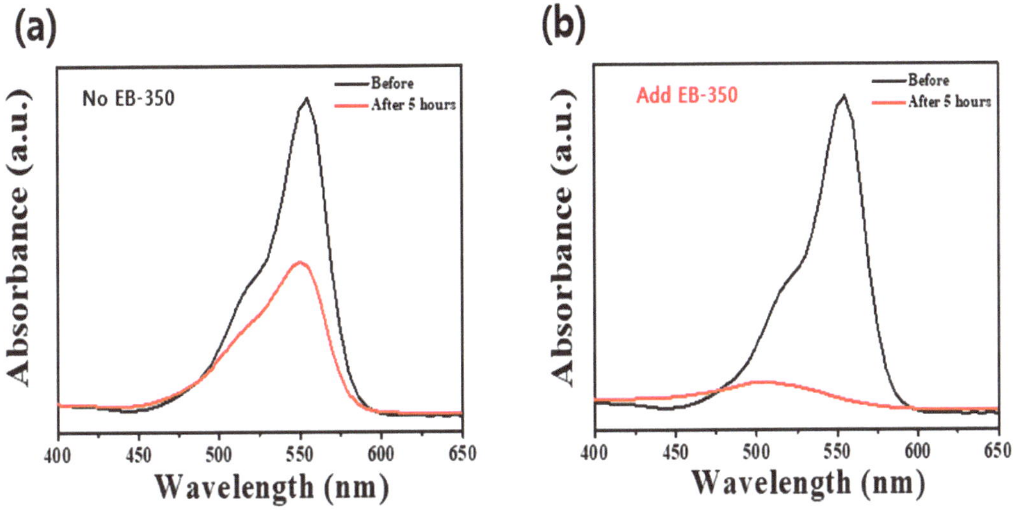

Figure 6. UV–Visible absorption spectra obtained before and after decolorization of RhB dye with the APTC coating layers on the ABS plastic filter substrates (**a**) without EB-350 and (**b**) with EB-350.

Figure 7 depicts the appearance of an air purifier with the filter of Figure 4b installed inside a vehicle. The product is expected to be released and sold as early as next year in gold and silver colors. We will also consider more colors and designs that reflect flow mechanics for future products to cater to the consumer's preference.

Figure 7. Images of air purifier adapted APTC filter.

3. Materials and Methods

3.1. Synthesis of Amorphous Ti-Based Hydroperoxo Complex (ATPC)

All the chemicals used in the synthesis of the amorphous Ti-based hydroperoxo complex (ATPC), namely titanium hydride (TiH_2, Sigma-Aldrich, St Louis, MO, USA) and hydrogen peroxide solution (H_2O_2, 30 wt.% in H_2O, Sigma-Aldrich), were used as starting agents without any further purification. The typical synthetic method for ATPC was very simple. First, TiH_2 powder (3.0 g) and distilled water (75 mL) were vigorously stirred at 25 °C. Second, 100 mL of H_2O_2 was slowly added into the solution to form a Ti–OOH surface with a TiH_2 core. After 1.5 h, metallic Ti reacted with H_2O_2, resulting in a gray colored gel state due to an exothermic reaction, followed by an increase in temperature up to 100 °C. Finally, 50 mL of H_2O_2 was added into the gray gel and stirred for 1 h to form a green colored gel state. After the gel was dried at 100 °C and ground, a fine powder was obtained. The details of our synthetic method as well as the characteristics of ATPC have already been reported elsewhere [21]. With the measurements of XPS and Raman spectra, we confirmed that our synthesized ATPC photocatalysts had large amounts of hydroperoxo groups and many oxygen vacancies on the ATPC surface that could induce very high photocatalytic activity under the visible light catalyst.

XPS shown in Figure 8 was measured to confirm the Ti oxidation number and oxygen related species as well asoxygen vacancy. As shown in the bottom figure of Figure 8a, the typical binding energy of Ti^{4+} was confirmed in the case of TiO_2. One noticeable thing is that the binding energy of metallic Ti and Ti^{4+} was confirmed through the data of TiH_2, because the surface of TiH_2 can easily oxidize even in the air. In the case of green and yellow colored ATPC, they showed the various Ti oxidation states such as Ti^{2+}, Ti^{3+}, and Ti^{4+}. This was because surface oxidation proceeded slowly from metallic Ti. In addition, Figure 8b shows the measurements of the O 1s peaks and the results are as follows. First, the green colored ATPC showed the oxygen-bonded binding energy between Ti^{4+}/Ti^{2+} and V_O (oxygen vacancy)/TiOH/TiOOH observed to be high. This could be attributed to our synthesized peroxo complexes, as oxidation and polymerization reactions took place on TiH_2 surface. In the case of yellow colored ATPC, it can be explained that the oxidation reaction proceeded by H_2O_2 from the green gel so that the Ti-OOH groups increased. Raman spectra were also measured to obtain supplement information for peroxo groups present in the synthesized ATPC. As shown in Figure 8c, there were two vibration modes (916 and 685 cm^{-1}) originating from the peroxo groups. In addition, strong Eg mode (286 cm^{-1}) derived from oxygen vacancy and typical TiO_2 vibration mode (531 cm^{-1}) was also measured [27].

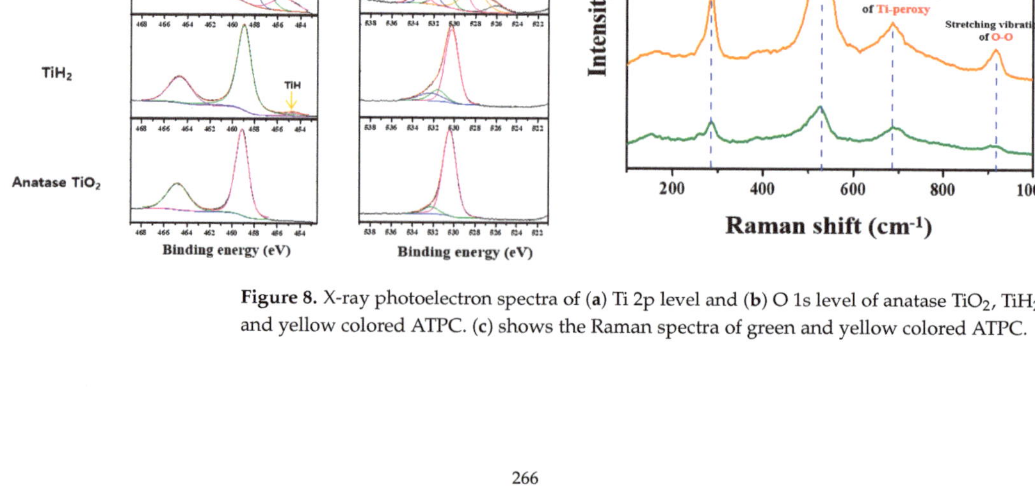

Figure 8. X-ray photoelectron spectra of (**a**) Ti 2p level and (**b**) O 1s level of anatase TiO_2, TiH_2, green and yellow colored ATPC. (**c**) shows the Raman spectra of green and yellow colored ATPC.

3.2. Characterization

Under a 200 W blue LED lamp, the photocatalytic activity of the synthesized ATPC was first evaluated by the decolorization of RhB. Figure 9 presents the detailed diagrams for the photocatalytic activity measurements for the removals of (a) RhB dye and (b) ACT VOC, respectively. In a typical photocatalytic activity measurement procedure, 20 mg of ATPC powder was added to a 70 mL aqueous solution of 10 ppm RhB and stirred for about 30 min to establish an adsorption–desorption equilibrium. As the first step, a blank test was conducted for about 30 min to confirm whether the RhB dye was physically adsorbed on the surface of the ATPC photocatalyst or decomposed without a visible light source. Next, under visible light conditions, the suspensions were irradiated at a certain distance. During light irradiation, every 5 min the suspensions (each 4 mL) were collected and separated via centrifugation. The separated solution was transferred to a 1 cm $\times$ 1 cm cuvette for absorbance measurement using a UV–Visible spectrophotometer. By checking the peak of absorbance at 550 nm corresponding to the $\pi \rightarrow \pi^*$ transition, the photocatalytic activities of each sample were measured. An application test for an airpurifier system was also conducted with the synthesized ATPC coating layers as follows. First, the samples were washed and centrifuged with distilled water after each photocatalytic activity measurement was finished to remove the RhB dye from the sample surface. The following procedure was repeated several times.

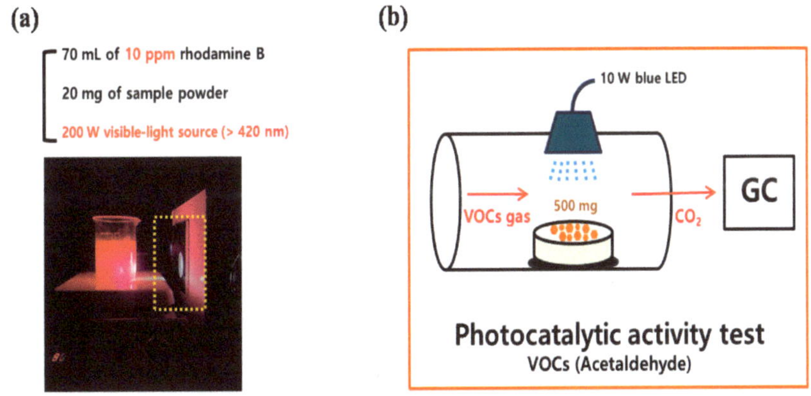

Figure 9. Schematic of the photocatalytic activity tests for the removal of (**a**) RhB dye and (**b**) ACT VOC.

4. Conclusions

In this study, an amorphous Ti-based hydroperoxo complex (ATPC) as the new photocatalyst involvingonly titanium hydride (TiH_2) and H_2O_2 was synthesized by using a facile method undermild conditions. Based on the results presented above, we conclude that all the synthesized samples showeda good degradation efficiency owing to the active carriers formed from the peroxo groups and oxygen vacancies of ATPC. Overall data suggest that this study presents a very good approach to solving air pollution problems such as VOCs. The reason is that in the process of synthesizing ATPC, not only low-cost precursors were used but also a facile synthesis method at room temperature was proposed. Moreover, both the synthesis of ATPS and the fabrication of ATPC-based coating layer are expected to be applicable to various fields such as the removal of both organic pollutants and hazardous wastes as well as a scale-up synthesis method. In addition, since our synthetic method does not need any additional heat or pressure during the synthesis process, this mild synthetic method will be very valuable if it can be applied to other metallic materials besides TiH_2. Furthermore, it will be easy to build a mass synthesis system in an industrial aspect later.

Author Contributions: Conceptualization, J.W.L. and J.-H.B.; methodology, J.W.L.; software, R.H.J.; validation, J.W.L., R.H.J. and I.S.; formal analysis, J.W.L.; investigation, J.W.L.; resources, J.-H.B.; data curation, R.H.J.; writing—original draft preparation, J.W.L.; writing—review and editing, R.H.J., I.S. and J.-H.B.; visualization, J.W.L.; supervision, J.-H.B.; project administration, J.-H.B.; funding acquisition, J.-H.B. All authors have read and agreed to the published version of the manuscript.

Funding: This work was supported by Basic Science Research Program through the National Research Found This ation of Korea (NRF) funded by the Ministry of Education (No. NRF-2019R1A6A1A10073079).

Data Availability Statement: Not applicable.

Conflicts of Interest: The authors declare no conflict of interest.

References

1. Liu, Y.; Shao, M.; Fu, L.; Lu, S. Source profiles of volatile organic compounds (VOCs) measured in China: Part I. *Atmos. Environ.* **2008**, *42*, 6247–6260. [CrossRef]
2. Phillips, M.; Herrera, J.; Krishnan, S. Variation in volatile organic compounds in the breath of normal humans. *J. Chromatogr. B* **1999**, *729*, 75–88. [CrossRef]
3. Fan, Z.; Lioy, P.; Weschler, C.; Fiedler, N.; Zhang, J. Ozone-Initiated Reactions with Mixtures of Volatile Organic Compounds under Simulated Indoor Conditions. *Environ. Sci. Technol.* **2003**, *37*, 1811–1821. [CrossRef] [PubMed]
4. Calfapietra, C.; Fares, S.; Manes, F.; Morani, A.; Sgrigna, G.; Loreto, F. Role of Biogenic Volatile Organic Compounds (BVOC) emitted by urban trees on ozone concentration in cities: A review. *Environ. Pollut.* **2013**, *183*, 71–80. [CrossRef]
5. He, J.; Zou, Z.; Yang, X. Measuring whole-body volatile organic compound emission by humans: A pilot study using an air-tight environmental chamber. *Build. Environ.* **2019**, *153*, 101–109. [CrossRef]
6. Yang, S.; Gao, K.; Yang, X. Volatile organic compounds (VOCs) formation due to interactions between ozone and skin-oiled clothing: Measurements by extractionanalysis-reaction method. *Build. Environ.* **2016**, *103*, 146–154. [CrossRef]
7. Yu, L.; Wang, L.; Xu, W.; Chen, L. Adsorption of VOCs on reduced graphene oxide. *J. Environ. Sci.* **2018**, *67*, 171–178. [CrossRef]
8. Kamal, M.S.; Razzak, S.A.; Hossain, M.M. Catalytic oxidation of volatile organic compounds (VOCs)—A review. *Atmos. Environ.* **2016**, *140*, 117–134. [CrossRef]
9. Yang, C.; Qian, H.; Li, X.; Cheng, Y.; He, H.; Zeng, G.; Xi, J. Simultaneous Removal of Multicomponent VOCs in Biofilters. *Trends Biotechnol.* **2018**, *36*, 673–685. [CrossRef]
10. Ostyn, N.R.; Steele, J.A.; Prins, M.D. Low-temperature activation of carbon black by selective photocatalytic oxidation. *Nanoscale Adv.* **2019**, *1*, 2873–2880. [CrossRef]
11. Sheng, Z.; Ma, D.; He, Q.; Wu, K. Mechanism of photocatalytic toluene oxidation with $ZnWO_4$: A combined experimental and theoretical investigation. *Catal. Sci. Technol.* **2019**, *9*, 5692–5697. [CrossRef]
12. Shayegan, Z.; Haghighat, F. Effect of surface fluorination of P25-TiO_2 on adsorption of indoor environment volatile organic compounds. *Chem. Eng. J.* **2018**, *346*, 578–589. [CrossRef]
13. Erdogan, N.; Jongee, P.; Woohyuk, C.; Soo, Y.K.; Abdullah, O. Alkaline hydrothermal synthesis, characterization, and photocatalytic activity of TiO_2 nanostructures: The effect of initial TiO_2 phase. *J. Nanosci. Nanotechnol.* **2019**, *19*, 1511–1519. [CrossRef] [PubMed]
14. Gao, X.X.; Ge, Q.Q.; Xue, D.J.; Ding, J.; Ma, J.Y. Tuning the fermi-level of TiO_2 mesoporous layer by lanthanum doping towards efficient perovskite solar cells. *Nanoscale* **2016**, *8*, 16881–16885. [CrossRef] [PubMed]
15. Giannakopoulou, T.; Papailias, I.; Todorova, N.; Boukos, N.; Liu, Y.; Yu, J.; Trapalis, C. Tailoring the energy band gap and edged' potentials of g-C_3N_4/TiO_2 composite photocatalysts for NO_X removal. *Chem. Eng. J.* **2017**, *310*, 571–580. [CrossRef]
16. Lopez, R.; Ricardo, G. Band-gap energy estimation from diffuse reflectance measurements on sol-gel and commercial TiO_2: A comparative study. *J. Sol-Gel Sci. Technol.* **2012**, *61*, 1–7. [CrossRef]
17. Mara, M.W.; Bowman, D.N.; Buyukcakir, O.; Shelby, M.L. Electron injection from copper diimine sensitizers into TiO_2: Structural effects and their implications for solar energy conversion devices. *J. Am. Chem. Soc.* **2015**, *137*, 9670–9684. [CrossRef]
18. Ovsyannikov, S.V.; Xiang, W.; Vladimir, V.S.; Alexander, E.K.; Natalia, D.; Gaston, G.; Leonid, D. Structural stability of a golden semiconducting orthorhombic polymorph of Ti_2O_3 under high pressures and high temperatures. *J. Phys. Condens. Matter* **2010**, *22*, 375402–375411. [CrossRef]
19. Panayotov, D.A.; John, T.Y.J. Depletion of conduction band electrons in TiO_2 by water chemisorptions-IR spectroscopic studies of the independence of Ti-OH frequencies on electron concentration. *Chem. Phys. Lett.* **2005**, *410*, 11–17. [CrossRef]
20. Qiu, M.; Yuan, T.; Zhangxian, C.; Zeheng, Y.; Wenming, L.; Kai, W.; Lei, W.; Kun, W.; Weixin, Z. Synthesis of Ti^{3+} self-doped TiO_2 nanocrystals based on Le Chatelier's principle and their application in solar light photocatalysis. *RSC Adv.* **2016**, *6*, 74376–74383. [CrossRef]
21. Lee, J.W.; Jeong, R.H.; Kim, D.I.; Yu, J.-H.; Nam, S.-H.; Boo, J.-H. Facile synthesis of amorphous Ti-peroxo complex for photocatalytic activity under visible-light irradiation. *J. Clean. Prod.* **2019**, *239*, 118013. [CrossRef]
22. Gligorovski, S.; Strekowski, R.; Barbati, S.; Vione, D. Environmental implications of hydroxyl radicals. *Chem. Rev.* **2015**, *115*, 13051–13092. [CrossRef]

23. Shu, Y.; Ji, J.; Zhou, M.; Liang, S.; Xie, Q.; Li, S.; Liu, B.; Deng, J.; Cao, J.; Liu, S.; et al. Selective photocatalytic oxidation of gaseous ammonia at ppb level over Pt and F modified TiO_2. *Appl. Catal. B Environ.* **2022**, *300*, 120688. [CrossRef]
24. Karmakar, S.; Barman, S.; Rahimi, F.A.; Biswas, S.; Nath, S.; Maji, T.K. Developing post-modified Ce-MOF as a photocatalyst: A detail mechanistic insight into CO_2 reduction toward selective C2 product formation. *Energy Environ. Sci.* **2023**, *16*, 2187–2198. [CrossRef]
25. Talaiekhozani, A.; Rezani, S.; Kim, K.-H.; Sanaye, R.; Amanie, A.M. Recent advances in photocatalytic removal of organic and inorganic pollutants in air. *J. Clean. Prod.* **2021**, *278*, 123895. [CrossRef]
26. Papailias, I.; Todorova, N.; Giannakopoulou, T.; Dvoranová, D.; Brezová, V.; Dimotikali, D.; Trapalis, C. Selective removal of organic and inorganic air pollutants by adjusting the g-C_3N_4/TiO_2 ratio. *Catal. Today* **2021**, *361*, 37–42. [CrossRef]
27. Zheng, W.; Liu, X.; Yan, Z.; Zhu, L. Ionic liquid-assisted synthesis of large-scale TiO_2 nanoparticles with controllable phase by hydrolysis of $TiCl_4$. *ACS Nano* **2009**, *3*, 115–122. [CrossRef]

MDPI AG
Grosspeteranlage 5
4052 Basel
Switzerland
Tel.: +41 61 683 77 34

Catalysts Editorial Office
E-mail: catalysts@mdpi.com
www.mdpi.com/journal/catalysts

www.ingramcontent.com/pod-product-compliance
Lightning Source LLC
Chambersburg PA
CBHW041418210926
51850CB00013B/3303